Natural Science in Archaeology

Series Editors

Günther A. Wagner, Hirschhorn, Germany

Christopher E. Miller, Institut für Naturwissenschaftliche, Eberhard-Karls-Universität Tübingen, Tübingen, Germany

Holger Schutkowski, School of Applied Sciences, Talbot Campus, Bournemouth University, Poole, Dorset, UK

The last three decades have seen a steady growth of application of natural scientific methods to archaeology. The interdisciplinary approach of archaeometry has found increasing appreciation by the archaeologists and is now considered indispensable and an integral part of archaeological studies. Interdisciplinary collaboration requires a multidisciplinary background. It is becoming increasingly difficult for the individual to grasp the whole field of archaeometry with its rapid developments. The aim of the series *Natural Science in Archaeology* is to bridge this information gap at the interface between archaeology and science. The individual volumes cover a broad spectrum of physical, chemical, geological, and biological techniques applied to archaeology as well as to palaeoanthropology with the interested nonspecialist in mind.

The single monographs cover:

- large fields of research
- specific methods of general interest (archaeometric methods of dating, material analysis, environmental reconstruction, geophysical prospecting, remote sensing and data processing)
- materials of interest to the archaeologist, such as sediments, soils, metal and nonmetal artifacts, animal and plant remains and other organic residues
- practical aspects such as sampling and data interpretation
- case studies, to demonstrate the potential and limitations of the various techniques.

Ariane Ballmer • Albert Hafner •
Willy Tinner
Editors

Prehistoric Wetland Sites of Southern Europe

Archaeology, Dendrochronology, Palaeoecology and Bioarchaeology

 Springer

Editors
Ariane Ballmer
Independent Researcher
Bern, Switzerland

Institute of Archaeological Sciences
Oeschger Centre for Climate Change
Research (OCCR)
University of Bern
Bern, Switzerland

Albert Hafner
Institute of Archaeological Sciences
Oeschger Centre for Climate Change
Research (OCCR)
University of Bern
Bern, Switzerland

Willy Tinner
Institute of Plant Sciences
Oeschger Centre for Climate Change
Research (OCCR)
University of Bern
Bern, Switzerland

ISSN 1613-9712
Natural Science in Archaeology
ISBN 978-3-031-52779-1   ISBN 978-3-031-52780-7   (eBook)
https://doi.org/10.1007/978-3-031-52780-7

This Springer imprint is published by the registered company Springer Nature Switzerland AG
The registered company address is: Gewerbestrasse 11, 6330 Cham, Switzerland

If disposing of this product, please recycle the paper.

Preface and Acknowledgements

While the desideratum concerning a southern European perspective on prehistoric wetland settlements has been latent for several decades, the need for the present collection of papers has increasingly emerged in the context of the editors' work within the ERC Synergy project 'Exploring the dynamics and causes of prehistoric land use change in the cradle of European farming' (EXPLO). The project tackles the adaptation strategies applied by early European farming communities to changing climate and environmental conditions. The Southwestern Balkans, in particular the border triangle between northern Greece, North Macedonia, and Albania, is not only one of the first areas where agriculture from Western Asia reached Europe more than 8000 years ago, it also features lakes which offer unique and underexplored archives of cultural and environmental (pre-)history. There, archaeological excavations of prehistoric settlements and on- and off-site bioarchaeological and palaeoecology investigations are conducted, especially in the wetland areas of the Lakes Ohrid, Prespa, and Kastoria as well as in smaller lakes such as Volvi or Zazari. With the participation of the Universities of Bern (cPI Albert Hafner, Institute of Archaeological Sciences; PI Willy Tinner, Institute of Plant Sciences; both Oeschger Centre for Climate Change Research (OCCR)), Oxford (PI Amy Bogaard, School of Archaeology), and Thessaloniki (PI Kostas Kotsakis, School of History and Archaeology), the project was implemented in 2019 and will run for a total of six years until 2025. Researchers from the fields of archaeology, biology, and climate sciences are involved.

Based on the desideratum mentioned above, the editors planned a conference in 2020 in Bern, entitled *Prehistoric Wetland Sites of Southern Europe: Archaeology, Dendrochronology, Palaeoecology and Bioarchaeology* with invited experts from Europe, the USA as well as China. The interest in a pan-Southern European scope of the archaeological wetland settlement phenomenon—from the Iberian Peninsula to the Balkan Peninsula—and consequently in promoting the exchange among the relevant experts—has been reflected not least in the positive response of numerous sponsors which showed willingness to generously co-finance the conference. We take this opportunity to thank these funders for their favourable support. Due to the pandemic-related cancellation of the workshop, the event was held online in a closed setting over two days at the end of April 2021. Despite its virtual mode, the conference was a success, with 23 papers presented and at times up to 80 colleagues attending online. In the present book, 16 of the 23 presented papers are published.

The editors thank the authors for their lively participation in the conference, the preparation of the manuscripts, and their patience during the preparation phase of the publication. We also wish to express our gratitude to the experts (in alphabetical order) who critically peer reviewed the essays: Cyrille Billard (Service Régional d'Archéologie (SRA), Direction Régionale des Affaires Culturelles (DRAC) Normandie, Caen, France), Felix Bittmann (Natural Sciences Department, Lower Saxony Institute for Historical Coastal Research (NIhK), Wilhelmshaven, Germany), Enno Giele (Institute of Chinese Studies, Heidelberg University, Heidelberg, Germany), Maja Gori (Institute of Heritage Science, National Research Council of Italy (ISPC-CNR), Rome, Italy), Kristof Haneca (Flemish Heritage Institute, Flanders Heritage Agency Department, Brussels, Belgium), Maria Ivanova-Bieg (Vienna Institute for Archaeological Science (VIAS), University of Vienna, Vienna, Austria), Peter Kuniholm (Laboratory of Tree-Ring Research and School of Anthropology, University of Arizona, Tucson, USA), and Oliver Nelle (Tree-ring laboratory Hemmenhofen, State Office for Cultural Heritage Baden-Württemberg, Gaienhofen-Hemmenhofen, Germany). Finally, we thank the Editors of the series *Natural Science in Archaeology*, Günther A. Wagner, Christopher E. Miller, and Holger Schutkowski for their interest in our publication project, as well as the Publishing Editor Annett Büttner and the Production Coordinator Chandra Sekaran Arjunan for their guidance through the publication process.

The prepress of the book was co-financed by the Burgergemeinde Bern and foundations which wish to remain unnamed. The scientific and editorial supervision of this book, coordinated by Ariane Ballmer, as well as the papers tagged with the keyword 'EXPLO' were prepared within the framework of the ERC Synergy project 'Exploring the dynamics and causes of prehistoric land use change in the cradle of European farming' (EXPLO), funded by the European Union's Horizon 2020 research and innovation program under the grant agreement No. 810586.

Bern, Switzerland   Ariane Ballmer
                                                                               Albert Hafner
                                                                               Willy Tinner

About This Book

Unique in its scope, this book provides for the first time a Southern European perspective on prehistoric wetland settlements and their natural environment. These are dwellings originally built in humid locations, i.e. on shores and in shallow water areas of lakes, bogs, marshes, rivers, estuaries, and lagoons. The relevant archaeological remains are in most cases waterlogged and offer outstanding preservation conditions for organic materials and are moreover in close proximity to uninterrupted natural archives (e.g. lake or mire sediments), which allows for a broad range of transdisciplinary research approaches. The sites discussed in this book date from the Neolithic and the Bronze Age (c. 5500–1000 BC) and are located in nine countries of Southern Europe, i.e. Spain, France, Italy, Slovenia, Croatia, North Macedonia, Albania, Greece and Bulgaria.

Four dimensions of prehistoric wetland settlements are explored in the book—the archaeological, the dendroarchaeological, the palaeoecological and the bioarchaeological: Part I is dedicated to archaeology, i.e. the excavation of settlement remains, their transdisciplinary exploration as well as their interpretation; Part II deals with dendroarchaeology and its contribution to the understanding of occupation sequences and regional chronologies; and Part III concerns uninterrupted off-site palaeoecological records of past ecosystem change, including human–environment interactions, as well as bioarchaeological on-site approaches to subsistence strategies and land use practices.

Prehistoric Wetland Sites of Southern Europe showcases how different disciplines and areas of expertise from the humanities and the natural sciences meet on an equal footing to elaborate coherent pictures of the past. Besides a cross-section of research statuses of different archaeological sites, currently ongoing research as well as novel, hitherto unpublished case studies and findings are made accessible to the international research community. Drawing on a wide range of expert contributions from both archaeology and the natural sciences, this book targets scholars, professionals, and students from the fields of prehistoric archaeology and palaeosciences and is furthermore of interest to cultural heritage stakeholders.

Contents

Contributors

Vasiliki Andreaki Department of Prehistory, Universitat Autònoma de Barcelona, Barcelona, Spain

Ferran Antolín Department of Natural Sciences, German Archaeological Institute, Berlin, Germany;
Integrative Prehistory and Archaeological Science (IPAS), Department of Environmental Sciences, University of Basel, Basel, Switzerland

Marco Baioni Museo Archeologico della Valle Sabbia, Gavardo, Italy

Ariane Ballmer Independent Researcher, Bern, Switzerland;
Institute of Archaeological Sciences and Oeschger Centre for Climate Change Research (OCCR), University of Bern, Bern, Switzerland

Daria G. Banchieri Varese, Italy

Joan Anton Barceló Department of Prehistory, Universitat Autònoma de Barcelona, Barcelona, Spain

Valeska Becker Department for Prehistoric and Protohistoric Archaeology, University of Münster, Münster, Germany

Amy Bogaard School of Archaeology, University of Oxford, Oxford, UK

Matthias Bolliger Institute of Archaeological Sciences and Oeschger Center for Climate Change Research (OCCR), University of Bern, Bern, Switzerland;
Dendrochronological Laboratory, Archaeological Service of the Canton of Bern, Bern, Switzerland

Sarah Brechbühl Institute of Plant Sciences and Oeschger Centre for Climate Change Research (OCCR), University of Bern, Bern, Switzerland

Mike Charles School of Archaeology, University of Oxford, Oxford, UK

Katarina Čufar University of Ljubljana, Biotechnical Faculty, Department of Wood Science and Technology, Ljubljana, Slovenia

Kalin Dimitrov National Archaeological Institute with Museum, Bulgarian Academy of Sciences, Sofia, Bulgaria;
Centre for Underwater Archaeology, Ministry of Culture of the Republic of Bulgaria, Sozopol, Bulgaria

John Francuz Institute of Archaeological Sciences, University of Bern, Bern, Switzerland

Kathrin Ganz Institute of Plant Sciences and Oeschger Centre for Climate Change Research (OCCR), University of Bern, Bern, Switzerland

Patrick Gassmann Neuchâtel, Switzerland

Sylvia Gassner Institute of Plant Sciences and Oeschger Centre for Climate Change Research (OCCR), University of Bern, Bern, Switzerland

Pavel Georgiev Centre for Underwater Archaeology, Ministry of Culture of the Republic of Bulgaria, Sozopol, Bulgaria;
Centre for Maritime Archaeology, University of Southampton, Southampton, UK

Tryfon Giagkoulis Department of Archaeology, School of History and Archaeology, Aristotle University of Thessaloniki, Thessaloniki, Greece

Erika Gobet Institute of Plant Sciences and Oeschger Centre for Climate Change Research (OCCR), University of Bern, Bern, Switzerland

Albert Hafner Institute of Archaeological Sciences and Oeschger Centre for Climate Change Research (OCCR), University of Bern, Bern, Switzerland

Amy Holguin School of Archaeology, University of Oxford, Oxford, UK

Marco Hostettler Institute of Archaeological Sciences and Oeschger Centre for Climate Change Research (OCCR), University of Bern, Bern, Switzerland

Katarina Jerbić Institute of Archaeological Sciences, University of Bern, Bern, Switzerland;
Flinders University, Adelaide, South Australia, Australia

Ana Jesus Integrative Prehistory and Archaeological Science (IPAS), Department of Environmental Sciences, University of Basel, Basel, Switzerland

Kostas Kotsakis Department of Archaeology, School of History and Archaeology, Aristotle University of Thessaloniki, Thessaloniki, Greece

Thibault Lachenal UMR 5140 Archéologie des Sociétés Méditerranéennes (ASM), Université Paul-Valéry, CNRS, MC, Montpellier, France

Petrika Lera Institute of Archaeology, Academy of Albanological Studies, Tirana, Albania

André F. Lotter Institute of Plant Sciences and Oeschger Centre for Climate Change Research (OCCR), University of Bern, Bern, Switzerland

Oriol López-Bultó Museu d'Arqueologia de Catalunya, Barcelona, Spain

Andrej Maczkowski Institute of Archaeological Sciences and Oeschger Center for Climate Change Research (OCCR), University of Bern, Bern, Switzerland

Claudia Mangani Museo Civico Archeologico 'G. Rambotti', Desenzano del Garda, Italy

Nicoletta Martinelli Laboratorio Dendrodata, Verona, Italy

Héctor Martínez Grau Integrative Prehistory and Archaeological Science (IPAS), Department of Environmental Sciences, University of Basel, Basel, Switzerland

Roberto Micheli Soprintendenza Archeologia, Belle Arti e Paesaggio del Friuli Venezia Giulia, Trieste, Italy

Mario Mineo Museo Delle Civiltà and Museo Preistorico Etnografico 'Luigi Pigorini', Rome, Italy

César Morales-Molino Institute of Plant Sciences and Oeschger Centre for Climate Change Research (OCCR), University of Bern, Bern, Switzerland

Cécile Oberweiler CNRS, UMR 7041 ArScAn, Université Paris I Panthéon Sorbonne, Université Paris Nanterre, Paris, France

Antoni Palomo Department of Prehistory, Universitat Autònoma de Barcelona, Barcelona, Spain

Raquel Piqué Department of Prehistory, Universitat Autònoma de Barcelona, Barcelona, Spain

Nayden Prahov National Archaeological Institute with Museum, Bulgarian Academy of Sciences, Sofia, Bulgaria;
Centre for Underwater Archaeology, Ministry of Culture of the Republic of Bulgaria, Sozopol, Bulgaria

Johannes Reich Institute of Archaeological Sciences and Oeschger Centre for Climate Change Research (OCCR), University of Bern, Bern, Switzerland

Mauro Rottoli Laboratorio di Archeobiologia, Musei Civici di Como, Como, Italy

Carolina Senn Institute of Plant Sciences and Oeschger Centre for Climate Change Research (OCCR), University of Bern, Bern, Switzerland

Raül Soteras Department of Natural Sciences, German Archaeological Institute, Berlin, Germany

Bigna L. Steiner Integrative Prehistory and Archaeological Science (IPAS), Department of Environmental Sciences, University of Basel, Basel, Switzerland

Elizabeth Stroud School of Archaeology, University of Oxford, Oxford, UK

Guoping Sun Zhejiang Provincial Institute of Relics and Archaeology, Hangzhou, Zhejiang Province, China

Xavier Terradas Archaeology of Social Dynamics, Spanish National Research Council (CISC)—Mila y Fontanals Institute on Humanities Research (IMF), Barcelona, Spain

Antoine Thévenaz Institute of Plant Sciences and Oeschger Centre for Climate Change Research (OCCR), University of Bern, Bern, Switzerland

Willy Tinner Institute of Plant Sciences and Oeschger Centre for Climate Change Research (OCCR), University of Bern, Bern, Switzerland

Gilles Touchais CNRS, UMR 7041 ArScAn, Université Paris I Panthéon Sorbonne, Université Paris Nanterre, Paris, France

Lieveke van Vugt Institute of Plant Sciences and Oeschger Centre for Climate Change Research (OCCR), University of Bern, Bern, Switzerland

Anton Velušček Research Centre of the Slovenian Academy of Sciences and Arts, Institute of Archaeology, Ljubljana, Slovenia

Haowei Wo Department of Archaeology, Hangzhou City University, Hangzhou, Zhejiang Province, China

Prehistoric Wetland Sites of Southern Europe: Archaeological Matter, Environmental Context, Research Potential, and Threats to Preservation

Ariane Ballmer, Albert Hafner, and Willy Tinner

Abstract

Archaeological remains of dwellings that were originally built in wetland environments and today in many cases are waterlogged, offer rich materials and data due to their outstanding preservation. At the same time, off-site deposits in wetlands bear detailed information on palaeoenvironmental conditions. The unique methodological possibility to correlate archaeological settlement sequences with temporally uninterrupted palaeoenvironmental records in a high temporal resolution, and thus to reconstruct coherent long-term human–environment relationships, is of particular significance. In this opening chapter, the authors introduce the basic parameters of an overarching, contextual perspective to prehistoric wetland settlements of Mediterranean Europe, not only in geographical terms, but also in (inter-) disciplinary, or methodological terms, respectively. Sites from eastern Spain, southern France, Italy, Slovenia, the Balkan Peninsula, and the Bulgarian Black Sea coast are discussed by archaeologists, dendrochronologists, bioarchaeologists, and palaeoecologists. Whereas the waterlogging of the anthropogenic remains and environmental data allow for advanced archaeological and palaeoenvironmental research, at the same time the in situ-preservation of the relevant sites, deposits and findings is at stake due to natural erosion processes and human interventions, as well as increasingly to climate change. To preserve this exceptional cultural heritage, the authors underline the pressing necessity and importance to record, inventory, and protect, or professionally excavate and document these sites.

A. Ballmer (✉)
Independent Researcher, Bern, Switzerland
e-mail: mail@arianeballmer.com

A. Ballmer · A. Hafner
Institute of Archaeological Sciences, Oeschger Centre for Climate Change Research (OCCR), University of Bern, Bern, Switzerland
e-mail: albert.hafner@unibe.ch

W. Tinner
Institute of Plant Sciences, Oeschger Centre for Climate Change Research (OCCR), University of Bern, Bern, Switzerland
e-mail: willy.tinner@unibe.ch

Keywords

Wetland archaeology · Waterlogging · Wetland settlements · Southern Europe · Mediterranean Europe · Dendroarchaeology · Dendrochronology · Palaeoecology · Bioarchaeology · Transdisciplinary research · Cultural heritage protection · EXPLO

A. Ballmer et al. (eds.), *Prehistoric Wetland Sites of Southern Europe*, Natural Science in Archaeology, https://doi.org/10.1007/978-3-031-52780-7_1

1.1 Wetland Sites: Archaeological Evidence, Methodological Potential and Epistemological Value, and Preservation Bias

Wetland environments have been frequented, used, and occupied worldwide since the Palaeolithic, through the ages, and still are in recent times (Menotti and O'Sullivan 2013a, b). Constituting a specific site category in prehistoric archaeology, **'wetland settlements' are dwellings which were originally built on humid, occasionally submerged, or water-affected subsoils—lake shores, bogs, rivers, estuaries, and lagoons**. Thus, they display architectural features intended to keep the floors dry: the buildings are raised from the ground on wood piled platforms, or feature underlays of wood grids, branches, and twigs (for an overview see Menotti 2004, 2013). Compared to terrestrial sites they tend to occur more rarely in the archaeological record, which is due to their state of preservation, the chances of their discovery and the applied research strategies. The notion that wetland environments may have been less frequently settled both in space and time is a hypothesis, which needs to be carefully evaluated on a regional scale and for specific time periods.

Archaeological sites under water and in waterlogged sediments are among the most **important archives for studying human prehistory as well as human–environment relationships** of the Holocene (Maarleveld 2014; Rey et al. 2017; Hafner et al. 2020). Unlike in terrestrial soils, organic material (plant remains such as wood, fibres, grains, seeds or fruits, microfossils such as pollen and spores, as well as ancient environmental DNA) is preserved over millennia in these contexts, offering an exceptional range and quality of information. Thus, an abundance of organic artefacts has survived in these contexts, from wooden architectural elements and tool components to products from plant fibres such as woven fabrics. The specific conditions in waterlogged milieus also offer opportunities particularly to three interdisciplinary research fields: **dendroarchaeology**

(establishing chronologies and deducing climatic conditions based on archaeological wood) (Schweingruber 1988; Menotti 2004; Haneca et al. 2009), **bioarchaeology** (gaining insights into subsistence economy and life ways based on on-site plant and animal remains) (Jacomet 2004, 2013; Jacomet and Brombacher 2005; Schibler 2004), and **palaeoecology** (reconstructing the local and regional vegetation and land use practices through off-site palaeoecological records, using palynology, macrofossil and aDNA approaches, and drawing conclusions on climate) (Gobet et al. 2003; Rey et al. 2017; van Vugt et al. 2022).

The peculiarity of wetland sites is that under the best conditions the **settlement data can be correlated with ecological off-site data in a very high temporal resolution**, allowing the most accurate possible integrative reconstruction of past human–environment relationships (Moore et al. 1991; Rey et al. 2017). This advantage is due to the site-specific possibilities of using tree-ring and varve chronology combined with radiocarbon dating (wiggle-matching), an elaborate approach that is routinely and successfully applied in archaeology (Bronk Ramsey et al. 2001; Galimberti et al. 2004). As a matter of fact, off-site waterlogged deposits with palaeoenvironmental data can also be found in the environs of terrestrial settlements, allowing for the reconstruction of the palaeoenvironment there. However, two limitations need to be considered: first, due to preservation conditions, terrestrial settlements barely ever provide absolute occupation chronologies that are comparably detailed to wetland settlements. Hence, the chronological correlation of settlement and environmental data is not possible with comparable precision. Second, in terrestrial settlements the preservation of botanical remains is limited and heavily relies on charring. This impairs the knowledge about the spectrum of exploited and used plant and animal resources, which would be necessary for the reconstruction of a comprehensive picture of human–environment relations. Thus, the **methodological and epistemological value of** wetland sites clearly lies in the increased complexity and diversity of information provided by the favourable preservation conditions on the one hand, and the potential of highly resolved dating of both on- and off-site data, and consequently their temporally accurate confrontation, on the other. Thus, their information value regarding questions on human–environment interactions, subsistence economy and land use enhances that from concurrent archaeological sites on mineral soils (which in turn occur in larger numbers). Consequently, wetland settlements refer to more than an architectural phenomenon and specific ways of life. In fact, they offer a unique archival quality, which can even be tentatively used for the complementary understanding of terrestrial settlements, in particular regarding absolute chronologies and economic aspects, as well as the spectrum of (organic) material culture, including wood building techniques.

Despite the specific preservation conditions offered by the waterlogged sediments, these also lead to a taphonomic bias and thus, to a certain degree, to a distorted perception of the bigger picture. From a methodological point of view, it is therefore not surprising that wetland settlements tend to be attributed key roles in larger-scale scenarios of past processes and developments. For instance, in the case of La Marmotta at Lake Bracciano (Rome, Lazio, Italy) (see Chap. 4 in this book), the site's preservation conditions significantly impact the interpretation of the settlement's position within the supra-regional context and the reconstruction of diffusion dynamics. Whereas crop remains have been recorded in a whole series of Early Neolithic sites along the European Mediterranean coast, the wetland settlements of La Marmotta, La Draga (Lake Banyoles, Girona, Spain) (Chap. 11), and Isolino Virginia (Lake Varese, Lombardy, Italy) (Chap. 16), in particular allow for a deeper insight into the Early Neolithic spectrum of cultivated plants and hence are considered reference nodes within a conceptual network of spreading dynamics in the Western Mediterranean. Thus, for all the positivism regarding the preservation conditions waterlogged wetland settlements

offer, possible taphonomic biases need to be considered when discussing contexts beyond individual sites.

1.2 Southern Europe: No Blank Zone on the Map of Prehistoric Wetland Sites

Currently, the highest and best-researched concentration of prehistoric wetland settlements is found around the Alps, especially north and south of the Alpine range. Their waterlogged remains on lake shores, rivers or in bogs are dated between c. 5000 and 500 cal BC, i.e. are mainly from the Neolithic and Bronze Age, and in significantly fewer cases from the Iron Age (Menotti 2004). Established in 2011, the serial UNESCO Cultural Heritage site 'Prehistoric Pile Dwellings around the Alps' includes 111 of the over 900 known sites in six countries around the Alps (Switzerland, Austria, France, Germany, Italy, and Slovenia) (Dunning and Hafner 2005; Hafner 2022).

The present book deals with prehistoric wetland settlements south of this cluster. Focusing on Southern Europe, i.e. Mediterranean Europe, the discussed sites spread from eastern Spain across southern France, Italy, Slovenia, then down the Balkan Peninsula and as far as the Black Sea coast (Fig. 1.1). They are roughly dated between 5500 and 1000 cal BC, i.e. the Neolithic and the Bronze Age. The additional case study from China is to be understood as an illustrative excursus, which should exemplarily open a window to further regions of the world featuring prehistoric wetland settlements.

The state of research on Southern European wetland sites is quite diverse, though regional research discrepancies over such a large geographical, multinational area are not surprising. Being in close exchange with the circum-alpine research community, the Italian, Spanish and Slovenian sites have always had access to tight networks of specialised expertise. At the same time, wetland settlements in many other parts of Southern Europe have been standing in the shadows of other site categories, research

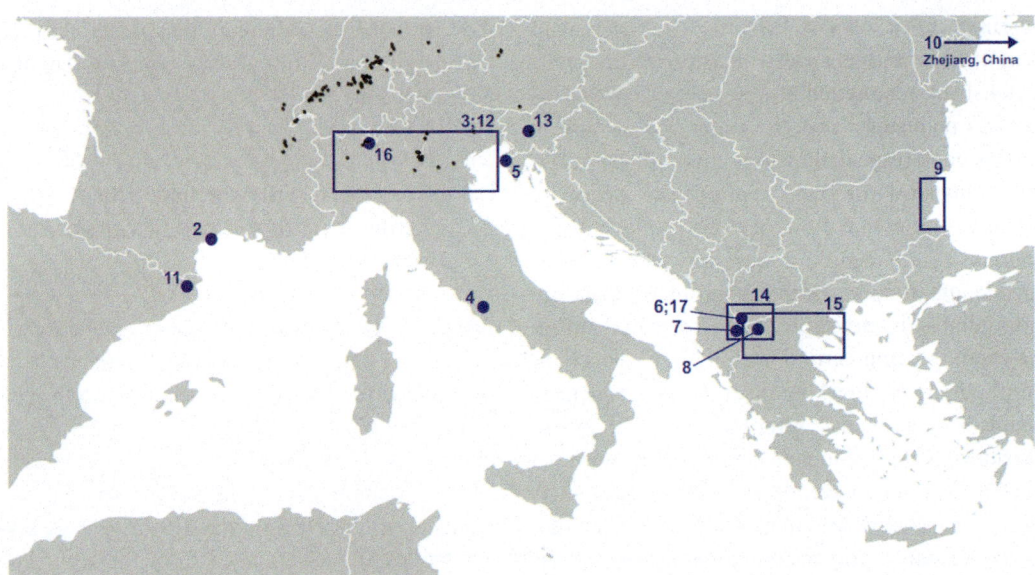

Fig. 1.1 Prehistoric wetland sites of Southern Europe. **Blue dots and squares**: sites and study areas discussed in the book; the **numbers** correspond to the Chapter numbers: (**2**) La Motte, Agde, Occitania (France); (**3**) Northern Italy (**4**) La Marmotta, Rome, Lazio (Italy); (**5**) Zambratija, Umag, Istria (Croatia); (**6**) Ploča Mičov Grad, Lake Ohrid (North Macedonia); (**7**) Sovjan, Korçë Basin (Albania); (**8**) Dispilio and Amindeon Basin, West Macedonia (Greece); (**9**) Black Sea Coast (Bulgaria); (**10**) Zhejiang Province (China); (**11**) La Draga, Lake Banyoles, Girona (Spain) (**12**) Northern Italy; (**13**) Ljubljansko Barje, Ljubljana (Slovenia); (**14**) Southwestern Balkans; (**15**) Southern Balkans and Greece; (**16**) Isolino Virginia, Lake Varese, Lombardy (Italy); (**17**) Ploča Mičov Grad, Lake Ohrid (North Macedonia). **Black dots**: sites registered as UNESCO Cultural Heritage 'Prehistoric Pile Dwellings around the Alps' (https://www.palafittes.org/homepage.html) (A. Ballmer)

questions and action priorities for some time. On the one hand, **expert knowledge, and material resources necessary for conducting research in waterlogged archaeological contexts has been less available in the regions concerned. There, the practice of recognising, excavating, and protecting archaeological wetland sites was lesser established** on the other hand—the two factors obviously being interdependent.

This is especially the case for the Balkans, where for a long time wetland settlements have only been excavated very selectively and in many cases with less specialised approaches (for an overview see Naumov 2020). While prehistoric wetland sites are known from the entire Balkan peninsula, a few have made international headlines in the recent past, of which two are especially highlighted as follows. The pile-dwellings on the North Macedonian shore of Lake Ohrid have gained a certain international prominence through the previous work of the North Macedonian

archaeologist P. Kuzman (Kuzman 2013, 2016, 2017), not least significantly due to the open-air museum at the 'Bay of Bones' near Gradište—designed by him—, which is built on the premises of the archaeological site of Ploča Mičov Grad itself and counts as a top regional destination for both national and international tourists since 2008 (cf. Chap. 6 in this book). Another exceptional case concerns the large-scale excavations in the Amindeon Basin in the Florina region (Northwestern Greece), where since 2000 multiple wetland settlements have been excavated at unprecedented expense (Chrysostomou et al. 2015; Giagkoulis 2020) (cf. Chap. 8 in this book). These showcase excavations were possible due to the polluter-pays principle: since the lignite extraction work posed a significant threat to vast archaeological remains, the Greek state mining company had to bear the costs of their professional excavation (not of the post-excavation work though, leading to the majority of the findings being unevaluated and unpublished to date). Further wetland settlements have been extensively researched and represent key sites in Balkan prehistory: several of them are discussed in this book (especially Chaps. 7, 8, 9 and 14). In the framework of the currently ongoing ERC Synergy-project *Exploring the dynamics and causes of prehistoric land use change in the cradle of European farming* (EXPLO),[1] highly specialised interdisciplinary on- and off-site research is being carried out around Lake Ohrid (on both the North Macedonian and the Albanian side), and on several shores in Greece, such as those of Lake Kastoria and Great Lake Prespa, as well as in smaller Greek lakes such as for instance Volvi or Zazari (Gassner et al. 2020; Hafner et al. 2021; Reich et al. 2021; Bolliger et al. 2023; Brechbühl et al. 2023; Chaps. 6, 8, 14, 15 and 17 in this book).

As a matter of fact, the Southern European facies of prehistoric wetland settlements, albeit with smaller-scale perspectives, has always been on the radar of European archaeology:

The Conference Proceedings *WES'04. Wetland Economies and Societies* (Trachsel and Della Casa 2005) do not only involve all relevant regions around the Alps, and especially contrast the northern alpine wetland settlements with the southern alpine ones, but for instance also include the Albanian wetland site of Sovjan (Touchais et al. 2005), arguing that:

> (…) dealing with economy and society inevitably lead to discussions on regions, landscapes, habitats and their particular features, as being part of the cultural background of Neolithic and Bronze Age communities throughout Europe. (Della Casa 2005, 12)

This wide-ranging survey additionally brought the multi- and interdisciplinary research practice of wetland archaeology into the limelight. Regarding the contributing authors, sites and methods, the present book, in parts, can certainly be understood as a logical consequence

of some 20 years of continued research, shifting the geographical focus, and relying on new discoveries and methodological advances.

When the nearly thousand-page *Oxford Handbook of Wetland Archaeology* (Menotti and O'Sullivan 2013a) discusses wetland settlements in Europe, the Americas, Africa, the Middle East, Asia and Oceania, the prehistoric examples of Southern Europe are interestingly only mentioned in passing but are not discussed further:

> Prior to the development of the lake-dwelling tradition in the Alpine region, lakeside settlements were mainly located in the Mediterranean area. Some of the best examples are those of Sovjan (c. 7000 cal BC), former Lake Maliq in Albania, Dispilio (c. 5300–3500 cal BC) on Lake Kastoria, Greece, La Marmotta (c. 5500–5200 cal BC) on Lake Bracciano, Italy, La Draga (c. 5400–5000 cal BC) on Lake Banyoles, Spain, and Isolino Virginia (Lake Varese)'

and

> Bronze Age wetland settlements are not found anywhere else in the Mediterranean except on Lake Ohrid on the border between Albania and Macedonia, where (especially on the Macedonian shores of the lake) pile dwelling communities became established and thrived throughout this period. A Bronze Age anthropogenic layer, dating to the end of the third millennium cal BC, is also found at Sovjan (former Lake Maliq, Albania, not far from Lake Ohrid), a site already occupied in the Early Neolithic. (Menotti and O'Sullivan 2013b, 16–17; see also Pétrequin 2013, 258 in the same Handbook, referring to the sites of La Marmotta, La Draga, Isolino Virginia and Dispilo).

In the 23rd/24th issue of *Plattform* (2014/15), the annual publication of the Southern German *Association of Pile-Dwelling and Local History* (Verein für Pfahlbau- und Heimatkunde) an entire part is dedicated to 'Pile-dwellings in Europe'. In the foreword the editor G. Schöbel states:

> With contributions on early pile-dwellings in the Balkans, i.e. from Southeastern Europe, a new research approach becomes apparent that has only received little attention so far,

and

> In contrast, the pile-dwelling settlements of Greece, Macedonia, and Albania – although known and much older – have received little attention. (Translated from German) (Schöbel 2014/15, 1).

---

[1] ERC Synergy-project *Exploring the dynamics and causes of prehistoric land use change in the cradle of European farming* (EXPLO) (2019–25), hosted by the Universities of Bern, Oxford, and Thessaloniki (cf. Preface of this Book).

Hence, in addition to an article on Isolino Virginia at Lake Varese (Italy) (Banchieri 2014/15), the issue of *Plattform* also includes prehistoric wetland sites from northern Greece (Chrysostomou and Jagoulis 2014/15), North Macedonia (Naumov 2014/15), and Albania (Oberweiler et al. 2014/15).

In the recently published edited book *Settling waterscapes in Europe. The archaeology of Neolithic and Bronze Age pile-dwellings* (Hafner et al. 2020), the geographical focus is on the northern Alpine region (France, Switzerland, Germany and Austria) as well as the Baltic States and Russia, but it also includes two contributions dealing with wetland sites on the Balkan Peninsula (Giagkoulis 2020; Naumov 2020).

In this respect, with regard to prehistoric wetland settlements, **Southern Europe is by no means to be understood as a blank zone, but rather as less intensively and systematically researched, less internationally published, and above all not overarchingly approached**. This book intends to satisfy this desideratum.

1.3 Prehistoric Wetland Sites of Southern Europe: Archaeology, Dendrochronology, Palaeoecology and Bioarchaeology

The aim of this book is not to offer a complete compilation of Southern European wetland sites, but rather to present a cross-section of research statuses of different sites, currently ongoing research as well as novel, hitherto unpublished findings from relevant disciplines that have not been accessible to the international research community. By bringing together a selection of Mediterranean European wetland sites (cf. Fig. 1.1), a contextual perspective is opened for the first time. When multidisciplinary approaches have become a matter of course in archaeological sciences, this book is a testimony of transdisciplinary research: the presented studies showcase how disciplines and areas of expertise from the humanities and the natural sciences meet on an

equal footing and gain advanced knowledge through dialogue. In the process, the high degree of specialisation, and the conceptual, methodological, and practical overcoming of disciplinary boundaries becomes equally visible.

The book is divided into three parts, the first being dedicated to archaeology in the integral sense, i.e. the excavation of sites, their interdisciplinary exploration, as well as their interpretation, the second to dendroarchaeology and its contribution to the understanding of occupation sequences and regional chronologies, and the third to off-site palaeoecological research on past environments and people's impact on it, as well as bioarchaeological on-site approaches to subsistence strategies and land use practices, based on plant remains.

1.3.1 Archaeology

Part I of this book assembles eight contributions on selected archaeological sites and features, focusing on cases whose results have been underrepresented in the published literature to date. Based on new or previously only partly accessible research results, the chapters provide an impression of the broad variety of Neolithic and Bronze Age settlements in Southern Europe, characterised by diverse features, site chronologies and environmental settings. The pivotal impact of transdisciplinary research to the understanding of these sites—from specialised excavation and object recovery techniques, to documentation methods like underwater photogrammetry (Structure from Motion), dating methods (dendrochronology and radiocarbon dating), to palaeoecological off-site research on environmental conditions, to bioarchaeological analyses of on-site plant and faunal macroremains—is also made clear and will be further developed subsequently in parts II and III of the book.

Thibault Lachenal (Chap. 2) presents the Late Bronze Age site of La Motte 1 near Agde on the French Mediterranean coast. Now flooded by the river Hérault, the settlement was originally situated on the edge of a coastal lagoon. Due to its exposure to a highly variable natural environment the settlement was provided with elaborate bank fortification structures, which were intended to protect the village from water erosion. The example testifies to the perception and handling of both environmental opportunities and hazards by the settlement's inhabitants.

Marco Baioni, Claudia Mangani and Roberto Micheli (Chap. 3) provide a synthetic overview of the northern Italian wetland settlements, a large part of which are UNESCO Cultural Heritage sites. Spreading from western Piedmont to Veneto, i.e. from Trentino to the lower Lombardy and Veneto plains, particular clusters are located around Lakes Varese and Garda. Whereas they are settled from the Early Neolithic to the Late Bronze Age, a particularly high number between the Early and Middle Bronze Age (c. 2000–1400 cal BC). The authors provide the latest state of research, including new results—such as from the new investigations on the settlements of Palù di Livenza and Lucone di Polpenazze.

Occupied from as early as 5600 cal BC, the pile-dwelling of La Marmotta at Lake Bracciano (Rome, Lazio, Italy) belongs to the earliest-dated wetland settlements in this book's geographical scope. **Ariane Ballmer, Mario Mineo and Valeska Becker (Chap. 4)** review the current state of research and highlight selected aspects, such as the site chronology, as well as outstanding topics related to the settlers' way of life, in particular concerning agriculture but also navigation. With its early occupation dates and the fascinating spectrum of finds, the site seems to refer to a 'pioneer settlement' of a group of early farmers, and might have given substantial impetus to the Neolithisation of the western Mediterranean region.

Katarina Jerbić (Chap. 5) explores the wetland settlement of Zambratija Bay near Umag on the Adriatic coast of northern Croatia. The submerged site contains the remains of a pile-dwelling settlement from around 4000 cal BC. The author especially points out the potential of the site for future research, which she mainly sees in dendrochronology as well as palaeoenvironmental analyses.

Johannes Reich, Marco Hostettler, Ariane Ballmer and Albert Hafner (Chap. 6) present the latest results from the Neolithic and Bronze Age (middle of 5th millennium cal BC, and 18th–14th centuries cal BC) pile-dwelling settlement at Ploča Mičovgrad Grad near Gradište on the North Macedonian shore of Lake Ohrid. The latest research on this site represents a milestone in Balkan wetland settlement archaeology, especially in terms of excavation techniques, absolute dating, and interdisciplinary research approaches (see Chaps. 14, 15 and 17 in this book).

The essay by **Gilles Touchais, Cécile Oberweiler and Petrika Lera (Chap.** 7) introduces the findings of the French-Albanian excavations undertaken between 1993 and 2006 at Sovjan (Korçë Basin, Albania) and complements them with the very latest results, with a focus on the Bronze Age occupation. With the help of the waterlogged stratigraphy as well as the corresponding features and find material from the occupation layers, the authors are able to set decisive reference points for the Bronze Age in the Southwestern Balkans.

Kostas Kotsakis and Tryfon Giagkoulis (Chap. 8) summarise the results from the rescue excavations project in the Four Lakes Region in the Amindeon Basin in north-western Greece, and furthermore present the latest findings from the ongoing investigations at the Neolithic pile-dwelling settlement at Dispilio on the southern shore of Lake Kastoria. Besides the chronology of the wetland occupations, yielding dates as early as c. 5600 cal BC (Dispilio), they also pursue clues on the house construction and the internal structure of the different settlements, and furthermore discuss the sites' natural environment.

Ariane Ballmer, Kalin Dimitrov, Nayden Prahov and Pavel Georgiev (Chap. 9) provide an overview of the Bulgarian version of prehistoric wetland settlements from palaeolake shores, marshes, riverbanks, and lagoons, which according to current knowledge concentrate along the Black Sea coast. Evidence of Eneolithic and Early Bronze Age pile-dwelling settlements is available from the Varna Lakes area, as well as from the sites of Sozopol harbour,

Ropotamo and Urdoviza. The authors relate the deliberate settling of the wetlands to strategic roles within the natural and economic landscape, especially to the exploitation of aquatic resources, but possibly also in connection with the copper ore distribution network.

Finally, a geographical excursus to eastern China serves as an exemplary illustration of the globality of the phenomenon of prehistoric wetland settlements: **Haowei Wo and Guoping Sun (Chap.** 10) offer insights into the wetland settlements of the Zhejiang Province, the best-preserved archaeological features from prehistoric China. Here, wetland settlements first appear in the 8th and 7th millennia cal BC with the Early Neolithic Shangshan culture, followed by wetland settlements of the Kuahuqiao and Hemudu cultures. Rice cultivation and the exploitation of maritime resources are established. The Liangzhu city, dated to the 4th and 3rd millennium cal BC (Late Neolithic), shows characteristics of an 'early state', featuring complex social organisation, monumental architecture, and elaborate infrastructural buildings.

1.3.2 Dendrochronology

The chapters in Part II are dedicated to the dendrochronology of prehistoric Southern Europe, partly related to individual sites and their dating, but also concerning the methodological construction of regional standard curves. The fundamental research for absolute chronologies of the relevant regions relies on the advantages of waterlogged remains, especially construction timbers. While dendroarchaeology is firmly established in Spain, Italy and Slovenia, and circumalpine chronologies can even be correlated with one another (Billamboz and Martinelli 2015), the tree-ring chronologies of the Mediterranean zone still show major gaps and therefore rely on basic research, i.e. the construction of regional standard chronologies through systematic sampling and dating of archaeological wood from prehistoric settlement contexts (previous relevant works are by Kuniholm and Striker 1987; Manning et al. 2006;

Čufar et al. 2008, 2010; Westphal et al. 2010; Pearson et al. 2014; Ważny et al. 2014; Roibu et al. 2021). Especially in recent years, important gaps have been filled thanks to targeted, systematic sampling and elaborate analyses (Hafner et al. 2021; Maczkowski et al. 2021; Andreaki et al. 2022; Bolliger et al. 2023).

Oriol López-Bultó, Vasiliki Andreaki, Patrick Gassmann, Joan Anton Barceló, Ferran Antolín, Antoni Palomo, Xavier Terradas and Raquel Piqué (Chap. 11) discuss the latest results on the absolute dating of the Early Neolithic lakeshore settlement of La Draga (Lake Banyoles, Girona, Spain). Based on tree-ring analyses of the construction timbers, radiocarbon-dating, and wiggle-matching, three building phases are proposed, corresponding to settlement stages between c. 5300 and 4800 cal BC. Furthermore, a minimum lifetime of the wooden constructions of 27 years is established.

The general dendrochronology of the northern Italian lakeshore settlements is presented by **Nicoletta Martinelli (Chap.** 12). The tree-ring chronologies are dated by radiocarbon and wiggle-matching and (with interruptions) cover Bronze Age occupation phases from the beginning of the Early Bronze Age around 2200 cal BC to the 14th century cal BC (Middle Bronze Age, beginning of Late Bronze Age). Interestingly, the single sequences covering the time span between the 16th century and the end of the 14th century cal BC coincide with a period from which no wetland settlements are evidenced in the northern Alpine foreland.

Anton Velušček and Katarina Čufar (Chap. 13) look back on 25 years of dendrochronological research at the Ljubljana marsh (Slovenia). Between 1995 and 2017, nearly 8800 samples of waterlogged wood from 16 archaeological sites, mainly from vertical piles the dwellings were built on, were collected, and examined. Oak and ash tree-ring chronologies were established for most of the sites. Absolute dating was carried out by means of radiocarbon measurements and wiggle-matching, and for the 4th millennium cal BC settlements with the help of teleconnection with the German-Swiss reference chronology. Eventually, the authors point to the Slovenian oak chronologies' potential to be teleconnected with those from other regions.

For the Southwestern Balkans, a region which until recently has shown significant deficits in the absolute dating of prehistoric time periods, **Andrej Maczkowski, Matthias Bolliger and John Francuz (Chap.** 14) present tree-ring chronologies for the archaeological sites of Sovjan (Korçë Basin, Albania), Ploča Mičov Grad (Lake Ohrid, North Macedonia) and Dispilio (West Macedonia, Greece). These are the first prehistoric, centennial and multi-centennial tree-ring chronologies for the region—anchored in a high temporal resolution by means of radiocarbon dating and wiggle-matching—covering various periods of the Neolithic, Chalcolithic and the Bronze Age.

1.3.3 Palaeoecology and Bioarchaeology

Finally, three essays from the fields of palaeoecology and bioarchaeology address the ecological and subsistence economic dimensions of prehistoric wetland settlements. Palaeoecological and bioarchaeological research has a long and strong tradition in Southern Europe (Palaeoecology: e.g. Mercuri et al. 2010; Mercuri 2014; Ucchesu et al. 2015a, b, 2017a, b; Ramos-Román et al. 2016; Pérez-Jordà et al. 2017; Sarigu et al. 2017, 2022; Mora-González et al. 2018; López-Sáez et al. 2020; Florenzano et al. 2022; bioarchaeology, or archaeobotany respectively: e.g. Colledge and Conolly 2007; Marinova 2007; Rottoli and Pessina 2007; Marinova and Popova 2008; Valamoti et al. 2008, 2020; Bogaard et al. 2016; Antolín 2016; Kreuz and Marinova 2017; Kotzamani and Livarda 2018; Bouby et al. 2020; de Vareilles et al. 2020; Piqué et al. 2021; Vaiglova et al. 2021—just to mention a few publications from the past 20 years). While bioarchaeology has the advantage of reaching very high taxonomical resolutions for plants and animals used by humans and brought to the settlements, palaeoecology has the advantage of providing uninterrupted ecological time-series including long-term information on abundances of crops and weeds. Such time-series are particularly

useful to reconstruct land use activity trends and developments (e.g. crop production, or harvest success and failure) over centuries and millennia, independently from archaeological excavations, and thus in fact complement the bioarchaeological and archaeological evidence. Recent developments in the field of palaeoecology include probabilistic approaches to assess past land use intensities as well as a more detailed categorisation of organisms that are either native or introduced to a region. This problem is particularly pronounced for Southern Europe since many crops and weeds or their relatives (e.g. wild cereal species) have their origins in the Mediterranean region, including the Fertile Crescent. Thus, until recently, the detailed disentanglement of human and natural impacts on vegetation posed quite a challenge (Deza-Araujo et al. 2020, 2022a, b), making thorough off-site land use reconstructions difficult. Recent progress also includes the development of highly precise off-site chronologies that allow a better comparison between off-site and quite accurately dated on-site records, specifically the application of wiggle-matching techniques to off-site records (Rey et al. 2019a, b). These advances hold out the prospect of corresponding applications and results in Southern European case studies.

Based on vegetation shifts, **César Morales-Molino, Lieveke van Vugt, Ariane Ballmer, Sarah Brechbühl, Kathrin Ganz, Sylvia Gassner, Erika Gobet, Albert Hafner, André F. Lotter, Carolina Senn, Antoine Thévenaz and Willy Tinner (Chap.** 15) discuss the impact the first farmers in the Southern Balkans had on their natural environment. They refer to recent methodological developments in palaeoecology that contribute to tightening its linkages with archaeology and bioarchaeology. Two case studies are in the centre: Limni Zazari (Macedonia, Northern Greece) and Ploča Mičov Grad (Lake Ohrid, North Macedonia). As an example, the correlation of highly precise and accurate lake-sediment chronologies with dendrochronologically-dated archaeological settlements is carried out. Furthermore, using corresponding data from pollen assemblages from lake surface samples, the authors show that modern pollen–vegetation relationships can be reliably used to interpret fossil pollen records. Continuous pollen records of cereals and ruderal plants from Limni Zazari (Greece) allow the dating of early farming activities to around 6250 cal BC. At Ploča Mičov Grad (North Macedonia), pollen evidence of cereals and weeds hint at an onset of agriculture at 5500–5100 cal BC, i.e. significantly earlier than the tree-ring inferred beginning of the settlement occupation.

Bigna L. Steiner, Ferran Antolín, Raül Soteras, Mauro Rottoli and Daria G. Banchieri (**Chap.** 16) present the results of archaeobotanical analysis and radiocarbon dating performed on drilling cores extracted from the site of Isolino Virginia at Lake Varese (Lombardy, Northern Italy). At least two phases of occupation are identified, one between 5000 and 4700 cal BC, and another one between 4250 and 3650 cal BC, yielding an impressive number of plant macro-remains. The main crops during the 5th millennium cal BC appear to be naked wheat, naked barley, flax, and poppy, and possibly also pea. For the time being, this crop assemblage is rather comparable with some from the Western Mediterranean area than with the ones from the Eastern Italian sites, where glume wheats seem to have been the most important crops.

The book concludes with a Chapter by **Amy Holguin, Ferran Antolín, Mike Charles, Ana Jesus, Héctor Martínez Grau, Raül Soteras, Bigna L. Steiner, Elizabeth Stroud and Amy Bogaard** (**Chap.** 17), dedicated to the abundant, well-preserved botanical macroremains from the Late Neolithic occupation layers of the pile-dwelling settlement at Ploča Mičov Grad at Lake Ohrid (North Macedonia). Within an occupation duration of a few decades, the inhabitants of the Late Neolithic pile-dwelling settlement apparently made use of a range of cultivated cereals (particularly einkorn, emmer and barley), pulses (including lentil, pea and bitter vetch) and oilseed crops (flax and opium poppy), as well as wild fruit and nuts (such as almond, pistachio, blackberries and wild strawberry), thought to have been grown, or collected respectively in the vicinity of the village. Based on weed remains,

the authors assume high-input agrarian practices, featuring small-scale and labour-intensive cultivation. Thus, they add an essential dimension to our picture of the everyday life at Ploča Mičov Grad and initiate discussions about the strategic use and alteration of the wetland settlement's environment.

1.4 The Prehistoric Wetland Sites of Southern Europe at Stake

Evidently, waterlogged anthropogenic and natural deposits bear an enormous methodological and research potential and epistemological value for archaeology. Yet, in addition to investigating the sites, their protection and preservation is equally crucial. This is because cultural and natural heritage in waterlogged sediments, and thus the possibility of recording and documenting the corresponding findings, is in a race against time: the in situ-preservation of archaeological wetland sites worldwide is threatened by resource extraction, land reclamation through drainage and filling, construction works and landscaping, water management, and finally by the consequences of climate change.

When in the past, a series of archaeological wetland sites in the Amindeon Basin in Northern Greece had been severely threatened by mining activities (cf. *supra*), they were detected in time and could be safeguarded by means of extensive rescue excavations. However, others risk being unrecognised or remain unprotected, and thus are in danger of possibly being destroyed, or insufficiently documented due to a lack of resources, time pressure and economic priorities. Aware that there are many more, we would like to address two specific situations within the EXPLO-study area to illustrate some common threats to which archaeological wetland sites are exposed: an example of damage of waterlogged archaeological remains through man-made drainage measures and land reclamation was recently recorded on the Albanian shore of Lake Ohrid, in the vicinity of the village of Udënisht (cf. Andoni et al. 2017). The archaeological layers of the Neolithic pile-dwelling settlement of 'Lin 3' have been preserved both underwater as well as on the adjacent terrestrial shore. Farming activities in the land area involves the improvised extension of the agricultural fields towards the lake into the reed belt by means of slash-and-burn and subsequent earth fill, as well as the maintenance of a system of irrigation canals between the lake and the cultivated plots. Whereas ploughing seems unlikely to reach the archaeological layers, the water canals are clearly more invasive, as is evident from the frequent recovery of waterlogged, archaeological timbers in these trenches. A striking example of recent water level change, potentially exposing flooded archaeological remains of lake shore settlements to the air, is observed at Great Lake Prespa, located on the tripoint of Albania, Greece, and North Macedonia. Its water level has dropped by c. 8 m since 1987, likely due to climatically induced reduction in precipitation, as well as excessive water extraction for agricultural purposes (van der Schriek and Giannakopoulos 2017). As a result, today, the formerly submerged remains of at least a part of the Neolithic shore settlement of Kallamas on the Albanian lake shore are largely on dry land (Westphal et al. 2010; Oberweiler et al. 2014). Obviously further, so far unidentified, sites around Lake Prespa may be affected, too.

The **archaeological waterlogged sites' exposure to erosion, oxidation, and biological decay in response to rising water temperatures, lake and sea-level changes, or wetland drying, inevitably lead to an irreversible loss of unique archives to human and environmental (pre-) history**. Specifically, climate change-induced risks to archaeological wetland sites have been addressed and discussed by experts for years (e.g. Menotti and O'Sullivan 2013b; Van de Noort 2013a, b). This threat cannot be emphasised enough and, moreover, is becoming increasingly urgent. At the same time, the focus of previous studies by international initiatives on climate change and cultural heritage was mostly on the effects of climate change on historic buildings and urban settlement landscapes, as well as on technical preventive measures to future climate change induced issues

(Sabbioni et al. 2010; Bertolin and Camuffo 2014; Dunkley 2015; Dastgerdi et al. 2019; Sesana et al. 2021; Hafner 2022; Hafner and Underwood 2022). To bring the precarious situation described above into focus, the impact of climate change on the particularly vulnerable waterlogged archaeological relics in lakes, bogs, marshes, rivers, and oceans needs to be further identified, recorded, assessed, and monitored in detail. While a significant number of circum-alpine wetland sites benefit from intense monitoring and protection measures due to their UNESCO-status, in the future, special attention must be paid to regions exhibiting archaeological wetland sites where they are still under-explored, barely valorised, and less specifically protected. In the Balkans, for instance, the protection efforts at Lake Ohrid are particularly noteworthy in this regard: because of numerous endemic species and an important series of cultural heritage (mostly from historical times), the North Macedonian part of Lake Ohrid was classified as a mixed natural and cultural UNESCO World Heritage property in 1980 (*Natural and Cultural Heritage of the Ohrid region*) (https://whc.unesco.org/en/list/99/). In 2019, the World Heritage property was extended to the Albanian part of the lake. This promises to entail an inventory of the sites, which should lead to specific protection efforts in the future.

When the findings from wetland sites that are continuously obtained and presented by researchers further emphasise their cultural value, two strategies for preservation in particular have been established: on the one hand, immediate protection through the conservation of the archaeological and environmental contexts in situ (e.g. Ramseyer 2013; Olivier 2013; Hafner and Underwood 2022), and on the other hand the safeguarding approach by means of recovery and documentation of the features and finds to the highest possible standards and conditions.

In this sense, this book shall set an example: the presented research not only captures the phenomenon of prehistoric wetland settlements in Southern Europe but also demonstrates the extent to which state-of-the-art wetland archaeology, including cutting-edge interdisciplinary approaches can provide novel insights of the highest precision, thereby doing justice to the ephemeral cultural asset and its overarching significance.

Acknowledgements We would like to thank all the authors for participating in this book project, sharing their experience, knowledge, and research findings, and hence contributing to the archaeology of Southern European wetland sites. The preparation of this chapter was funded by the European Research Council (ERC) under the European Union's Horizon 2020 research and innovation programme, grant agreement No. 810586.

References

Andoni E, Hasa E, Gjipali I (2017) Neolithic settlements on the Western Bank of Lake Ohrid: Pogradec and Lin 3. In: Përzhita L (ed) Proceedings of the international conference 'new archaeological discoveries in the Albanian Region', Tirana, 30–31 Jan 2017. Botimet Albanologjike, Tirana, pp 45–61

Andreaki V, Barceló JA, Antolín F, Gassmann P, Hajdas I, López-Bultó O, Martínez-Grau H, Morera N, Palomo A, Piqué R, Revelles J, Rosillo R, Terradas X (2022) Absolute chronology at the waterlogged Site of La Draga (Lake Banyoles, NE Iberia): Bayesian chronological models integrating tree-ring measurement, radiocarbon dates and micro-stratigraphical data. Radiocarbon 64(5):907–948. https://doi.org/10.1017/RDC.2022.56

Antolín F (2016) Local, intensive and diverse? Early farmers and plant economy in the North-East of the Iberian Peninsula (5500–2300 cal BC). Barkhuis Publishing, Groningen

Banchieri DG (2014/15) Isolino Virginia im Lago di Varese, Italien. Der älteste Pfahlbau der Alpensüdseite und seine überregionalen Verbindungen. Plattform 23/24:32–43. https://www.pfahlbauten.de/wp-content/uploads/2020/06/Plattform_23_24.pdf

Bertolin C, Camuffo D (2014) Climate change impact on movable and immovable cultural heritage throughout Europe: damage risk assessment, economic impact and mitigation strategies for sustainable preservation of cultural heritage in the times of climate change. Clim Cult Deliv 5(2). https://www.climateforculture. eu/index.php?inhalt=download&file=pages/user/ downloads/project_results/D_05.2_final_publish. compressed.pdf

Billamboz A, Martinelli N (2015) Dendrochronology and Bronze Age pile-dwellings on both sides of the Alps: from chronology to dendrotypology, highlighting settlement developments and structural woodland changes. In: Menotti F (ed) The end of the lake-dwellings in the Circum-Alpine region. Oxbow Books, Oxford, Philadelphia, pp 68–84

Bogaard A, Hodgson J, Nitsch E, Jones G, Styring A, Diffey C, Pouncett J, Herbig C, Charles M, Ertuğ F, Tugay O, Filipovic D, Fraser R (2016) Combining functional weed ecology and crop stable isotope ratios to identify cultivation intensity: a comparison of cereal production regimes in Haute Provence, France and Asturias, Spain. Veg Hist Archaeobotany 25:57–73. https://doi.org/10.1007/s00334-015-0524-0

Bolliger M, Maczkowski A, Francuz J, Reich J, Hostettler M, Ballmer A, Naumov G, Taneski B, Todoroska V, Szidat S, Hafner A (2023) Dendroarchaeology at Lake Ohrid: 5th and 2nd millennia BC tree-ring chronologies from the waterlogged site of Ploča Mičov Grad, North Macedonia. Dendrochronologia:126095. https://doi.org/10.1016/j.dendro.2023. 126095

Bouby L, Marinval P, Durand F, Figueiral I, Briois F, Martzluff M, Perrin T, Valdeyron N, Vaquer J, Guilaine J, Manen C (2020) Early Neolithic (ca. 5850–4500 cal BC) agricultural diffusion in the Western Mediterranean: an update of archaeobotanical data in SW France. PLOS ONE 15(4):e0230731. https://doi.org/10.1371/journal.pone.0230731

Brechbühl S, van Vugt L, Gobet E, Morales-Molino C, Volery J, Lotter AF, Ballmer A, Brugger SO, Szidat S, Hafner A, Tinner W (2023) Vegetation dynamics and land-use change at the Neolithic lakeshore settlement site of Ploča Mičov Grad, Lake Ohrid, North Macedonia. Vegetation History and Archaeobotany

Bronk Ramsey C, van der Plicht J, Weninger B (2001) 'Wiggle matching' radiocarbon dates. Radiocarbon 43 (2A):381–389. https://doi.org/10.1017/S00338222 00038248

Chrysostomou P, Jagoulis T (2014/15). Land- und Seeufersiedlungen der 'Kultur der vier Seen', Griechenland. Neue archäologische Einblicke in die prähistorischen Siedlungen im Amindeon-Becken. Plattform 23/24:4–9. https://www.pfahlbauten.de/wp-content/uploads/2020/06/Plattform_23_24.pdf

Chrysostomou P, Jagoulis T, Mäder A (2015) The 'culture of four lakes'. Prehistoric lakeside settlements (6th–2nd mill. BC) in the Amindeon Basin, Western Macedonia, Greece. Archäologie Schweiz 38(3):24–32

Colledge S, Conolly J (2007) The Neolithisation of the Balkans: a review of the archaeobotanical evidence. In: Spataro M, Biagi P (eds) A short walk through the Balkans: the first farmers of the Carpathian Basin and adjacent Regions. Proceedings of the conference held at the Institute of archaeology UCL on June 20th–22nd, 2005, Società per La Preistoria e Protostoria Della Regione Friuli-Venezia Giulia. Quaderno, Trieste, pp 25–38

Čufar K, de Luis M, Zupančič M, Eckstein D (2008) A 548-year tree-ring-chronology of oak (*Quercus* spp.) for southeast Slovenia and its significance as a dating tool and climate archive. Tree-Ring Res 64(1):3–15. https://doi.org/10.3959/2007-12.1

Čufar K, Kromer B, Tolar T, Velušček A (2010) Dating of 4th millennium BC pile-dwellings on Ljubljansko barje, Slovenia. J Archaeol Sci 37:2031–2039. https://doi.org/10.1016/j.jas.2010.03.008

Dastgerdi A, Sargolini M, Pierantoni I (2019) Climate change challenges to existing cultural heritage policy. Sustainability 11:5227. https://doi.org/10.3390/su11195227

Della Casa P (2005) Introduction. Wetland economies and societies: 150 years of research on prehistoric economy and society in lake dwellings. In: Trachsel M, Della Casa P (eds) WES'04. Wetland economies and societies. Proceedings of the international conference Zurich. Collectio archaeologica, vol 3. Chronos, Zurich, 10–13 Mar 2004, pp 11–16

de Vareilles A, Bouby L, Jesus A, Martin L, Rottoli M, Linden MV, Antolín F (2020) One sea but many routes to Sail. The early maritime dispersal of Neolithic crops from the Aegean to the western Mediterranean. J Archaeol Sci Rep 29:102140. https://doi.org/10.1016/j.jasrep.2019.102140

Deza-Araujo M, Morales-Molino C, Tinner W, Henne PD, Heitz C, Pezzatti GB, Hafner A, Conedera M (2020) A critical assessment of human-impact indices based on anthropogenic pollen indicators. Quatern Sci Rev 236:106291. https://doi.org/10.1016/j.quascirev.2020.106291

Deza-Araujo M, Morales-Molino C, Conedera M, Henne PD, Krebs P, Hinz M, Heitz C, Hafner A, Tinner W (2022a) A new indicator approach to reconstruct agricultural land use in Europe from sedimentary pollen assemblages. Palaeogeogr Palaeoclimatol Palaeoecol 599:111051. https://doi.org/10.1016/j.palaeo.2022.111051

Deza-Araujo M, Morales-Molino C, Conedera M, Pezzatti GB, Pasta S, Tinner W (2022b) Influence of taxonomic resolution on the value of anthropogenic pollen indicators. Veg Hist Archaeobotany 31(1):67–84. https://doi.org/10.1007/s00334-021-00838-x

Dunkley M (2015) Climate is what we expect, weather is what we get. Managing the potential effects of oceanic climate change on underwater cultural heritage. In: Willems WJH, van Schaik HPJ (eds) Water & Heritage. Material, conceptual and spiritual connections. Sidestone Press, Leiden, pp 217–229. https://www.sidestone.com/openaccess/9789088902789.pdf

Dunning C, Hafner A (2005) Das Projekt 'Pfahlbauten des Alpenraums als UNESCO-Kulturerbe'. Informationen zur Nominierung auf die 'liste indicative' der Schweizerischen Bundesregierung vom Dezember 2004. In: Trachsel M, Della Casa P (eds) WES'04. Wetland economies and societies. Proceedings of the international conference Zurich. Collectio Archaeologica, vol 3. Chronos, Zurich, 10–13 Mar 2004, pp 297–298

Florenzano A, Zerboni A, Carter JC, Clo E, Mariani GS, Mercuri AM (2022) Environmental and land use changes in a Mediterranean landscape: palynology and geoarchaeology at ancient Metapontum (Pantanello, Southern Italy). Quatern Int 635:105–124. https://doi.org/10.1016/j.quaint.2022.01.004

Galimberti M, Bronk Ramsey C, Manning S (2004) Wiggle-match dating of tree-ring sequences. Radiocarbon 46(2):917–924. https://doi.org/10.1017/S0033822200035967

Gassner S, Gobet E, Schwörer C, van Leeuwen J, Vogel H, Giagkoulis T, Makri S, Grosjean M, Panajiotidis S, Hafner A, Tinner W (2020) 20,000 years of interactions between climate, vegetation and land use in Northern Greece. Veget Hist Archaeobot 29:75–90. https://doi.org/10.1007/s00334-019-00734-5

Giagkoulis T (2020) On the edge. The pile-field of the Neolithic Lakeside Settlement Anarghiri IXb (Amindeon, Western Macedonia, Greece) and the non-residential Wooden structures on the periphery of the habitation. In: Hafner A, Dolbunova E, Mazurkevich A, Pranckenaite E, Hinz M (eds) Settling waterscapes in Europe. The archaeology of neolithic and Bronze Age pile-dwellings. Open series in prehistoric archaeology, vol 1. Propylaeum, Bern, Heidelberg, pp 137–155. https://doi.org/10.11588/propylaeum.714

Gobet E, Tinner W, Hochuli PA, van Leeuwen JFN, Ammann B (2003) Middle to Late Holocene vegetation history of the Upper Engadine (Swiss Alps): the role of man and fire. Veg Hist Archaeobotany 12:143–163. https://doi.org/10.1007/s00334-003-0017-4

Hafner A (2022) UNESCO world heritage sites under water: archaeological places of outstanding universal value. In: Hafner A, Öniz H, Semaan L, Underwood CJ (eds) Heritage under water at risk. Challenges, threats and solutions. International Council on Monuments and Sites (ICOMOS), Paris, pp 26–30. https://openarchive.icomos.org/id/eprint/2488

Hafner A, Underwood CJ (2022) Introduction to the impact of climate change on underwater cultural heritage and the decade of ocean science for sustainable development 2021–2030. In: Hafner A, Öniz H, Semaan L, Underwood CJ (eds) Heritage under water at risk. Challenges, threats and solutions. International Council on Monuments and Sites (ICOMOS), Paris, pp 118–125. https://openarchive.icomos.org/id/eprint/2488

Hafner A, Dolbunova E, Mazurkevich A, Pranckenaite E, Hinz M (eds) (2020) Settling waterscapes in Europe. The archaeology of neolithic and Bronze Age pile-dwellings. In: Open series in prehistoric archaeology, vol 1. Propylaeum, Bern, Heidelberg, pp 1–6. https://doi.org/10.11588/propylaeum.714

Hafner A, Reich J, Ballmer A, Bolliger M, Antolín F, Charles M, Emmenegger L, Fandré J, Francuz J, Gobet E, Hostettler M, Lotter AF, Maczkowski A, Morales-Molino C, Naumov G, Stäheli C, Szidat S, Taneski B, Todoroska V, Bogaard A, Kotsakis K, Tinner W (2021) First absolute chronologies of neolithic and Bronze Age settlements at Lake Ohrid based on dendrochronology and radiocarbon dating. J Archaeol Sci Rep 38:103107. https://doi.org/10.1016/j.jasrep.2021.103107

Haneca K, Čufar K, Beeckman H (2009) Oaks, tree-rings and wooden cultural heritage: a review of the main characteristics and applications of oak dendrochronology in Europe. J Archaeol Sci 36:1–11. https://doi.org/10.1016/j.jas.2008.07.005

Jacomet S (2004) Archaeobotany: a vital tool in the investigation of lake-dwellings. In: Menotti F (ed) Living on the Lake in prehistoric Europe. 150 years of lake-dwelling research. Routledge, Oxon, New York, pp 162–177

Jacomet S (2013) Archaeobotany: analyses of plant remains from waterlogged archaeological sites. In: Menotti F, O'Sullivan A (eds) Oxford handbook of Wetland archaeology. Oxford University Press, Oxford, pp 497–514. https://doi.org/10.1093/oxfordhb/9780199573493.013.0030

Jacomet S, Brombacher C (2005) Reconstructing intra-site patterns in Neolithic lakeshore settlements: the state of archaeological research and future prospects. In: Trachsel M, Della Casa P (eds) WES'04. Wetland economies and societies. Proceedings of the international conference Zurich. Collectio Archaeologica, vol 3. Chronos, Zurich, 10–13 Mar 2004, pp 69–94

Kotzamani G, Livarda A (2018) People and plant entanglements at the dawn of agricultural practice in Greece. An analysis of the Mesolithic and Early Neolithic archaeobotanical remains. In: Urem-Kotsou D, Tasic N, Buric M, Papageorgopoulou C (eds) The meolithic of Northern Greece and the Balkans: the environmental context of cultural transformation. Quat Int Spec Issue 496:80–101

Kreuz A, Marinova E (2017) Archaeobotanical evidence of crop growing and diet within the areas of the Karanovo and the linear pottery cultures: a quantitative and qualitative approach. Veg Hist Archaeobotany 26:639–657. https://doi.org/10.1007/s00334-017-0643-x

Kuniholm PI, Striker CL (1987) Dendrochronological investigations in the Aegean and neighboring regions, 1983–1986. J Field Archaeol 14(4):385–398

Kuzman P (2013) Praistoriski palafitni naselbi vo Makedonija. In: Dimitrova E, Donev J (eds) Makedonija: Mileniumski Kulturno-Istoriski Fakti. Skopje, pp 297–430

Kuzman P (2016) Od Zlastrana do Penelopa: Neolitskite lokaliteti vo Ohridsko (I). In: Fidanoski L, Naumov G (eds) Neolit vo Makedonija: Novi Soznanija i Perspektivi. Centar za istražuvanje na predistorijata, Skopje, pp 23–40

Kuzman P (2017) Chronological and geographic routes of Ohrid's oldest population. Arheoloski Informator 1:147–156

López-Sáez JA, Carrasco RM, Turu V, Ruiz-Zapata B, Gil-García MJ, Luelmo-Lautenschlaeger R, Pérez-Díaz S, Alba-Sánchez F, Abel-Schaad D, Ros X, Pedraza J (2020) Late Glacial-early holocene vegetation and environmental changes in the western Iberian central system inferred from a key site: the Navamuño record, Béjar range (Spain). Quatern Sci Rev 230:106167. https://doi.org/10.1016/j.quascirev.2020.106167

Maarleveld TJ (2014) Underwater sites in archaeological conservation and preservation. In Smith C (ed) Encyclopedia of global archaeology. Springer, New York. https://doi.org/10.1007/978-1-4419-0465-2_522

Maczkowski A, Bolliger M, Ballmer A, Gori M, Lera P, Oberweiler C, Szidat S, Touchais G, Hafner A (2021) The Early Bronze Age dendrochronology of Sovjan (Albania): a first tree-ring sequence of the 24th–22nd c. BC for the southwestern Balkans. Dendrochronologia 66:125811. https://doi.org/10.1016/j.dendro.2021.125811

Manning SW, Bronk Ramsey C, Kutschera W, Higham T, Kromer B, Steier P, Wild EM (2006) Chronology for the Aegean Late Bronze Age 1700–1400 BC. Science 312:565–569. https://doi.org/10.1126/science.1125682

Marinova E (2007) Archaeobotanical data from the Early Neolithic of Bulgaria. In: Colledge S, Conolly J, Shennan S (eds) The origins and spread of domestic plants in Southwest Asia and Europe. Taylor & Francis Group, Walnut Creek, pp 93–109

Marinova E, Popova T (2008) Cicer arietinum (chick pea) in the Neolithic and Chalcolithic of Bulgaria: implications for cultural contacts with the neighbouring regions? Veg Hist Archaeobotany 17:73–80. https://doi.org/10.1007/s00334-008-0159-5

Menotti F (ed) (2004) Living on the lake in prehistoric Europe: 150 years of lake-dwelling research. Routledge, London

Menotti F (2013) Wetland occupations in prehistoric Europe. In: Menotti F, O'Sullivan A (eds) Oxford handbook of Wetland archaeology. Oxford University Press, Oxford, pp 11–25. https://doi.org/10.1093/oxfordhb/9780199573493.013.0003

Menotti F, O'Sullivan A (eds) (2013a) Oxford handbook of Wetland archaeology. Oxford University Press, Oxford

Menotti F, O'Sullivan A (2013b) General introduction to the handbook. In: Menotti F, O'Sullivan A (eds) Oxford handbook of Wetland archaeology. Oxford University Press, Oxford, pp 1–6. https://doi.org/10.1093/oxfordhb/9780199573493.013.0001

Mercuri AM (2014) Genesis and evolution of the cultural landscape in central Mediterranean: the 'where, when and how' through the palynological approach. Landscape Ecol 29:1799–1810. https://doi.org/10.1007/s10980-014-0093-0

Mercuri AM, Florenzano A, N'Siala IM, Olmi L, Roubis D, Sogliani F (2010) Pollen from archaeological layers and cultural landscape reconstruction: case studies from the Bradano valley (Basilicata, southern Italy). Plant Biosyst 144:888–901. https://doi.org/10.1080/11263504.2010.491979

Moore PD, Webb JA, Collison ME (1991) Pollen analysis. Blackwell Scientific Publications, Oxford

Mora-González A, Delgado-Huertas A, Granados-Torres A, Contreras Cortés F, Pavón Soldevilla I, Duque Espino D (2018) Complex agriculture during the second millennium BC: isotope composition of carbon studies (δ^{13}C) in archaeological plants of the settlement Cerro del Castillo de Alange (SW Iberian Peninsula, Spain). Veg Hist Archaeobotany 27:453–462. https://doi.org/10.1007/s00334-017-0634-y

Naumov G (2014/15) Prähistorische Pfahlbauten im Ohrid-See, Republik Mazedonien. Plattform 23/24: 10–20. https://www.pfahlbauten.de/wp-content/uploads/2020/06/Plattform_23_24.pdf

Naumov G (2020) Neolithic wetland and lakeside settlements in the Balkans. In: Hafner A, Dolbunova E, Mazurkevich A, Pranckenaite E, Hinz M (eds) Settling waterscapes in Europe. The archaeology of neolithic and Bronze Age Pile-Dwellings. Open series in prehistoric archaeology, vol 1. Propylaeum, Bern and Heidelberg, pp 111–135. https://doi.org/10.11588/propylaeum.714

Oberweiler C, Lera P, Kurti R, Touchais G, Aslaksen OC, Blein C, Elezi G, Gori M, Krapf T, Maniatis Y, Wagner S (2014) Mission archéologique franco-albanaise du bassin de Korçë. Bull Arch Écoles Fr L'étranger Balkans. https://doi.org/10.4000/baefe.1660

Oberweiler C, Touchais G, Lera P (2014/15) Prähistorische Siedlungen im Seeuferbereich von Korça, Albanien. Plattform 23/24:22–31. https://www.pfahlbauten.de/wp-content/uploads/2020/06/Plattform_23_24.pdf

Olivier A (2013) International and national Wetland management policies. In: Menotti F, O'Sullivan A (eds) Oxford handbook of Wetland archaeology. Oxford University Press, Oxford, pp 687–702. https://doi.org/10.1093/oxfordhb/9780199573493.013.0041

Pearson CL, Ważny T, Kuniholm PI, Botić K, Durman A, Seufer K (2014) Potential for a new multimillennial tree-ring chronology from Subfossil Balkan River Oaks. Radiocarbon 56:51–59. https://doi.org/10.2458/azu_rc.56.18342

Pérez-Jordà G, Peña-Chocarro L, García Fernández M, Vera Rodríguez JC (2017) The beginnings of fruit tree cultivation in the Iberian Peninsula: plant remains from the city of Huelva (southern Spain). Veg Hist Archaeobotany 26:527–538. https://doi.org/10.1007/s00334-017-0610-6

Pétrequin P (2013) Lake-dwellings in the Alpine Region. In: Menotti F, O'Sullivan A (eds) Oxford handbook of Wetland archaeology. Oxford University Press, Oxford, pp 253–267. https://doi.org/10.1093/oxfordhb/9780199573493.013.0016

Piqué R, Alcolea M, Antolín F, Berihuete-Azorín M, Berrocal A, Rodríguez-Antón D, Herrero-Otal M, López-Bultó O, Obea L, Revelles J (2021) Mid-holocene palaeoenvironment, plant resources and human interaction in Northeast Iberia: an archaeobotanical approach. Appl Sci 11:5056. https://doi.org/10.3390/app11115056

Ramos-Román MJ, Jiménez-Moreno G, Anderson RS, García-Alix A, Toney JL, Jiménez-Espejo FJ, Carrión JS (2016) Centennial-scale vegetation and North Atlantic Oscillation changes during the Late Holocene in the southern Iberia. Quatern Sci Rev 143:84–95. https://doi.org/10.1016/j.quascirev.2016.05.007

Ramseyer D (2013) Preservation against erosion: protecting Lake Shores and Coastal environments. In: Menotti F, O'Sullivan A (eds) Oxford handbook of Wetland archaeology. Oxford University Press, Oxford, pp 615–661. https://doi.org/10.1093/oxfordhb/9780199573493.013.0039

Reich J, Steiner P, Ballmer A, Emmenegger L, Hostettler M, Stäheli C, Naumov G, Taneski B, Todoroska V, Schindler K, Hafner A (2021) A novel Structure from Motion-based approach to underwater pile field documentation. J Archaeol Sci Rep 39:103120. https://doi.org/10.1016/j.jasrep.2021.103120

Rey F, Gobet E, van Leeuwen J, Gilli A, van Raden U, Hafner A, Wey O, Rhiner J, Schmocker D, Zünd J, Tinner W (2017) Vegetational and agricultural dynamics at Burgäschisee (Swiss Plateau) recorded for 18,700 years by multi-proxy evidence from partly varved sediments. Veg Hist Archaeobotany 26(6):71–586. https://doi.org/10.1007/s00334-017-0635-x

Rey F, Gobet E, Schwörer C, Wey O, Hafner A, Tinner W (2019a) Causes and mechanisms of synchronous succession trajectories in primeval Central European mixed Fagus sylvatica forests. J Ecol 107:1392–1408. https://doi.org/10.1111/1365-2745.13121

Rey F, Gobet E, Szidat S, Lotter AF, Gilli A, Hafner A, Tinner W (2019b) Radiocarbon wiggle matching on laminated sediments delivers high-precision chronologies. Radiocarbon 61:265–285. https://doi.org/10.1017/RDC.2018.47

Roibu CC, Ważny T, Crivellaro A, Mursa A, Chiriloaei F, Ştirbu MI, Popa I (2021) The Suceava oak chronology: a new 804 years long tree-ring chronology bridging the gap between central and south Europe. Dendrochronologia 68:125856. https://doi.org/10.1016/j.dendro.2021.125856

Rottoli M, Pessina A (2007) Neolithic agriculture in Italy: an update of archaeobotanical data with particular emphasis on northern settlements. In: Colledge S, Conolly J, Shennan S (eds) The origins and spread of domestic plants in Southwest Asia and Europe. Taylor & Francis Group, Walnut Creek, pp 141–154

Sabbioni C, Brimblecombe P, Cassar M (2010) The Atlas of climate change impact on European cultural heritage. Scientific analysis and management strategies. Noah's ark. In: Global climate change impact on built heritage and cultural landscapes. Anthem Press, London

Sarigu M, Grillo O, Lo Bianco M, Ucchesu M, d'Hallewin G, Loi MC, Venora G, Bacchetta G (2017) Phenotypic identification of plum varieties (*Prunus domestica* L.) by endocarps morpho-colorimetric and textural descriptors. Comput Electron Agric 136:25–30. https://doi.org/10.1016/j.compag.2017.02.009

Sarigu M, Sabato D, Ucchesu M, Loi MC, Bosi G, Grillo O, Torres SB, Bacchetta G (2022) Discovering Plum, Watermelon and Grape cultivars founded in a Middle Age Site of Sassari (Sardinia, Italy) through a computer image analysis approach. Plants (Basel) 11(8):1089. https://doi.org/10.3390/plants11081089

Schibler J (2004) Bones as a key for reconstructing the environment, nutrition and economy of the lake-dwelling societies. In: Menotti F (ed) Living on the Lake in prehistoric Europe. 150 years of lake-dwelling research. Routledge, Oxon, New York, pp 144–161

Schöbel G (2014/15) Editorial. Plattform 23/24:1. https://www.pfahlbauten.de/wp-content/uploads/2020/06/Plattform_23_24.pdf

Schweingruber FH (1988) Tree rings: basics and applications of dendrochronology. Kluwer Academic Publishers, Dordrecht. https://doi.org/10.1007/978-94-009-1273-1

Sesana E, Gagnon AS, Ciantelli C, Cassar JA, Hughes JJ (2021) Climate change impacts on cultural heritage: a literature review. Wires Clim Change 12(4):e710. https://doi.org/10.1002/wcc.710

Touchais G, Lera P, Oberweiler C (2005) L'habitat protohistorique lacustre de Sovjan (Albanie): dix ans de recherches franco-Albanaises (1993–2003). In: Trachsel M, Della Casa P (eds) WES'04. Wetland economies and societies. In: Proceedings of the international conference Zurich. Collectio Archaeologica, vol 3. Chronos, Zurich, 10–13 Mar 2004, pp 255–258

Trachsel M, Della Casa P (eds) (2005) WES'04. Wetland economies and societies. In: Proceedings of the international conference Zurich. Collectio Archaeologica, vol 3. Chronos, Zurich, 10–13 Mar 2004

Ucchesu M, Orru M, Grillo O, Venora G, Usai A, Serreli PF, Bacchetta G (2015a) Earliest evidence of a primitive cultivar of *Vitis vinifera* L. during the Bronze Age in Sardinia (Italy). Veg Hist Archaeobotany 24:587–600. https://doi.org/10.1007/s00334-014-0512-9

Ucchesu M, Pena-Chocarro L, Sabato D, Tanda G (2015b) Bronze Age subsistence in Sardinia, Italy: cultivated plants and wild resources. Veg Hist Archaeobotany 24:343–355. https://doi.org/10.1007/s00334-014-0470-2

Ucchesu M, Sarigu M, Del Vais C, Sanna I, d'Hallewin G, Grillo O, Bacchetta G (2017a) First finds of Prunus domestica L. in Italy from the Phoenician and Punic periods (6th–2nd centuries BC). Veg Hist Archaeobotany 26:539–549. https://doi.org/10.1007/s00334-017-0622-2

Ucchesu M, Sau S, Luglie C (2017b) Crop and wild plant exploitation in Italy during the Neolithic period: new data from Su Mulinu Mannu, Middle Neolithic site of Sardinia. J Archaeol Sci Rep 14:1–11. https://doi.org/10.1016/j.jasrep.2017.05.026

Vaiglova P, Coleman J, Diffey C, Melanie F, Pappa M, Halstead P, Valamoti S, Cavanagh W, Renard J, Buckley M, Bogaard A (2021) Exploring diversity in Neolithic agropastoral management in mainland Greece using stable isotope analysis. Environ Archaeol 28(2):62–85. https://doi.org/10.1080/14614103.2020.1867292

Valamoti SM, Samuel D, Bayram M, Marinova E (2008) Prehistoric cereal foods from Greece and Bulgaria: investigation of starch microstructure in experimental and archaeological charred remains. Veg Hist Archaeobotany 17:265–276. https://doi.org/10.1007/s00334-008-0190-6

Valamoti SM, Chondrou D, Bekiaris T, Ninou I, Alonso N, Bofill M, Ivanova M, Laparidou S, McNamee C, Palomo A, Papadopoulou L, Prats G, Procopiou H, Tsartsidou G (2020) Plant foods, stone tools and food preparation in prehistoric Europe: an integrative approach in the context of ERC funded project PLANTCULT. J Lithic Stud 7(3):21. https://doi.org/10.2218/jls.3095

Van de Noort R (2013a) Climate change archaeology. Building resilience from research in the world's coastal wetlands. Oxford University Press, Oxford. https://doi.org/10.1093/acprof:osobl/9780199699551.001.0001

Van de Noort R (2013b) Wetland archaeology in the 21st century: adapting to climate change. In: Menotti F, O'Sullivan A (eds) Oxford handbook of Wetland archaeology. Oxford University Press, Oxford, pp 719–732. https://doi.org/10.1093/oxfordhb/9780199573493.013.0043

van der Schriek T, Giannakopoulos C (2017) Determining the causes for the dramatic recent fall of Lake Prespa (southwest Balkans). Hydrol Sci J 62(7):1131–1148. https://doi.org/10.1080/02626667.2017.1309042

Van Vugt L, Garcés-Pastor S, Gobet E, Brechbühl S, Knetge A, Lammers Y, Stengele K, Greve Alsos I, Tinner W, Schwörer C (2022) Pollen, macrofossils and sedaDNA reveal climate and land use impacts on Holocene mountain vegetation of the Lepontine Alps, Italy. Quat Sci Rev 296:107749. https://doi.org/10.1016/j.quascirev.2022.107749

Ważny T, Lorentzen B, Köse N, Akkemik Ü, Boltryk Y, Güner T, Kyncl J, Kyncl T, Nechita C, Sagaydak S, Kamenova Vasileva J (2014) Bridging the gaps in tree-ring records: creating a high-resolution dendrochronological network for Southeastern Europe. Radiocarbon 56(4):39–50. https://doi.org/10.2458/azu_rc.56.18335

Westphal T, Tegel W, Heussner KU, Lera P, Rittershofer KF (2010) Erste dendrochronologische Datierungen historischer Hölzer in Albanien. Archaeol Anz 2:75–95. https://doi.org/10.34780/8ib4-c616

Part I
Archaeology

From Lagoon to River: Bank Management at the Submerged Late Bronze Age Settlement of La Motte (Agde, France)

2

Thibault Lachenal

Abstract

The La Motte 1 site (Agde), submerged in the bed of the Hérault River, corresponds to a settlement from the end of the Bronze Age initially established on the edge of a lagoon that is now clogged. The site is characterised by more than 500 wooden piles divided into two main groups located on shoals. The use of varied species attests to the exploitation of distinct environments such as Mediterranean mixed oak forests and riparian forests. Moreover, the spatial analysis illustrates the presence of several coherent alignments that use a specific species. The excavation trenches dug support the interpretation thereof as systems for bank maintenance and protection via different techniques: wattle made from the strawberry tree on oak piles, and possible live stakes made of willow. The study of the goods and radiocarbon dating show the successive installation of these systems between the 10th and the beginning of the 8th centuries BC. The goal of the various technical solutions was to protect the areas around the settlement from erosion. Their use must therefore be examined in the light of the changes in their immediate surroundings, which saw a rise in the sea level coupled with a progressive filling of the delta of the Hérault. This case study thus illustrates the capacity to adapt and the vulnerability of the populations of the Bronze Age in the face of changes in a particularly sensitive environment.

Keywords

Southern France · Late Bronze Age · Lagoon · Piles · Wattle · Bank reinforcement

2.1 Introduction

Currently submerged in the bed of the Hérault River, the La Motte 1 site was discovered in 2001 during a subaquatic survey. In 2004, a dig revealed a deposit of objects, mostly made of bronze, corresponding to a rich ornament with feminine connotations dating to the eighth century BC (Moyat et al. 2005, 2007; Verger et al. 2007). Although this exceptional discovery was the focus of the work during this dig, the presence of a settlement from the end of the Bronze Age, embodied by wooden piles and numerous remains of a domestic nature, was also brought to light. The context in which this settlement was located remained unclear: low bank of a river, island, or shoal in the riverbed, or even a village built on a lagoon (Moyat et al. 2007, 80)?

T. Lachenal (✉)
UMR 5140 Archéologie des Sociétés
Méditerranéennes (ASM), Université Paul-Valéry,
CNRS, MC, Montpellier, France
e-mail: thibault.lachenal@cnrs.fr

© The Author(s) 2025
A. Ballmer et al. (eds.), *Prehistoric Wetland Sites of Southern Europe*,
Natural Science in Archaeology, https://doi.org/10.1007/978-3-031-52780-7_2

Likewise, its extent and duration of occupation were not able to be determined. New subaquatic excavation campaigns were thus undertaken in 2011, accompanied by a research program on the palaeoenvironment and the geomorphology of the territory around Agde from the beginning of the Holocene[1] (Gascó et al. 2014; Gascó et al. 2015; Deviller et al. 2019; Lachenal et al. 2020). These multidisciplinary studies revealed a permanent settlement set up on the edge of an ancient lagoon, with an economy mainly based on agriculture and livestock (Bouby et al. 2016; Lespes et al. 2019). Both the lagoon and the alluvial plain were exploited, although land resources seem to be prevalent. Several signs point to the elevated social status of the group (or of certain individuals) who occupied the site. Herd management, which indicates the search for quality meat, supposes a privileged social environment (Lespes et al. 2019). The presence of several deposits of metal objects, including a rich ceremonial garment, indicates the same (Verger et al. 2007; Lachenal 2022a). Metallurgy seems to have been a dynamic craft activity at the site, as attested by the presence of several moulds, which raises the question of the role of the settlement in the production and distribution of objects made from copper alloy, potentially manufactured using the copper ore from the Cabrières district located in the hinterland (Lachenal 2022b). The location of the site near the coast and the natural landmark formed by the Cap d'Agde potentially gave it a strategic position in the organisation of exchanges. The founding of this community does not therefore seem to be the result of chance. However, such interface areas between the land and sea, reputed to be unsanitary, present numerous constraints. Bronze Age lagoon settlements around the Gulf

of Lion were thus considered for a long time to be sites occupied temporarily, during the dry season (Py 1990).

Indeed, adaptation to this environment required the implementation of specific structures by the settlement's occupants. Made of wood, these structures were exceptionally preserved due to the submersion of the site. Here, we present the various types of structures identified during digs, their supposed function, and their chronology.

2.2 From Lagoon to River: Location and Palaeogeography of the Site

The La Motte site is currently located in the bed of the Hérault River, 5 km from its mouth and a little more than 500 m north of the modern city of Agde, in a portion of the river intersected by the Canal du Midi (Fig. 2.1). The site is located downstream of a meander formed by the river where it meets the northern flow of the Mont Saint-Loup volcanic complex.

The archaeological remains, which appear on the surface of the river bottom, are located on shoals preserved from erosion (Fig. 2.2), at a depth of between 4 and 6 m. It should be noted, however, that the bottom water level is artificially raised by 1.83 m by the La Pansière dam, 400 m downstream.

The location of the settlement at La Motte in potentially very dynamic coastal surroundings immediately raised the question of the site's environment and the evolution thereof. A set of 34 boreholes associated with 61 radiocarbon datings and biological and geochemical analyses allowed the geomorphological evolution of the lower Hérault valley from the beginning of the Holocene to be described (Devillers et al. 2019). Until 6500 cal BP, the morphology was dominated by waves and retrogradation. During this phase, successive ephemeral river mouths formed and were rapidly submerged by the rising sea level. A progradation phase started after 6500 cal BP. It progressively filled the estuary with the advance of the ria, and several shallow

[1] This program, which is still underway, ran in 2011 and from 2013 to 2015 under the supervision of Jean Gascó (CNRS UMR 5140 ASM), then from 2016 to 2021 under Thibault Lachenal (CNRS UMR 5140 ASM), in collaboration with the association Ibis (Agde). The work was supported by the LabEx ARCHIMEDE via the Investissement d'Avenir program, ANR-11-LABX-0032–01, and the projects DYLITAG (dir. B. Devillers) and SiLÂB (dir. T. Lachenal).

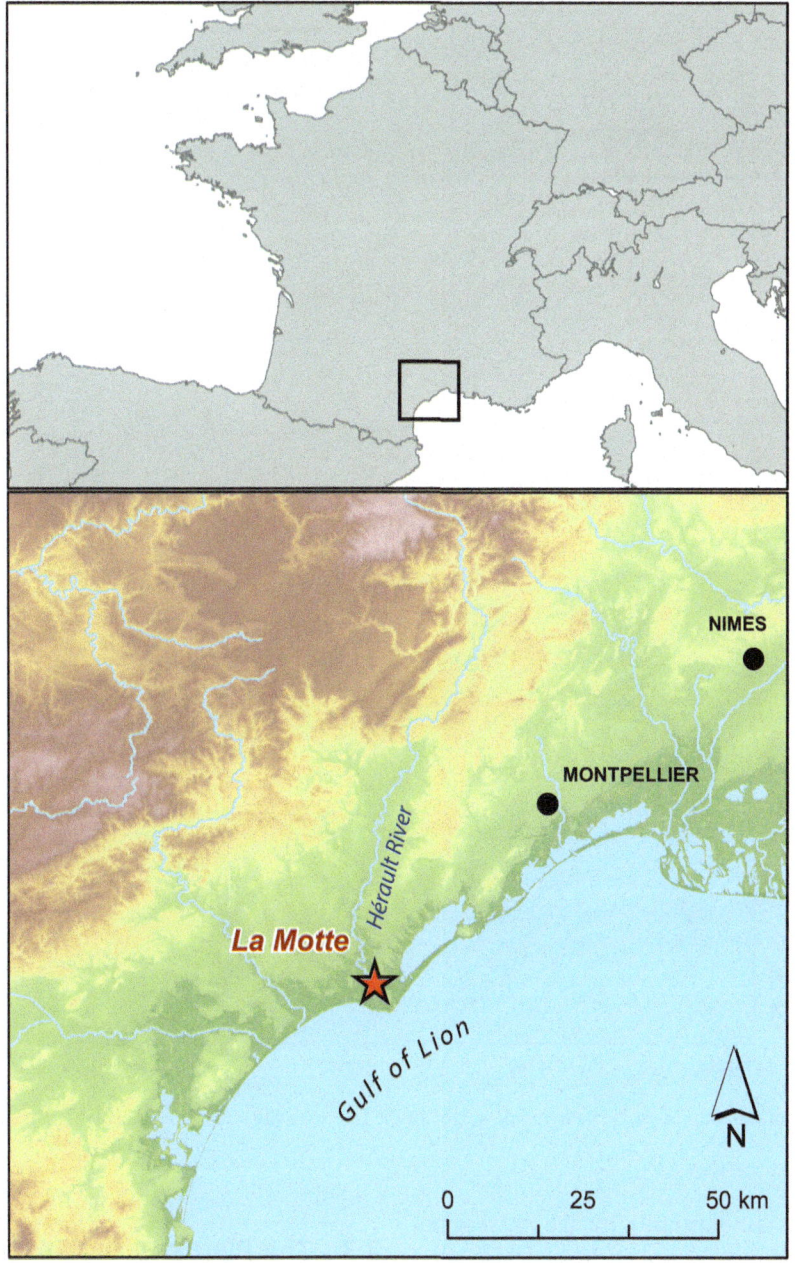

Fig. 2.1 Location of the La Motte site in south-eastern France (GIS by T. Lachenal; background BD by ALTI®)

lagoons associated with beach ridges have been distinguished. For the Bronze Age, the high density of information points to a delta morphology with a pronounced fluvial dominance. It appears that the site of La Motte was established on the edge of an ancient lagoon, near the mouth of the river. During the site's occupation, the fluvial sedimentation dynamics, associated with anthropic waste, led to the filling in of the lagoon before the sediments transported by the river clogged the lower part of the Hérault valley (Devillers et al. 2019). This mosaic of

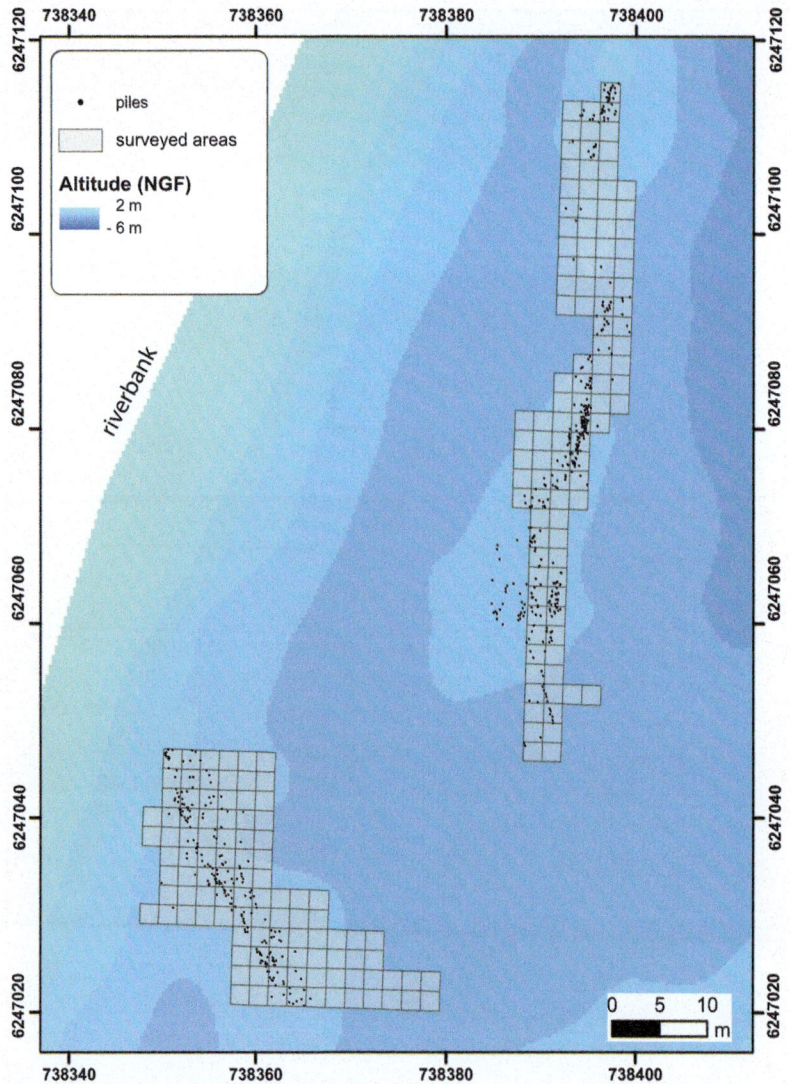

Fig. 2.2 Map of the features on the bathymetric survey of the La Motte site (GIS T. Lachenal; bathymetry by F. Yung)

environments is illustrated by the palaeoento-mological and carpological analyses, which attest to the nearby presence of Mediterranean and riparian forests, the river, and the biotopes found along coastal lagoons, such as *sansouïres* (salt marshes) and reedbeds (Bouby et al. 2016, 85–86). The site was thus located in Bronze Age lagoon habitats that are well known along the coast of the Gulf of Lion, in particular around the Étang de l'Or and Étang de Thau (Dedet and Py 1985; Leroy 2010).

2.3 The Plan and the Structures

The study of the La Motte site uses two complementary approaches: an extensive overhead survey of all the structures directly visible on the surface and excavation trenches in several key sectors to document the stratigraphy of the site and its evolution. This strategy was adopted in response to the time and cost constraints specific to subaquatic operations.

The La Motte site is characterised by more than 500 wooden piles divided into two main groups. The first, to the west and near the right bank, has a northwest-southeast orientation and includes more than 200 pieces of architectural wood. The second, near the left bank to the east, is more roughly oriented north–south and contains 323 piles (Figs. 2.2 and 2.3). In these two sectors, located on shoals, the wood was associated with remains from domestic activities (pottery, fauna, grinding goods, etc.), as well as numerous basalt blocks. The latter very probably come from the outcrops located farther south, from the volcanic flow on which the modern city of Agde is built. Riprap consisting of large blocks of basalt reaching a width of 50 cm also borders the remains to the east and to the north. An excavation trench carried out in sector 1 (Fig. 2.3) revealed, however, that they were installed after the Bronze Age occupation and must date to between antiquity and the modern era.

2.3.1 The Piles

The wood identified corresponds for the most part to vertical elements, piles and stakes, some of which may have collapsed. The specimens analysed reveal the use of varied species, which indicates a diversified exploitation of distinct environments: Mediterranean mixed oak forest, alluvial forest (floodplain) and riparian forest (streambed). Oak (*Quercus* sp.) and elm (*Ulmus* sp.) are highly dominant, but willow (*Salix* sp.), holm oak (*Quercus ilex*), alder (*Alnus* sp.), and ash (*Fraxinus* cf. *angustifolia/ornus*) are also well represented. The diameters of the wood, which range from 3 to 20 cm, are for the most part between 6 and 10 cm (Fig. 2.4). These were therefore mostly young trees. The piles were made from whole trunks of green wood, which is evidenced by the deformation of some pile tips after their staking and were not debarked. They seem to reflect a simple exploitation of a forest environment in which wood was plentiful. The points of these piles were sharpened to a single or multiple bevel/s (Lachenal et al. 2020).

2.3.2 Spatial Distribution of the Species of Wood

In the sectors in which the anatomic identification of the wood was carried out exhaustively, it is possible to perform a spatial analysis of the structuring of the site based on the nature of the wood used (Lachenal et al. 2020).

In sectors 1 and 2 on the left bank, among the wood that was subject to an anatomical determination, elm (*Ulmus* sp.) and oaks (*Quercus* sp. and *Quercus* cf. *coccifera/ilex*) represent the large majority (Fig. 2.5). Several coherent arrangements can be distinguished, with a line of fallen oak specimens to the south that seems to divide into two branches, which are also present farther north. One of them is superimposed on an alignment of elm posts.

In sectors 3 and 4, to the west on the right bank, there is also a large variety of species, with ten tree species identified (Fig. 2.6). Three of them, however, are clearly superior in number: oak (82 pieces), elm (78 piles), and willow (*Salix* sp.: 26). The groupings of the wood show several alignments using a specific species. To the southwest, a line of piles corresponds for the most part to specimens made of elm. In parallel, three alignments of oak piles can also be distinguished farther east, towards the middle of the river. Finally, in the northwest of this sector, three small sections principally use stakes made from willow. The coherence of these various arrangements allows them to be interpreted as several successive structures. The use of different wood could thus be explained as the result of the exploitation of varied environments: deciduous oak forest, alluvial forest for elm and riparian forest for willow.

2.3.3 Interpretation of the Alignments of Piles: Results of the Excavation Trenches

2.3.3.1 Sector 2

An excavation trench passing through squares D44, D63, D87, and D118 was dug in the southern part of the line of oak piles of sector 2 in

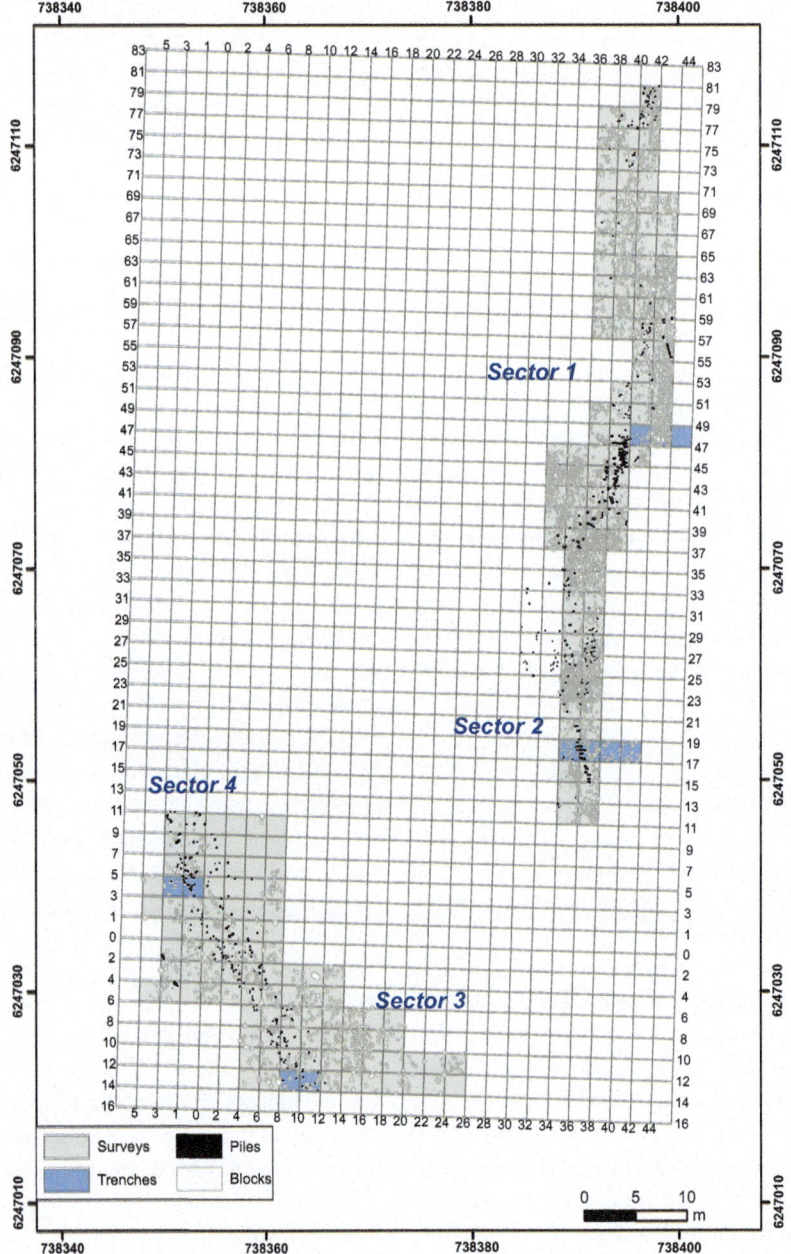

Fig. 2.3 General mapping of the features of the La Motte site and location of the trial trenches (GIS by T. Lachenal)

order to interpret these alignments of piles (Fig. 2.3). The dig revealed the presence of wattle (US 3) inclined to the east (Fig. 2.7b). The posts supporting it are oak piles on which the bark has been partially preserved, with irregular, non-rectilinear shapes between 2 and 2.5 m long

(Fig. 2.7a). They support a lattice of unpruned strawberry-tree (*Arbutus unedo*) branches that are twisted and woven between the piles (Fig. 2.8). The presence of oysters in a living position at the tips of the piles suggests that the base of this structure was underwater when it was

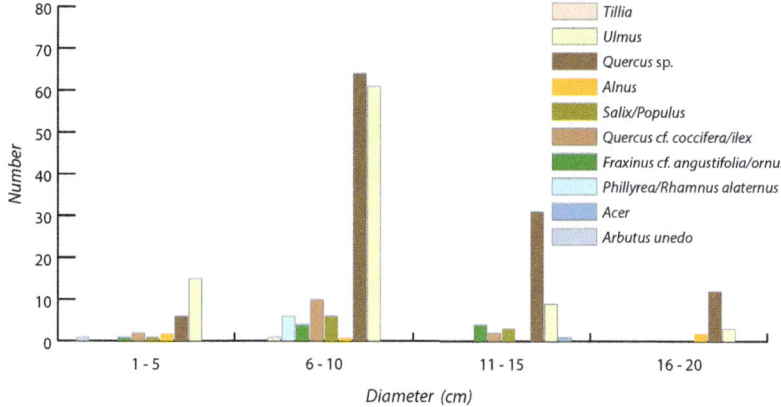

Fig. 2.4 Distribution of the wood species of the piles by diameter class (Xylological analysis by S. Greck)

in use (Fig. 2.7a). This structure can be interpreted as a bank reinforcement system. The incline observed can thus correspond either to the collapse of the structure or to its installation on an inclined slope. A level of basalt blocks, some of which had been colonised by oysters (US 23), was present at the base of the wattle (Fig. 2.7b). This corresponds to a deliberate laying of stones that may also have been used to reinforce the bank and act as a submerged foundation for the wattle, according to a mixed solution currently used in vegetation engineering[2] (Adam et al. 2008, 152).

The stratigraphy associated with this wattle features the alternation, under more recently formed units (US 17, 14), of layers of sandy and clay loam grey in colour, rich in ceramic goods, fauna, and shells (US 11–15, 10, 18, 1, 2, 7, 22), with browner layers very rich in organic matter but poorer in terms of goods (US 4, 6, 8, 13, 12, 19). All these levels are, just like the wattle, inclined towards the east. They thus tend to confirm the presence of a bank in this sector, located in contact with the protohistoric settlement and the palaeolagoon. The various types of levels identified can be interpreted as fill of

different origins: waste from the kitchen and other domestic activities for the levels rich in goods, and waste from construction phases or deposits of plants transported by the current for those rich in organic matter. Either the wattle arrangement was used to protect the banks from erosion, or it was deliberately placed as formwork to hold this fill and reclaim land from the lagoon, similar to the systems of artificial islands in the village of Ganvié, located on Lake Nokoué in Benin (Pétrequin and Pétrequin 1984).

2.3.3.2   Sector 3

In Sector 3, the trench dug in D26 (Gascó et al. 2014, 2015; Lachenal et al. 2020), is located at the southern end of the alignments of oak piles identified to the west (Figs. 2.3 and 2.6). The stratigraphy encountered here also has a marked incline to the southwest, perpendicular to the alignments of piles, indicating the presence of a bank. It consists of heterogeneous levels of domestic waste (US 11, 12, 13, 17) and others especially rich in stratified or compacted plant organic matter (US 2, 3, 4, 6, 8, 9, 18), while sterile sandy clay layers may have been deposited by episodes of flooding (US 5, 7, 10). In US 15, an arrangement of wattle supported by posts, partly collapsed, was identified (Fig. 2.9). The presence of oysters fastened onto its base indicates its installation in partly submerged surroundings. It can be interpreted as wattle

---

[2] The French term for vegetation engineering, 'génie végétal', is defined as the use of live plants, parts thereof, and seeds in order to solve engineering problems in the mechanical fields of protection against erosion and land stabilisation and regeneration (Adam et al. 2008, 43).

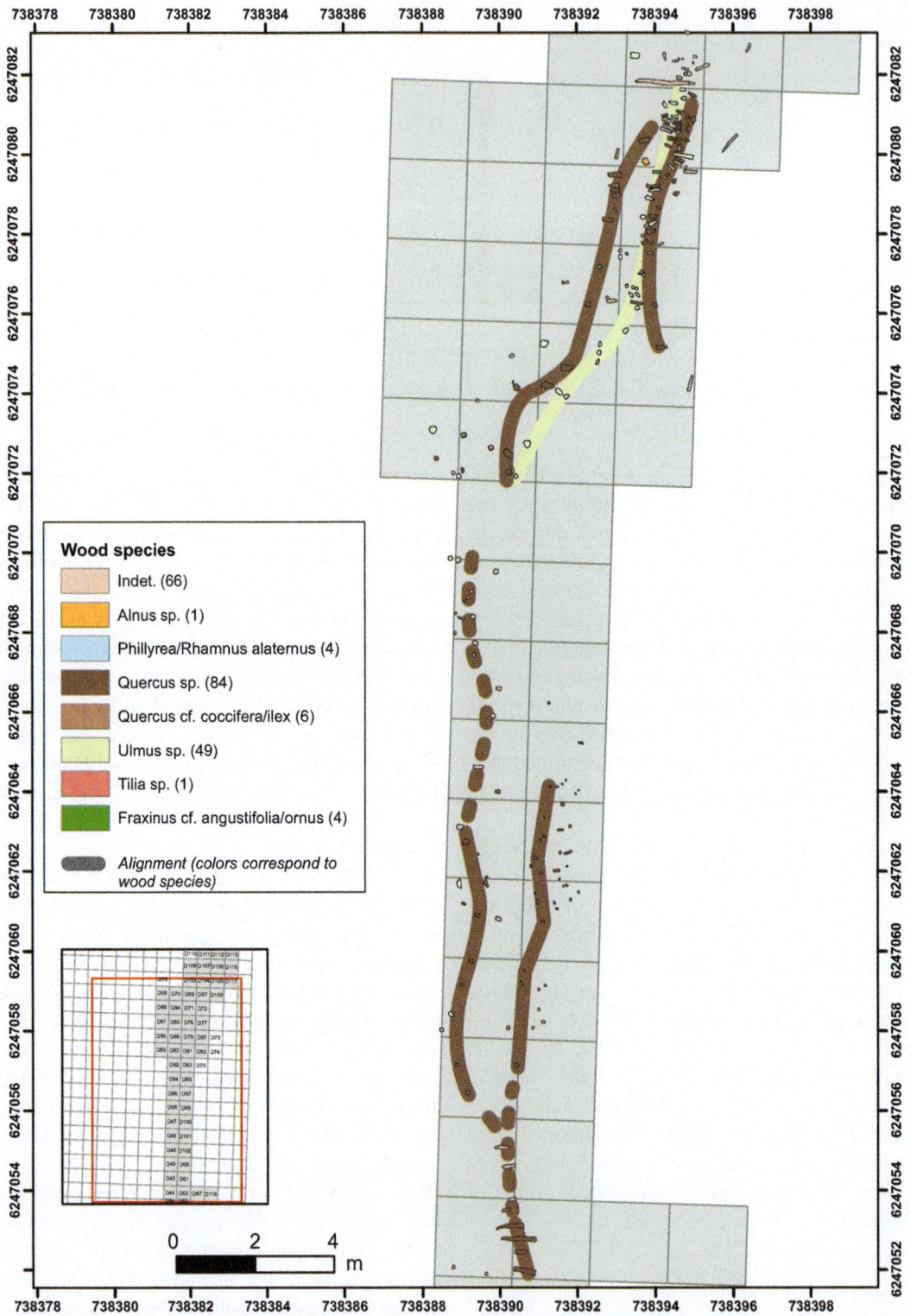

Fig. 2.5 Interpretation of the distribution plan of pile wood species in sectors 1 and 2 (GIS by T. Lachenal; xylological analysis by S. Greck)

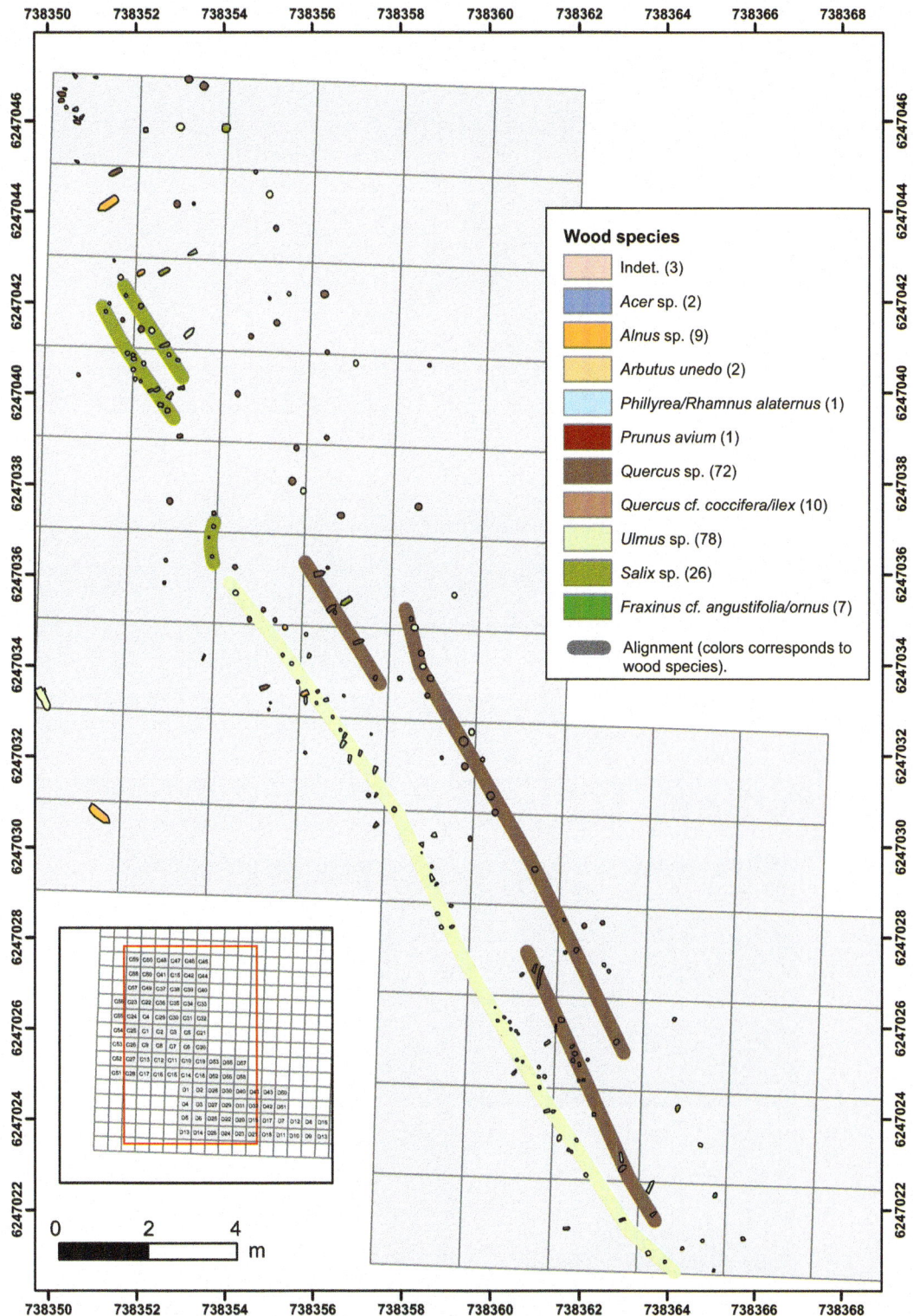

Fig. 2.6 Interpretation of the distribution plan of pile wood species in sectors 3 and 4 (GIS T. Lachenal; xylological analysis by S. Greck)

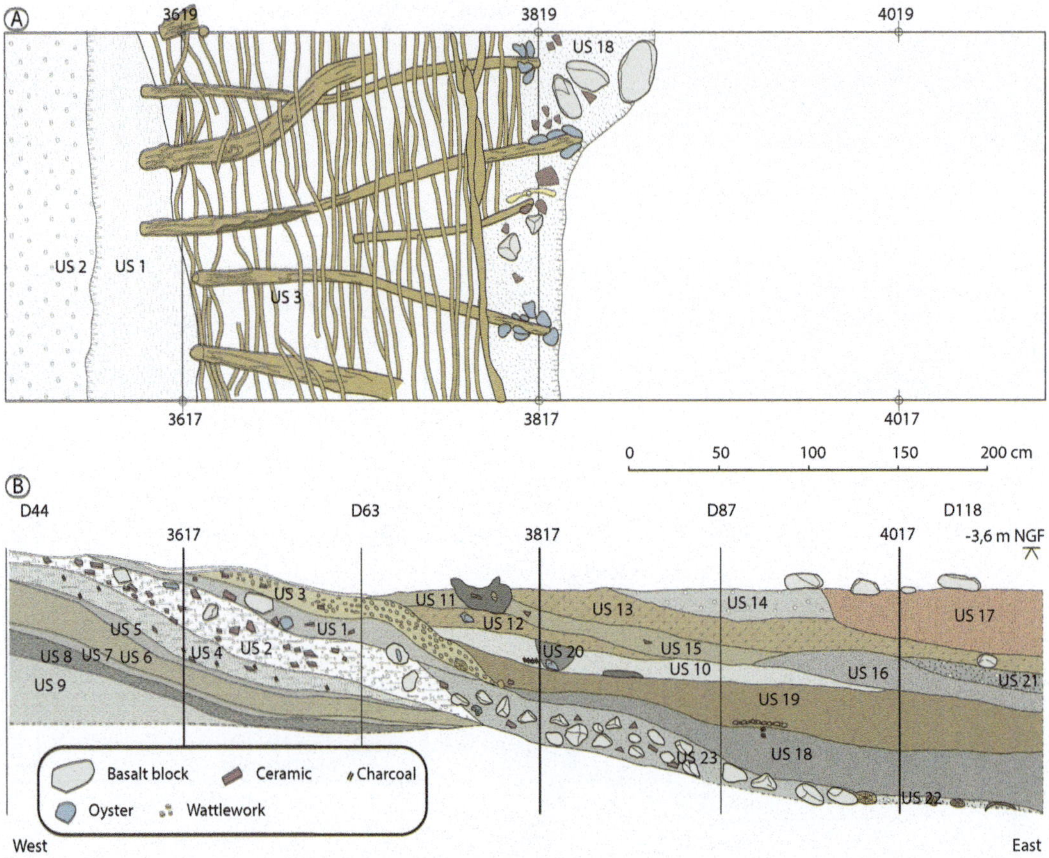

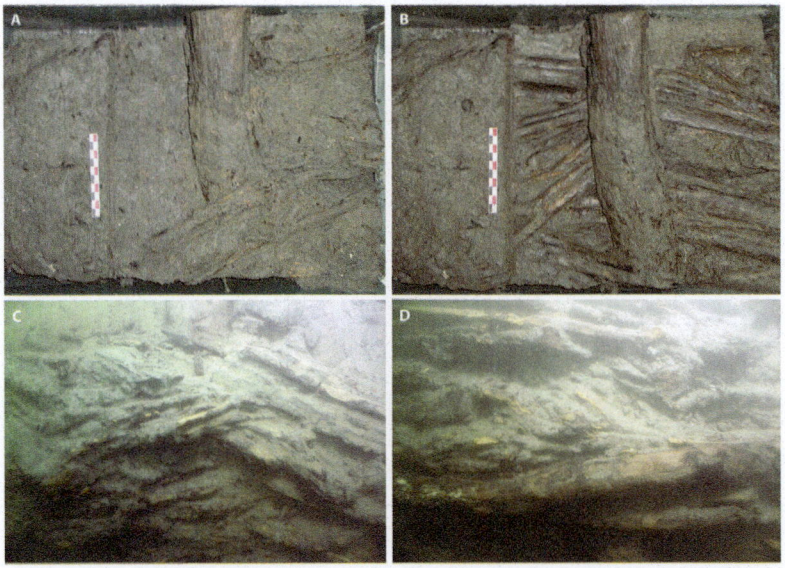

Ⓐ

US 18

US 2 US 1

US 3

3619   3819   4019

3617   3817   4017

0   50   100   150   200 cm

Ⓑ

D44   D63   D87   D118

3617   3817   4017   -3,6 m NGF

US 3   US 11
US 5   US 1   US 12   US 13   US 14
US 8 US 7 US 6   US 4 US 2   US 20   US 15   US 17
US 9   US 10   US 16   US 21
                                                US 19
                                        US 23   US 18
                                                US 22

Basalt block   Ceramic   Charcoal

Oyster   Wattlework

West   East

Fig. 2.7 **a** Plan of the wattle in the sector 2 excavation (D44–D63–D87–D118). **b** Stratigraphic section (drawing by F. Laurent, CAD by T. Lachenal)

Fig. 2.8 **a, b** Strips 2 and 3 of the excavation of the wattle block in the laboratory (photo L. Chabal). **c, d** Details of the wattle in the excavation (Photo by J.-C. Iché)

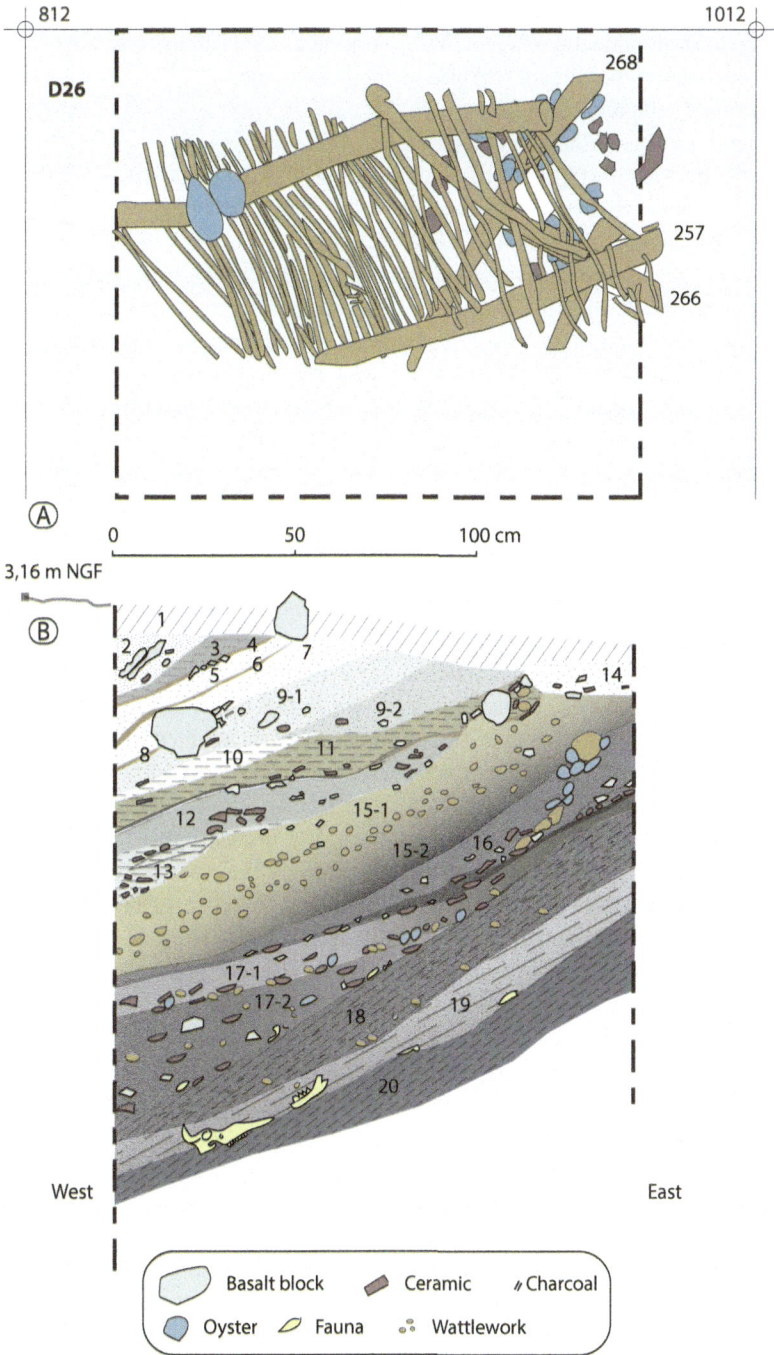

Fig. 2.9 **a** Plan of wattle remains in the trial trench of sector 3 (D26). **b** Stratigraphic section (drawing by B. Debrand, CAD by J. Gascó)

supporting a bank, having an opposite orientation with respect to that of sector 2, but a similar installation and function.

2.3.3.3   Sector 4

In this sector, a trench was dug in squares C22 and C23, where an alignment of willow piles was set up (Figs. 2.6 and 2.10a). This is also the zone in which the deposit of ornamental elements was discovered. The excavation once again revealed a stratigraphy strongly inclined towards the southwest, perpendicularly to the alignment of piles identified in this sector (Fig. 2.10b). This sequence is thus also located in a bank area on the edge of the protohistoric occupation. Here again, there are anthropic deposits corresponding to domestic waste or fill (8, 11, 15, 17, 19, 22, 23, 24), levels of accumulation of small plant debris transported by the current (US 12, 14, 18, 21), and flood episodes marked by pure clay sediment (1/13). The excavation also uncovered the piles embedded on the edge of this bank, which are not associated with wattle here. They consist of willow and elm with a small diameter, between 8 and 10 cm, and with the bark preserved (Fig. 2.11). The willows have the particularity of having conserved small branches, and a fine development of roots was sometimes observed above their tips, even though their points were sharpened. It is thus possible to put forward the hypothesis of these being live stakes, a technique still in use today for bank stabilisation. The use of willow is revealing because it is the preferred species in vegetation engineering today due to its pioneering nature and its aptitude for vegetative regeneration (Adam et al. 2008, p. 97). This species is also particularly resistant to submersion and flooding; therefore, its use suggests good knowledge of its ecology on the part of the site's inhabitants. Another hypothesis is that the branches were left behind to facilitate the installation of the piles in the ground.

At the base of this trench, six fallen piles were discovered, belonging to a first collapsed palisade on sterile lagoon levels (US 25). These piles still have a part of their bark and have a sharpened tip, always directed towards the upper part of the bank (Fig. 2.12). This preoccupation with reinforcing the banks thus required regular structural works by the site's occupants. The presence of oysters and shipworm holes on these piles indicate a brackish lagoon environment. The willow piles installed later, however, perform better in riverine surroundings. The sequence thus appears to record the progressive passage from a lagoon edge to a fluvial environment, as also indicated by the geomorphological studies.

2.3.4 The Question of the Settlement Structures

The alignments of piles can for the most part be interpreted as systems for bank reinforcement and maintenance at the centre of which there was a settlement, as suggested by the levels of domestic waste identified in the excavations, which produced a very large amount of goods. Unfortunately, the part of the site located between these supposed banks was very highly eroded, as shown by the bathymetric survey (Fig. 2.2). A transect of subaquatic boreholes confirmed the absence of preserved strata in this part of the river. It is probable that the wattle systems acted as a framework for the archaeological layers, in accordance with their initial goal of bank maintenance, and maintained them in place. On the contrary, when moving away from these structures, the levels were washed away by the action of the river, a phenomenon marked by the presence of erosional ledges. It is possible, however, to detect the remains of settlement structures in the immediate vicinity of the banks. In particular in sector 4, in the northwest of the site, posts with a large diameter that are not integrated into the alignments of piles stand out. They allow the reconstruction of a structure on load-bearing posts, having a subrectangular shape and ending in an apse to the north (Fig. 2.13). A characteristic of this structure is the use of highly varied types of wood, whereas the bank structures generally show a relative homogeneity in the species used. Other pieces of wood with a large diameter present in various sectors of the site raise questions about the presence of other buildings. However, at present, in the spaces studied, no other complete and coherent floorplan was discerned. The presence of settlement structures is, however, also suggested by the discovery of wall elements made of daub (Lachenal et al. 2020).

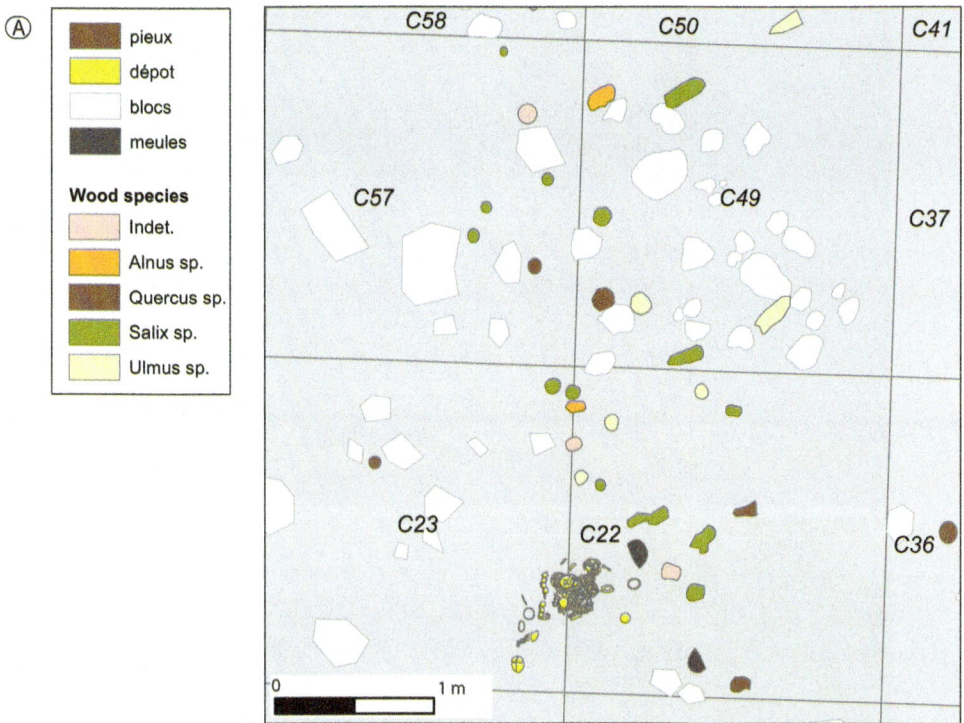

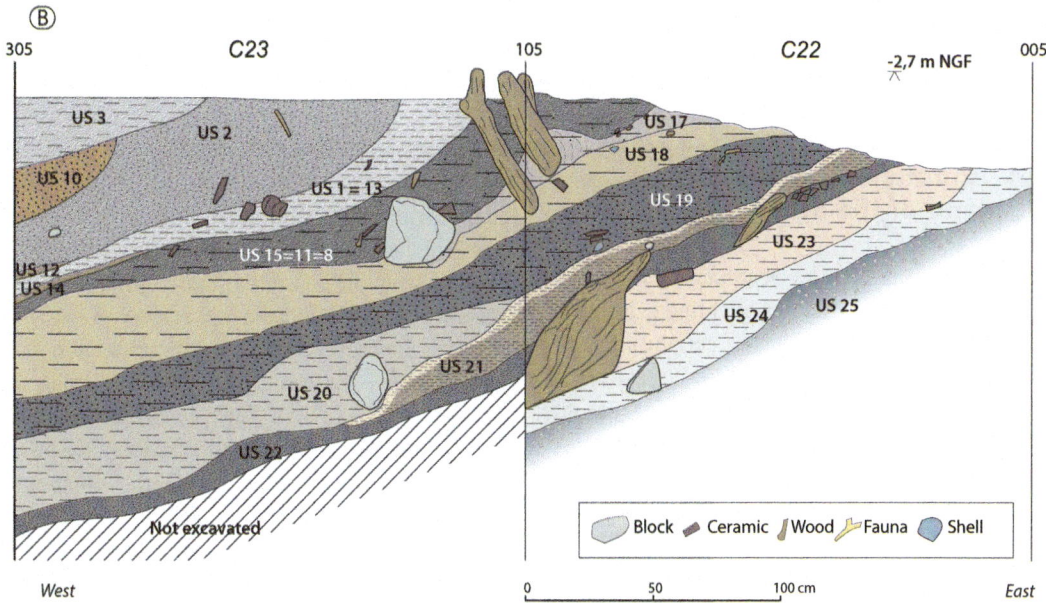

Fig. 2.10 **a** Plan of the trial trench in sector 4 (C22–C23). **b** Stratigraphic section (drawing by A. Sebastia and T. Lachenal, CAD by T. Lachenal)

Fig. 2.11 **a** View of the piles appearing in the trial trench of sector 4 (C22–C23) during the excavation (photo by J. Montès). **b** Detail of a branch preserved on a trunk (photo by T. Lachenal). **c** Willow pile sample (Photo by J.-C. Iché)

2.4 Chronology of the Site and of the Structures

In order to determine the absolute chronology of the deposits and structures identified, radiocarbon dating was carried out on certain piles and on material from the excavation trenches. At present, twenty dates are available for the site (Table 2.1). These dates are associated with other chronological information provided by the stratigraphy, the typology of the goods, and dendrochronological synchronisation. As a result, they lend themselves to Bayesian modelling (Bronk Ramsey 2009; Lanos and Dufresne 2012), through which the validity of the periodisation proposed can be verified and the limits and duration of each phase can be suggested. The model was created using the ®ChronoModel software (Lanos and Philippe 2017, 2018). Two types of periodisation were carried out,

sometimes including the same dates, relating on the one hand to the infrastructure and on the other hand to the chronotypological data indicated by the ceramic goods (Figs. 2.14 and 2.15). The information on stratigraphic anteriority and posteriority are indicated by arrows (from the oldest to the most recent). Certain piles were also synchronised by dendrochonological analysis (Lachenal et al. 2020). Thus, piles 215 and 254 were shown to be contemporary. Pile 23, however, was cut down after pile 318, which was installed at the same time as pile 315, during the construction of the wattle in sector 2. This structure is located in the stratigraphy between US 2 and 12 in the excavation trench dug in this sector.

Here, we have distinguished a phase prior to the installation of the wattle in sectors 2 and 3, a phase corresponding to this installation, and a later phase. Two other periods correspond to the installation of the two levels of piles and of

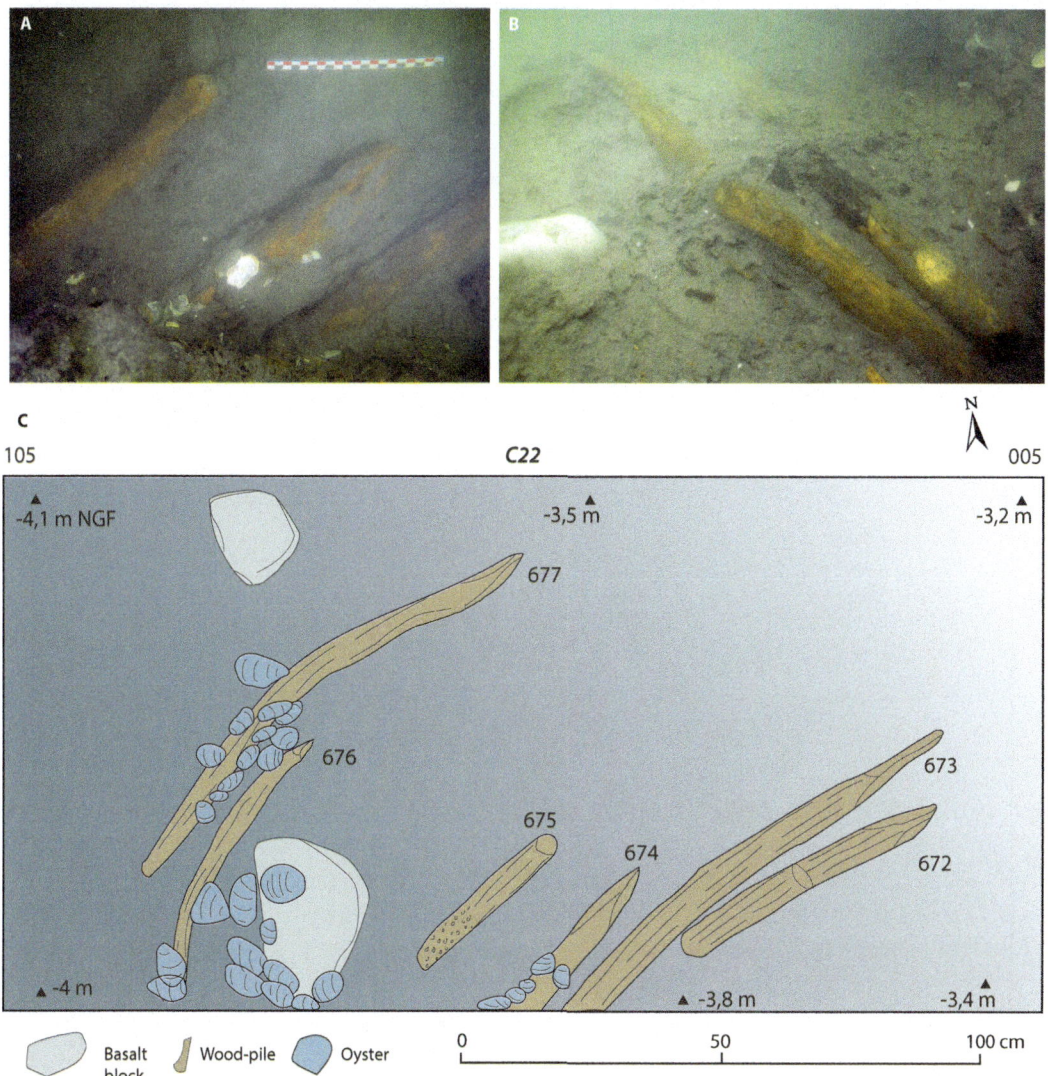

Fig. 2.12 Views and plan of the piles collapsed at the base of the trial trench of sector 4 (photo by T. Lachenal and J.-C. Iché; drawings and CAD by T. Lachenal)

willow stakes uncovered in the excavation trench of sector 4 (Figs. 2.14 and 2.15). This analysis shows that the collapsed palisade in this zone dates to the 10th century BC (Fig. 2.16). The associated goods correspond to the Late Bronze Age IIIA of the regional chronology (Dedet 2014), which can be correlated with the classic north-alpine Hallstatt B1-period (David-Elbiali and Dunning 2005). These corpora are characterised by an abundance of convex, more rarely carinate, bowls and dishes, frequently decorated

with patterns incised with a bifid comb with blunt teeth (Fig. 2.17, No. 1–4, 6, 9–10). The large jars with a neck sometimes underlined with a fingertip-impressed band (Fig. 2.17, No. 8), or with a cylindrical neck decorated with fluting and a rim, are also characteristic of this period (Fig. 2.17, No. 13).

Later, the installation of the wattle in sectors 2 and 3 would have been carried out at the end of the 10th century or at the beginning of the 9th century BC (Fig. 2.16). The levels preceding the

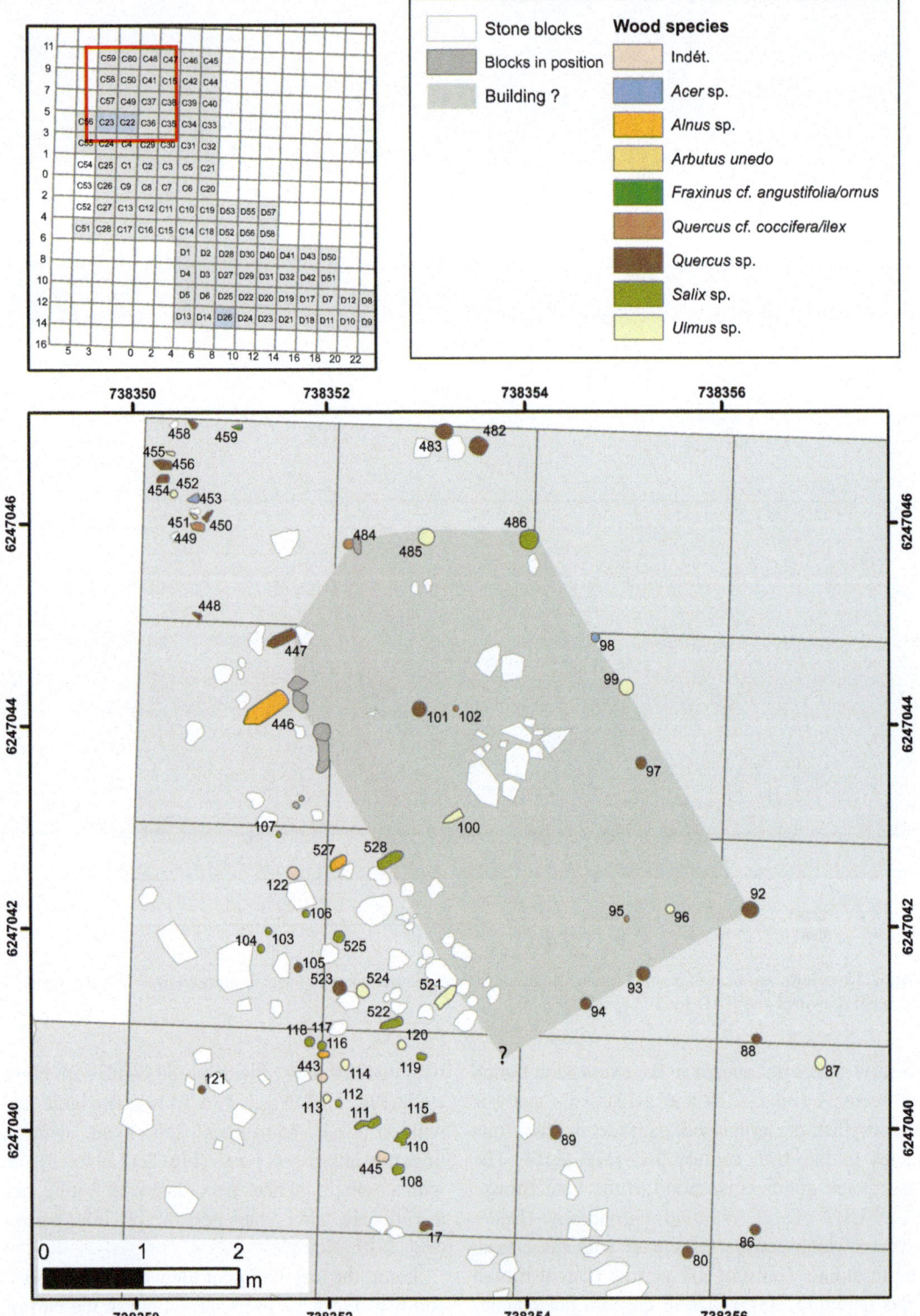

Fig. 2.13 Plan of a possible building identified in sector 4 (GIS by T. Lachenal; xylological analysis by S. Greck)

Table 2.1 Radiocarbon dates of the La Motte site

Sample	Material	Lab. code	Age BP	±	Cal BC (Intcal20)	Phase: chronotypology	Phase: bank protection	Phase: dendrochronology
C22 US24 Pile 673	Wood (pile)	Poz-143291	2790	30	**1011–891 (83%)** 879–835 (12%)	BF IIIa	Collapsed fence	
US105.17	Wood (branch)	Poz-129419	2795	30	1040–1036 (1%) **1013–893 (86%)** 877–836 (9%)	BF IIIb early		
C22 US 22	Wood (branch)	Poz-129417	2730	30	**926–809 (95%)**	BF IIIb early		
D26 US 16	Wood (branch)	Ly-11446	2710	30	**909–807 (95%)**	BF IIIb early	Before wattle	
D87 US 23	Carbonised residue	Poz-96358	2700	35	**911–803 (95%)**	BF IIIb early	Before wattle	
D26 US 4	*Triticum dicoccum*	Ly-9038	2765	30	**994–830 (95%)**	BF IIIb late	After wattle	
US105.11	Wood (Mould tenon)	Poz-129413	2710	30	**909–807 (95%)**	BF IIIb late		
D26 US 11	*Triticum dicoccum*	Ly-9039	2705	30	**906–806 (95%)**	BF IIIb late	After wattle	
D26 US 13	Wood (branch)	Ly-11444	2700	30	**903–805 (95%)**	BF IIIb late	After wattle	
D87 US12	Charcoal (wood branch)	Poz-96359	2690	35	**911–803 (95%)**	BF IIIb late	After wattle	
C22 US19	Wood (branch)	Poz-111143	2655	35	898–857 (19%) **847–785 (76%)** 783–779 (1%)	BF IIIb late		
Pile 113	Wood of *Salix* sp. (last rings)	Poz-96228	2605	35	**819–761 (95%)**	Bronze/Iron transition	Willow piles	
C23 US15	Wood (branch)	Poz-96229	2555	30	**800–748 (95%)**	Bronze/Iron transition		
C23 US13	Wood (branch)	Poz-96360	2525	35	**791–718 (92%)** 707–700 (20%)	Bronze/Iron transition		
US105.03	Wood (board)	Poz-129414	2510	30	785–782 (1%) **779–716 (88%)** 709–700 (6%)	Bronze/Iron transition		
Pile 215	Wood of *Quercus* sp. (last rings)	Ly-11448	2780	30	**1006–889 (76%)** 881–834 (20%)			Synchro 1
D26 US 15 Pile 257	Wood of *Quercus* sp. (last rings)	Ly-11443	2730	30	**926–809 (95%)**		Wattle	
Pile 315	Wood of *Quercus* sp. (last rings)	Ly-11445	2710	30	978–948 (8%) **936–817 (88%)**		Wattle	
Pile 254	Wood of *Quercus* sp. (last rings)	Ly-11447	2695	30	**901–804 (95%)**			Synchro 1
Pile 23	Wood of *Quercus* sp. (last rings)	ARC2445	2620	45	896–867 (7%) **839–755 (88%)**			Synchro 2

BF bronze final (Late Bronze Age). Calibration made with ®ChronoModel using the IntCal20 curve (Reimer et al. 2020). The highest probabilities are highlighted in bold

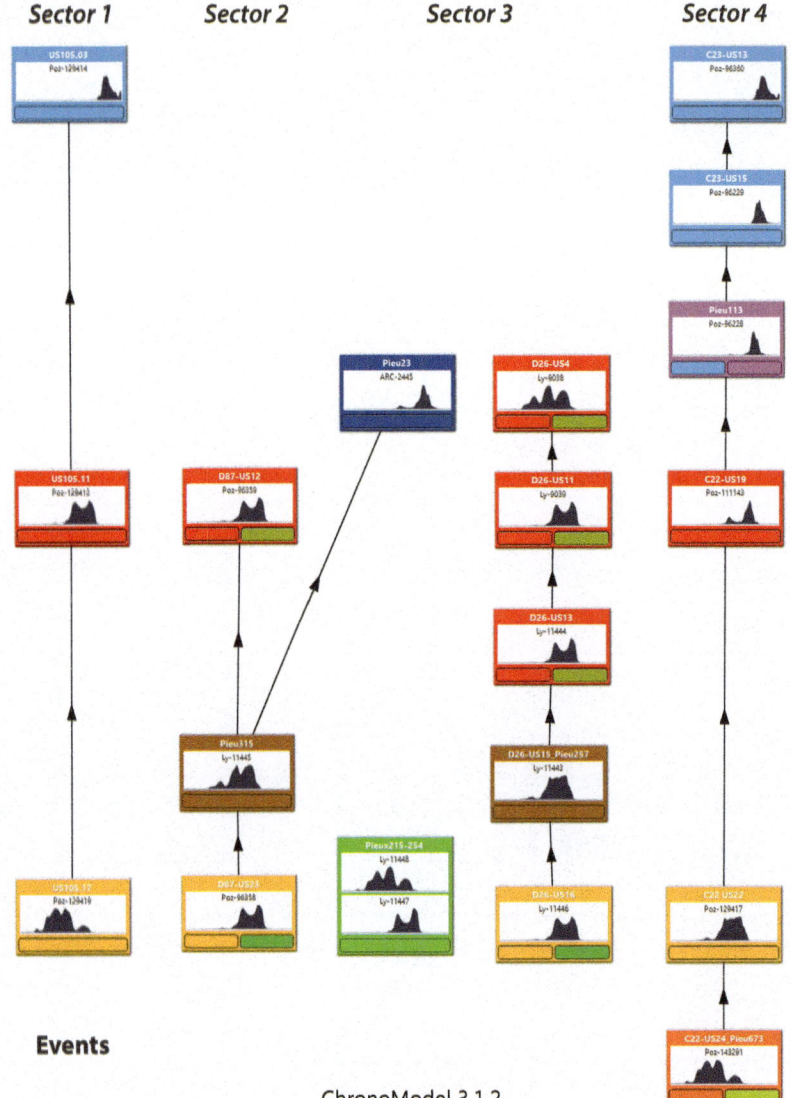

Fig. 2.14 Chronological modelling based on events dated by radiocarbon (®ChronoModel)

installation of these structures contained ceramic goods from the beginning of the Mailhac I style of the Late Bronze Age IIIB (Guilaine 1972). This style saw the appearance of rectilinear beakers (Fig. 2.17, No. 21) and cups with internal incised decoration (Fig. 2.17, No. 16). The decoration was still most often geometric, in the form of meanders or broken lines (Fig. 2.17, No. 21–24). This stage could correspond to the Hallstatt B2-period.

The strata covering the wattle systems date to the second half of the 9th century BC and are possibly contemporary to the Hallstatt B3-period (Fig. 2.16). The associated goods feature the appearance of more complex decoration, whether figurative, zoomorphic, or anthropomorphic (Fig. 2.18).

Finally, the levels uncovered in sector 4 together with the alignment of willow piles would illustrate the beginning of the 8th century BC (Fig. 2.16). This period corresponds to the

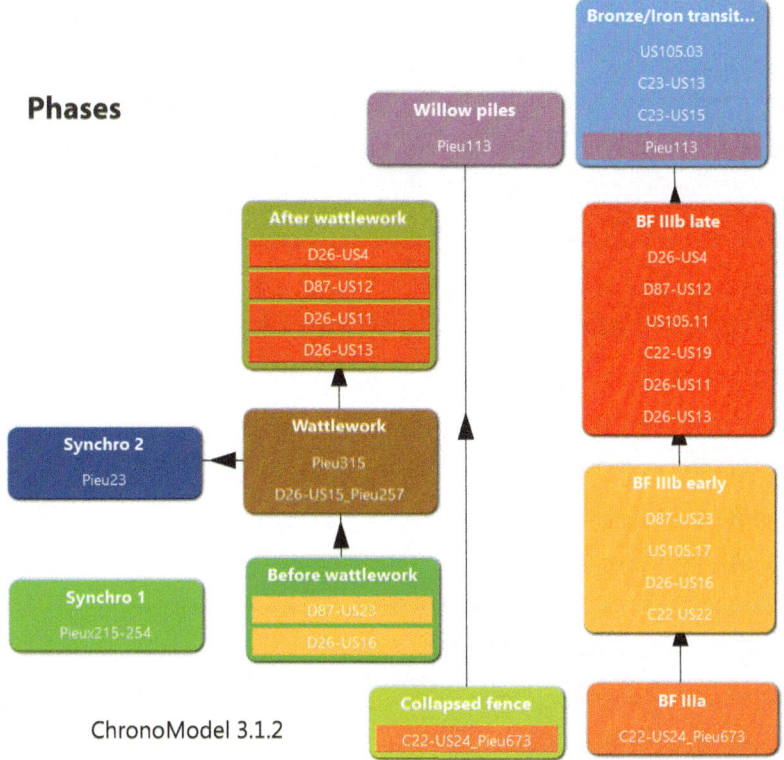

Fig. 2.15 Chronological model of event phases (®ChronoModel)

phase of transition between the Bronze Age and the Iron Age defined regionally (Janin 1992), contemporary to the beginning of the Hallstatt C-period in continental Europe. This period saw a continuation of the Mailhac I style, together with the appearance of new forms such as low sinuous carinated bowls (Fig. 2.18, No. 12–14). The latter are characteristic of the 8th century BC in the necropolises of Languedoc (Janin 1992; Taffanel and Janin 1998).

2.5   Conclusion

Deliberately settled at the interface between two environments, lagoon and alluvial plain, the occupants of the La Motte site had to carry out works in order to protect their settlement for at least two centuries (Fig. 2.19). The excavation trenches dug in three sectors of the site provide information on the implementation of the alignments of piles revealed by the spatial analysis while taking into account the xylological determination. They all point to systems for bank maintenance and protection via different techniques: simple wooden piles, strawberry-tree wattle supported by oak posts, and possible live willow stakes. This implies a comprehension of the natural functioning of the ecosystems, with the selection and implementation of species of wood and techniques suitable for the management and restoration of the area bordering the lagoon. The occupants of the site sought long-term solutions capable of protecting their settlement and apparently tested several methods. The various parallel alignments identified could thus correspond to successive works, as the domestic waste created near the homes allowed land to be reclaimed from the lagoon. Their goal appears to have been to protect the area around the settlement from erosion. Their implementation must thus be viewed in the light of the

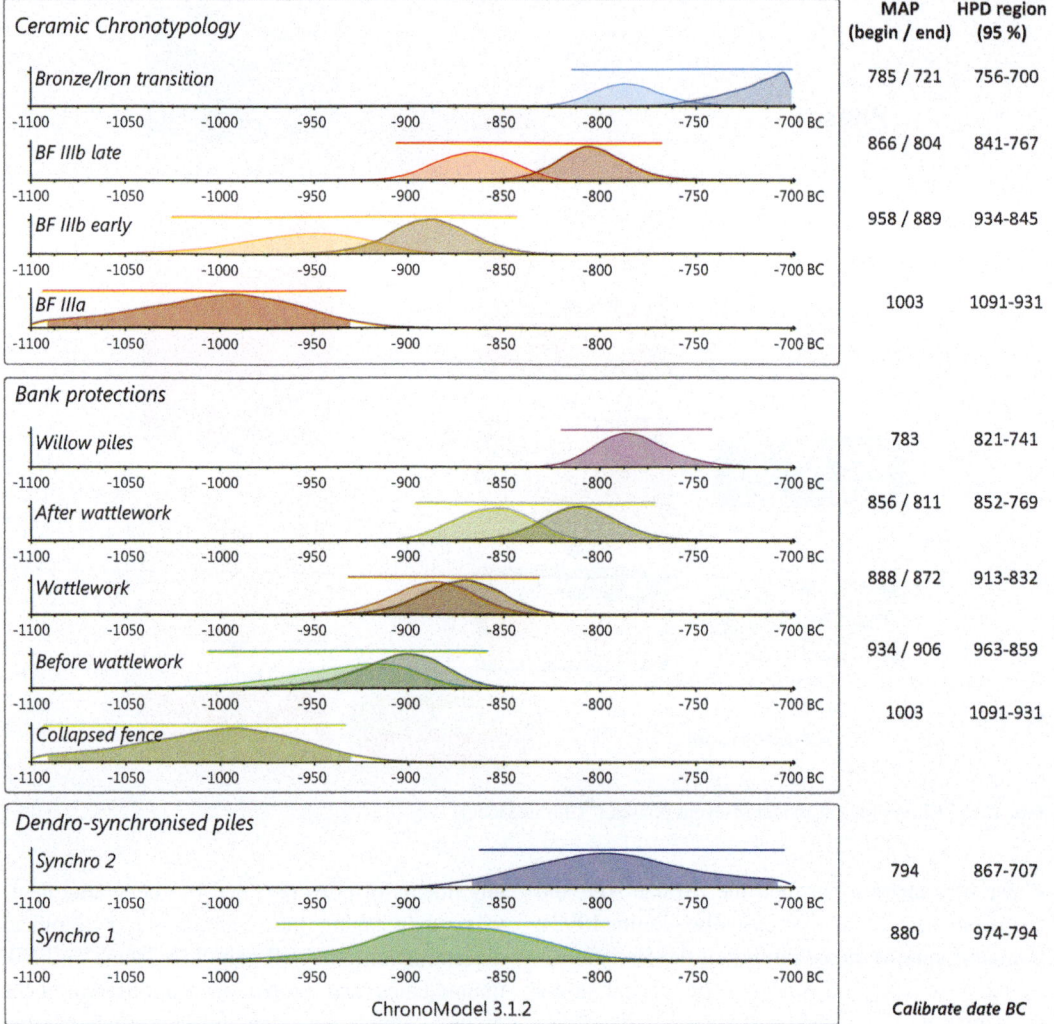

Fig. 2.16 Histograms of the phases modelling made by using ®ChronoModel. MAP, Modes a posteriori of event start and end. HPD, Highest Probability Density region at 95% of event start, end and duration (Author)

geomorphological evolution of the surroundings, which involved a rise in the sea level coupled with a progressive filling of the Hérault delta. The sedimentological studies reveal an increase in the influence of the river during the last phase of occupation of the site, which was dated to the beginning of the 8th century BC. The abandonment of the settlement could thus be connected to an aggradation of the river. This phenomenon must have destabilised the surroundings in which the site's occupants lived, with an increase in floods that threatened the settlement or the beginning of a filling of the lagoon to which it

was connected. The settlement potentially moved onto the basalt relief located 500 m downstream, where the modern city of Agde is situated. Indeed, it was in this sector that the Peyrou necropolis, dating to the 7th century BC, was excavated (Nickels et al. 1989).

The bank reinforcement systems uncovered at La Motte are unprecedented for the Bronze Age in a lagoon setting. They illustrate the vulnerability and the adaptation capacity of a community established in an unstable environment and thus provide a faraway echo of problems that are just as relevant today.

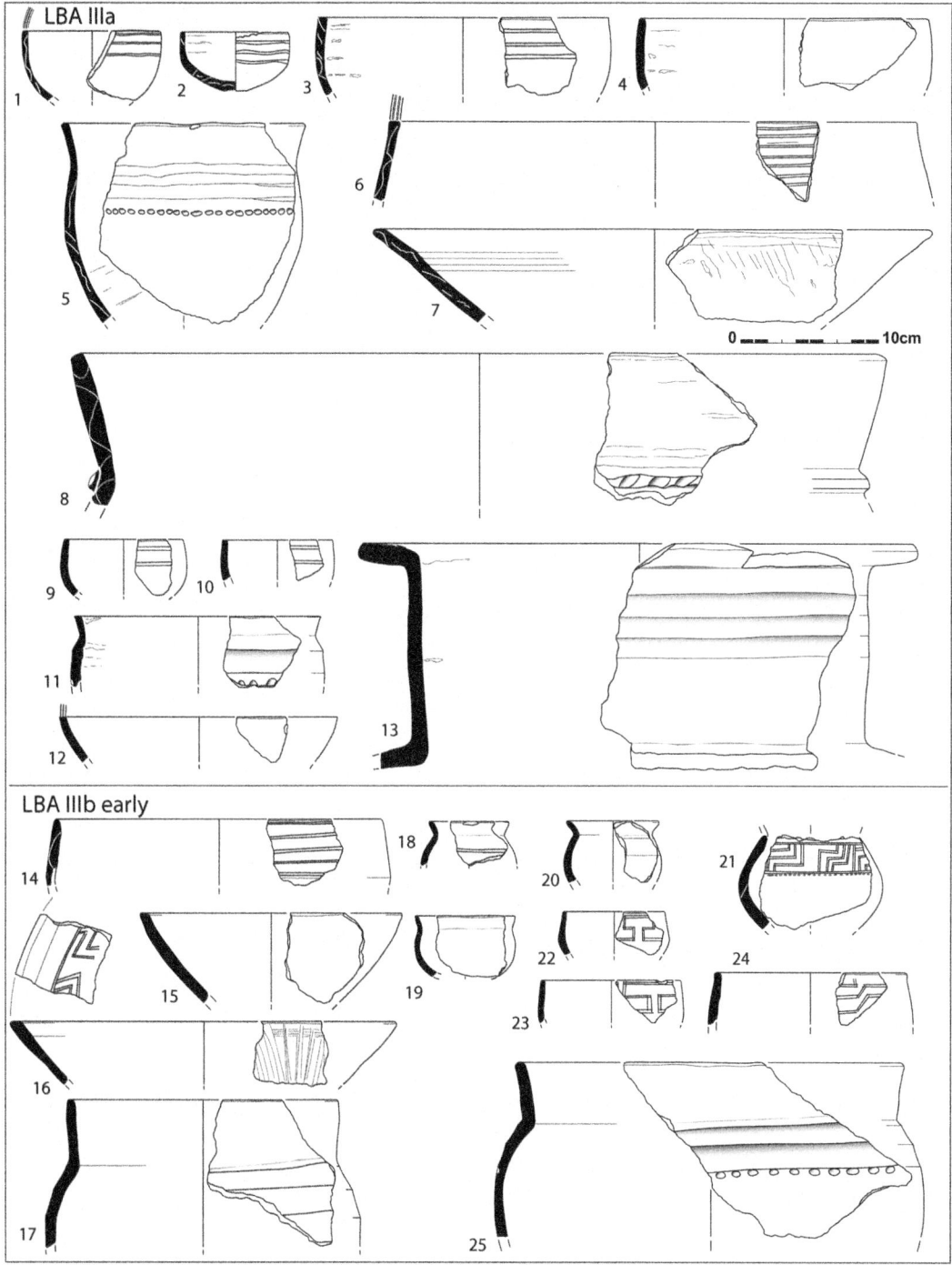

Fig. 2.17 Selection of pottery vessels dated to the LBA IIIa and early LBA IIIb (drawing and CAD by T. Lachenal)

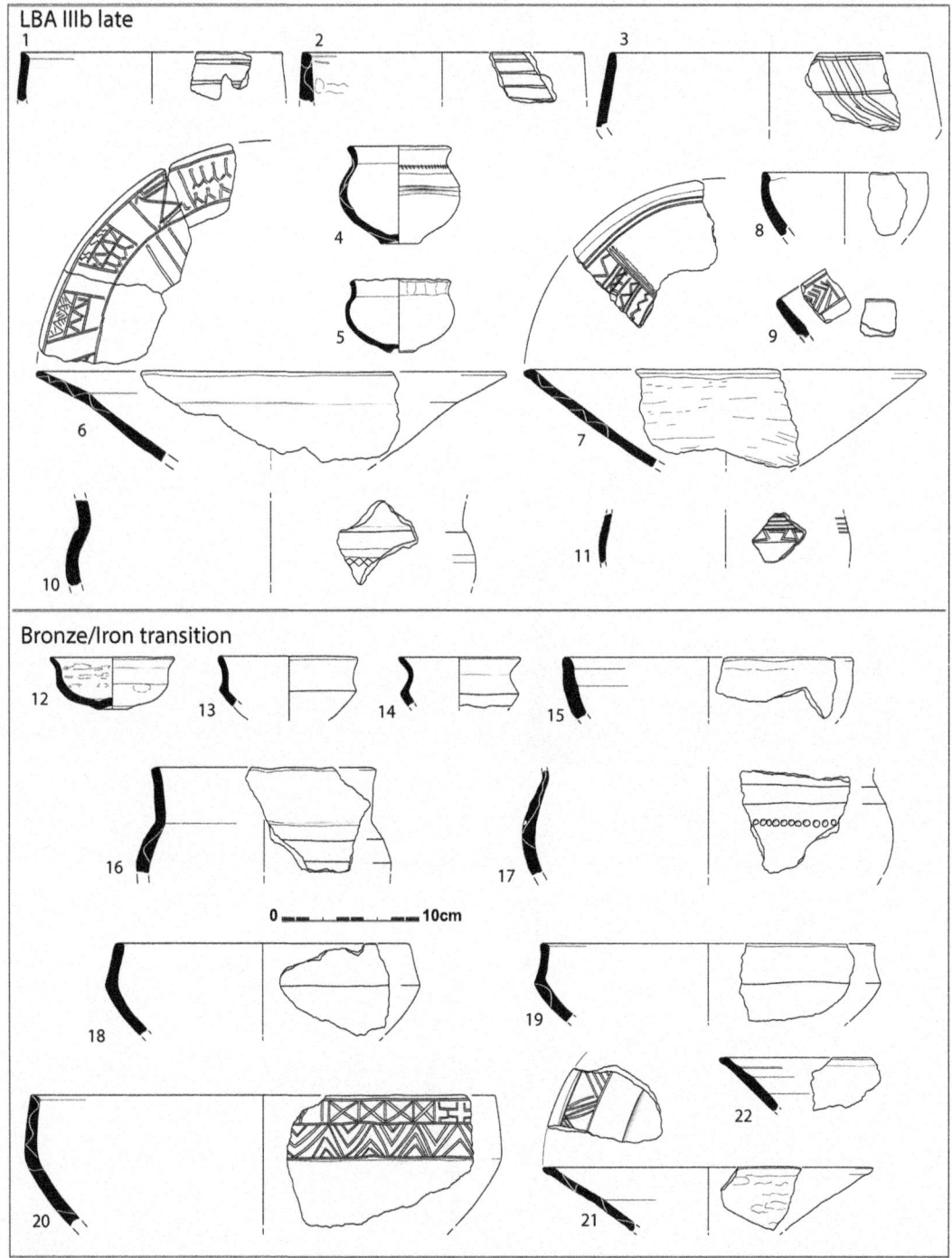

Fig. 2.18 Selection of pottery vessels dated to the late LBA IIIb and to the transition phase between Bronze and Iron Ages (drawing and CAD by T. Lachenal)

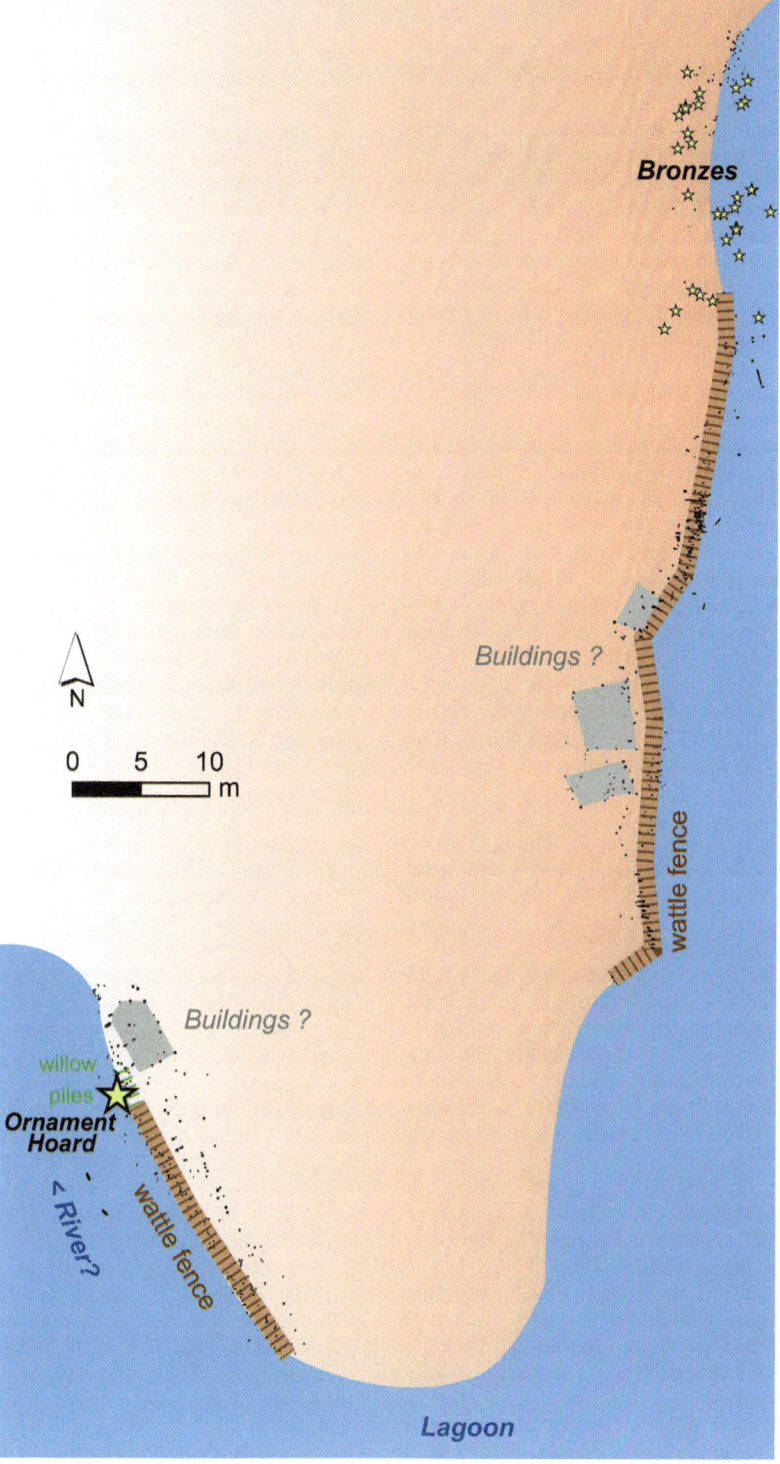

Fig. 2.19 Proposed interpretation of the structures of the La Motte site (Author)

References

Adam P, Debiais N, Gerber F, Lachat B (2008) Le génie végétal. Un manuel technique au service de l'aménagement et de la restauration des milieux aquatiques. La documentation française, Paris

Bouby L, Ponel P, Girard V, Chia TC, Garnier L, Tillier M, Devillers B, Lachenal T, Tourrette C, Gascó J (2016) Premiers résultats carpologiques et entomologiques sur le site subaquatique Bronze final de la Motte (Agde, Hérault). In: Dietsch-Sellami M-F, Hallavant C., Bouby L, Pradat B (eds) Plantes, produits végétaux et ravageurs. Aquitania, Pessac, pp 65–87

Bronk Ramsey C (2009) Bayesian analysis of radiocarbon dates. Radiocarbon 51(1):337–360

David-Elbiali M, Dunning C (2005) Le cadre chronologique relatif et absolu au nord-ouest des Alpes entre 1060 et 600 av. J.-C. In: Bartoloni G, Delpino F (eds) Oriente e Occidente: metodi e discipline a confronto. Riflessioni sulla cronologia dell'età del Ferro italiana. Mediterranea 1. Istituti editoriali e poligrafici Internazionali, Pisa, Roma, pp 145–195

Dedet B (2014) Le style céramique du Bronze final IIIa en Languedoc oriental. Doc D'archéol Mérid 35(12):85–126

Dedet B, Py M (1985) L'occupation des rivages de l'étang de Mauguio (Hérault) au Bronze final et au premier âge du Fer. Synthèses et annexes, vol 3. ARALO, Caveirac

Devillers B, Bony G, Degeai J-P, Gascò J, Lachenal T, Bruneton H, Yung F, Oueslati H, Thierry A (2019) Holocene coastal environmental changes and human occupation of the lower Hérault River, southern France. Quatern Sci Rev 222:105912. https://doi.org/10.1016/j.quascirev.2019.105912

Gascó J, Borja G, Tourrette C, Verdier J-L, Bouby L, De Villers B, Greck S, Yung F (2014) Le site subaquatique de la Motte (Agde, Hérault) à la fin de l'âge du Bronze. In: Sénépart I, Léandri F, Cauliez J, Perrin T, Thirault É (eds) Chronologie de la Préhistoire récente dans le Sud de la France. Actualité de la recherche. Archives d'Écologie Préhistorique, Toulouse, pp 625–630

Gascó J, Borja G, Tourrette C, Yung F, Verdier J-L, Bouby L, Devillers B, Greck S, Baisse F, Barthelemy C, Chabbert J, Constant D, Debrand B, Dez J, Iche J-C, Laurent F, Puech J-P, Rouvet P, Rolland C, Sabastia A (2015) Une occupation lagunaire palafittique aux IXe–VIIIe s. a. C.: La Motte (Agde) au fond du fleuve Hérault. In: Olmer F, Roure R (eds) Les Gaulois au fil de l'eau. Ausonius, Bordeaux, pp 69–86

Guilaine J (1972) L'âge du Bronze en Languedoc occidental, Roussillon, Ariège. Klincksieck, Paris

Janin T (1992) L'évolution du Bronze final IIIb et la transition Bronze-Fer en Languedoc occidental d'après la culture matérielle et les nécropoles. Doc D'archéol Mérid 15:243–259

Lachenal T (2022a), Dépôts d'objets métalliques sur les berges du site de la Motte (Agde, Hérault), au tournant des âges du Bronze et du Fer. In: Lourdaux-Jurietti S, Colas G (eds) Des épées pour la Saône? Les dépôts de l'âge du Bronze en milieu humide. Musée Vivant Denon, Chalon-sur-Saône, pp 102–103

Lachenal T (2022b) Enter the Matrix. Late Bronze Age Casting Moulds from "La Motte" (Agde, Dép. Hérault/F) in their Context. Archäologisches Korrespondenzblatt 52(1):65–90

Lachenal T, Gascó J, Devillers B, Bouby L, Chabal L, Girard V, Greck S, Guibal F, Lespes C, Liottier L, Ponel P, Tourrette C (2020) Un habitat de la fin de l'âge du Bronze entre lagune et fleuve: le site immergé de la Motte à Agde (Hérault, France). In: Billaud Y, Lachenal T (eds) Entre terres et eaux. Les sites littoraux de l'âge du Bronze: spécificités et relations avec l'arrière-pays. Société préhistorique française, Paris, pp 217–255. http://www.prehistoire.org/offres/file_inline_src/515/515_P_48188_5e6226de06f09_12.pdf

Lanos P, Dufresne P (2012) Modélisation statistique bayésienne des données chronologiques. In: De Beaune SA, Francfort H-P (eds) L'archéologie à découvert. Hommes, objets, espaces et temporalités. CNRS, Paris, pp 238–248

Lanos P, Philippe A (2017) Hierarchical Bayesian modeling for combining dates in archaeological context. J Soc Franç Stat 158(2):72–88

Lanos P, Philippe A (2018) Event date model: a robust Bayesian tool for chronology building. Commun Stat Appl Methods 25(2):131–157

Leroy F (2010) Les habitats littoraux protohistoriques des côtes de Méditerranée nord-occidentale. In: Delestre X, Marchesi H (eds) Archéologie des rivages méditerranéens: 50 ans de recherche. Errance, Paris, pp 137–148

Lespes C, Lachenal T, Gardeisen A, Gascó J (2019) New perspectives on the Lagoon sites of the Late Bronze Age in the South of France Revealed by Animal exploitation at the La Motte I Site (Hérault). J Archaeol Sci 25:206–216. https://doi.org/10.1016/j.jasrep.2019.04.005

Moyat P, Dumont A, Verger S, Mariotti JF, Greck S, Janin T (2005) Note d'information. Un habitat et un dépôt d'objets métalliques protohistoriques découverts dans le lit de l'Hérault à Agde. C R Séances Acad Inscr Belles Lett 149(1):371–394

Moyat P, Dumont A, Mariotti JF, Janin T, Greck S, Bouby L, Ponel P, Verdin P, Verger S (2007) Découverte d'un habitat et d'un dépôt métallique non funéraire du VIIIe s. av. J.-C. dans le lit de l'Hérault à Agde, sur le site de la Motte. Jahrb Römisch-Germanischen Zentralmuseums Mainz 54:53–84

Nickels A, Marchand G, Schwaller M (1989) Agde, la nécropole du premier Age du Fer. CNRS, Paris

Pétrequin AM, Pétrequin P (1984) Habitat lacustre du Bénin. Une approche ethnoarchéologique. Recherches sur les civilisations, Paris

Py M (1990) Culture, économie et société protohistoriques dans la région nîmoise. Ecole française de Rome, Rome.

Reimer PJ, Austin WEN, Bard E, Bayliss A, Blackwell PG, Bronk Ramsey C, Butzin M, Cheng H, Edwards RL, Friedrich M, Grootes PM, Guilderson TP, Hajdas I, Heaton TJ, Hogg AG, Hughen KA, Kromer B, Manning SW, Muscheler R, Palmer JG, Pearson C, van der Plicht J, Reimer RW, Richards DA, Scott EM, Southon JR, Turney CSM, Wacker L, Adolphi F, Büntgen U, Capano M, Fahrni SM, Fogtmann-Schulz A, Friedrich R, Köhler P, Kudsk S, Miyake F, Olsen J, Reinig F, Sakamoto M, Sookdeo A, Talamo S (2020) The IntCal20 Northern Hemisphere Radiocarbon Age Calibration Curve (0–55 cal kBP). Radiocarbon 62 (4):725–757

Taffanel O, Taffanel J, Janin T (1998) La nécropole du Moulin à Mailhac (Aude). ARALO, Lattes

Verger S, Dumont A, Moyat P, Mille B (2007) Le dépôt de bronzes du site fluvial de La Motte à Agde (Hérault). Jahrb Römisch-Germanischen Zentralmuseums Mainz 54:85–171

Wetland Archaeology in Northern Italy: An Overview

3

Marco Baioni, Claudia Mangani, and Roberto Micheli

Abstract

Northern Italy has furnished abundant evidence of the 'pile-dwelling phenomenon' which is found throughout the Alpine area. Neolithic pile-dwellings are known for northern Italy starting from the Early Neolithic (Isolino di Varese). However, lake-dwelling villages were most widespread in Northern Italy between about 2000 and 1400 BC, between the Early and Middle Bronze Age, and continued in the Late Bronze Age. At a time when in other Alpine regions pile-dwellings became rarer, in a large part of northern Italian territory they became the main settlement model. Pile-dwelling villages spread from western Piedmont to Veneto, from Trentino to the lower Lombardy and Veneto plains, along the banks of lakes, peat bogs and river depressions. The main concentration of pile-dwelling settlements was in the Varese lakes and around Lake Garda. In recent decades, in part thanks to the introduction of dendrochronology, our knowledge of these sites has deepened. New excavations, scientific analysis and dendrochronology have helped to increase our knowledge of these settlements.

Keywords

Northern Italy · Pile-dwellings · Neolithic · Bronze Age · Wooden elements · Structural typology

M. Baioni (✉)
Museo Archeologico della Valle Sabbia, Gavardo, Italy
e-mail: marco.baioni.archeologo@gmail.com

C. Mangani
Museo Civico Archeologico 'G. Rambotti', Desenzano del Garda, Italy
e-mail: claudia.mangani@comune.desenzano.brescia.it

R. Micheli
Soprintendenza Archeologia, Belle Arti e Paesaggio del Friuli Venezia Giulia, Trieste, Italy
e-mail: roberto.micheli@cultura.gov.it

3.1 Introduction

Northern Italy is home to numerous of the 'pile-dwelling' sites that have been found throughout the Alpine region since the 19th century. The first fortuitous discoveries of pile-dwellings in Italy occurred between 1830 and 1850 in Peschiera (province of Verona) on Lake Garda, when the Austrian engineer corps was reorganising the defences of the local fortress.[1] However, the first real interest in the pile-dwellings of northern Italy occurred after the discovery of Swiss pile-

---

[1] This military fort, which originated in Roman times, reached its greatest strength under the Republic of Venice in the 16th century. In the 19th century it was part of the defensive system organised by the Austrian Empire to defend Lombardy-Venetia.

© The Author(s) 2025
A. Ballmer et al. (eds.), *Prehistoric Wetland Sites of Southern Europe*, Natural Science in Archaeology, https://doi.org/10.1007/978-3-031-52780-7_3

dwelling villages during winter 1853–54 (Ruoff 2004; Marzatico 2004; Pétrequin 2013; Menotti 2013). On 25 April 1863, Swiss researchers presented the results of their investigations into Lake Zurich pile-dwellings in Milan to the Società Italiana di Scienze Naturali, which conducted a visit to the Varese lakes in Lombardy, thought then to be especially suitable for 'stilt-house' settlements. Three days later, in the presence of Antonio Stoppani, Édouard Desor and Gabriel De Mortillet, the first underwater searches were conducted in Lake Varese, which led to the discovery of the Bodio centrale and Isolino di Varese sites (De Marinis 1982). These discoveries constituted the starting point of studies on the prehistoric pile-dwellings of northern Italy.

Research into pile-dwellings was carried out in the late nineteenth and early 20th century using the methods available at that time. In Italy, after an initial 19th century interest in pile-dwellings influenced by romantic notions and interpretations framed according to national identity, in the period between two world wars attention to them—and to prehistory in general—lost momentum in favour of exaltation of the Roman civilisation as the origin of Italian history and culture (Marzatico 2004). Only in the 1970s did scientifically accurate research resume, with the work of Renato Perini in Fiavé in Trentino (Perini 1984, 1994) and Lavagnone in Lombardy (Perini 1981, 1988), leading to an appreciation of the complexity of prehistoric pile-dwellings.

The paper offers a brief summary of Neolithic and Bronze Age wetland sites, focussing on two notable localities, Palù di Livenza for the Neolithic and Lucone di Polpenazze for the Bronze Age. Both sites were inscribed on the UNESCO World Heritage List in 2011 and are under investigation by the authors as part of multidisciplinary projects aimed at understanding the dynamics of the prehistoric peoples who settled in the wetlands.

3.2 Geographical Distribution

Approximately 200 known pile-dwelling sites are distributed in wetland zones along the entire southern belt of the Alps, from western Piedmont to Friuli (Fig. 3.1). Their main wetland features and landscape contexts can be summarised as follows:

- Small lakes and peat bogs in the Alpine belt, well represented by the sites of Fiavé and Ledro in Trentino;
- Large Prealpine lakes, especially Lake Varese and Lake Garda in Lombardy and Veneto;
- Small lakes and peat bogs in moraine areas, distributed throughout the belt running between the Alps and the Po Plain, becoming more numerous in the Lake Garda area in Lombardy and Veneto;
- Small lakes and peat bogs in the Berici and Euganei hills in Veneto;
- Rivers, springs, and marshy areas in the lowlands. These pile-dwellings are found mainly in the lowlands of eastern Lombardy and Veneto.

3.3 Chronology

Although there the presence of forager groups is evidenced during the early Holocene in northern Italian wetlands, the appearance of stable settlements dates to the Early Neolithic,[2] but is most documented later during the Middle and Late Neolithic,[3] with the Square-mouthed Pottery culture[4] (*Cultura dei Vasi a Bocca Quadrata*) and then the Lagozza culture.

The greatest diffusion of pile-dwelling sites, however, occurred at the beginning of the Early Bronze Age,[5] between the end of the 3rd and beginning of the 2nd millennium BC, when they became the predominant type of settlement within the Polada culture (Fasani 1982; Baioni et al. 2018; Martinelli 2019), although other

---

[2] The Early Neolithic is hereafter indicated as EN.

[3] The Middle and Late Neolithic are hereafter referred to as MN and LN respectively.

[4] Hereafter SMP culture.

[5] Bronze Age subdivisions are hereafter indicated as follows: EBA (Early Bronze Age), MBA (Middle Bronze Age), LBA (Late Bronze Age), FBA (Final Bronze Age).

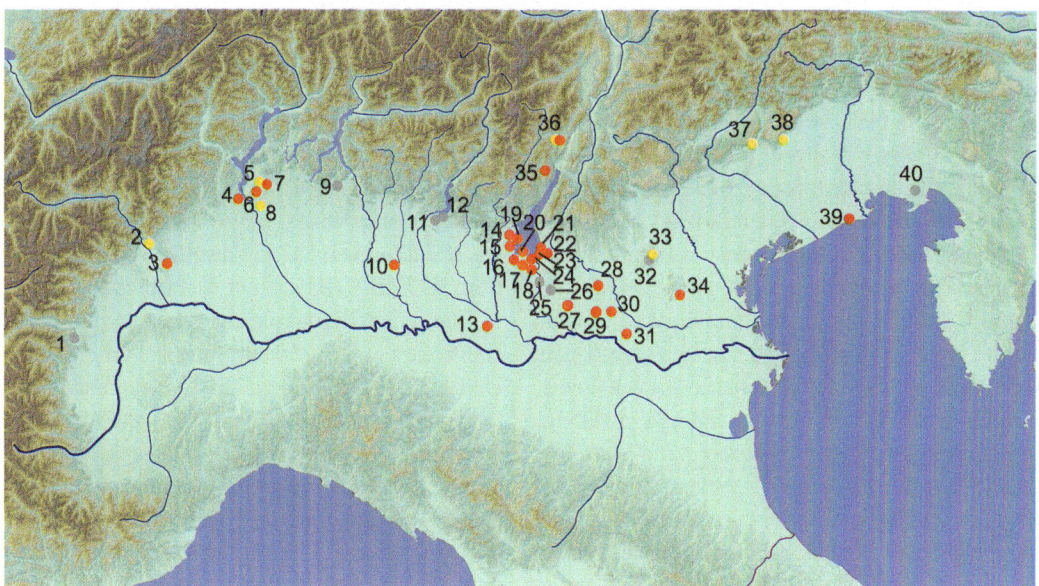

Fig. 3.1 Distribution map of the main Italian pile dwellings. Neolithic sites in yellow; Bronze Age sites in red; sites not mentioned in the texts in grey. For reasons of scale, some points indicate several settlements. (**1**) Avigliana (Torino), Torbiera di Trana; (**2**) Montalto Dora (Torino), Lago Pistono; (**3**) Viverone (Biella)/Azeglio (Torino), Lago di Viverone (sites Vi1, Vi2, Vi3); (**4**) Arona (Novara), Lagone di Mercurago; (**5**) Biandronno (Varese), Isolino Virginia; (**6**) Cadrezzate (Varese), sites in Lake Monate; (**7**) Bodio Lomnago (Varese), Bodio Centrale, Desor Maresco, Keller-Gaggio; Bardello (Varese), Palude Ranchet, Bardello Stoppani; Cazzago Brabbia (Varese), Ponti or Cazzago; (**8**) Besnate (Varese), Lagozza; (**9**) Bosisio Parini (Lecco/Como), Cascina del Pascolo; (**10**) Sergnano (Cremona); (**11**) Corte Franca (Brescia), Valle delle Paiole and other sites; (**12**) Iseo (Brescia), Torbiere di Iseo (several sites); (**13**) Piadena Drizzona (Cremona), Lagazzi del Vho; (**14**) Polpenazze del Garda (Brescia), Lucone; (**15**) Moniga del Garda (Brescia), Porto; Padenghe del Garda (Brescia), West Garda and La Cà; Desenzano del Garda (Brescia), Corno di Sotto; (**16**) Desenzano del Garda (Brescia), Lavagnone; (**17**) Cavriana (Mantova), Bande-Corte Carpani; Solferino (Mantova), Barche; (**18**) Monzambano (Mantova), Castellaro Lagusello; (**19**) Manerba del Garda (Brescia), San Sivino-Gabbiano; (**20**) Sirmione (Brescia), San Francesco, Porto Galeazzi, Lugana Vecchia, La Maraschina; (**21**) Cisano (Verona), Porto; (**22**) Cavaion Veronese (Verona), Cà Nova; (**23**) Lazise (Verona), La Quercia and Pacengo (Porto, Bor, Bosca); (**24**) Peschiera del Garda (Verona), Frassino, Belvedere and historical Peschiera sites (Imboccatura del Mincio, Bacino Marina; palafitta del Mincio or Setteponti); (**25**) Volta Mantovana (Mantova), Isolone del Mincio; (**26**) Roverbella (Mantova), Prestinari and minor sites; (**27**) Isola della Scala (Verona) sites; (**28**) Oppeano (Verona), Feniletto, 4C, 4D, 4E sites; (**29**) Nogara (Verona), Dossetto; (**30**) Cerea (Verona), Tombola; (**31**) San Pietro Polesine (Rovigo), Canàr; (**32**) Arcugnano (Vicenza), Fondo Tomellero, and other sites; (**33**) Arcugnano (Vicenza), Molino Casarotto, Le Fratte; (**34**) Arquà Petrarca (Padova), Laghetto della Costa; (**35**) Ledro (Trento), Molina di Ledro; (**36**) Fiavé (Trento), Torbiera Carera; (**37**) Colmaggiore di Tarzo (Treviso), Revine lago; (**38**) Polcenigo/Caneva (Pordenone), Palù di Livenza; (**39**) Caorle (Venezia), San Gaetano; (**40**) Terzo di Aquileia (Udine), Canale Anfora (Authors)

types are also known such as those on hilltops and in caves. Between the late EBA and early MBA there occurred perhaps the greatest spread of this type of settlement: the Polada culture, starting from the Lake Garda area, spread along the main rivers in the central Po Plain reaching the River Po (Rapi 2013), while further west in Piedmont and western Lombardy the Bodio-Mercurago-Pollera phase of the North-western Bronze Age culture developed (Grassi and Mangani 2014).

At the beginning of the MBA, alongside pile-dwellings, the first dry-ground villages surrounded by ditches with house structures still on posts emerged; the construction system for buildings on stilts successfully tested in wet environments was thus extended to dry-land settlements in the case of the *terramare* (Bernabò Brea et al. 1997; Vanzetti 2013). Pile-dwellings continued to be used during the LBA and FBA, but their incidence progressively decreased. Although the LBA was a phase of major demographic expansion, only a few pile-dwellings are known, and the general abandonment of wetland settlements is documented. In the subsequent FBA the number of sites became even lower, especially in the Po Plain, and pile-dwellings became extremely rare (Billamboz and Martinelli 2015). Similar evidence of settlement reduction is also documented in the southern Po Plain, where geohydrological data reveal reduced water availability in the LBA and FBA that seems to mark the end of the *terramare* phenomenon (Cremaschi et al. 2006).

A difference in the incidence of the pile-dwelling phenomenon can be observed between the north and south of the Alps: in northern Italy there were few pile-dwellings in the late 4th millennium and the first half of the 3rd millennium BC, while their number increased between the end of the 3rd and the first half of the 2nd millennium BC (Fig. 3.2). In comparison, the period of maximum development north of the Alps is found during the Neolithic, with a peak during the late phases, and then again in the LBA and FBA. Thus, the spread of pile-dwelling settlements on the two sides of the Alps seems to have had an asynchronous development in prehistoric times, with different origins and motivations.

3.4 Environmental Studies

In recent years, several research projects have been undertaken to study the environmental conditions of these settlements and their diachronic variations. With regard to Neolithic sites, reconstructive environmental analyses are underway in Palù di Livenza and Isolino Virginia. As for the Bronze Age, numerous projects have already been completed or are still in progress in Viverone, Oppeano, Ledro (Joannin

et al. 2014), Sergnano, Bodio (Grassi and Mangani 2014), Lucone di Polpenazze (Valsecchi et al 2006) and Lavagnone (Ravazzi et al. 2018). Based on the studies carried out in the latter two sites, a synthesis has recently been presented for the Lake Garda area, which highlights phases of anthropic impact, as well as reforestation after the abandonment of pile-dwelling sites starting from 1200 BC (Ravazzi et al. in press).

Further synthesis work has been carried out in connection with the drafting of the Italian UNESCO World Heritage List Management Plan (Ruggiero et al. 2022). In this case, the focus was on the state of conservation of the archaeological deposits and assessment of the erosion phenomena that damage archaeological deposits in the World Heritage List site components (Baioni et al. 2015). The objective here was to evaluate the extent of erosion phenomena damaging structures and strata.

3.5 Definition of Pile-Dwellings and Their Structural Typology

Although the concept of pile-dwellings may appear to be linked to specific structural characteristics of buildings raised on posts above the ground or water surface,[6] the use of this term is often not based on detailed analysis of the settlement structures, but instead it has a more generic meaning linked to the environmental conditions of the archaeological deposit. The Italian term *palafitta*, from the original meaning of 'piling', was used in the 19th century to translate the German *Pfahlbau* introduced by F. Keller in 1854. As research continued and the complexity of these settlements was better understood, the term appeared ambiguous, since the 'fields of poles' could derive either from shore settlements in dry conditions or settlements on decks.[7]

---

[6] Essentially, houses on poles overlooking water.

[7] In German, the use of two terms, *Moorsiedlungen* for shore settlements on the ground and *Ufersiedlungen* for settlements on decks, has become widespread (Billamboz and Schlichtherle 1984; Schlichtherle 2004).

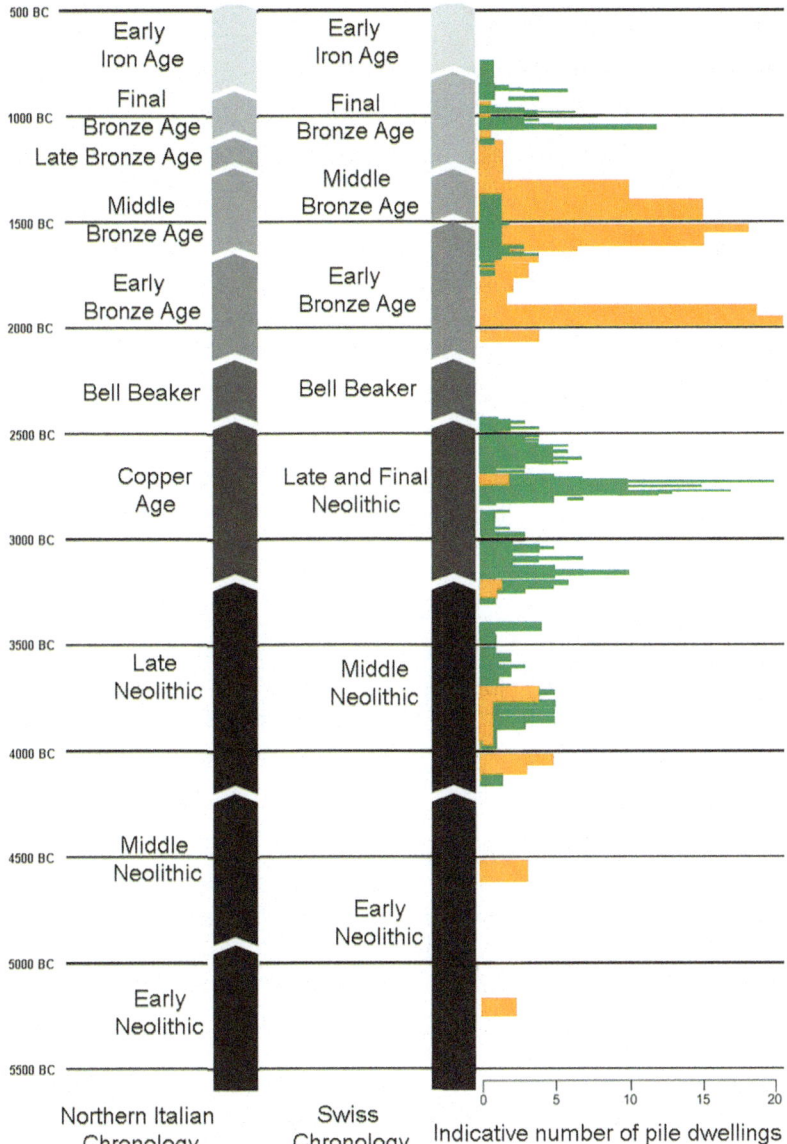

Fig. 3.2 Chart showing the chronological distribution of pile dwellings in Italy and the rest of the circum-Alpine region (Graphics by M. Baioni after similar charts present in the literature, starting with Corboud and Petrequin 2004, Fig. 4)

In Italy, on the other hand, an attempt was made to distinguish between pile-dwellings sensu stricto, with a deck of planks supported by posts, and villages built on layers of timber, branches, stones, and earth, based directly on the ground (in Italian *bonifiche* or reclaimed land). This distinction was formalised by Balista and Leonardi (1996), who also attempted a classification of the two structural types. This classification, taken up in a recent overview (Baioni et al. 2018), was developed in the context of studies on pile-dwellings of the Bronze Age, but can now also be extended to the Neolithic.

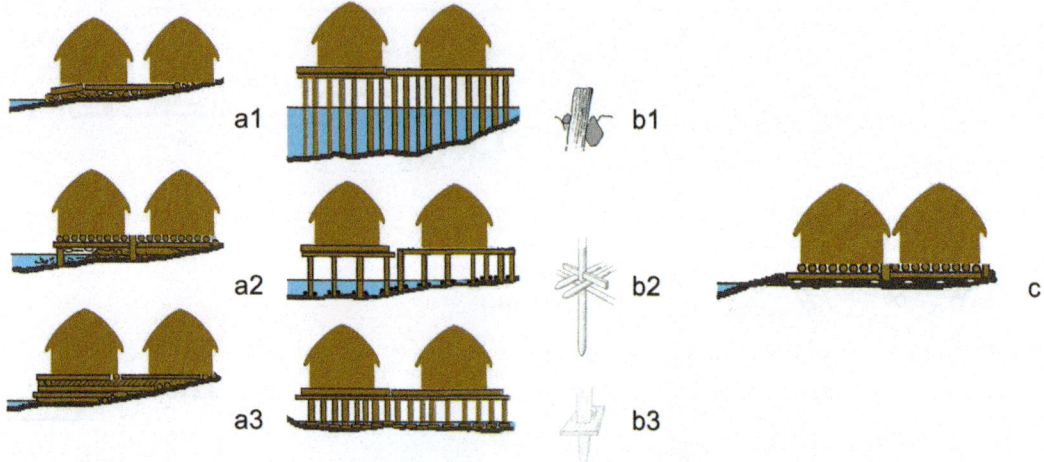

Fig. 3.3 Summary diagram of the various types of pile dwellings (M. Baioni after Balista and Leonardi 1996; De Marinis et al. 2005; Rapi 2013)

According to this classification, based on the remains found at some Bronze Age sites (Fig. 3.3), northern Italian wetland settlements may be divided into:

- *Settlements built directly on the ground surface or soil, or timber structures laid on it (bonifica)*, i.e. houses built at ground level, often with wooden floors lying on a framework of logs and small branches or more complex wooden structures. Within this group it is possible to recognise a further subdivision into constructions built on the ground: on layers of timber, branches, stones and/or soil (Type A1: Lavagnone 4); or on wooden structures, such as large oak beams placed side by side and supported by small posts driven into the lake silt (Type A2: Arquà Petrarca; Isolino; Molino Casarotto); or on *caissons*, rectangular structures made of wooden beams and filled with various materials (Type A3: Barche di Solferino); or directly on the ground surface (Type C: Ledro).
- *Dwellings on raised decks*, i.e. houses built on structures that raise them to varying heights above ground or water level. This group includes settlements with constructions built: on high raised decks supported by isolated poles (Type B1: Lucone D, Lavagnone 2, Fiavé 3); on high raised decks supported by posts held by a grid structure (*podium*) of horizontal beams (Type B2: Fiavé 6); on low raised decks supported by poles stabilised with pile shoes (B3: Lavagnone 3).

3.6 Neolithic Occupation of Wetland Zones

The first stable occupation of wetlands south of the Alps started in the second half of the 6th millennium BC and increased later during the Neolithic, in the 5th millennium BC. The initial phase corresponds to a later period of the EN that followed the first expansion of the Impresso-Cardial complex groups. The submerged site of La Marmotta in Lake Bracciano (see Chap. 4 in this book) in central Italy constitutes an interesting example of this phenomenon; other cases are also documented in northern Italy on Lake Varese. A similar occupation pattern is also found at La Draga in Catalonia (see Chap. 11 in this book) and Dispilio in Greece (see Chap. 8 in this book). All these examples indicate that wetlands attracted settlement within a few centuries after the initial early farmers' expansion and the first transformation of forests and woodlands into agricultural landscapes in southern Europe. The ecological variety associated with the availability of different soils, compared to those previously managed by the Neolithic farmers, certainly favoured the occupation of these areas (Menotti 2013; Pétrequin 2013; Hafner et al. 2016). Wetlands have the ecological characteristic of comprising a great variety of habitats present in limited spaces where different ecosystems coexist, with remarkable biodiversity and a high productivity of plant biomass which attracts wild animals. They are thus rich in natural resources and particularly favourable for subsistence. Unfortunately, few of the Neolithic settlements investigated to date in northern Italy preserve archaeological deposits with reliable stratigraphies; furthermore, the trenches excavated are often limited in size and, unlike other areas in the Alpine region, little information is available concerning the internal organisation of Neolithic villages and their spatial extension in wetlands.

The main Neolithic remains are wooden platforms or areas of reclaimed land which suggest the presence of dwellings on the ground, while only in very few cases is there evidence of features on stilts. For this reason, Palù di Livenza constitutes a very interesting case study of a Neolithic pile-dwelling settlement.

3.6.1 Palù di Livenza: A Multiphase Pile-Dwelling Settlement

Palù di Livenza is a wetland zone at the foot of the Cansiglio plateau near Pordenone in north-east Italy, extending into the territory of Caneva and Polcenigo where the karst springs of the River Livenza are located. The presence of wooden posts and other archaeological finds has been known since the 19th century. The numerous investigations carried out in the last forty years have led to the discovery of a large number of archaeological finds which revealed the exceptional nature of the site. Systematic investigation by means of surveys, excavations, and coring, focussed along the drainage channel in Sectors 1 and 2, have furnished information about a Neolithic pile-dwelling settlement (Fig. 3.4 (1)).

Archaeological excavations have uncovered wooden features with aerial platforms and land reclamation works, demonstrating various phases of site occupation (Peretto and Taffarelli 1973; Corti et al. 1998; Vitri et al. 2002).

A new phase of research has been launched in Sector 3: although the excavation area was limited to only 48 m^2, corresponding to just a small fraction of the archaeological site's total extension, the investigations outlined a diachronic sequence of five main Neolithic structural phases in which four different pile-dwelling features alternate with abandonment of the area (Micheli et al. 2018, 2019, in press) (Fig. 3.4 (2)). Hundreds of wooden elements were identified, consisting of both horizontal beams and fixed vertical poles. The construction systems identified show technical similarities and employ the same type of foundation structures: sleeper beams and plinths made of worked oak logs supported standing buildings, which may have been both residential in function and used for other purposes, such as granaries or other storehouses (Fig. 3.4 (3, 4)).

Proceeding from below, structural phases 1, 4 and 5 are represented by the remains of the foundations on sleeper beams of three buildings that may be interpreted as dwellings on stilts based on the dimensions of wooden features and the accumulations of waste deposits with abundant potsherds, stone tools and unfinished artefacts, baked clay fragments, organic and animal remains. Phase 2 revealed the remains of a smaller structure on *Pfahlschuh*-type plinths, probably rectangular in shape, interpreted as an elevated silos held up by fixed wooden posts (Fig. 3.4 (2 and 4)); this structure was destroyed by fire, as indicated by the abundance of burnt remains of cereals, including fragments of whole ears. The remains of Phase 3 comprised a foundation structure on sleepers of smaller dimensions than those of Phases 1, 4 and 5 which can be interpreted as a deposit or warehouse, also destroyed by fire. Inside this construction were preserved abundant grain, numerous dried small wild apples, shapeless blocks of flint and some unfinished wooden objects (a small spoon, agricultural tools and other objects that are difficult to interpret). The wooden tools found include a digging stick with a thickened, pointed end—a

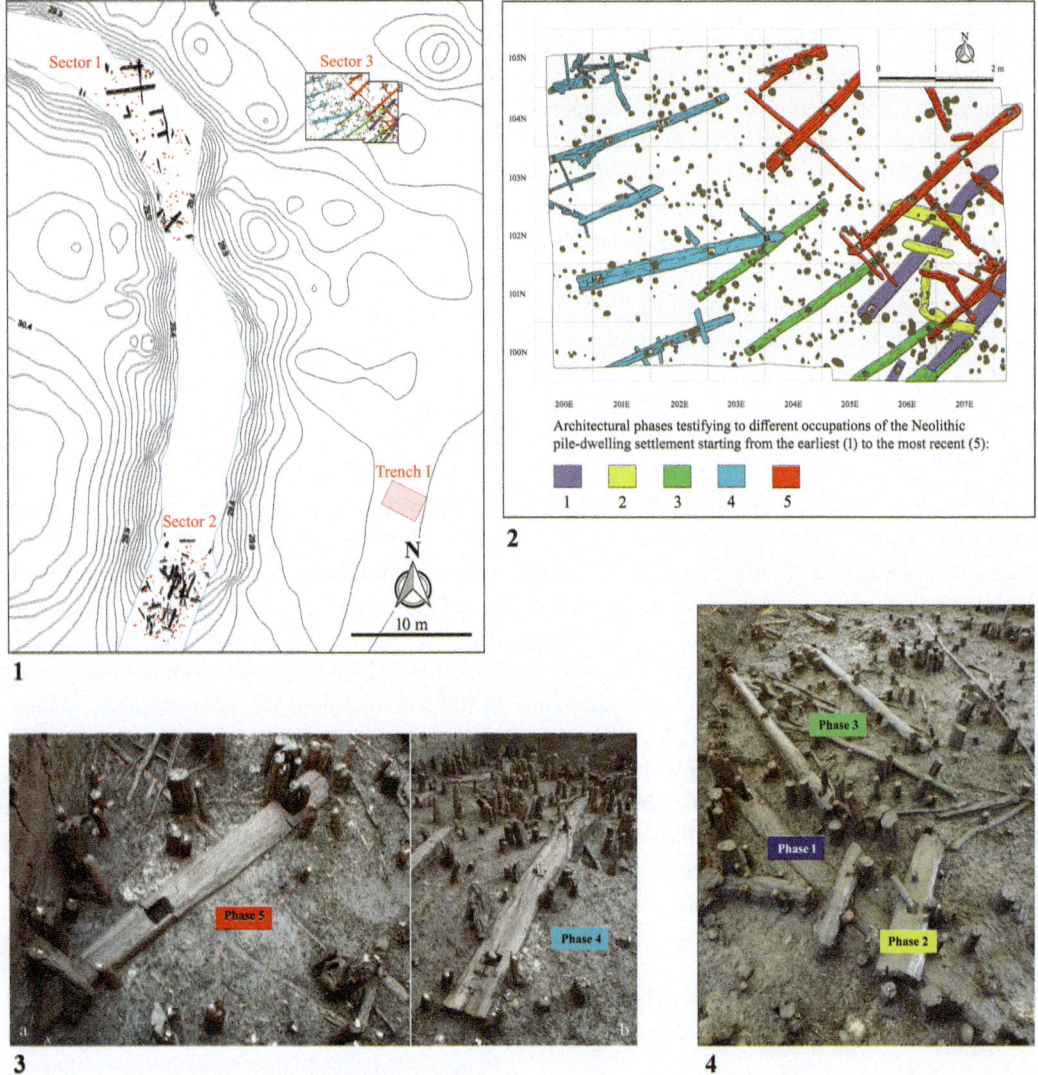

Fig. 3.4 Palù di Livenza, a Neolithic multiphase pile-dwelling settlement: (**1**) The artificial channel with the location of Sectors 1–3; (**2**) Palimpsest of the wooden foundation structures identified in Sector 3; (**3a**) Detail of the phase 5A sleeper beam; (**3b**) Detail of the phase 4 sleeper beam; (**4**) Palimpsest of the wooden foundation structures identified in phases 1–3 in the south-eastern part of Sector 3 (SABAP-FVG archive)

kind of a proto-spade, which we may imagine was used for tilling the soil and digging holes in the ground. Over time, through the various phases, there is a progressive rotation of the orientation of the foundations of the overlapping Neolithic pile-dwellings in relation to the changing conformation of the land and seasonal variations in the water level (Micheli et al. 2023).

Oak sleepers between 3 and 4 m long (Fig. 3.4 (3)), with regularly-spaced rectangular through-holes, constituted the main foundation elements that supported the posts that held up the raised structures by means of mortise-and-tenon joints in Phases 1, 3, 4 and 5. These structures are also associated with numerous piles which lack any clear order, but some of which presumably functioned together with the plinth systems and sleeper beams; they were constructed at the same time and/or after the pile-dwellings were built. The same foundation mechanism based on mortise-and-tenon joints underwent progressive improvement over time, passing from 2 holes in Phase 1 beams to 3 holes in those of Phase 3, up to 4 holes in Phases 4 and 5. This suggests refinement of the construction techniques used to support the aerial features, giving greater stability to the buildings, that was based on local experience acquired in situ by the Neolithic people living at Palù di Livenza as an adaptive response to the particular environmental conditions of the wetland—where water was in the past, as it is today, the natural element that regulates everything. The data from Sector 3 reveal the complexity of the architectural mechanisms developed by the Neolithic farmers to preserve their dwellings from the threat of water level fluctuations.

The artefacts found suggest that the lower Phases 1–4 can be ascribed to the SMP culture in some of its various aspects that evolved in the second half of the 5th millennium BC, starting from 4400 to 4300 cal BC. The last Phase 5 may be attributed to the LN Alpine groups and dates to between c. 3950 and 3650 cal BC (Micheli 2018; Micheli et al. 2018, 2019, 2023).

3.6.2 Neolithic Remains in Northern Italian Wetlands

Although Italian Neolithic settlements in wetlands are generally not numerous, those investigated so far reveal the complexity of the architectural solutions devised by the Neolithic groups to adapt their constructions to water level fluctuations, the instability of the waterlogged soil and the prolonged damp environmental conditions. Evidence from northern Italy suggests that Neolithic dwellings were built on the ground, directly on wooden platforms or planking on the surface of reclaimed land or on preparation layers, as seen in the examples dating to the EN at Isolino Virginia and Pizzo di Bodio, to the MN at Fimon-Molino Casarotto, to the RN at Lagozza di Besnate and Colmaggiore di Tarzo, and lastly to the

LN at Fiavé 1. The data emerging from these sites suggest the existence of Neolithic perilacustrine settlements that developed along the outer belt of the lakes' maximum flooding zones, or in areas along the shores or on islands, in any case in areas of the wetlands at higher altitudes.

Elevated constructions, on an aerial deck supported by poles above water, are therefore currently documented with certainty only at Palù di Livenza, where in fact oscillations of the water table in the basin were, as today, frequent, and irregular. The same foundation system seen in the aerial deck constructions was used for the erection of the larger buildings in Phases 1, 3, 4 and 5 (houses and a warehouse), while a system based on *Pfahlschuhe*-type plinths was adopted for the Phase 2 structure, interpreted as a probable elevated silo (Fig. 3.4 (2 and 4)). This archaeological evidence reveals the substantial continuity over time of the architectural solutions adopted, which must have been particularly effective in limiting the problems arising from the presence of water and persistent dampness, while at the same time giving stability to the raised buildings.

At Isolino Virginia (see Chap. 16 in this book) the site extends over an island, where a deposit about 4 m thick preserves a succession of Neolithic wooden platforms (Fig. 3.5 (1)), and the remains of subsequent repeated occupations throughout the Neolithic.[8] The buildings were erected not later than the early phase of the Neolithic on reclaimed land directly overlying lake sediments, which over time led to subsidence and the collapse of these structures. This situation resulted in repeated interventions and reclamation operations with deposits of tree trunks and mixed branches, together with sand and stones (Fusco 1976–77; Guerreschi 1978–79; Guerreschi et al. 1992; Baioni et al. 2005). At Pizzo di Bodio on Lake Varese a deeply-stratified settlement was present during the E/MN, and later in the Copper Age dwellings were built on reclaimed land from the early stages of occupation (Banchieri and Balista 1991).

---

[8] The site was occupied during the EN by people of the Isolino group, throughout the MN by local SMP culture-related groups, and in the LN by Lagozza culture groups.

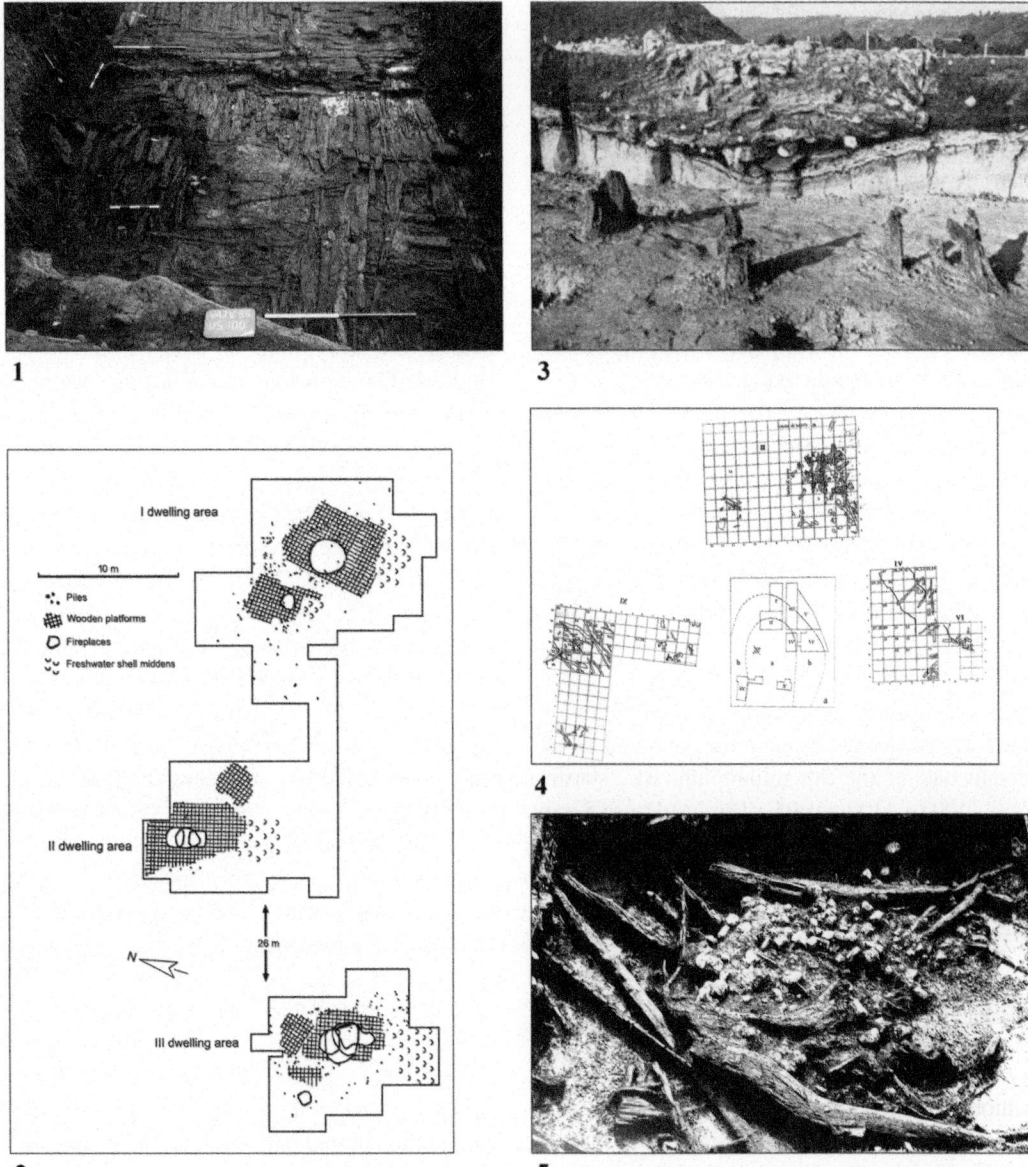

Fig. 3.5 Neolithic settlements in wetland zones in northern Italy: (**1**) EN wooden platform found during the 1995 excavations at Isolino Virginia; (**2**) Plan of dwelling areas with wooden structures found in the MN site of Fimon-Molino Casarotto; (**3**) Detail of a multi-phase hearth in dwelling area 1 at Fimon-Molino Casarotto; (**4**) Plan of the wooden structures found in the LN site of Fiavé 1; (**5**) Detail of wooden platform remains in the area north of Sector IX at Fiavé 1 (1 after Baioni et al. 2005; 2–3 after Bagolini 1983–84; 4–5 after Perini 1984)

At the Fimon–Molino Casarotto site (Vicenza) three main features attributed to the first phase of the SMP culture were investigated in the early 1970s (Bagolini et al. 1973). Wooden platforms on reclaimed land located on the edge of a lake basin rich in natural resources were composed of parallel horizontal elements fixed to the ground by posts driven into the silt; each platform had a large hearth in the centre, which appeared to have been rebuilt over time during diverse occupation phases (Fig. 3.5 (2, 3)). The site was occupied at various times during the MN between 4900–4800 and 4300–4200 cal BC. In the same basin, recent investigations at Le Fratte (Bianchin Citton 2016) brought to light the remains of a LN settlement consisting of alignments of wooden poles without associated floor levels.

Investigations at Lake Pistono near Montalto Dora (Torino) in Piedmont identified Neolithic remains located on a peninsula overlooking the lake basin (Padovan et al. 2019). Roughly perpendicular features consisting of two rows of post-holes were interpreted as the remains of two dwellings, maybe on stilts, referable to the SMP culture (phase 2) dating to around the mid-5th millennium BC.

Land reclamation features used to consolidate the ground on which stood raised dwellings, probably on stilts, have been identified since the late 19th century at Lagozza di Besnate (Varese) (Guerreschi 1966; 67; Borrello 1984), location of the eponymous RN culture site, related to the Chassey and Cortaillod groups. Other reclamation works consisting of alignments of wooden poles, burnt wooden planks and stone embankments were identified at Colmaggiore di Tarzo (Treviso) in Veneto, about 20 km from Palù di Livenza, where a multi-phase LN settlement extended over an area situated between two lakes (Bianchin Citton 1994). The first occupation of the famous multi-phase Bronze Age site of Fiavé in Trentino is attributed to the LN and dated to about 3900 and 3600 cal BC. It developed on a natural island (Fig. 3.5 (4)) and on areas where land reclamation (Fig. 3.5 (5)) had been carried out using artificial networks of larch and pine trunks, with twigs, stones, gravel, and earth (Perini 1984, 29–51).

To find other similar Neolithic pile-dwellings with foundations such as sleeper beams or *Pfahlschuhe*-type plinths as documented at Palù di Livenza we must look beyond Italy, remaining however in the Alpine region. There are various wetland sites in southern Germany, south-eastern France, and Switzerland, dating from the second half of the 5th to the 4th millennium BC, where variations on *Pfahlschuhe*-type plinths are known, as well as long sleeper beams with multiple holes or other mortise-and-tenon joint systems for sustaining the weight of stilt buildings (Pétrequin and Pétrequin 1988; Schlichtherle and Billiamboz 2015; Ebersbach 2013; Hofmann et al. 2016). Archaeological data reveal that similar solutions were adopted by groups that inhabited different places at various times during the Neolithic. In many cases, the Neolithic groups that developed these methods did not have direct relations or contacts with each other, as the case of Palù di Livenza also seems to suggest, even when the other known Italian cases are considered. It is probable that ingenuity, empirical and technological knowledge acquired through experience, and the need to find effective solutions to contingent problems posed by the various wetland environments occupied by Neolithic communities led the various groups to independently devise similar apparently simple mechanisms that sufficed to give stability to aerial constructions and limit the threat posed by water in lakeside villages.

3.7 Bronze Age Occupation of Wetland Zones

During the Bronze Age, the pile-dwelling phenomenon spread throughout northern Italy, starting from the pre-Alpine belt and its numerous intra-moraine lakes. The phenomenon also characterises both peat bogs and small lakes in western Piedmont and the belt between the provinces of Lecco, Como, and Bergamo, with a concentration in the southern area of Lake Iseo and its peat bogs (Poggiani Keller et al. 2005). In the Veneto area, the presence of pile-dwellings on the lagoon near the sea is also known, such as the LBA site of San Gaetano (Caorle–Venice).

However, the areas where Bronze Age pile-dwellings are particularly concentrated are essentially three:

- between Lake Garda and the River Po, including western Trentino, eastern Lombardy and western Veneto;
- between the Berici and Euganei hills in Veneto;

- between Lake Varese and Lake Viverone (western Piedmont-eastern Lombardy).

Undoubtedly the richest and most complex of these is the Garda area, where the main ongoing research is also concentrated. These three areas are briefly described below, with reference to the sites of greatest interest.

3.7.1 Area Between Lake Garda and the River Po, Including Western Trentino, Eastern Lombardy and Western Veneto

Two important sites are known: Ledro (Molina di Ledro–Trento, IT-TN-01) and Fiavé (Fiavé Carera–Trento, IT-TN-02). Ledro is an important pile-dwelling settlement that was excavated between the 1930s and 1960s, unfortunately without recording which layers most of the finds came from (Rageth 1975).

In the area of the ancient Lake Carera at Fiavé,[9] seven consecutive pile-dwellings have been brought to light: they occupied different areas of the basin during a long period of time, from LN (*Fiavé 1* village) to LBA (*Fiavé 7* village). In what is called 'area 2' we can still see the dense vertical pile field of *Fiavé 3* and *Fiavé 5* villages (erected between the EBA and MBA) (Fig. 3.6 (2)). The more interesting structures are those of *Fiavé 6* village (advanced MBA): built on a peninsula and enclosed by a palisade, the village spread out from the shore area, where an aerial deck supported by posts was held by a grid (*podium*) of horizontal poles (Type B2) (Fig. 3.6 (3)). The shape and dimensions of the houses have been hypothesised (with two different proposals), without resorting to dendrochronological dating (Perini 1984, 1994) (Fig. 3.6 (1)).

Between 1986 and 1993 a new area (area 4) was identified: it has characteristics similar to area 2 (Marzatico 1996).

There are numerous submerged sites along the southern shores of Lake Garda. The sites where investigations are in progress include San Sivino–Gabbiano (Manerba del Garda–Brescia, IT-LM-02), identified in 1971, where the first underwater survey was carried out by L.H. Barfield in 1978. Recently a new survey has been undertaken, accompanied by sampling of the piles in order to reconstruct, thanks to dendrochronology, the organisation of the settlement.[10]

The Moniga Porto site, investigated in 1981, featured preserved in situ structures and stratigraphy (Pia 1982), while that at Corno di Sotto, identified in 1966, has recently (2013) been the object of a research project that detected a strong erosive process damaging anthropogenic deposits (Baioni et al. 2015).

There is little information regarding the submerged pile-dwellings around the Sirmione peninsula (Mangani and Ruggiero 2018), while a study has recently been published that attempts to understand the complex topography of the Peschiera del Garda sites (Albertini and Martinelli 2018).

Recent research has furnished detailed information concerning a submerged pile-dwelling at Belvedere (Peschiera del Garda-VR, IT-VN-04), where dendrochronology revealed the presence of posts from both an older village (2060–2040 cal BC) and a more recent one (1751, 1655–1635 cal BC) (Capulli et al. 2014; see Chap. 12 in this book).

Another important submerged site is La Quercia (Lazise–Verona), investigated between 1984 and 1994. It is the only settlement in Lake Garda where stratigraphic underwater excavations have been carried out, together with a topographic survey which revealed more than 6000 piles in an area of almost 2 ha (Fozzati et al. 2015).

---

[9] The chronological sequence of the Fiavé site was drawn up by Perini (1981, 1984), who identified seven phases from Fiavé 1 (LN) to Fiavé 7 (LBA).

[10] Research direction: Soprintendenza ABAP-BG-BS, in collaboration with Manerba del Garda municipality and economic support from the Lombardy Regional Authority.

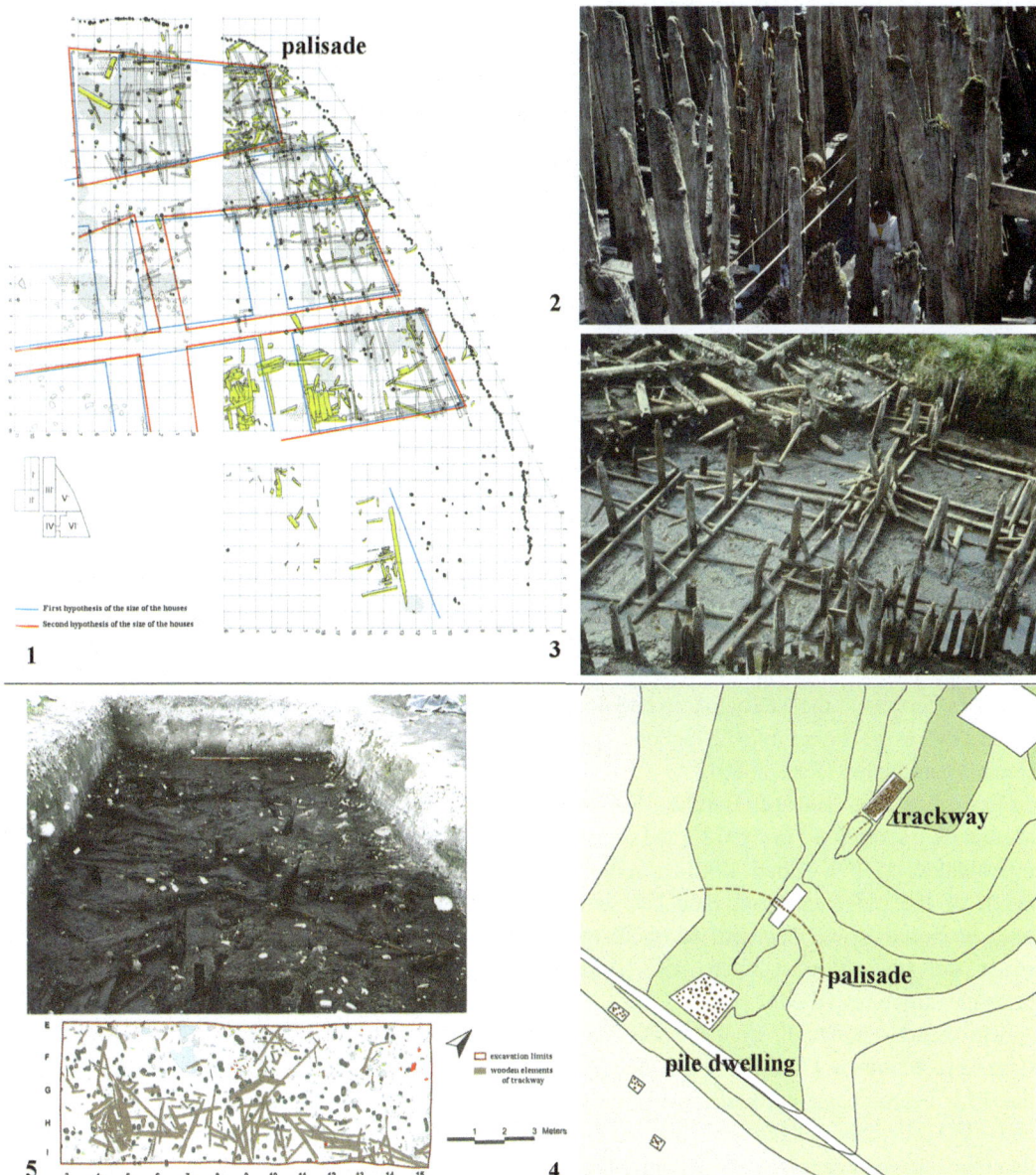

Fig. 3.6 Fiavé-Carera (Trento): (**1**) Plan of the Fiavé 6 phase with indication of the two house-shape hypotheses; (**2**) Posts of Fiavé 3–5 pile dwelling; (**3**) Beam grid supporting the posts of Fiavé 6. Desenzano del Garda– Lavagnone (Brescia); (**4**) Areas excavated at Lavagnone with EBA structures (from Rapi 2013 with modifications); (**5**) EBA trackway in Sector B (After De Marinis et al. 2005)

A particularly rich picture emerges from the numerous small lakes and intra-moraine basins surrounding the southern part of Lake Garda. The Lucone basin (Polpenazze del Garda–Brescia, IT-LM-05) is on the eastern side; known since the 1960s, with 4 different sites, it was occupied from the beginning of the EBA until the late MBA (Poggiani Keller et al. 2005). Only Lucone A and Lucone D have been excavated; pile-dwelling settlements were found in both. Since 2007, annual excavation campaigns aimed at thoroughly documenting an EBA pile-dwelling at Lucone D have been conducted. For this reason, it has been chosen as a representative site (see infra, Sect. 3.7.4).

The Lavagnone basin (Desenzano del Garda/ Lonato del Garda–Brescia, IT-LM-01), known since the 19th century, was first systematically excavated by Perini (1981, 1988). The University of Milan[11] has continued research at Lavagnone since 1989 (De Marinis 2007; Rapi 2013). The area has been divided into various excavation sectors (A, B, C, D, E); the settlement identified underwent topographical shifts and changes in architectural techniques during the Bronze Age (Fig. 3.6 (4)). Different settlement areas, a timber trackway (a road leading to the settlement), and a small part of a palisade have been identified (De Marinis 2007; De Marinis et al. 2005) (Fig. 3.6 (5)). Research is now focussed on a residential nucleus (Sector E) located towards the centre of the basin, dating to the MBA, which has yielded interesting wooden finds such as a pirogue and a yoke. In addition, the study of archaeological objects from the layers of 'Lavagnone 3' was recently completed (Rapi 2020).

Regarding the site of Barche (Solferino– Mantova), identified in 1918 and partially investigated in 1938 (Zorzi 1940), a reinterpretation of the old excavation data has recently been proposed by G. Leonardi to try to reconstruct the interesting wooden box (*cassone*) drainage structures (Baioni et al. 2018).

The nearby basin of Bande–Corte Carpani (Cavriana–Mantova, IT-LM-07) was identified in the 19th century and excavated between 1952 and 1983 (Piccoli 1982a). This too is a pile-dwelling site characterised by topographic and structural variations over time. Recently, N. Martinelli published a study of the plan of part of the settlement (excavations 1981–1983) based on dendrochronological analyses, in which two walkways dated between 1999 and 1994 cal BC, and a house built in 1980 and renovated in

1974 cal BC, with spaces organised in two 'naves' (Baioni et al. 2018), were identified (Fig. 3.7 (1)).

At the small lake of Castellaro Lagusello (Monzambano–Mantua, IT-LM-08) two settlements dating between the MBA and LBA have been identified. The first excavations (1977–78) recognised a sequence of settlements on drained land (Piccoli 1982b). The most recent research has identified a large deck set on the ground (Fasani 2002).

At Lake Frassino (Peschiera del Garda–Verona, IT-VN-05), a submerged pile-dwelling was identified in the early 1980s (Fozzati et al. 2015; Gonzato et al. 2019) and investigated by underwater excavation campaigns carried out by the Soprintendenza between 1989 and 2000. Research was resumed in 2014. The finds date to both the EBA and MBA. Dendrochronological and radiocarbon dates of samples taken from the lake range from 1776 to 1703 cal BC (see Chap. 12 in this book).

Numerous settlements in marshy areas are documented outside the moraines around southern Lake Garda, near springs, rivers, and ancient riverbeds. Two new settlements were discovered during archaeological controls in a wetland area near Vallese (Oppeano–Verona). The first settlement (named 4C) dates to the EBA and the second (4D) to the MBA. The oldest (site 4C), a short-lived pile-dwelling abandoned after a flooding event, preserves different types of wooden structures; its overall development occurred in the late EBA 1 and early EBA 2 (Gonzato et al. 2021a). In site 4D, four phases have been identified. Phase 2 featured the best-preserved wooden structures, marked by standing wooden partitions delimiting frequently restored hearths and floor sequences (with intercalated finely bedded hearth rake-out layers). These structures were rectangular in shape and bounded by horizontal wooden elements and a wattle and daub wall (Gonzato et al. 2021b) (Fig. 3.7 (2)).

Tombola (Cerea–Verona, IT-VN-18), already known from surveys in 1955, was investigated in 1999. Vertical piles and horizontal timbers forming two rectangular structures were identified. The site dates to between the late MBA and

---

[11] Director of research: M. Rapi. The first dendrochronological analyses carried out by P. I Kuniholm (Griggs et al. 2007) are being extended by a new dating project by Ivana Pezzo, which is currently under way.

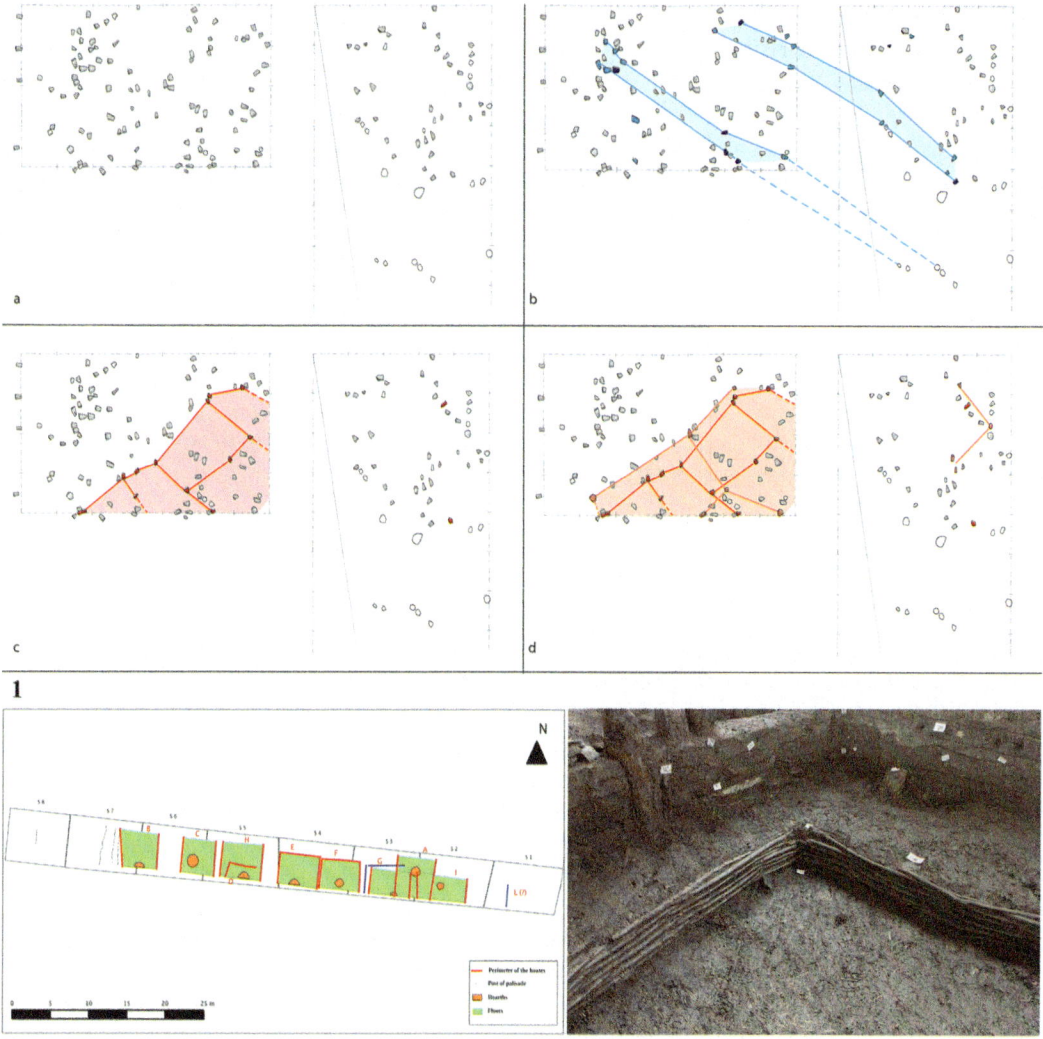

Fig. 3.7 (**1**) Bande di Cavriana (Mantova)—Plan of building reconstructions based on dendrochronology; (**a**) Reconstruction of post positions at insertion levels in the 1981 (right) and 1983 (left) excavation areas; (**b**) Identification of walkways made from wood felled between 1999 and 1994 cal BC; (**3**) Identification of load-bearing posts of a building, probably residential, erected in 1980 cal BC; (**4**) Identification of posts related to a probable renovation of the building using trees felled in 1974 cal BC (N. Martinelli, graphics E. Cavallo). (**5**) Oppeano (Verona)–Vallese, site 4D, plan of excavation area with row of dwellings (C. Mangani) and detail of a wooden structure (After Gonzato et al. 2021b)

LBA. Dendrochronology has revealed two different phases of tree felling (Salzani et al. 2018; see Chap. 12 in this book).

The site of Canàr (San Pietro Polesine–Rovigo), excavated numerous times in the 20th century (1984–88, 1990, 1994), presents two main occupation phases. Alongside a ditch, both a palisade and fence were found, made up of boards inserted vertically into the ground. These were contemporary with remains related to residential structures; unfortunately, it was not possible to ascertain their possible connection with some pole shoes present in the investigated area (Salzani et al. 1996).

3.7.2 Area Between the Berici and Euganei Hills (Veneto)

The hilly areas of the Berici and Euganei, where there are various peat bogs and small lakes, are located in the central part of the Veneto region. The numerous sites include Laghetto della Costa (Arquà Petrarca–Padua, IT-VN-07), a settlement on drained land identified in 1885, that was investigated in the late nineteenth and early 20th century (Martinelli 2015).

3.7.3 Eastern Piedmont and Western Lombardy

The second area in terms of the number of Bronze Age pile-dwelling settlements known is the zone around Lake Varese, comprising various smaller lakes and peat bogs. Many pile-dwellings (De Marinis 2009) were identified during 19th century research along the shores of Lake Varese, Lake Monate and nearby marshy areas. In general, these settlements have yielded finds dating from the late EBA to early MBA, together with a few LBA artefacts. More recent research has been conducted at only a few of these sites, such as Sabbione (Cadrezzate–Varese, IT-LM-12) and Bodio centrale (Bodio Lomnago–Varese, IT-LM-10). The Sabbione pile-dwelling, investigated between 1980 and 1996, shows a progressive enlargement of the village area marked by the presence of several palisades; at least two houses have been identified (Fig. 3.8 (1, 2)). They are both rectangular in plan and are built on three rows of load-bearing posts; the larger covers an area of about 32 square metres, the smaller about 15 square metres (absolute dates between 1633 and 1625 cal BC). The settlement, erected between 1674 and 1605 BC, was located on the subaerial shore of the lake (Martinelli 2017).

New research was conducted at Bodio centrale between 2006 and 2012; this was not limited to the identification of wooden structures but included an underwater excavation of approximately 107 square metres (Fig. 3.8 (3)). A probable dwelling structure was identified, dating

between 1693 and 1674 cal BC ± 22. The finds, among the few discovered during controlled archaeological excavations in the Varese lakes area, date to the early MBA and belong to the north-western *facies* 'Bodio-Mercurago-Pollera' phase (Grassi and Mangani 2014).

Three settlements have been identified since 1966 on the bottom of Lake Viverone. Site Vi.1-Emissario (Viverone/Azeglio-Turin, IT-PM-01) has been investigated since 1977 by means of topographic survey campaigns of piles and dendrochronological sampling, which led to the definition of the complex organisation of this circular settlement with a diameter of about 70 m and a complex system of palisades with an access walkway and central walkway. The dendrochronological study of 149 wooden elements of the perilacustrine settlement Vi1-Emissario on Lake Viverone, with the use of wiggle-matching, allowed the construction phases of the structures to be dated to between 1424 and 1401 ± 41 cal BC (2σ) (Rubat Borel et al. 2022).

Among the research that has continued in recent decades, that at the site of Lucone D has been chosen as a case study to illustrate a settlement with structures with a high raised deck supported by isolated posts (Type B1) dating to the EBA.

3.7.4 Lucone D: An EBA Pile-Dwelling

Since 2007, research has been resumed[12] in the Lucone di Polpenazze pile-dwelling site D (Polpenazze del Garda-BS, IT-LM-05) by the Archaeological Museum of Valle Sabbia (Gavardo-BS).[13] At present the excavation area consists of two contiguous sectors (1 and 2), with a total of area of 343 m^2 (Baioni et al. 2020, 2021a, b, 2022) (Fig. 3.9 (1, 2)).

Lucone D was a short-lived settlement (under 1 hectare) with two main phases: it was founded in 2034–2033 BC and the latest felling of trees

---

[12] The first research dates back to the second half of the last century. For an overview see Baioni et al. (2007).
[13] Director of research: M. Baioni.

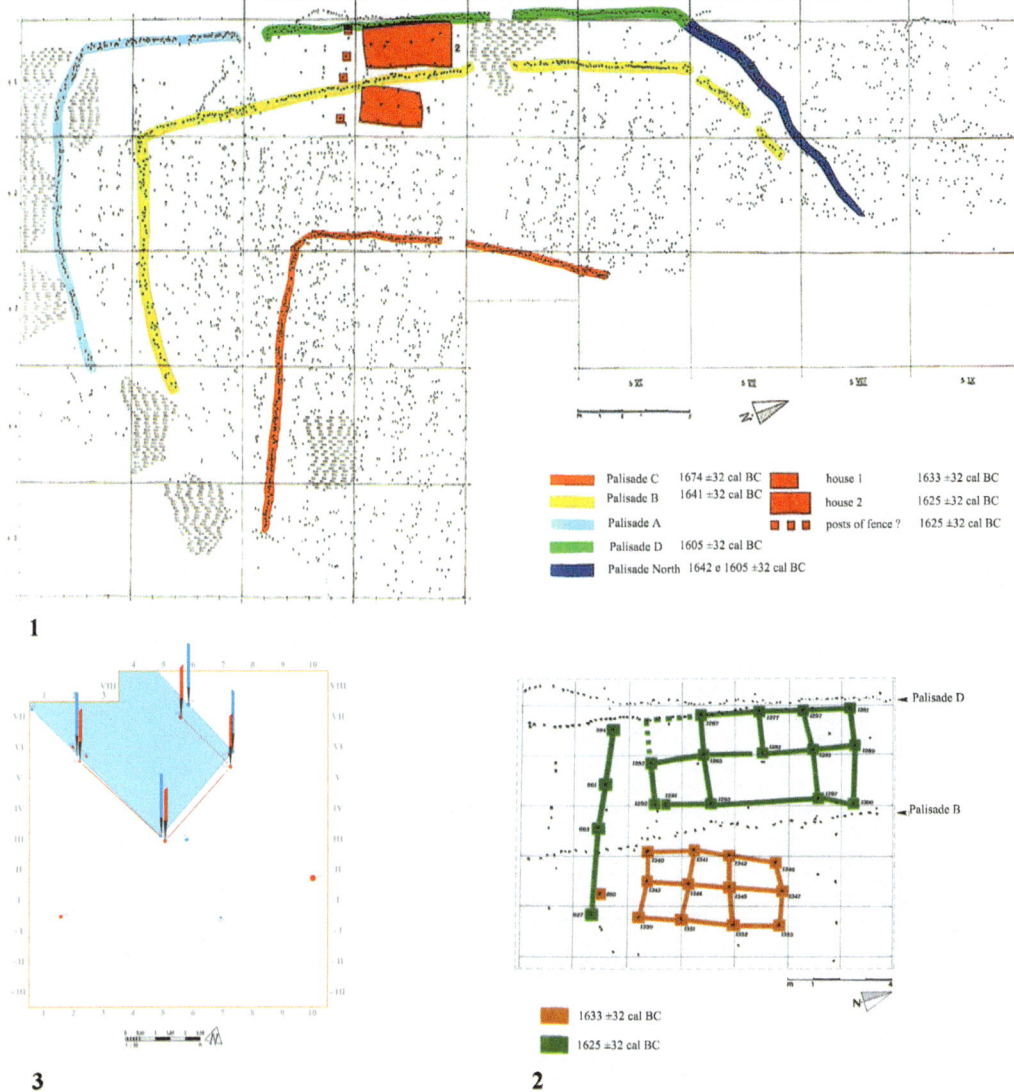

Fig. 3.8 Cadrezzate (Varese), Lake Monate, 'Il Sabbione' pile dwelling: (**1**) Dating and localisation of village structures based on dendrochronology (N. Martinelli; graphic F. Orsi); (**2**) Dating and reconstruction based on dendrochronology of two buildings in the village (N. Martinelli; graphic F. Orsi). (**3**) Bodio (Varese), Bodio Centrale pile dwelling, hypothetical reconstruction of a rectangular building made of elm posts (blue) and then reinforced with oak posts (red) (N. Martinelli, graphics by C. Mangani)

used for the rebuild occurred in 1967 BC (dendrochronological dates, see Chap. 12 in this book). From a stratigraphic perspective, two macro-phases—LUD1 and LUD2—may be clearly recognised:

Phase LUD1 represents the first period of use of the settlement and ended with a fire (LUD 1b)

which affected the entire excavated part of the village. *Phase LUD2* represents the reoccupation of the area after the fire.

The stratigraphy is characterised by the contrast between areas of continuous deposition of layers with a plant-based matrix and extensive dumps of waste material, that in this phase

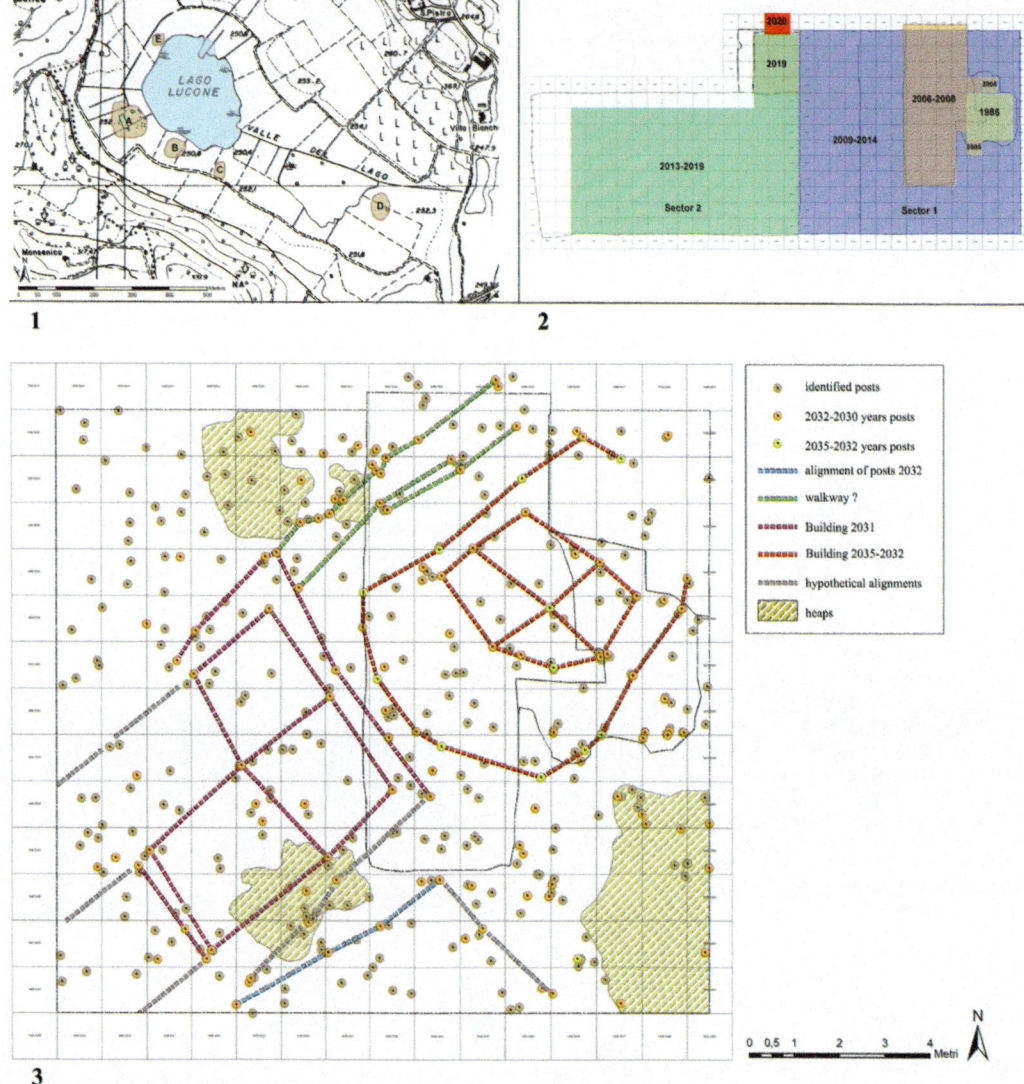

Fig. 3.9 Polpenazze del Garda (Brescia)—Lucone: (**1**) Map of Lucone basin with main archaeological sites (A, B, C, D, E) and excavation areas (green) (CTR Regione Lombardia, T. Quirino). (**2**) Lucone D: Excavation area showing excavation campaigns and Sectors 1 and 2 (M. Baioni). (**3**) Lucone D, Sector 1: Hypothetical reconstruction of village layout using 2035–2032 BC and 2032–2030 BC posts: older building (red), more recent building (purple), connecting walkway (green) (N. Martinelli, E. Saletta, graphics by T. Quirino)

become large and complex, containing dozens of interleaved lenses and patches of various sorts.

The number of archaeological finds is considerable, as is usual in prehistoric pile-dwellings. Lucone D also offers a good opportunity for multidisciplinary studies of the ancient

environment, palaeoeconomy and palaeodiet (Badino et al. 2011; Perego 2017; Bona 2018; Perego and Jacomet 2013).

The excavation of area D provided a lot of data concerning the structures of the pile-dwelling village. These are still being studied and include more than 1300 wooden elements overall, plus hundreds of fragments of heat-altered clay from hearths and walls. In Sector 1 about 460 vertical poles have been identified, of which about 300 have been dated by dendrochronology. As often happens in these 'post field' pile-dwelling sites of the Italian EBA, the posts identified belong to various construction phases and only by combining dendrochronological dating (which identifies posts felled at the same time) and spatial analysis using GIS is it possible to discern which posts were functionally related.

At present, spatial analysis has been carried out only for part of Sector 1. From this first study, it has proved possible by means of careful in-depth analysis and dendrotypological observations, to identify a small rectangular building encircled by a row of posts dating to the years 2034 and 2033 BC. For the following years 2032 and 2031 BC, other structures have been identified for which we were not always able to define complete plans, because parts of them lie beyond the excavation area. This was a larger rectangular house, surrounded by a perimeter structure to which a walkway seems to be connected. The houses are of different sizes, but their width is around 5 m (Fig. 3.9 (3)).

The fire deposit is the main source of the numerous wooden elements found that belonged to the wooden deck of the houses, the walls and probably the roof.

There are many wooden boards, generally not fully preserved, whose remains are 20–30 cm wide and occasionally more than 2 m long (average thickness 5–8 cm) (Fig. 2.10). Sometimes there are preserved mortises and joints for the insertion of the planks between the poles (Fig. 3.10 (3)). There are also numerous beams that must have belonged to the standing structure of the house, sometimes with pile shoes (Fig. 3.10 (2)).

Fig. 3.10 Polpenazze del Garda (Brescia)—Lucone D, Sector 2: (**1**) Plan with two beams with 25 rectangular holes found together with tapered posts (C. Mangani, T. Quirino); (**2**) Photographic plan of second beam with tapered posts (C. Mangani); (**3**) Detail of first beam with post inserted in one of the rectangular holes (Photo by M. Baioni)

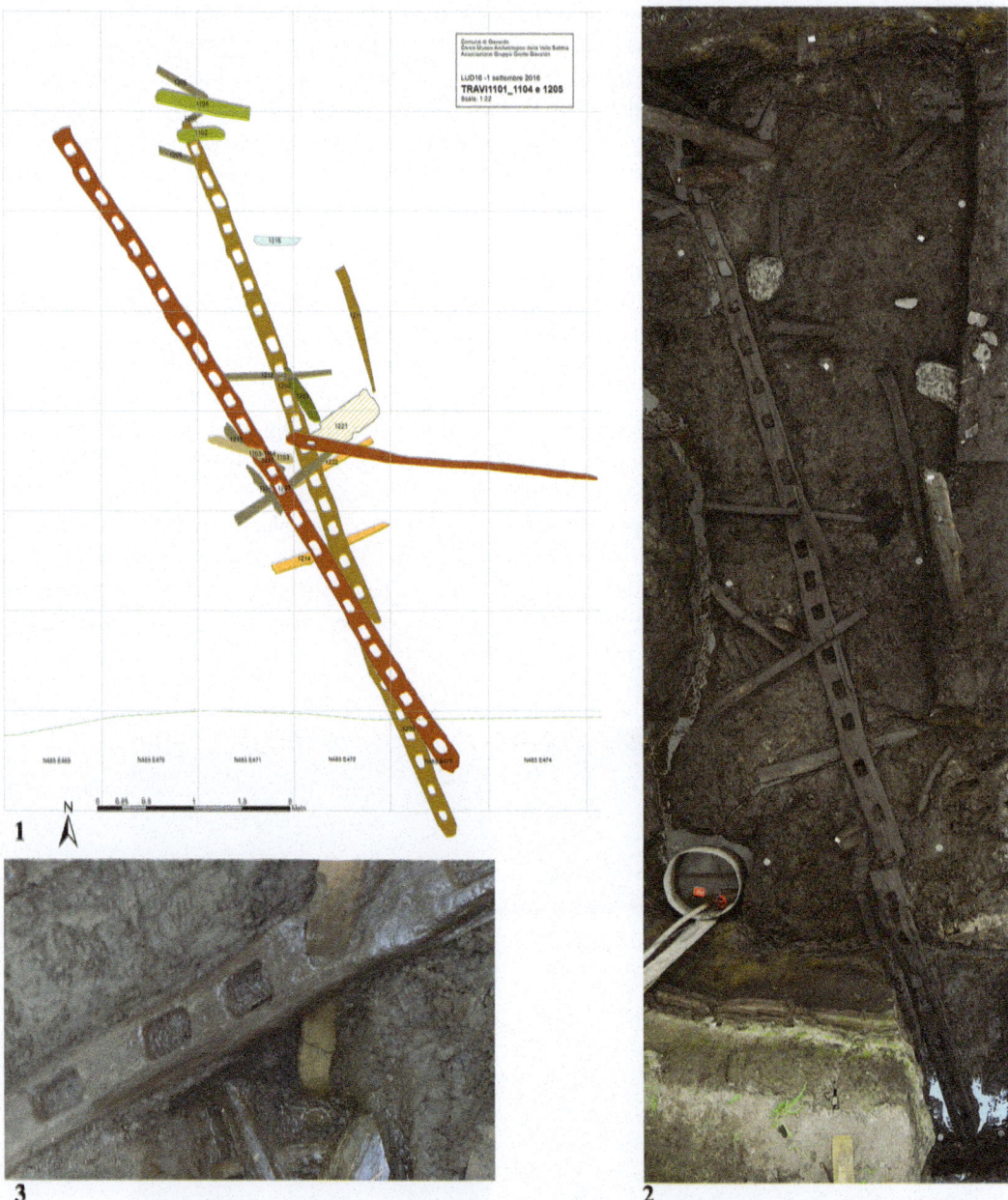

Fig. 3.11 Polpenazze del Garda (Brescia)—Lucone D: (**1**) Sector 1, photographic plan of fire deposit surface with boards and beams at different burning stages (C. Mangani); (**2**) Sector 2, photographic detail of tapered pole inserted into its pile-shoe (Photograph M. Baioni); (**3**) Sector 2, photograph of wooden door made of three boards and joists (Photo by M. Baioni)

Of particular interest are two beams, almost 8 m long, which were found in the peaty levels towards the centre of the basin (Fig. 3.11 (1, 2)). These are twin elements, perhaps made from the same oak trunk, with 25 square holes in corresponding positions, maybe mortises for the insertion of joists. The use of these two structural timbers is still unclear; perhaps they were elements that connected the wall and the roof. In 2020 in Sector 2, among the material involved in the collapse that occurred after the fire, the first wooden door from an Italian context was discovered. The door consists of three wooden planks, two of which were made by literally carving trunks, resulting in two large handles. The third central one is simply a smooth board, made from a different wood. Beams that held the structure together were passed through the handles (Fig. 3.11 (3)). The artefact is the first discovery in Italy of a type of wooden door that is rather rare, but known in Switzerland, at Zurich Opera House car park (Bleicher and Ruckstuhl 2015, 76–79) and at Pfäffikon-Burg (Eberli 2010).

Certain wooden elements may have belonged to the roof structures, such as a wooden beam with two oblique ends and a not-through-passing mortise on the side opposite the sloping surfaces. It could be a support element for a kind of roof truss. The discovery of numerous rectangular-shaped bark fragments, in some cases with a hole, suggests the presence of bark shingles on the roof. In the fire layers especially, parts of the houses made of clay are preserved. These are often parts of a box-type hearth, but fragments of clay bearing imprints of poles or branches, perhaps parts of walls or floors, are also present.

3.8 Conclusions

In northern Italy, human occupation of the wetlands can be dated from the EN and continued until the LN. This first occupation may be explained by the wide availability of natural resources and water that these places offered all year round, and the greater variety of land that could also be exploited for agriculture; they were thus particularly favourable for Neolithic subsistence. Subsequent archaeological evidence is scarce until the EBA, when in northern Italy there was an explosion of pile-dwelling settlements in the area affected by the Polada culture; it may in fact be noted that the wetland sites were not of particular interest to Copper Age groups.[14]

Although Neolithic wetland sites are not yet numerous, they already reveal the appearance of complex architectural strategies – which then became more complex in the Bronze Age. From the Neolithic onwards, pile-dwelling communities show remarkable ability in their choice of wood and its processing. In the Bronze Age more sophisticated techniques appeared that led to the manufacture of wooden structures of considerable length. Excavations in progress and the application of dendrochronology to old research material have also begun to provide information on settlement layouts and house plans in Italy.

References

Albertini I, Martinelli N (2018) Nuove ricerche per una topografia dei siti palafitticoli di Peschiera del Garda (Verona). Bollettino Del Museo Civico Di Storia Naturale Di Verona 42:89–106

Badino F, Baioni M, Castellano L, Martinelli N, Perego R, Ravazzi C (2011) Foundation, development and abandoning of a Bronze Age pile dwelling ("Lucone D", Garda lake) recorded in the palynostratigraphic sequence of the pond offshore the settlement. Il Quaternario – Italian J Quat Sci special number 24:177–179

Bagolini B (1983–84) Recenti ricerche sulle palafitte del Veneto. Alcune considerazioni sugli insediamenti in ambienti umidi. Sibrium XVII:57–81

Bagolini B, Barfield LH, Broglio A (1973) Notizie preliminari delle ricerche sull'insediamento neolitico di Fimon-Molino Casarotto (Vicenza). Rivista di Scienze Preistoriche XXVIII:161–215

Baioni M, Binaghi Leva MA, Borrello MA (2005) L'Isolino di Varese. Alcuni dati da recenti interventi. In: Della Casa P, Trachsel M (eds) WES '04 wetland economies and societies: 150 years of research on prehistoric economy and society in lake dwellings. Collectio Archæologica, vol 3. Chronos, Zurich, pp 209–214

Baioni M, Bocchio G, Mangani C (2007) Il Lucone di Polpenazze: storia delle ricerche e nuove prospettive. In: Atti del XVI Convegno Archeologico Benacense. Annali Benacensi, vol XIII–XIV. Tipolitografia Fratelli Geroldi, Brescia, pp 83–102

Baioni M, Furlanetto G, Grassi B, Longhi C, Mangani C, Martinelli N, Nicosia C, Ravazzi C, Ruggiero MG, Voltolini D (2015) Due palafitte a confronto: Bodio

---

[14] There is very little evidence of wetland occupation dating to the Copper Age in northern Italy, which corresponds to the Central European Final Neolithic:

possible cases are the Fiavé 2 phase and the presence of traditional bell-shaped ornaments and lithic technology resembling late Copper Age models such as in Porto Galeazzi and Corno di Sotto (Baioni et al. 2015).

centrale (Varese – Italia, IT-LM-10) e Corno di Sotto (Desenzano del Garda – Italia). Considerazioni sui processi d'erosione e su problemi di conservazione. In: Archéologie et Erosion 3, Atti della tavola rotonda (Arenenberg-Hemmenhofen, 8–10 ottobre 2014). Mêta Jura, Lons-le-Saunier, pp 175–182

Baioni M, Leonardi G, Fozzati L, Martinelli N (2018) Le palafitte: definizione e caratteristiche di un fenomeno complesso attraverso alcuni casi di studio. In: Baioni M, Mangani C, Ruggiero MG (eds) Le Palafitte: ricerca, conservazione, valorizzazione, Atti dell'incontro internazionale (Desenzano del Garda, 6–8 ottobre 2011), Palafitte. Collana di studi sui siti preistorici in ambiente umido, vol 0. SAP Società Archeologica, Quingentole (MN), pp 27–42

Baioni M, Mangani C, Gleba M (2020) Spinning and weaving in a pile dwelling of four thousand years ago: data from the excavations at Lucone di Polpenazze del Garda (Brescia): In: Redefining ancient textile handcraft structures, tools and production processes, proceedings of the VIIth international symposium on textiles and dyes in the Ancient Mediterranean World (Granada, Spain 2–4 Oct 2019). Universidad de Granada, Granada, pp 87–200

Baioni M, Bona F, Mangani C, Martinelli N, Nicosia C, Perego R, Quirino T, Saletta E (2021a) Daily life in a north Italian Early Bronze Age pile dwelling: Lucone di Polpenazze del Garda (Italy–Brescia). In: Jallot L, Peinetti A (eds) Use of space and domestic areas: functional organisation and social strategies, Proceedings of XVIII UISPP world congress (4–9 June 2018, Paris, France), Session XXXII-1, UISPP Proceedings Series, vol 18. Archaeopress, Oxford, pp 53–66

Baioni M, Mangani C, Martinelli N (2021b) The site of Lucone di Polpenazze and the chronology of Early Bronze Age pile dwellings in the Lake Garda area (Italy – Lombardy/Veneto). In: Marcigny C, Mordant C (eds) Bronze 2019, 20 ans de recherches. Actes du colloque international anniversaire de l'APRAB, Bayeux (19–22 juin 2019). Bulletin de l'APRAB, suppl no 7. APRAP, Dijon, pp 373–383

Baioni M, Mangani C, Bona F, Gulino F, Longhi C, Martinelli N, Nicosia C, Perego R, Quirino T, Redolfi Riva F (2022) Il sito D del Lucone di Polpenazze del Garda (BS): un breve quadro di sintesi. In: Atti LII Riunione Scientifica IIPP "Preistoria e Protostoria in Lombardia e nel Canton Ticino". Rivista di Scienze Preistoriche LXXII: 477-491.

Balista C, Leonardi G (1996) Gli abitati di ambiente umido nel Bronzo Antico dell'Italia settentrionale. In: Cocchi Genick D (ed) L'antica età del Bronzo in Italia, Atti del Congresso di Viareggio, 9–12 gennaio 1995. Octavo, Franco Cantini Editore, Firenze, pp 199–228

Banchieri DG, Balista C (1991) Note sugli scavi di Pizzo di Bodio (Varese) 1985–88. Preistoria Alpina 27:197–242

Bernabò Brea M, Cardarelli A, Cremaschi M (eds) (1997) Le terramare: la più antica civiltà padana. Electa, Milano

Bianchin Citton E (1994) Il sito umido di Colmaggiore di Tarzo (TV). In: Friuli-Venezia Giulia e Istria,

Proceedings of the XXIX scientific conference of Italian Institute of Prehistory and Protohistory. Firenze, pp 201–217

Bianchin Citton E (ed) (2016) Nuove ricerche nelle Valli di Fimon. L'insediamento del tardo Neolitico de Le Fratte di Arcugnano. Editrice Veneta, Vicenza

Billamboz A, Schlichtherle H (1984) "Pfahlbauten". Urgeschichtliche Ufer- und Moorsiedlungen, Neue Forschungen in Südwestdeutschland. Gesellschaft für Vor- und Frühgeschichte in Württemberg und Hohenzollern e.V., Stuttgart

Billamboz A, Martinelli N (2015) Dendrochronology and Bronze Age pile-dwellings on both sides of the Alps: from chronology to dendrotypology, highlighting settlement developments and structural woodland changes. In: Menotti F (ed) The end of the lake-dwellings in the Circum-Alpine region. Oxbow Books, Oxford, pp 68–84

Bleicher N, Ruckstuhl B (2015) VII Die Archäologischen Befunde. In: Blecher N, Harb C (eds) Zürich-Parkhaus Opéra, Eine neolithische Feuchtbodenfundstelle, Band I: Befunde, Schichten und Dendroarchäologie. Monographien der Kantonsarchäologie Zürich, vol 48. Kantonsarchäologie, Dübendorf, pp 50–99

Bona F (2018) The Early Bronze Age Pile Dwelling of Lucone lake (Site D). Preliminary report on archaeozoological data. In: Baioni M, Mangani C, Ruggiero MG (eds) Le Palafitte: ricerca, conservazione, valorizzazione, Atti dell'incontro internazionale (Desenzano del Garda, 6–8 ottobre 2011), Palafitte. Collana di studi sui siti preistorici in ambiente umido, vol 0. SAP Società Archeologica, Quingentole (Mn), pp 185–192

Borrello MA (1984) The Lagozza culture in Northern and Central Italy. Studi Archeologici, vol 3. Istituto Universitario di Bergamo, Bergamo

Capulli M, Fozzati L, Martinelli N, Pellegrini A (2014) La palafitta sommersa di Peschiera Belvedere sul lago di Garda (VR). Le ricerche archeologiche subacquee e l'utilizzo della tecnologia GIS come supporto per le analisi spaziali e la ricostruzione planimetrica delle strutture palafitticole. In: Leone D, Turchiano M, Volpe G (eds) Atti del III Convegno di Archeologia Subacquea, Manfredonia 4–6 ottobre 2007. Edipuglia, Bari, pp 103–110

Corboud P, Petrequin P (2004) Les sites préhistoriques littoraux du Léman et leurs relations avec le Jura français. Archäologie Schweiz 27(2):54–64

Corti P, Martinelli N, Micheli R, Montagnari Kokelj E, Petrucci G, Riedel A, Rottoli M, Visentini P, Vitri S (1998) Siti umidi tardoneolitici: nuovi dati da Palù di Livenza (Friuli-Venezia Giulia, Italia). In: Proceedings of the XIII congress of International Union of Prehistoric and Protohistoric Sciences, vol 6(II). A. B.A.C.O., Forlì, pp 1379–1391

Cremaschi M, Pizzi C, Valsecchi V (2006) Water management and land use in the terramare and a possible climate co-factor in their abandonment: the case study of the terramare of Poviglio Santa Rosa (Northern Italy). Quatern Int 151:87–98

De Marinis RC (1982) Storia della scoperta delle palafitte varesine. In: Palfitte: mito e realtà. Museo Civico di Storia Naturale, Verona, pp 71–83

De Marinis RC (ed) (2007) Studi sull'abitato dell'età del Bronzo del Lavagnone, Desenzano del Garda. Notizie Archeologiche Bergomensi, vol 10. Civico Museo Archeologico di Bergamo, Bergamo

De Marinis RC (2009) L'età del Bronzo nelle palafitte del Lago di Varese. In: De Marinis, RC, Massa S, Pizzo M (eds) Alle origini di Varese e il suo territorio. Le collezioni del sistema archeologico provinciale. L'Erma di Bretschneider, Roma, pp 124–139

De Marinis RC, Rapi M, Ravazzi C, Arpenti E, Deaddis, M, Perego R (2005) Lavagnone (Desenzano del Garda). New excavations and palaeoecology of a Bronze Age pile dwelling site in northern Italy. In: Della Casa P, Trachsel M (eds) WES'04 Wetland Economies and Societies. Proceedings of the international conference in Zurich, 10–13 March 2004. Collectio Archaeologica, vol 3. Chronos, Zurich, pp 221–232

Eberli U (2010) Holzkonstruktionen. In: Eberli U, Die horgenzeitliche Siedlung Pfäffikon-Burg. Monographien der Kantonsarchäologie Zürich, vol 40. Kantonsarchäologie, Dübendorf, pp 66–69

Ebersbach R (2013) Houses, households, and settlements: architecture and living space. In: Menotti F, O'Sullivan A (eds) The Oxford handbook of wetland archaeology. Oxford University Press, Oxford, pp 283–301

Fasani L (1982) Gli insediamenti palafitticoli italiani. Distribuzione geografica e inquadramento cronologico-culturale. In Palfitte: mito e realtà. Museo Civico di Storia Naturale, Verona, pp 33–40

Fasani L (2002) Età del Bronzo. In: Aspes A (ed) Preistoria Veronese, contributi e aggiornamenti, Memorie del Museo Civico di Storia Naturale di Verona, 2 serie, Sezione Scienze dell'Uomo, vol 5. Museo civico di storia naturale, Verona, pp 107–153

Fozzati L, Leonardi G, Martinelli N (2015) Wetlands. Palafitte e siti umidi nell'Età del bronzo del Veneto: territori e cronologia assoluta. In: Leonardi G, Tiné V (eds) Preistoria e Protostoria del veneto. Studi di Preistoria e Protostoria, vol 2. Istituto Italiano di Preistoria e Protostoria, Firenze, pp 241–250

Fusco V (1976–77) La stazione preistorica dell'Isolino di Varese. Sibrium XIII:1–27

Gonzato F, Baldo M, Mangani C (2021a) "Symbolic" limits and functional embankments: interactions with the landscape in the MBA site of Vallese di Oppeano (Verona-Italy). In: Marcigny C, Mordant C (eds) Bronze 2019. 20 ans de recherche, Actes du colloque International anniversaire de l'APRAB Bayeux (19–22 juin 2019). Bulletin de l'APRAB, suppl no 7. APRAP, Dijon, pp 387–398

Gonzato F, Mangani C, Salzani L (2019) Plain, mountain and lake: the Frassino pile-dwelling site in the middle of a network. In: Billaud Y, Lachenal T (eds) Entre terres et eaux Les sites littoraux de l'âge du Bronze: spécificités et relations avec l'arrièrepays Actes de la séance de la Société préhistorique française d'Agde

(20–21 octobre 2017), Dijon, Association pour la promotion des recherches sur l'âge de Bronze. Société préhistorique française, Paris, pp 115–140

Gonzato F, Mangani C, Martinelli N, Nicosia C (2021b) Different ways to handle the domestic space by comparison: the case of Bronze Age villages in Vallese di Oppeano (Verona –ITA). In: Jallot L, Peinetti A (eds) Use of space and domestic areas: functional organisation and social strategies, Proceedings of XVIII UISPP world congress (4–9 June 2018, Paris, France), Session XXXII-1, UISPP Proceedings Series, vol 18. Archaeopress, Oxford, pp 67–76

Grassi B, Mangani C (2014) Storie sommerse. Ricerche archeologiche alla palafitta di Bodio centrale a 150 anni dalla scoperta. Fantigrafica, Cremona

Griggs CB, Kuniholm PI, Newton MW (2007) Lavagnone di Brescia in the Early Bronze Age: Dendrochronological Report. Notizie Archeologiche Bergomensi 10:19–33

Guerreschi G (1966–1967) La Lagozza di Besnate e il Neolitico superiore padano. Rivista archeologica dell'antica provincia e diocesi di Como 148–149:5–352

Guerreschi G (1978–79) Nuovi scavi archeologici all'Isolino di Varese. Sibrium XIV:287–295

Guerreschi G, Catalani P, Ceschin N (1992) I nuovi scavi all'Isolino di Varese (1977–1986). Sibrium XXI:9–64

Hafner A, Pétrequin P, Schlichtherle H (2016) Ufer- und Moorsiedlungen. Chronologie, kulturelle Vielfalt und Siedlungsformen. In: 4000 Jahre Pfahlbauern. Thorbecke, Ostfildern, pp 59–64

Hofmann D, Ebersbach R, Doppler T, Whittle A (2016) The life and times of the house: multi-scalar perspectives on settlement from the neolithic of the Northern Alpine Foreland. Eur J Archaeol 19(4):596–630

Joannin S, Magny M, Peyron O, Vannière B, Galop D (2014) Climate and land-use change during the late Holocene at Lake Ledro (Southern Alps, Italy). Holocene 24(5):591–602

Mangani C, Ruggiero MG (2018) 2. Le palafitte. In: Roffia E (ed) Sirmione in età antica. Il territorio del comune dalla Preistoria al Medioevo. Edizioni Et, Milano, pp 59–78

Martinelli N (2015) La palafitta di Arquà Petrarca: note sulla topografia, la tipologia e la cronologia assoluta delle strutture rinvenute sulle sponde del Laghetto della Costa. In: Bianchin Citton E, Rossi S, Zanovello P (eds) Dinamiche insediative nel territorio dei Colli Euganei dal Paleolitico al Medioevo, Atti del convegno di studi, Este-Monselice, Padova. Monselice, pp 121–132

Martinelli N (2017) Gli insediamenti palafitticoli del lago di Monate. Il contributo della dendrocronologia allo studio dell'antica e media età del Bronzo. In: Harari M (ed) Il territorio di Varese in età preistorica e protostorica. Nomos Edizioni, Busto Arsizio (VA), pp 173–195

Martinelli N (2019) Prehistoric pile-dwellings in northern Italy: an archaeological and dendrochronological overview. In: Shindo L, Edouard J-L, Sumera F, Bailly M, Hartamann-Virnich A (eds) ARCADE.

Approche diachronique et Regards croisés: Archéologie, Dendrochronologie et Environnement, *Actes du colloque*. Aix-en-Provence, pp 69–78

Marzatico F (1996) La fine del Bronzo Antico sulla base delle recenti ricerche a Fiavé, zona 4 (scavi 1986–1993). In: Cocchi Genick D (ed) L'antica età del bronzo, Atti del Congresso Nazionale. All'Insegna del Giglio, Firenze, pp 247–256

Marzatico F (2004) 150 years of lake-dwelling research in Northern Italy. In: Menotti F (ed) Living on the lake in Prehistoric Europe. 150 years of lake-dwelling research. Routledge, London, pp 83–97

Menotti F (2013) Wetland occupations in Prehistoric Europe. In: Menotti F, O'Sullivan A (eds) The Oxford handbook of wetland archaeology. Oxford University Press, Oxford, pp 11–25

Micheli R (2018) Abitare le aree umide della pedemontana veneto-friulana alla fine del Neolitico: nuovi dati dal Palù di Livenza. In: Arnosti G, Riviera G, Schincariol F (eds) Dalla preistoria all'Alto medioevo nell'Antico Cenedese. Antichità Altoadriatiche, vol 89. Editreg, Trieste, pp 75–96

Micheli R, Bassetti M, Degasperi N (2019) Nuove indagini e prospettive della ricerca nella palafitta preistorica del Palù di Livenza. Quaderni Friulani di Archeologia XXIX:37–48

Micheli R., Bassetti M., Degasperi N. (2023) Palù di Livenza: un insediamento pluristratificato del Neolitico nella Pedemontana pordenonese. In: Caramella L.A.R. (ed.) Dall'acqua alla terra: cambiamenti nell'occupazione del territorio, Sibrium – Atti, 1. Varese, Centro di Studi Preistorici e Archeologici, pp. 26-61.

Micheli R, Bassetti M, Degasperi N, Fozzati L, Martinelli N, Rottoli M (2018) Nuove ricerche al Palù di Livenza: lo scavo del Settore 3. In: Borgna E, Cassola Guida P, Corazza S (eds) Preistoria e Protostoria del Caput Adriae. Studi di Preistoria e Protostoria, vol 5. Istituto Italiano di Preistoria e Protostoria, Firenze, pp 481–490

Padovan S, Rubat Borel F, Berruti G, Daffara S, Mancusi VG, Zunino M (2019) Il sito perilacustre vbq di Montalto Dora nel quadro del Neolitico del Piemonte. In: Maffi M, Bronzoni L, Mazzieri P (eds) Le quistioni nostre paletnologiche più importanti... Trent'anni di tutela e ricerca preistorica in Emilia occidentale, Atti del Convegno di Studi in onore di Maria Bernabò Brea. Archeotravo Cooperativa Sociale, Piacenza, pp 11–23

Perego R (2017) Contribution to the development of the Bronze Age plant economy in the surrounding of the Alps: an archaeobotanical case study of two Early and Middle Bronze Age sites in Northern Italy (Lake Garda region). Ph.D. thesis, University of Basel (CH)

Perego R, Jacomet S (2013) New finds of the new glume wheat type from Early Bronze Age pile-dwellings in northern Italy. In: 16th symposium of the international work group for palaeoethnobotany, IWGP. Thessaloniki, Greece 17–22 June 2013, Abstract volume, p 189

Peretto C, Taffarelli C (1973) Un insediamento del Neolitico Recente al Palù di Livenza (Pordenone). Rivista di Scienze Preistoriche XXVIII:235–260

Perini R (1981) La successione degli orizzonti culturali dell'abitato dell'età del Bronzo nella torbiera del Lavagnone. Bullettino Di Paletnologia Italiana 82:117–166

Perini R (1984) Scavi archeologici nella zona palafitticola di Fiavé-Carera. Parte I. Campagne di scavo 1969–1976. Situazione dei depositi e dei resti strutturali. Patrimonio storico e artistico del Trentino, vol 8. Trento

Perini R (1988) Gli scavi nel Lavagnone. Sequenza e tipologia degli abitati dell'età del Bronzo. Annali Benacensi 9:109–154

Perini R (1994) Scavi archeologici nella zona palafitticola di Fiavé-Carera. Parte III. Campagne di scavo 1969–1976. Resti della cultura materiale. Ceramica, vol 1, Patrimonio storico e artistico del Trentino, vol 10. Trento

Pétrequin A-M, Pétrequin P (1988) Le Neolithique des lacs. Préhistoire des lacs de Chalain et de Clairvaux (4000–2000 av. J.-C.), Editions Errance, Paris

Pétrequin P (2013) Lake-dwelling in the Alpine Region. In: Menotti F, O'Sullivan A (eds) The Oxford handbook of wetland archaeology. Oxford University Press, Oxford, pp 253–267

Pia GE (1982) Moniga (Brescia). In: Palafitte: mito e realtà. Grafiche Fiorini, Verona, pp 162–164, fig 26

Piccoli A (1982a) Bande di Cavriana (MN) – 1982 indagine d'emergenza. Sibrium XVI:51–68

Piccoli A (1982b) Saggio esplorativo nell'insediamento perilacustre di Castellaro Lagusello. Studi in onore di Rittatore Vonwiller, vol I. Como, pp 443–485

Poggiani Keller R, Binaghi Leva MA, Menotti EM, Roffia E, Pacchieni T, Baioni M, Martinelli N, Ruggiero MG, Bocchio G (2005) Siti d'ambiente umido della Lombardia: rilettura di vecchi dati e nuove ricerche. In: Della Casa P, Trachsel M (eds) WES'04 Wetland Economies and Societies. Proceedings of the international conference in Zurich, 10–13 March 2004. Collectio Archaeologica, vol 3. Chronos, Zurich, pp 233–250

Rageth J (1975) Der Lago di Ledro im Trentino und seine Beziehungen zu den alpinen und mitteleuropäischen Kulturen. Bericht der Römisch-Germanischen Kommission 55:1–259

Rapi M (2013) Dall'età del Rame all'età del Bronzo. I primi villaggi palafitticoli e la cultura di Polada. In: De Marinis, RC (ed) L'età del Rame. La pianura padana e le alpi al tempo di Ötzi. Compagnia della stampa, Brescia, pp 525–544

Rapi M (2020) Il complesso del Lavagnone 3. Scavi dell'Università degli Studi di Milano (1989–2006), Palafitte. Collana di studi sui siti preistorici in ambiente umido, vol 2(1). SAP, Quingentole

Ravazzi C, Badino F, Castellano L, de Nisi, D, Furlanetto G, Perego R, Zanon M, Dal Corso M, De Amicis M, Monegato G, Pini R, Valle F (2018) Introduzione allo studio stratigrafico e paleoecologico dei laghi intramorenici del Garda. In: Baioni M, Mangani C, Ruggiero MG (eds) Le Palafitte: ricerca, conservazione, valorizzazione, Atti dell'incontro internazionale (Desenzano del Garda, 6–8 ottobre 2011), Palafitte. Collana di studi sui siti preistorici in ambiente umido, vol 0. SAP Società Archeologica, Quingentole (MN), pp 167–184

Ravazzi C, Artioli G, Badino F, Baioni M, Banino R, Castellano L, Castelletti L, Chiesa S, Colombaroli D, Comolli R, Cremaschi M, Croce E, Dal Corso M, Dal Sasso G, Deaddis M, De Amicis M, Ferrario F, Fontana F, Furlanetto G, Garozzo L, Livio F, Mangani C, Marchetti M, Martinelli E, Michetti AM, Motella De Carlo S, Nicosia C, Perego P, Peresani M, Pini R, Poggiani Keller R, Quirino T, Rapi M, Rottoli M, Ruggiero M G, Tinner W, Tramelli A, Trentacoste A, Vallé F, Visentin D, Wick L, Zanon M, Zerboni A (in press) Scenari di ricostruzione delle interazioni uomo-ambiente-clima in Lombardia (N-Italia) dal Paleolitico medio all'età del Ferro. In: LII Riunione Scientifica IIPP "*Preistoria e Protostoria in Lombardia e nel Canton Ticino*". Rivista di Scienze Preistoriche

Ruoff U (2004) Lake-dwelling studies in Switzerland since "Meilen 1854". In: Menotti F (ed) Living on the lake in Prehistoric Europe. 150 years of lake-dwelling research. Routledge, London, pp 9–21

Rubat Borel F, Martinelli N, Köninger J, Menotti F (2022) Un contributo per la cronologia del Bronzo Medio: il sito perilacustre di Viverone e l'Italia nordoccidentale. In: LII Riunione Scientifica IIPP "*Preistoria e Protostoria in Lombardia e nel Canton Ticino*". Rivista di Scienze Preistoriche LXXII: 441-457.

Ruggiero MG, Baioni M, Mangani C (2022), Sito UNESCO Seriale Transnazionale "Siti palafitticoli preistorici dell'arco alpino". Elaborazione della parte nazionale del Piano di Gestione del sito: prime azioni e studi di fattibilità, Aligraphis

Salzani L, Martinelli N, Bellintani P (1996) La palafitta di Canar di San Pietro Polesine (Rovigo). In: Cocchi Genick D (ed) L'antica età del bronzo, Atti del Congresso Nazionale. All'Insegna del Giglio, Firenze, pp 282–285

Salzani L, Balista C, Butta P, Martinelli N, Torri P, Bosi G, Mazzanti M, Mercuri AM, Accorsi CA, Bertolini M, Thun Hohenstein U (2018) La palafitta di

Tombola di Cerea (VR). Lo scavo 1999. IpoTESI Di Preistoria 10:51–142

Schlichtherle H (2004) Lake-dwellings in South-western Germany. History of research and contemporary perspectives. In: Menotti F (ed) Living on the lake in prehistoric Europe. 150 years of lake-dwelling research. Routledge, London, New York, pp 22–35

Schlichtherle H, Billiamboz A (2015) Architecturale, sociale, écologique: les trois dimensions de la maison palafittique dans les villages lacustres et palustres néolithiques du sud-ouest de l'Allemagne. In: Rey P-J, Dumont A (eds) L'Homme et son environment: des lacs, des montagnes et des rivières. Bulles d'archéologie offerte à André Marguet. Suppléments à la Revue archéologique de l'Est, 40. ARTEHIS Éditions, Dijon, pp 99–114

Valsecchi V, Tinner W, Finsinger W, Ammann B (2006) Human impact during the Bronze Age on the vegetation at Lago Lucone (Northern Italy). Veg Hist Archaeobotany 15:99–113

Vanzetti A (2013) 1600? The rise of the Terramara system (Northern Italy). In: Meller H, Bertemes F (eds) 1600 – Kultureller Umbruch im Schatten des Thera-Ausbruchs? 4. Mitteldeutscher Archäologentag vom 14. bis 16. Oktober 2011 in Halle (Saale). 1600 – Cultural change in the shadow of the Thera-Eruption? 4th Archaeological Conference of Central Germany, October 14–16, 2011 in Halle (Saale). Tagungen des Landesmuseums für Vorgeschichte Halle, vol 9. Landesamt für Denkmalpflege und Archäologie in Sachsen-Anhalt, Landesmuseum für Vorgeschichte, Halle (Saale), pp 267–282

Vitri S, Martinelli N, Čufar K (2002) Dati cronologici dal sito di Palù di Livenza. In: Ferrari A, Visentini P (eds) Il declino del mondo neolitico. Ricerche in Italia centro-settentrionale fra aspetti peninsulari, occidentali e nord-alpini. Museo delle Scienze, Pordenone, pp 187–198

Zorzi F (1940) La palafitta di Barche di Solferino. (1ª Relazione). Bullettino di Paletnologia Italiana N.S. IV:41–82

The Early Neolithic Lake-shore Settlement of La Marmotta at Lake Bracciano (Anguillara Sabazia, Rome, Italy): a Critical Overview of the Current State of Research, and a Discussion of Selected Aspects

4

Ariane Ballmer, Mario Mineo, and Valeska Becker

Abstract

Occupied from around 5600 BC, the lake-shore settlement of La Marmotta at Lake Bracciano (Anguillara Sabazia, Rome, Italy) counts as one of the earliest permanently occupied wetland settlements in southern Europe. The first settlers were a Neolithic group, pursuing crop cultivation and stock breeding. The archaeological site with its fascinating spectrum of finds is not only of great interest per se but also plays a key role in the understanding of the supra-regional spread of the Neolithic along the northern Mediterranean coast. Thus, the settlement seems to have been a kind of gateway from which the western Mediterranean region, and possibly also the western Alpine region, would have received essential inputs, especially in terms of agricultural resources and practices. The pending evaluation of the extensive underwater excavations between 1992 and 2006 currently only allows for a provisional and relatively diffuse knowledge of the site. In the present chapter, the site chronology, selected aspects of the settlers' way of life, in particular concerning agriculture but also navigation, as well as the site's potential role in the Neolithisation of the western Mediterranean are dealt with in detail.

Keywords

Western mediterranean · Early Neolithic Italy · Neolithisation · Pile-dwelling · Tyrrhenian impressed ware · Painted ware · Ceramica lineare

A. Ballmer (✉)
Independent Researcher, Bern, Switzerland
e-mail: mail@arianeballmer.com

Institute of Archaeological Sciences and Oeschger Centre for Climate Change Research (OCCR), University of Bern, Bern, Switzerland

M. Mineo (✉)
Museo Delle Civiltà and Museo Preistorico Etnografico 'Luigi Pigorini', Rome, Italy
e-mail: mario.mineo@cultura.gov.it

V. Becker
Department for Prehistoric and Protohistoric Archaeology, University of Münster, Münster, Germany
e-mail: valeska.becker@uni-muenster.de

4.1 Introduction

The settlement of La Marmotta belongs to the most intriguing sites of the initial Mediterranean Neolithic, while simultaneously raising many questions. In addition to an impressive inventory of spectacular organic finds due to waterlogging, the site's utmost importance lies in its chrono-spatial setting: first occupied in the middle of the 6th millennium BC, i.e. the Early Neolithic, the site is one of the first permanently occupied wetland settlements of the 'pile dwelling'-type in Europe. From La Marmotta, impetus for the Neolithisation of the western Mediterranean would have been given.

It is therefore not surprising that the site is referred to on a regular basis. At the same time, its archaeological materials remain largely unevaluated and hence unpublished. Preliminary reports by the responsible excavators as well as by specialists from the various disciplines were already submitted after the initial intervention in 1989 and published in 1993 (archaeology in general, incl. structures and finds: Fugazzola Delpino et al. 1993; archaeobotany: Rottoli 1993; dendrochronology: Martinelli 1993; archaeozoology: Cassoli and Tagliacozzo 1993; human tooth: Macchiarelli and Salvadei 1993). In 1995, an article was published on a dugout canoe (Fugazzola Delpino and Mineo 1995) and another on the faunal remains found in its context (Cassoli and Tagliacozzo 1995). From the late 1990s on, a series of general reviews on the site as well as specific find categories were published (review articles: Fugazzola Delpino 1998–1999; Fugazzola Delpino and Pessina 1999; Fugazzola Delpino and Mineo 2000; Fugazzola Delpino 2002; pottery: Delpino 2020; anthropomorphic figurine: Fugazzola Delpino 2000–01; wild boar tusk artefacts: Fiore et al. 2006; special plants: Rottoli 2001–02; Bernicchia and Gemelli 2004; Fugazzola Delpino 2004; Speroni 2004; Bernicchia et al. 2006; harvesting tools, especially sickles: Mazzucco et al. 2018, 2020; dugout canoes: Fugazzola Delpino and Mauro 2014; Mineo 2016; Bondioli et al. 2000; woodwork: Caruso Fermé et al. 2021; Mazzucco et al. 2022; textiles, basketry, and cordage: Mineo 2019; Mineo et al. 2023).

Although both specific aspects and larger contexts have been published, these mostly seem to be based on general observations, whereas the underlying information is largely inaccessible. With a few exceptions, practically no corresponding basic data (such as plans and trench sections of the excavation; systematic inventory lists featuring figures and further relevant data; drawings or photos of finds) is currently available, which means that numerous statements and

contexts cannot be verified or reassessed for discussion, respectively. With the intention of highlighting the significance of La Marmotta without repeating what has already been published in multiple general reviews, we chose to focus on selected topics, considering the latest state of research in order to ultimately discuss the site's further context, i.e. the site chronology, outstanding elements related to the settlers' way of life, in particular concerning agriculture but also navigation, and finally the site's potential role in the Neolithisation of the western Mediterranean.

4.2 Discovery and Investigation

In spring of 1989, when a pipeline was installed close to the town of Anguillara Sabazia (Rome), rich archaeological deposits were uncovered underneath the bed of Lake Bracciano, several metres below today's water level. It soon became apparent that these were the remains of a prehistoric settlement. Between 1992 and 2006, systematic underwater excavations were carried out under the supervision of the Soprintendenza Speciale of the Museo Nazionale Preistorico Etnogafico 'Luigi Pigorini' (today: Museo delle Civiltà) in Rome, directed by M. A. Fugazzola Delpino, and from 1993 to 1999 with the collaboration of M. Mineo (Fugazzola Delpino 2002).

4.3 Topographical Setting

The archaeological site of La Marmotta is located on the southeastern shores of a volcanic lake, the present-day Lake Bracciano, at the outflow of the River Arrone, in a bay sheltered on three sides by promontories (Fig. 4.1). It lies approximately 300 m seawards from the modern shoreline. The archaeological deposits are fully submerged at a depth of 9–12 m below today's water level. The site remains extend over an area of not less than 2 ha, of which 25% have been investigated. The archaeological relics owe their excellent preservation, on the one hand, to waterlogging, and, on the other hand, to a c. 2 m thick covering layer, which 'sealed' the settlement relatively quickly after its abandonment and thus protected it from later disturbances.

The choice of the settlement's location and its occupation for nearly 500 years may be related mainly to the favourable conditions of the surrounding area: the mineral-rich volcanic soils offered fertile arable land, ideal for farming. In addition, the locus is connected to the Tyrrhenian coast through the River Arrone, flowing from the southeastern shore of Lake Bracciano into the sea in the Fregene-region north of Ostia. If we assume an Etruscan river harbour in the hinterland of Fregene, like they are known from other Southern Etrurian rivers (Daum 2015; Michetti 2017), the navigability, or at least some sort of raftability of the Arrone in prehistoric times seems conceivable.

4.4 Site Chronology and Course of Events

4.4.1 Stratigraphical Sequence

According to the excavators, individual archaeological layers were hardly distinguishable from each other by eye, and moreover, there were no separating layers between them. Furthermore, the stratigraphy did not present itself identically in the different trenches. On a general level, the following observations seem valid: a 20–30 cm thick archaeological deposit was recorded (Layer 1), consisting of anthropogenic material successively accumulated on the lake bottom. Mainly based on the pottery assemblages, described in more detail in the following chapter, two levels (Levels I and II) could be distinguished within

Fig. 4.1 La Marmotta. Geographical setting of the archaeological site (Authors)

Layer 1 (Fugazzola Delpino et al. 1993; Fugazzola Delpino 2002):

- **Layer 1, Level II** corresponds to the very beginning of the settlement.
- **Layer 1, Level I** is characterised by a great number of findings and represents the most extensive village occupation. (As developed *infra*, based on the pottery, an older and a younger phase can be distinguished within this level, in the following referred to as 'bottom' or 'top', respectively).

Layer 1 was covered by a 2 m thick deposit of natural lake sediments with a characteristic component of snails...

- ... accordingly called **'Chiocciolaio'-layer**. The **bottom part** of this deposit consists of remains of the village's final occupation years as well as of its abandonment, including the collapsed and submerged structures. Since the pottery spectrum (cf. *infra*) from the Chiocciolaio-layer differs from that of the preceding anthropogenic Layer 1, Level I, these are to be understood as the eroded remains of a separate settlement phase.

4.4.2 Pottery Typology

The pottery recovered at La Marmotta constitutes an important part of the find material and allows to contextualise the settlement within the framework of archaeological cultural phenomena during the Italian Early Neolithic. Whereas the oldest Neolithic remains, including the most ancient pottery in Italy, are found in the regions of Apulia, Calabria, and Basilicata as well as on Sicily, the pottery from La Marmotta shows characteristics of a more advanced stage (*Neolitico antico II*). This stage refers to at least three different styles that display connections to various geographic regions. The pottery styles occurring at La Marmotta have been described and contextualised in several publications, especially by Fugazzola Delpino et al. 1993,

Fugazzola Delpino 2002, and Delpino 2020, yet not extensively laid out. In the following, the pottery will be presented in accordance with the published material and related to the layers it was found in. Moreover, it will be set in the overall framework of the Italian Early Neolithic and beyond.

The earlier archaeological level (**Layer 1, Level II**) contains particularly well-preserved pottery of high quality: it is made of purified clay and well-fired, and the surfaces are carefully smoothed or even polished. Fine ware shapes include bowls, simple and carinated cups as well as jars with and without handles. Amongst the coarse ware, flasks, bowls, and jars occur. The bottoms of both the fine and the coarse ware are either flat or rounded. In rare cases, they show stand rings or pedestals. Vertically and horizontally pierced lugs, knobs, and handles are present. The decorations on the pottery found in Layer 2, Level II are both painted and, to a rarer extent, impressed (Fugazzola Delpino et al. 1993).

The painted pottery found in Layer 1, Level II shows white, red, grey, and black painted ornaments (Fig. 4.2). The motives consist of horizontal bands of parallel or angular lines and ladder-like bands. In some cases, the inside of the vessels is decorated, too (Fig. 4.2, no. 4–5) (Fugazzola Delpino et al. 1993). In Italy, Early Neolithic painted pottery first occurs in the southeast from c. 5700 BC on, with the styles named *Lagnano da Piede*, *Masseria La Quercia* or *Passo di Corvo arcaico*, and *tipico*. These styles differ from each other with respect to motives and colours, but seem to occur almost contemporaneously at various sites, though with different micro-regional onsets, yet showing significant chronological overlaps (Tiné 2002). This also applies to the Catignano painted pottery, which resembles the one found at La Marmotta. The Catignano painted pottery is concentrated in the Abruzzo region, and, as a cultural phenomenon, is dated to around 5600/5500–4900 BC (Tozzi and Zamagni 2003; Radi and Tozzi 2009). The westernmost sites of this group are located about 140 km from Lake Bracciano.

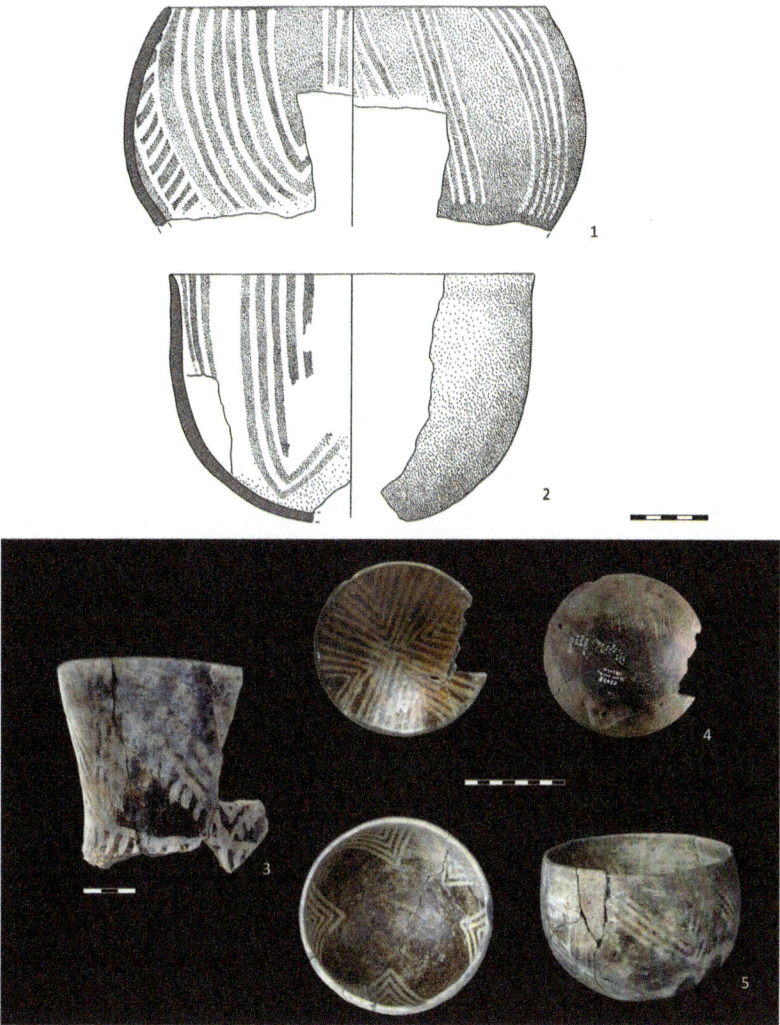

Fig. 4.2 La Marmotta. Pottery. **1–5**: Painted ware (Illustrations by L. Silenzi; first published by Fugazzola Delpino et al. 1993; photos: Museo delle Civiltà, Roma; composition by V. Becker)

The ceramics featuring impressions (Fig. 4.3, no. 1–10) belong to the so-called *Impresso* pottery culture (also referred to as *Impressa, Ceramica impressa-cardiale, Impresso-Cardial,* etc.). *Impresso* pottery constitutes the first pottery in Early Neolithic Italy, where it emerges around 6000 cal BC in the south of the country. Subsequently, it spread along the coasts, with regional styles named *Ceramica impressa adriatica, Ceramica impressa tirrenica* or *Ceramica impressa ligure*. At La Marmotta, we grasp the Tyrrhenian facies of this cultural phenomenon (ital. *Ceramica impressa tirrenica,* c. 5600–4900

BC) (Fugazzola Delpino et al. 1993). Sites attributed to the *Ceramica impressa tirrenica* are located in Lazio and Tuscany (Becker 2018) and have yielded pottery decorated with impressions and tremolo-like imprints mostly made with shells from molluscs like dog cockle (*Glycymeris glycymeris*) and others (*Cardium, Arca, Mytilus, Tellina, Pectunculus, Columbella, Conus, Cypraea*). In other cases, impressions are made with wooden or bone tools. They are arranged in shapes of triangles, lozenges, zigzag, and herringbone patterns. Bands filled with serrated impressions alternating with undecorated bands

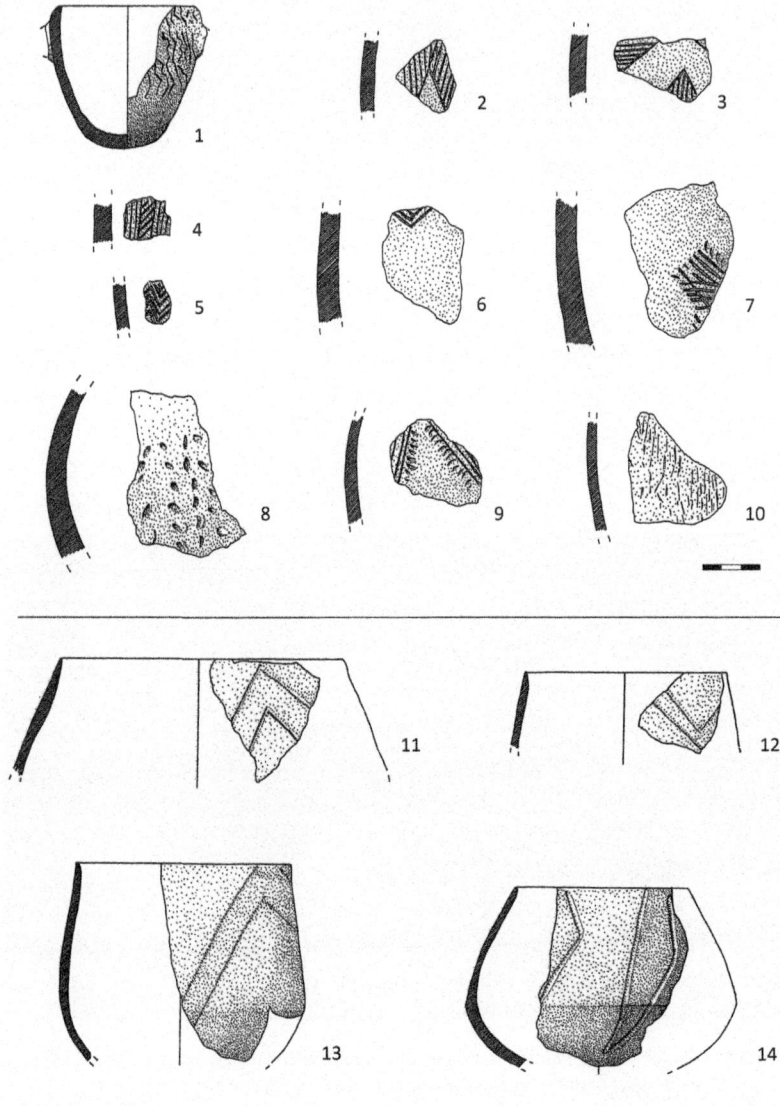

Fig. 4.3 La Marmotta. Pottery. **1–10**: *Impresso* ware, **11–14**: *Ceramica lineare* (Illustrations by L. Silenzi; first published by Fugazzola Delpino et al. 1993; composition by V. Becker)

are frequent. Sometimes they are arranged obliquely or vertically to the rim, sometimes hanging 'fringes' of punctures are featured below the rim. The very finely serrated or toothed punctures refer to a technique peculiar to the Tyrrhenian facies of Impressed ware. In some cases, the impressions were filled with white paste.

The subsequent level (**Layer 1, Level I**) allows for a distinction between an older and a younger phase, in the following referred to as 'bottom' or 'top', respectively:

Like Level II of Layer I, the **bottom part of level I** still contained both painted pottery and pottery decorated with impressions referring to the *Ceramica impressa tirrenica*, and the painted ware pointing to Catignano and southern Italy. On two bowls, a mixed decorative technique has been applied: their inner surface is painted, while the outer wall is decorated with imprints (cf. Binder et al. 2022).

The pottery in the **top part of level I** is made less carefully. A new type of decoration appears: incised lines. Whereas vessel shapes now encompass flat bowls, jars, and globular vessels, incised lines (mostly found on globular and semi-globular vessels) are arranged to form zigzags, crosses, and triangular motives. They are evidence of a new pottery style called 'pottery with incised lines' (Italian: *Ceramica a linee incise, Ceramica lineare*), which is typically found in Tuscany, Lazio, and Umbria at a more advanced stage of the early Neolithic (Fig. 4.3, no. 11–14). It dates around 5400/5300–4800/4700 BC and can be subdivided into three regional groups named after eponymous sites, *facies di Sasso, facies di Sarteano*, and *facies di Monte Venere* (Becker 2018). They only slightly differ from each other with respect to details in vessel shapes or decoration motives.

The most recent anthropogenic deposit, the so-called **Chiocciolaio-layer,** contained pottery decorated with incised lines only. It seems that at this stage, the *Ceramica lineare* would have completely replaced both the Tyrrhenian Impressed ware and the painted pottery. It is not quite clear whether this observation can be used as an argument for a more local orientation of the settlement and a decrease in supra-regional contacts.

Finally, some exceptional ceramic receptacles are worth mentioning, such as a vessel decorated with an incised 'sun'-motif. Incised and painted figural anthropomorphic and zoomorphic representations such as an incised motif resembling a butterfly are very rare or even unique (Fugazzola Delpino 2002).

The general chronological attribution of the present pottery styles, i.e. the Tyrrhenian Impressed ware, the painted wares of southern Italy and Catignano, and the *Ceramica lineare,* fits well with La Marmotta's absolute dates (*cf. infra*): the oldest pottery remains at La Marmotta belong to the Tyrrhenian Impressed complex, which is dated to around 5600–4800 BC. The painted ware occurs around the same time in southern Italy (Becker 2018). The same applies to the Catignano painted pottery, in which case a series of new dates point to its onset around 5600/5500 BC (Boschian and Colombo 2009). The relatively small number of absolute dates for the *Ceramica lineare* makes its chronological placement more difficult. Yet, the available radiocarbon data indicate its beginning at around 5400/5300 BC (Rosini et al. 2005), allowing us to assume a more recent dating than the previously mentioned styles or cultural phenomena.

4.4.3 Radiocarbon Dates

At an early stage of the site's excavation, the radiocarbon dating of a few construction timbers allowed the confirmation of the site's Early Neolithic context (cf. Fugazzola Delpino et al. 1993). Further ^{14}C-dates were specifically produced for dendrochronology (Fugazzola Delpino and Tinazzi 2010) (*cf.* Sect. 4.4). Though, from today's perspective, these values as such are less significant than those from recently measured short-lived botanical remains: first, the sampled timbers cannot be attributed to specific archaeological contexts at this stage; second, it is not always sufficiently clear which part of the timbers has been sampled, and third, the calibrated values show relatively wide probability ranges.

In recent years, radiocarbon ages of short-lived botanical remains (opium poppy capsules and cereal caryopsides) was performed (Table 4.1) (Salavert et al. 2020; Mazzucco et al. 2022), delivering much more accurate and comprehensible values than those previously available.

The ^{14}C-ages of the short-lived botanical remains roughly range between c. 5600 and 5200 cal BC: the charred grains from Layer 1, Level II covering a maximum timespan of between c. 5600 and 5400 cal BC, with those from Layer 1, Level I ranging between a little after 5500 and 5200 cal BC. Based on their ^{14}C-values, the opium poppy capsules appear to belong to the earlier occupation level (Layer 1, Level II). The close proximity of the dates, even across the levels, seems to confirm the stratigraphic finding, meaning that we are dealing with a continuous accumulation of Layer 1, possibly hinting at a more or less continuous site occupation.

Table 4.1 La Marmotta. Radiocarbon dates from short-lived botanical remains, sorted by sample context. Calibrated using OxCal 4.4 (Bronk Ramsey 2009) and the IntCal20 atmospheric calibration curve (Reimer et al. 2020) Mazzucco et al. (2022) have rightly pointed out, that the values of CNA-5289.1.2 and CNA-5291.1.1 might be the result of a secondary stratigraphical displacement of the relevant sample material (composition by A. Ballmer)

Lab no.	Sample material	^{14}C Age [yr BP]	cal BC 2σ (95.4%)	Sample context	Publication
ECHo2454	Opium poppy capsule (*Papaver somniferum*)	6600 ± 50	5622–5478	Layer 1	Salavert et al. (2020)
ECHo2657	Opium poppy capsule (*Papaver somniferum*)	6600 ± 30	5617–5480	Layer 1	Salavert et al. (2020)
CNA-5754.1.2	Charred caryopsis of emmer wheat (*Triticum dicoccum*)	6367 ± 34	5471–5220	Layer 1, Level I	Mazzucco et al. (2022)
CNA-5288.1.1	Charred caryopsis of emmer wheat (*Triticum dicoccum*)	6393 ± 40	5474–5230	Layer 1, Level I	Mazzucco et al. (2022)
CNA-5289.1.2	Charred caryopsis of emmer wheat (*Triticum dicoccum*)	6577 ± 43	5620–5475	Layer 1, Level I (?)	Mazzucco et al. (2022)
CNA-5290.1.1	Charred caryopsis of emmer wheat (*Triticum dicoccum*)	6534 ± 43	5614–5381	Layer 1, Level II	Mazzucco et al. (2022)
CNA-5291.1.1	Charred caryopsis of emmer wheat (*Triticum dicoccum*)	6431 ± 42	5476–5321	Layer 1, Level II (?)	Mazzucco et al. (2022)
CNA-5755.1.1	Charred caryopsis of emmer wheat (*Triticum dicoccum*)	6558 ± 31	5611–5475	Layer 1, Level II	Mazzucco et al. (2022)

4.4.4 Dendrochronology

About 3400 wooden piles were recorded within the excavated area, most of them inserted vertically into the former lakebed. They used to be part of a bank reinforcement as well as of house substructures. The dendrochronological analysis of these timbers was first started by N. Martinelli (preliminary report: Martinelli 1993) and later taken over by O. Tinazzi at the Laboratory of the Italian Institute of Dendrochronology (Fugazzola Delpino 2002; Fugazzola Delpino and Tinazzi 2010). A total of 2044 piles, mostly of oak, but also of laurel, ash, and alder, were examined. The tree ring chronologies were built with oaks and ash. The absence of a dendrochronological reference curve for the concerned region and period posed a particular challenge and attempts to find supra-regional matches for possible teleconnections were inconclusive (Fugazzola Delpino and Tinazzi 2010).

In an article from 2010, M. A. Fugazzola Delpino and O. Tinazzi present an absolutely dated mean curve ('Mean Curve 1'), covering 248 years between 5538 and 5290 cal BC. Based on the available data, it is not possible to identify particular clusters amongst the felling dates between 5457 and 5290 cal BC. An additional (floating) curve ('GR13') is presented by the authors, covering 154 years towards the end of the 6th millennium cal BC.

While the suggested tree felling dates (or construction dates, respectively) coincide with the conclusions drawn from other material from La Marmotta (stratigraphy, pottery, ^{14}C-ages), the unavailability of the underlying data prevents a deeper discussion. The same applies to the statements on the building architecture and settlement layout deriving from the dendroarchaeological analyses (Fugazzola Delpino and Tinazzi 2010).

4.4.5 A Brief History of the Village

Based on the combined information available from the stratigraphy, the pottery typology, and absolute dating, a course of events as presented in Table 4.2 can be established:

Table 4.2 La Marmotta. Proposal of a 'village history'. The phasing is based on available information from stratigraphy, pottery typology, and absolute dating (dendrochronological dating must be considered with reservations, cf. Sect. 4.4) (A. Ballmer)

Identifiable phases	Stratigraphical deposits	Pottery typology and chronological framework of cultural phenomena	Absolute dates (^{14}C, dendro)
Post-occupation (flooding and covering of site)	Top part of Chiocciolaio-layer and lake sediments	–	–
Final occupation, incl. abandonment	**Chiocciolaio-layer** (mix of anthropogenic deposit and lake sediments)	*Ceramica lineare* c. 5400–4800 BC	• Dendro (GR13-curve): end of 6th mill. cal BC (building activities)
Third occupation	**Layer 1, Level I, top** (anthropogenic deposit, rich in structures and finds)	Tyrrhenian Impressed ware (made less carefully); *Ceramica lineare* c. 5400–4800 BC	• ^{14}C: c. 5500–5200 cal BC • Dendro (Media 1-curve): 5457–5290 cal BC (building activities)
Second occupation	**Layer 1, Level I, bottom** (anthropogenic deposit, rich in structures and finds)	Painted ware; Tyrrhenian Impressed ware (both high quality) c. 5600–4800 BC	• ^{14}C: c. 5500–5200 cal BC • Dendro (Media 1-curve): 5457–5290 cal BC (building activities)
First occupation	**Layer 1, Level II** (anthropogenic deposit)	Painted ware; Tyrrhenian Impressed ware (both high quality) c. 5600–4800 BC	• ^{14}C: c. 5600–5400 cal BC • Dendro (Media 1-curve): 5457–5290 cal BC (building activities)
Pre-occupation	Sterile loam	–	–

There is substantial archaeological evidence of settlement activity from the 56th century BC onwards, which, mainly based on the pottery, but also on stratigraphic indications, as well as reference points from absolute dating, can be divided into four phases.

According to Fugazzola Delpino and Tinazzi (2010), the tree rings of the GR13-curve, referring to the village's latest construction phase, show signs of 'stress', possibly hinting at one or several regional environmental event/s: fire episodes, micro-climatic shifts, or volcanic eruptions. However, the causality between these several possible impacts and the abandonment of the settlement remains unclear at this stage. On the other hand, conspicuous burning traces in the Chiocciolaio-layer indicate a village fire during the latest occupation phase (Fugazzola Delpino et al. 1993). Whether the settlement had to be abandoned because of a conflagration, or whether the settlement was intentionally set on fire on its abandonment, or burned down shortly after its abandonment, is unknown. The signs of rapid sedimentation of the last archaeological layer (evidenced by the good preservation of the archaeological finds on the surface of the latest deposit as well as by the quality of the covering layer) hint at the role of the rising lake water level in the abandonment of the settlement.

4.5 Perspectives on the Way of Life: Subsistence Economy and Navigation

4.5.1 Subsistence Economy: Domesticated Animals, Cereals, and Opium Poppy

The waterlogged archaeological layers have delivered rich evidence of the subsistence economy prevalent at the time of La Marmotta's occupation, featuring a broad spectrum of plant and animal resources deriving from agricultural production as well as from foraging activities.

Both the faunal remains and a part of the plant remains from the 1989 campaign have been published in preliminary reports by Cassoli and Tagliacozzo (1993) and Rottoli (1993; see also Rottoli 2001–02; Rottoli and Pessina 2007). Some findings and conclusions have been generally presented by Fugazzola Delpino in 2002. Specific plant species from La Marmotta, in particular wheat and opium poppy, have been included in recent research projects (de Vareilles et al. 2020; Salavert et al. 2020; Mazzucco et al. 2022), leading to some of the relevant basic data being published.

Awaiting more data to be available, we would like to briefly address two aspects that are important for the overall assessment of the site: domestic animals on the one hand, and the presence of domesticated cereals and opium poppy on the other.

As has already been implied in 1993 by P. F. Cassoli and A. Tagliacozzo, and further developed on a larger database by Fugazzola Delpino (2002), from the earliest occupation of La Marmotta, domestic animal species show an advanced state of domestication. This is evidenced by the sizes of cattle and goats, as well as the absence of horns in ewes. The minimal number of individuals in ovicaprids (sheep prevailing over goat) is stated to be significantly higher than both cattle and pigs together.

Although the 1993 report deals with a part of the botanical macro-remains only, it shows clearly that the main plant components of the core 'Neolithic package' are present at La Marmotta, i.e. wheat, barley, lentil, and pea. Around 65% of the recovered remains of cultivated plants consisted of cereals, represented by substantial proportions of charred specimens including ears, spikelets, and individual caryopses. In addition to two-row hulled barley (*Hordeum distichum*), emmer (*Triticum dioccum* L.), and einkorn (*Triticum monococcum* S.), archaic free-threshing tetraploid naked wheat (*Triticum durum/turgidum*) was identified. For the time being, there is only tentative evidence of free-threshing hexaploid naked wheat (*Triticum aestivum*) (Rottoli 1993; Rottoli and Pessina 2007).

The crucial role of arable farming at La Marmotta, in particular of cereal cultivation, is indirectly evidenced by the presence of numerous tools employed in the cultivation, harvesting, processing, and storing of plants and plant products. Above all, the 52 wooden sickles with implemented flint blades, or parts thereof, indicate the practice of arable farming (Fig. 4.4) (Mazzucco et al. 2018, 2020; Mineo et al. 2018, 2021, 2022).

Interestingly, abundant remains of opium poppy (*Papaver setigerum/somniferum*) have been found at La Marmotta (Fig. 4.5). In fact, this is the oldest record of opium poppy in the Mediterranean to date. Whereas it is not possible to tell whether the finds belong to the wild or the cultivated species (Rottoli 1993, 2001–02; Rottoli and Pessina 2007; Salavert et al. 2020), the opium poppy's seeds and capsules were presumably used in food preparation and for its oil content, as well as for medical purposes.

4.5.2   Navigation

Five dugout canoes have been discovered at La Marmotta, made from different types of wood (Fugazzola Delpino and Mauro 2014). The two largest ones (up to 11 m long) could have served in more complex undertakings.

The biggest dugout, named 'La Marmotta 1', made from an oak log and measuring 10.4 m in length, is thought to be evidence of advanced boat building knowledge (Fig. 4.6): the bow has a tapered shape, whereas the stern presents itself as bulky. The bottom of the hull is partially flattened, possibly to improve its seaworthiness. At the bottom, four transversal ribs are carved out in regular intervals. Near the stern, wood had been hewn off, leaving a ridge that is reminiscent of a keel. Based on its find context, the 'La Marmotta 1' is assigned to the earliest settlement occupation (Fugazzola Delpino and Mineo 1995; Fugazzola Delpino et al. 1997; Bondioli et al. 2000). Its seaworthiness was practically demonstrated in an experimental expedition: in 1998, Czech archaeologists successfully navigated a replica from Sicily to Portugal in five stages (Tichý 2001, 2016). Furthermore, it has been suggested that by fastening two dugouts with planks (for instance by means of the wooden

Fig. 4.4 La Marmotta. Selection of Early Neolithic 'sickles' with wooden hafts and flint implements (Courtesy of Museo delle Civiltà, Roma; composition by A. Ballmer)

fixtures found on both of the large dugouts from La Marmotta [Fugazzola Delpino and Mineo 1995; Fugazzola Delpino and Mauro 2014]), a catamaran-like vehicle could be assembled, hypothetically suitable to transport substantial amounts of cargo, including not only humans but possibly also livestock and crops (see Vigne and Cucchi 2005; Zilhão 2014).

Several miniature clay models of boats were found within the earliest occupation layer of the settlement. They measure about 15–20 cm in length each and are light enough to float (Fugazzola Delpino et al. 1993; Fugazzola Delpino 2002). One of the published ones features a raised prow and a raised stern as well as convex long sides (Fig. 4.7), referring to a watercraft type going beyond a simply shaped dugout. Another one has an eyelet at each of the narrow ends. The function of these boat models may be linked to cultic-ritual practices.

Some boats from La Marmotta, both operative log-boats and miniature models with developed features, are significantly more elaborate than

rafts or simple dugout canoes. This could hint at the very particular predisposition of the residential community to navigate (besides lakes and rivers, one would like to imagine the navigation of the seas)—possibly an essential component of their identity.

4.6 Reference to the Past: The 'Venus' from La Marmotta

The entirely preserved small figurine of a naked female is highly interesting and enigmatic at the same time (Fig. 4.8). It was found under a collapsed house floor in Layer 1, Level II, corresponding to the settlement's first occupation phase (Fugazzola Delpino 2002). The sculpture is three-dimensionally carved from light green steatite and measures 4.8 cm in height (Fugazzola Delpino 2000–2001). Its round head shows no face; at the back of the head, implicitly represented hair hangs down. The arms are slightly bent and rest on the belly under large breasts.

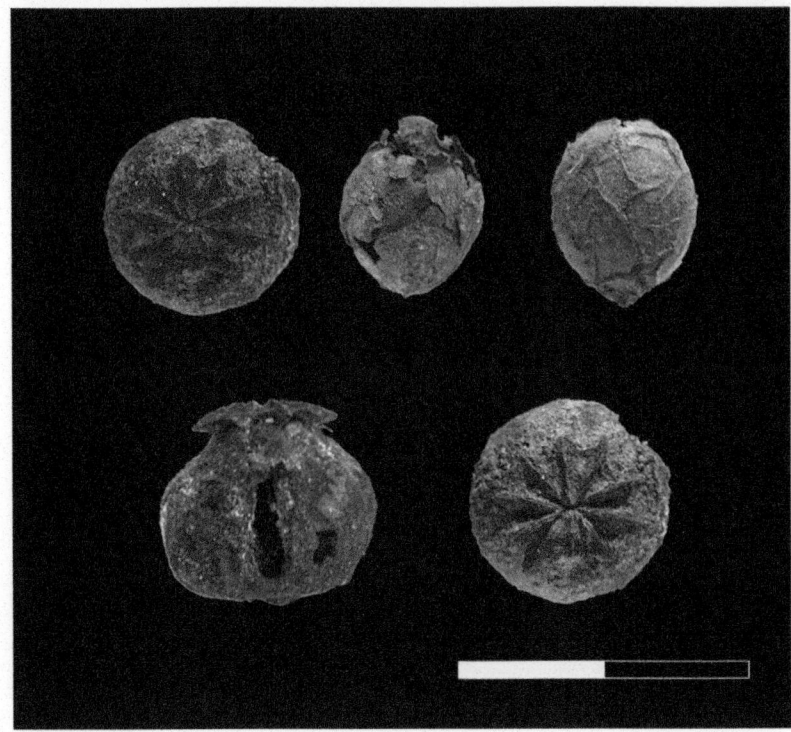

Fig. 4.5 La Marmotta. Opium poppy capsules (*Papaver setigerum/somniferum*) (Courtesy of Museo delle Civiltà, Roma)

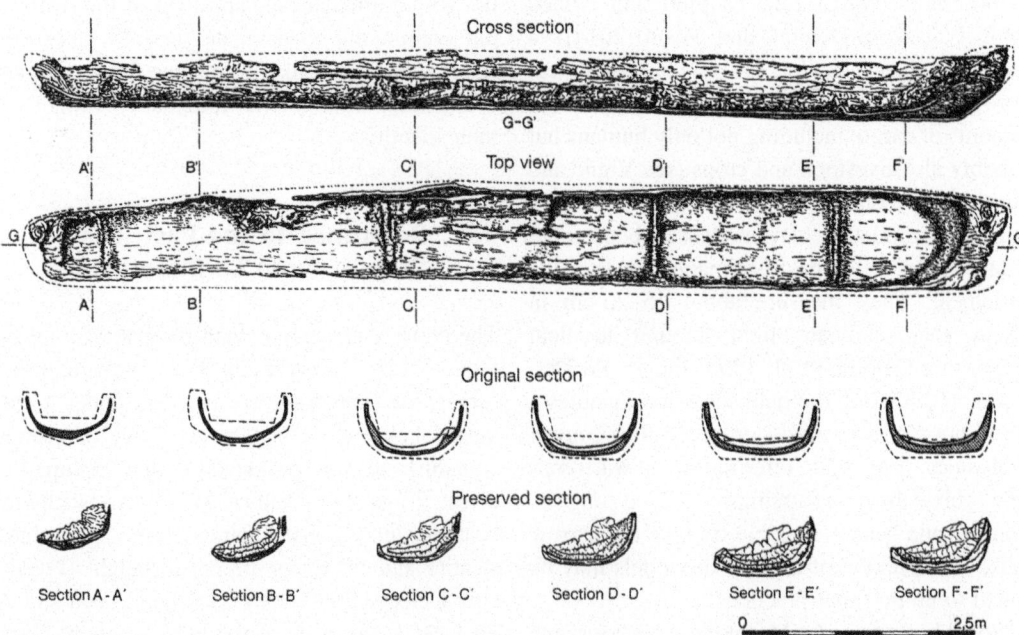

Fig. 4.6 La Marmotta. Dugout canoe 'La Marmotta 1' carved from a single oak log and featuring advanced boat building techniques (Illustration by Progetti and Sistemi; courtesy of the Museo Nazionale Preistorico Etnografico 'L. Pigorini', Rome; first published by Bondioli et al. 2000; modified by A. Ballmer)

Fig. 4.7 La Marmotta. Clay model of dugout canoe with raised prow and stern, and convex long sides (Courtesy of Museo delle Civiltà, Roma)

The abdominal region is corpulent; belly button and pubic triangle are carved. The thighs, buttocks, and legs are plastically sculpted and are equally full as the breasts and abdomen, while the lower legs are represented in a reduced way, even without feet.

From a stylistic point of view, the La Marmotta figurine must be attributed to the Upper Palaeolithic figurines of the Gravettian technocomplex (Hansen 2007; Skeates 2017). It shows striking resemblance to similar figurines from the Balzi Rossi/Grimaldi caves in Liguria (White and Bisson 1998) or even the famous 'Venus of Willendorf' (Wachau, Lower Austria) (Antl-Weiser 2008). Although M. A. Fugazzola Delpino (2000–2001) would like to establish a connection with Neolithic figurines from Thessaly, Anatolia, and the Levant, this is not obvious. Rather, the specimen from La Marmotta currently represents the southernmost finding point within the distribution area of the Gravettian figurines between the Pyrenees and the Don (Hansen 2007). The limestone raw material of the 'Venus from Willendorf' has recently been located in Lake Garda region in Northern Italy (Weber et al. 2022), evidencing the long-distance circulation of such pieces.

For all the sensations of the find, the temporal discrepancy between La Marmotta's settlement occupation and the stylistic context of the figurine cannot be ignored. J. Robb (2007) has plausibly argued that the La Marmotta figurine might in fact be a Palaeolithic product that would have been collected by the Early Neolithic settlers of La Marmotta from another site and absorbed into their own practices—and hence incorporated into the archaeological context. Due to the figurine's deposition in the context of a building interpreted as assembly house or communal building, it has been interpreted in relation to a 'founding ritual' of this specific building (Fugazzola Delpino 2002); it is of course tempting to expand this idea towards the initial founding of the entire settlement. Apart from the strange aesthetic and special status that the piece may have had for the community, it may also have been understood as a reference to a remote, absolute past, thus—as a symbol of a temporal continuity—creating a kind of legitimacy of the group and/or place.

4.7   La Marmotta and the Western Mediterranean Early Neolithic

At least three points allow an extended discussion of La Marmotta: the impressed pottery, the documented spectrum of cultivated plants, in particular the naked wheat, as well as the presence of opium poppy.

It is noticeable that the pottery from the earliest period of occupation has no local background, rather, the painted ware displays connections to southeastern and central-eastern

Fig. 4.8 La Marmotta. Figurine of a naked female made of green steatite. Likely made in the Upper Palaeolithic, it was collected, kept and deposited by the Early Neolithic settlers of La Marmotta (Courtesy of Museo delle Civiltà, Roma)

Italy. The concurrently present *Impresso* pottery belongs to the Tyrrhenian facies which is found in Lazio and Tuscany as well as in Sardinia and Corsica. It seems that in the course of the Neolithisation process, southeastern Italy would have been settled first (c. 6000 BC), followed by southern Italy, including Sicily (c. 6000/5900 BC), then quite rapidly afterwards Liguria (c. 5800 BC) and only thereafter Sardinia, Corsica, and the Tyrrhenian coast (c. 5600/5500 BC) (for a synopsis see Becker 2018). Based on the currently available data, La Marmotta would have taken a leading role in the spreading of the Tyrrhenian *Impresso* facies. Eventually, in the latest stage of La Marmotta's occupation, a more local pottery tradition is documented, evidenced by the presence of *Ceramica lineare*.

The evidence of a cultivated plant spectrum referring to the core 'Neolithic package' as well as of domesticated ovicaprids and other animals leads to the conclusion that La Marmotta must be understood as a 'pioneer' settlement of an already completely neolithisised group, quite possibly originating outside of the Tyrrhenian circle, for instance in southern Italy. The hypothesis of a 'migration' by sea can be supported by the presence of relatively elaborated dugout canoes (not only visible in actual

vehicles but also in their symbolic representation) in the first settlement phase of the village: authors like Guilaine and Manen (2007), Tresset and Vigne (2007), or Zilhão (2014) have put forward the realistic scenario according to which early farmers would have moved across the Mediterranean bringing with them seeds or seedlings of cultivated plants as well as livestock.

With this in mind (the presence of early Tyrrhenian *Impresso* pottery, domesticated plants and animals, as well as an occupation from as early as 5600 BC at the latest), La Marmotta can be considered as a catalyst of the Tyrrhenian Neolithisation and moreover of the Neolithisation of large areas of the western Mediterranean region (see, e.g. the southeastern Spanish site of La Draga, Banyoles, occupied from c. 5300 BC on: see Chap. 11 in this book), but also parts of central and Northern Italy (for instance, the waterlogged site of Isolino Virginia at Lake Varese, occupied from c. 5000 BC on, is also linked to the western Mediterranean tradition (see Chap. 16 in this book)), including the western Alpine sphere (Jacomet 2007). La Marmotta's role as prime mover is supported by the presence of opium poppy (*Papaver setigerum/somniferum*): A. Salavert and her team found that during the relevant period, opium poppy is only found in central Italy and the western Mediterranean region, with La Marmotta yielding one of the earliest pieces of evidence (Salavert et al. 2020; see also: Rottoli 1993, 2002; Rottoli and Pessina 2007). Based on its chrono-spatial distribution pattern, it is speculated that, in contrast to the cereals and pulses of Europe's first farmers, the domestication of the opium poppy could have taken place in central Italy from where it would have spread west, and eventually northwards (Salavert 2010, 2011, Salavert et al. 2018; de Vareilles et al. 2020; Jesus et al. in press).

Certainly, the reconstruction of the detailed spreading dynamics of the western Mediterranean Early Neolithic requires further research. Being inclined to grant La Marmotta the status of a hub in the Neolithisation process of the western Mediterranean and furthermore the western Alpine regions, the discussion underlines the urgency of evaluating and publishing the site.

4.8 Epilogue

Various find categories are currently being worked on by Italian and international specialists. Not yet published before the completion of this manuscript, but expected in 2023, the book *'The Submerged site of La Marmotta (Rome, Italy). Decrypting a Neolithic Society'* (Mineo et al. in press) will present a series of new, previously unpublished materials and promising analyses. Undoubtedly, the site will keep the research community further engaged.

Acknowledgements The authors thank all the researchers and specialists who have participated and/or collaborated in the excavation, documentation, restoration, study, determination, and further analysis of the materials of the archaeological site of La Marmotta since its discovery. Mario Mineo expresses special gratitude to the late Director of the Museo delle Civiltà, F. M. Gambari, and the Director General of the State Museums of Italy, Dr. M. Osanna, for their friendship and trust during the many years of research and collaboration; as well as to Dr. M. A. Fugazzola Delpino, the organiser and director of the excavations at La Marmotta between 1989 and 2009, who had a major influence on the exploration and publication of the site, and is largely responsible for the rich inventory of finds that is known today.

References

Antl-Weiser W (2008) The anthropomorphic figurines from Willendorf. Wiss Mitt Niederösterreichisches Landesmuseum 19:19–30

Becker V (2018) Studien zum Altneolithikum in Italien. Neolithikum und ältere Metallzeiten: Studien und Materialien, vol 3. Lit Verlag, Berlin

Bernicchia A, Gemelli V (2004) Studio micologico dei funghi. Bull Paletnologia Ital 95:17–20

Bernicchia A, Fugazzola Delpino MA, Gemelli V, Mantovani B, Lucchetti A, Cesari M, Speroni E (2006) DNA recovered and sequenced from an almost 7000 y-old Neolithic polypore Daedaleopsis Tricolor. Mycol Res 110(1):14–17. https://doi.org/10.1016/j.mycres.2005.09.012

Binder D, Gomart L, Huet Th, Kačar S, Maggi R, Manen C, Radi G, Tozzi C (2022) Le complexe de la Céramique Imprimée en Méditerranée centrale et nord-occidentale: une synthèse chronoculturelle (VIIe et VIe millénaires AEC). In: Binder D, Manen C (eds) Céramiques imprimées de Méditerranée occidentale (VIe millénaire AEC): données, approches et enjeux nouveaux. Western Mediterranean Impressed Wares (6th millennium BCE): New data, approaches and challenges. Actes de la séance de la Société préhistorique française de Nice (mars 2019). Société préhistorique française, Paris, pp 27–124

Bondioli L, Fugazzola MA, Macchiarelli R, Mineo M, Morigi G, Montanari M, Rossi L (2000) Soprintendenza Speciale al Museo Nazionale Preistorico Etnografico 'L. Pigorini'. Il restauro dei manufatti lignei provenienti da siti umidi. In: Schutz des Kulturerbes unter Wasser. Veränderungen europäischer Lebenskultur durch Fluss- und Seehandel. Beiträge zum Internationalen Kongress für Unterwasserarchäologie IKUWA 99, 18.–21. Februar 1999 in Sassnitz auf Rügen. Beiträge zur Ur- und Frühgeschichte Mecklenburg-Vorpommerns, vol 35. Archäologisches Landesmuseum für Mecklenburg-Vorpommern, Lübstorf, pp 475–480

Boschian G, Colombo M (2009) Infilling processes of large pit features at Catignano—Neolithic (Italy). In: Cavulli F (ed) Defining a methodological approach to interpret structural evidence. Proceedings of the XV World Congress (Lisbon, 4–9 September 2006). British Archaeological Reports International Series, vol 2045. Archaeopress, Oxford, pp 43–50

Bronk Ramsey C (2009) Bayesian analysis of radiocarbon dates. Radiocarbon 51(1):337–360. https://doi.org/10.1017/S0033822200033865

Caruso Fermé L, Mineo M, Ntinou M, Remolins G, Mazzucco N, Gibaja JF (2021) Woodworking technology during the early Neolithic: First results at the site of La Marmotta (Italy). Quat Int 593–594C:399–406

Cassoli PF, Tagliacozzo A (1993) La Marmotta, Anguillara Sabazia (RM). Scavi 1989. Analisi preliminare delle faune. Bull Paletnologia Ital N.S. II 84:323–337

Cassoli PF, Tagliacozzo A (1995) Appendice 3. I reperti ossei faunistici dell'area della piroga. In: Fugazzola

Delpino MA, D'Eugenio G, Pessina A, 'La Marmotta' (Anguillara Sabazia, RM). Scavi 1989. Un abitato perilacustre di età neolitica. Bull Paletnologia Ital N.S. IV 86:267–288

Daum J (2015) Hafenbau an der Küste des südlichen Etrurien. In: Schmidts T, Vučetić MM (eds) Häfen im 1. Millennium AD. Bauliche Konzepte, herrschaftliche und religiöse Einflüsse. Plenartreffen im Rahmen des DFG-Scherpunktprogramms 1630 'Häfen von der Römischen Kaiserzeit bis zum Mittelalter' im Römisch-Germanischen Zentralmuseum Mainz, 13.–15. Januar 2014. RGZM-Tagungen, vol 22. Interdisziplinäre Forschungen zu den Häfen von der Römischen Kaiserzeit bis zum Mittelalter in Europa, vol 1. Verlag des Römisch-Germanischen Zentralmuseums, Mainz, pp 9–22

Delpino C (2020) Il Neolitico antico dell'areale medio-tirrenico: Gli aspetti della ceramica impressa e della ceramica lineare nel Lazio e nel territorio di Roma. In: Anzidei AP, Carboni G (eds) Roma prima del mito. Abitati e necropoli dal Neolitico alla prima età dei metalli nel territorio di Roma (VI–III mill. a. C.). Archaeopress, Oxford, pp 3–24

Fiore I, Tagliacozzo A, Fugazzola Delpino MA (2006) L'utilizzo dei canini di suino nel villaggio neolitico de 'La Marmotta' (Anguillara Sabazia, Roma). Istituto Italiano di Preistoria e Protostoria, Firenze

Fugazzola Delpino MA (2004) Su alcuni funghi rinvenuti nel villaggio neolitico de 'La Marmotta' (Anguillara Sabazia, Roma) Considerazioni Preliminari. Bull Paletnologia Ital 95:1–20

Fugazzola Delpino MA, Mauro N (2014) La seconda imbarcazione monossile del villaggio neolitico de La Marmotta. In: Asta A, Caniato G, Gnola D, Medas S (eds) Archeologia, Storia, Etnologia Navale. Atti del II convegno nazionale Cesenatico Museo della Marineria (13–14 aprile 2012). Navis 5. Rassegna di studi di archeologia, etnologia e storia navale. Libreria Universitaria, Padova, pp 125–132

Fugazzola Delpino MA, Mineo M (1995) La piroga neolitica di Bracciano (La Marmotta 1). Bull Paletnologia Ital N.S. IV 86:197–266

Fugazzola Delpino MA, Mineo M (2000) Die frühneolithische Siedlung 'La Marmotta' (Italien) als Spiegelbild mittelmeerischer Kulturkontakte. In: Schutz des Kulturerbes unter Wasser. Veränderungen europäischer Lebenskultur durch Fluss- und Seehandel. Beiträge zum Internationalen Kongress für Unterwasserarchäologie IKUWA 99, 18.–21. Februar 1999 in Sassnitz auf Rügen. Beiträge zur Ur- und Frühgeschichte Mecklenburg-Vorpommerns, vol 35. Archäologisches Landesmuseum für Mecklenburg-Vorpommern, Lübstorf, pp 121–126

Fugazzola Delpino MA, Pessina A (1999) Le village submergé de La Marmotta (Lac de Bracciano, Rome). In: Vaquer J (ed) Le Néolithique du Nord-Ouest méditerranéen. XXIVe Congrès préhistorique de France: Le Néolithique du Nord-Ouest méditerranéen (Carcassonne, 26–30 Septembre 1994). Congrès Préhistorique de France, vol 24, 2. Société préhistorique française, Paris, pp 35–38

Fugazzola Delpino MA, Tinazzi O (2010) Dati di cronologia da un villaggio del Neolitico Antico. Le indagini dendrocronologiche condotte sui legni de La Marmotta (lago di Bracciano-Roma). Miscellanea in ricordo di Francesco Nicosia, Studia Erudita. Fabrizio Serra Editore, Pisa, Roma

Fugazzola Delpino MA, D'Eugenio G, Pessina A (1993) 'La Marmotta' (Anguillara Sabazia, RM). Scavi 1989. Un abitato perilacustre di età Neolitica. Bull Paletnologia Ital N.S. IV 84:181–342

Fugazzola Delpino MA (1998–99) La vita quotidiana del Neolitico. Il sito della Marmotta sul Lago di Bracciano. In: Pessina A, Muscio G (eds) Settemila anni fa il primo pane. Ambienti e culture delle società neolitiche. Museo Friulano di Storia Naturale, Udine, pp 185–192

Fugazzola Delpino MA (2000–01) La piccola 'dea madre' del lago Bracciano. Bull Paletnologia Ital 91–92:27–45

Fugazzola Delpino MA (2002) La Marmotta, Lazio. In: Fugazzola MA, Pessina A, Tinè V (eds) Le ceramiche impresse nel Neolitico antico. Italia e Mediterraneo. Istituto Poligrafico e Zecca dello Stato, vol 1. Studi di Paletnologia, Roma, pp 373–395

Guilaine J, Manen C (2007) From mesolithic to neolithic in the western mediterranean. Proc Brit Acad 144:20–51. https://doi.org/10.5871/bacad/9780197264140.003.0003

Hansen S (2007) Bilder vom Menschen der Steinzeit: Untersuchungen zur anthropomorphen Plastik der Jungsteinzeit und Kupferzeit in Südosteuropa. Archäologie in Eurasien, vol 20. P. von Zabern, Mainz

Jacomet S (2007) Neolithic Plant economies in the northern Alpine Foreland from 5500–3500 cal BC. In: Colledge S, Conolly J (2007) The origins and spread of domestic plants in southwest Asia and Europe. Left Coast Press, Walnut Creek, CA, pp 221–258

Jesus A, Bonhomme V, Evin A, Ivorra S, Soteras R, Salavert A, Antolín F, Bouby L (in press) A morphometric approach to track opium poppy domestication. Sci Rep 11:9778

Macchiarelli R, Salvadei L (1993) Appendice 4. 'La Marmotta', Anguillara Sabazia (RM). Scavi 1989. Molare deciduo umano da livelli neolici. In: Fugazzola Delpino MA, D'Eugenio G, Pessina A, 'La Marmotta' (Anguillara Sabazia, RM). Scavi 1989. Un abitato perilacustre di età neolitica. Bull Paletnologia Ital N.S. II 84:339–342

Martinelli N (1993) Appendice 2. La Marmotta (Anguillara Sabazia, RM): scavi 1989—indagini dendrocronologiche. Note preliminare. In: Fugazzola Delpino MA, D'Eugenio G, Pessina A, 'La Marmotta' (Anguillara Sabazia, RM). Scavi 1989. Un abitato perilacustre di età neolitica. Bull Paletnologia Ital N.S. IV 84:181–315

Mazzucco N, Ibáñez JJ, Capuzzo G, Gassin B, Mineo M, Gibaja JF (2020) Migration, adaptation, innovation: the spread of Neolithic harvesting technologies in the Mediterranean. PLoS ONE 15(4):e0232455. https://doi.org/10.1371/journal.pone.0232455

Mazzucco N, Mineo M, Arobba D, Caramiello R, Caruso Fermé LC, Ibáñez JJ, Morandi L, Mozota M, Pichon F, Portillo M, Rageaot M, Remolins G, Rottoli M, Gibaja JF (2022) Multiproxy study of 7500-year-old wooden sickles from the Lakeshore Village of La Marmotta Italy. Sci Rep 12(1):14976. https://doi.org/10.1038/S41598-022-18597-8

Mazzucco N, Capuzzo G, Petrinelli-Pannocchia C, Ibáñez JJ, Gibaja JF (2018) Harvesting tools and the spread of the neolithic into the central-western mediterranean area. Quatenary Int 470 (Part B):511–528. https://doi.org/10.1016/j.quaint.2017.04.018

Michetti LM (2017) Harbors. In: Naso A (ed) Etruscology. De Gruyter, Boston, Berlin, pp 391–405

Mineo M, Gibaja J, Mazzucco N (eds) (in press) The Submerged Site of La Marmotta (Rome, Italy). Decrypting a Neolithic Society. Oxbow Books, Oxford

Mineo M, Mazzucco N, Gibaja JF, Mozota M (2018) 'Sabres' from the Neolithic. The Sickles from La Marmotta. Awrana 2018. University of Nice Côte d'Azur, Nice

Mineo M, Mazzucco N, Rottoli M, Remolins G, Caruso-Fermé L, Gibaja J (2023) Textiles, basketry and cordage from the early Neolithic settlement of La Marmotta, Lazio. Antiquity First View, pp 1–17. https://doi.org/10.15184/aqy.2023.21

Mineo M (2016) Monossili d'Europa: costruite anche per le rotte marine? In: Guidi A, Cazzella A, Nomi F (eds) Ubi minor. Le isole minori del Mediterraneo centrale. Dal Neolitico ai primi contatti coloniali. Convegno di studi in ricordo di Giorgio Buchner, a 100 anni dalla nascita (1914–2014). Scienze dell'Antichità, vol 22(2). Edizioni Quasar, Sapienza Università di Roma, Roma, pp 453–475

Mineo M (2019) Tessiture e intrecci. In: Massussi M, Tucci S, Laurito R (eds) Trame di storia. Metodi e strumenti dell'archeologia sperimentale. Archeofest 2017. E.S.S. Editorial Service System, Roma, pp 119–146

Radi G, Tozzi C (2009) La ceramica impressa e la cultura di Catignano in Abruzzo. In: Barbaza M, Boissinot P, Briois F, Carrère I, Coularou J, Gascó J, Giraud P, Manen C, Marinval P, Midant-Reynes B, Perrin T, Vaquer J (eds) De Méditerranée et d'ailleurs… Mélanges offerts à Jean Guilaine. Les Archives d'Ecologie Préhistorique, Toulouse, pp 601–611

Reimer PJ, Austin WEN, Bard E, Bayliss A, Blackwell PG, Bronk Ramsey C, Butzin M, Cheng H, Edwards RL, Friedrich M, Grootes PM, Guilderson TP, Hajdas I, Heaton TJ, Hogg AG, Hughen KA, Kromer B, Manning SW, Muscheler R, Palmer JG, Pearson C, van der Plicht J, Reimer RW, Richards DA, Scott EM, Southon JR, Turney CSM, Wacker L, Adolphi F, Büntgen U, Capano M, Fahrni SM, Fogtmann-Schulz A, Friedrich R, Köhler P, Kudsk S, Miyake F, Olsen J, Reinig F, Sakamoto M, Sookdeo A, Talamo S (2020) The IntCal20 Northern hemisphere radiocarbon age calibration curve (0–55 cal kBP). Radiocarbon 62 (4):725–757. https://doi.org/10.1017/RDC.2020.41

Robb J (2007) The early mediterranean village: agency, material culture, and social change in Neolithic Italy. Cambridge University Press, Cambridge

Rosini M, Sarti L, Silvestrini M (2005) La ceramica del sito di Ripabianca di Monterado (Ancona) e le coeve produzioni dell'Italia centro-settentrionale. Riv Sci Preistoriche 55:225–263

Rottoli M, Pessina A (1984) Neolithic agriculture in Italy: an update of archaeobotanical data with particular emphasis on northern settlements. In: van Zeist W, Casparie WA (eds) Plants and ancient man: Studies in palaeoethnobotany. Proceedings of the Sixth Symposium of the International Work Group for Palaeoethnobotany, Groningen 30 May–3 June 1983. Balkema, Rotterdam, pp 141–153

Rottoli M (1993) Appendice 1. La Marmotta Anguillara Sabazia (RM). Scavi 1989, analisi paletnobotaniche: prime risultanze. In: Fugazzola Delpino MA, D'Eugenio G, Pessina A, 'La Marmotta' (Anguillara Sabazia, RM). Scavi 1989. Un abitato perilacustre di età neolitica. Bull Paletnologia Ital N.S. IV 84:305–315

Rottoli, M (2001–02) Zafferanone selvatico (Carthamus lanatus) e cardo della Madonna (Silybum marianum), piante raccolte o coltivate nel Neolitico antico a 'La Marmotta'? Bull Paletnol Ital 91–91:47–61

Salavert A (2011) Plant economy of the first farmers of central Belgium (Linearbandkeramik, 5200–5000 BC). Veg Hist Archaeobotany 20:321–332. https://doi.org/10.1007/s00334-011-0297-z

Salavert A, Zazzo A, Martin L, Antolín F, Gauthier C, Thil F, Tombret O, Bouby L, Manen C, Mineo M, Mueller-Bieniek A, Piqué R, Rottoli M, Rovira N, Toulemonde F, Vostrovská I (2020) Direct dating reveals the early history of opium poppy in western Europe. Nat Sci Rep 10:20263. https://doi.org/10.1038/s41598-020-76924-3

Salavert A, Martin L, Antolín F, Zazzo A (2018) The opium poppy in Europe: exploring its origin and dispersal during the Neolithic. Antiquity 92(364):e1. https://doi.org/10.15184/aqy.2018.154

Salavert A (2010) Le pavot (Papaver somniferum) à la fin du 6e millénaire av. J.-C. en Europe occidentale. Anthropobotanica 3(1):3–16

Skeates R (2017) Prehistoric figurines in Italy. In: Insoll T (ed) The Oxford handbook of prehistoric figurines. Oxford University Press, Oxford, pp 777–798

Speroni E (2004) Nota neurofarmacologica sui funghi. Bull Paletnologia Ital 95:19–20

Tichý R (2016) The earliest maritime voyaging in the mediterranean: view from Sea. Živá Archeologie REA 18:26–36

Tichý R (2001) Expedice Monoxylon. Pocházíme z mladší doby kamenné. Rekonstrukce a experiment v archeologii, Supplementum, vol 1. Společnost experimentální archeologie, Hradec Králové

Tiné V (2002) Le facies a ceramica impressa dell'Italia meridionale e della Sicilia. In: Fugazzola Delpino MA, Pessina A, Tiné V (eds) Le ceramiche impresse nel Neolitico antico. Italia e Mediterraneo. Studi di Paletnologia, vol I. Istituto Poligrafico e Zecca dello Stato, Roma, pp 131–165

Tozzi C, Zamagni N (eds) (2003) Gli scavi nel villagio Neolitico di Catignano (1971–1980). Istituto Italiano di Preistoria e Protostoria, Firenze

Tresset A, Vigne JD (2007) Substitution of species, techniques and symbols at the Mesolithic-Neolithic transition in Western Europe. Proc Br Acad 144:189–210

de Vareilles A, Bouby L, Jesus A, Martin L, Rottoli M, Linden MV, Antolín F (2020) One sea but many routes to Sail. The early maritime dispersal of Neolithic crops from the Aegean to the western Mediterranean. J Archaeol Sci Rep 29:102140. https://doi.org/10.1016/j.jasrep.2019.102140

Vigne JD, Cucchi T (2005) Premières navigations au Proche-Orient: les informations indirectes de Chypre. Paléorient 31(1):186–194. https://doi.org/10.3406/paleo.2005.4797

Weber G, Lukeneder A, Harzhauser M, Mitteroecker P, Wurm L, Hollaus LM, Kainz S, Haack F, Antl-Weiser W, Kern A (2022) The microstructure and the origin of the Venus from Willendorf. Nat Sci Rep 12:2926. https://doi.org/10.1038/s41598-022-06799-z

White R, Bisson M (1998) Imagerie féminine du Paléolithique. L'apport des nouvelles statuettes de Grimaldi. Gallia Préhistoire 40:95–132. https://doi.org/10.3406/galip.1998.2159

Zilhão J (2014) Early prehistoric navigation in the western Mediterranean: implications for the Neolithic transition in Iberia and the Maghreb. Eurasian Prehistory 11 (1–2):185–200

An Introduction to the Submerged Prehistoric Pile-Dwelling in Zambratija Bay on the Croatian Adriatic Coast

5

Katarina Jerbić

Abstract

Since its discovery in 2008, the submerged prehistoric pile-dwelling in Zambratija Bay has been a subject of handful discussions and papers. Although several small-scale surveys and test excavations were performed on the site immediately after the original discovery, targeted research has not been done there until 2017 as part of a small individual Ph.D. project, executed by the author of this paper. The aims of the paper are twofold: (1) to present preliminary data that was available prior to commencing the aforementioned Ph.D. and (2) to justify the fieldwork methods chosen for the Ph.D., which was successfully finished in 2020.

Keywords

Croatia · Adriatic coast · Pile-dwelling · Palaeolandscape

K. Jerbić (✉)
Institute of Archaeological Sciences,
University of Bern, Bern, Switzerland
e-mail: katarina.jerbic@unibe.ch

Flinders University, Adelaide, South Australia, Australia

5.1 A Modern Pile-Dwelling Fever

In 2008, a submerged prehistoric maritime pile-dwelling was discovered in Zambratija Bay in Croatia, at only around 100 kms distance from some of the most famous Alpine prehistoric pile-dwellings (Fig. 5.1). The site revealed an unusual opportunity for archaeologists in Croatia to experience what could be described as a 'modern-day' pile-dwelling fever (Benjamin et al. 2011; Koncani Uhač 2008, 2009; Koncani Uhač and Čuka 2015, 2018). This article represents an overview of the specific archaeological and environmental circumstances that surround the submerged site in Zambratija Bay. It is the first article derived from the most recent interdisciplinary research performed in 2017 on the site as part of an underwater archaeological Ph.D. at Flinders University in Adelaide, which took place between 2016 and 2020.

5.2 Site Discovery and Research up to 2016

Zambratija Bay is located on the northeastern Croatian Adriatic coast. The submerged pile-dwelling lies three metres underwater, and it was first recognised as a site in 2008 by underwater archaeologists from the Archaeological Museum of Istria in Pula (AMI) as part of an underwater inspection. Three features all found within a small area on the seabed were crucial for the

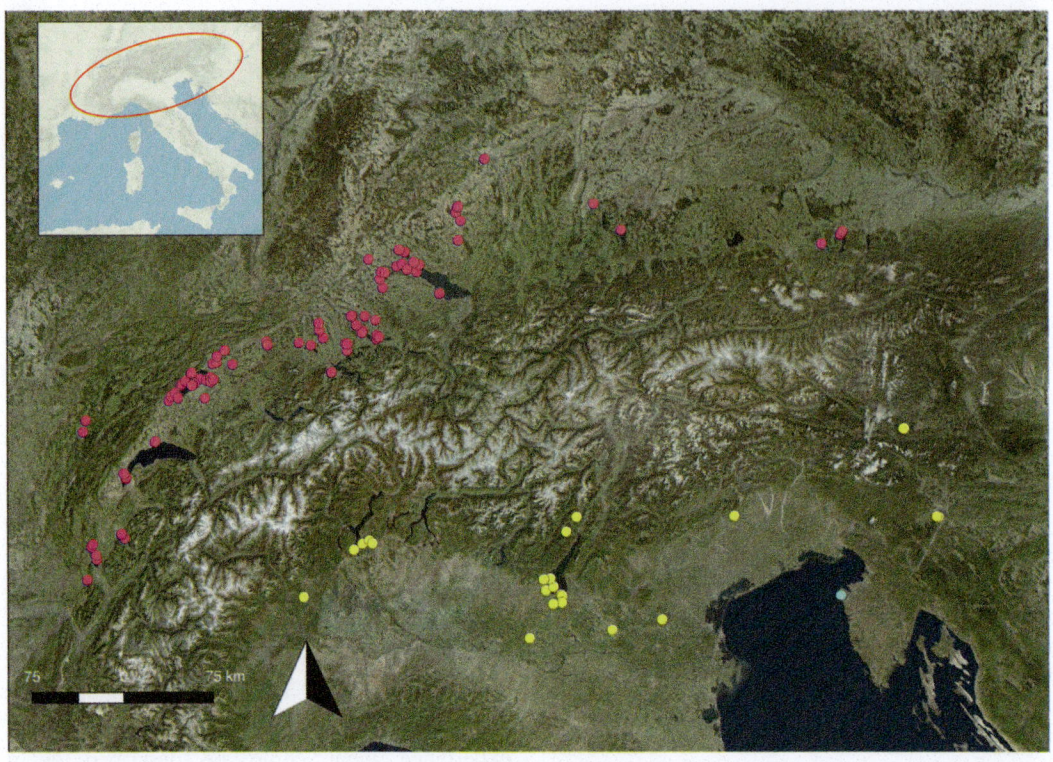

Fig. 5.1 Aqua: The location of Zambratija Bay. **Yellow**: Prehistoric pile-dwellings south of the Alps. **Purple**: Prehistoric pile-dwellings north of the Alps (Map edited by E. Aragon Nuñez based on UNESCO (2017–2018))

initial determination of the site as a potential prehistoric pile-dwelling: (1) wooden piles protruding out of the seabed, (2) an organic platform 30 by 67 m in size composed of peat and wooden material and (3) sherds of prehistoric pottery within the near vicinity of the first two features (Koncani Uhač 2008, 2009, 265). Together with two other underwater sites identified within the bay at the same time—the remains of a prehistoric laced boat (Boetto et al. 2015; Koncani Uhač 2009, 267; Koncani Uhač et al. 2017a; Koncani Uhač and Uhač 2012) and a submerged Roman Antiquity embankment/road (Koncani Uhač 2009, 264), the submerged prehistoric site was surveyed and excavated on several occasions between 2008 and 2017.

The initial archaeological campaign started immediately after discovery in 2008 as an investigation of the entire bay, resulting in the excavation of six trenches and a collection of pottery finds from the seabed (Koncani Uhač

2008, 2009). One of those trenches was placed within a location with a high concentration of wooden piles, revealing an undisturbed cultural context resembling a dwelling structure (Koncani Uhač and Čuka 2015), containing prehistoric pottery, superimposed by layers of peat and wood. The first total station survey was also conducted in 2008, documenting the excavated trenches, peat platform and 34 wooden piles protruding from the seabed on the north-western ridges of what was later identified as a submerged karstic sinkhole (Benjamin et al. 2011, 195; Koncani Uhač 2008, 397, 2009, 265). One wooden pile, later determined to be made of Oak (*Quercus L.*) (Koncani Uhač and Čuka 2015, 27), was fully excavated from the seabed and revealed traces of wood shaping technology, with a tapered base to allow for ease of placement into the natural ground. The pile was stored in the AMI laboratory until 2011, when a small sample taken from its surface was sent for radiocarbon

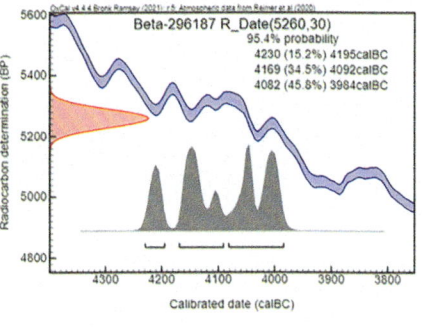

Fig. 5.2 Left: An updated calibration (Reimer et al. 2020) of the original radiocarbon date from Koncani Uhač and Čuka (2015). **Right**: The wooden pile that was extracted from the seabed. The sample for the radiocarbon analysis was taken from this pile (Original photos by AMI; image edited by E. Aragon Nuñez)

dating, revealing an age of 5260 ± 30 BP, calibrated to 4230–3984 cal BC (Fig. 5.2) (Beta-296187) (Koncani Uhač and Čuka 2015, 28; Reimer et al. 2020).[1] Further excavations and surveys followed in 2011, for the purposes of defining and marking more piles as well as determining the outer limitations of the settlement. A bathymetric survey was conducted in 2012 which resulted in recorded elevations of the seabed within the bay, providing the necessary data for the partial reconstruction of a submerged palaeolandscape where the aforementioned submerged karstic sinkhole was clearly identifiable (Fig. 5.3) (Koncani Uhač and Čuka 2015, 30). Later overlayed with the results from the total station survey, the bathymetric data revealed that the wooden piles and the peat platform were located around the outer elevated edges of the sinkhole (Koncani Uhač and Čuka 2015, 28). One more excavation campaign at the submerged settlement in Zambratija took place in 2014, when three trenches were excavated around the edges of the submerged sinkhole to further determine the site's archaeological and environmental stratigraphy, as well as to record additional piles (Koncani Uhač and Čuka 2015, 31).

Pottery sherds collected from around the site and in archaeological trenches from all archaeological campaigns mentioned above were typologically attributed to known prehistoric complexes in the Istrian Peninsula region, ranging from the Late Neolithic/Early Copper Age to the Bronze Age (Koncani Uhač and Čuka 2015). Although representing a large 3000-year timeframe, part of the contextualised pottery assemblage showed a resemblance to the so-called Nakovana-style, which in the Eastern Adriatic context represents the end of the Neolithic and the start of Early Copper Age (Forenbaher 1999–2000). On the Istrian Peninsula, Nakovana-style pottery is found in contextualised layers covering a radiocarbon age range between 4252–4048 and 3959–3797 cal BC (Forenbaher et al. 2013, 592). Together with the 4230–3984 cal BC radiocarbon date derived from the wooden pile, the finding of Nakovana-style pottery helped roughly set the occupation of the submerged settlement in Zambratija Bay to the Early Copper Age, potentially making it one of the earliest examples of the prehistoric pile-dwelling lifestyle within the broader Alpine-Adriatic region with closest comparable contemporary dates found at the Palù di Livenza and Hočevarica pile-dwellings in Italy and Slovenia (Corti et al. 1997; Čufar and Martinelli 2004; Čufar and Velušček 2004; Čufar et al. 2015; Tasca et al. 2019).

The archaeological and environmental findings in Zambratija Bay represented an unexpected discovery, and therefore, the initial

---

[1]This date was first published in Koncani and Uhač (2015) as 5260 ± 30 BP calibrated to 4230–3980 cal BC. The author had run the radiocarbon date through the most recent OxCal 4.4 IntCal20 calibration curve (https://c14.arch.ox.ac.uk/oxcal/OxCal.html) for the Northern Hemisphere.

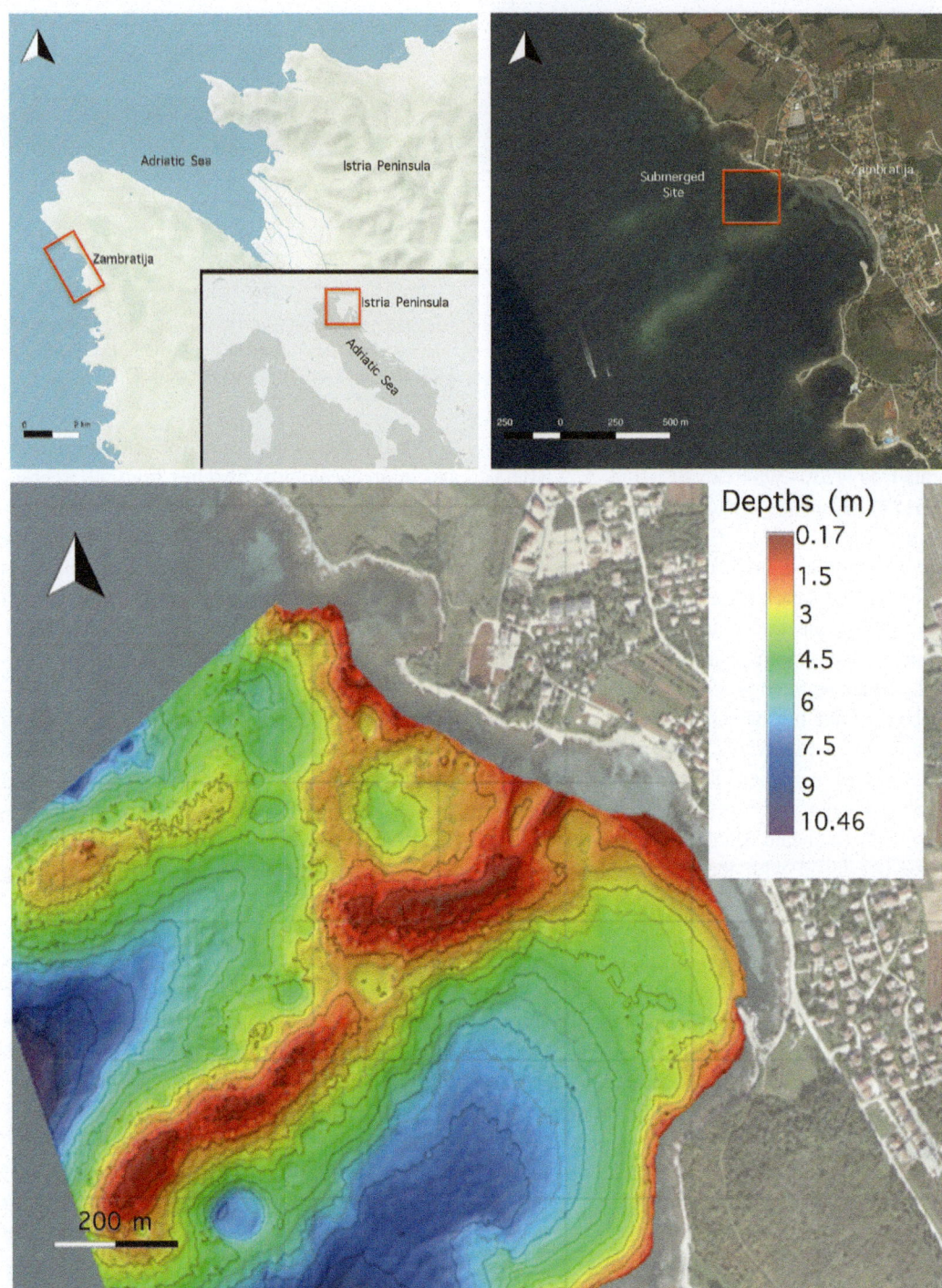

Fig. 5.3 Location of Zambratija Bay on the Istrian Peninsula and the bathymetric map of the bay (Image edited by E. Aragon Nuñez)

underwater research did not include a theoretical and methodological plan designed for the investigation of prehistoric pile-dwellings (Benjamin et al. 2011; Boetto et al. 2015; Koncani Uhač 2008, 2009; Koncani Uhač et al. 2017a; Koncani Uhač and Čuka 2015; Koncani Uhač and Uhač 2012). The duration of each those initial investigations did not exceed more than two weeks and consequently some of the larger questions, such as the spatial extents of the settlement, occupational timeline(s) and socio-economic developments of the settlement, as well as the environmental history and climatic

occurrences which triggered all the site formation processes leading to the conditions we see on the site in present time, therefore remained unknown.

5.3 Significance and Potential of the Submerged Site in Zambratija Bay

A Ph.D. research fieldwork was conducted in Zambratija Bay in May and June 2017. It was an interdisciplinary project divided into two fieldwork campaigns (seabed coring and waterlogged

Fig. 5.4 Up: One of the wooden pile samples from Zambratija Bay collected in 2017. **Down**: A small section of a seabed core from Zambratija Bay with clearly visible organic remains (wood and charcoal) as well as ash. This section of the core was positioned at 107–120 cm under the seabed, under a thick layer of marine sediments (Photos by K. Jerbić)

wood sample collection) which included an international team of maritime archaeologists and environmental scientists working together to address the site's archaeological and palaeoenvironmental significance and archaeological potential (Fig. 5.4). The significance and potential of the site can be divided into three main categories: (1) as a cultural indicator of past sea-level and environmental change; (2) as a potential 'Adriatic' version of the European prehistoric pile-dwelling phenomenon; and (3) as possibly being one of the archaeological sites to contribute to the connection of the Southern Alpine dendrochronological Oak sequence. A discussion about these three categories is presented below.

5.3.1 Zambratija Bay as an Archaeological Proxy for Past Sea Levels and Palaeoenvironment

Zambratija is an archaeological site with stratified evidence of human activities in a landscape that was once terrestrial but is now submerged three metres under water. The proximity of the site to the current shoreline provides an indication of former sea levels and a means to assess the local environmental history. After the most recent Ice Age around 20 million km^2 of the Earth's territory was inundated by the rising sea (Harff et al. 2016b, (1), leaving landscapes and natural habitats abandoned by their terrestrial flora, fauna and human populations. These submerged landscapes are now being investigated by collaborative research where existing records are updated with new available proxy data. Archaeologists worldwide are drawn to these investigations not only to find submerged archaeological sites, but also to reconstruct the ecosystems which the human populations were inhabiting and exploiting, as well as to find possible land bridges and passages used to occupy new territories (Flemming et al. 2003; Gearey et al. 2017; Harff et al. 2016a; Lewis Johnson and Stright 1992; Masters and Flemming 1983). According to the currently available data, the settlement in Zambratija Bay was in use around 6000 years ago. This

timeframe chronologically connects the site to interdisciplinary debates on the aftermath of the Pleistocene/Holocene transition visible in scientific records as climate and sea-level stabilisation (Felja et al. 2015; Smith et al. 2011; Stocchi et al. 2005; Šegota and Filipčić 1991), an event that partially influenced intense periods of prehistoric population migrations and the spread of farming and the Neolithisation of the European continent (Forenbaher and Miracle 2005; Forenbaher et al. 2013; McClure et al. 2014; Turney and Brown 2007). Due to its apparent preservation and the presence of waterlogged organic material, it is also highly significant as a very reliable archaeological indicator of past sea-level change and direct evidence of human resilience to extreme environmental circumstances.

5.3.2 Zambratija Bay as a Potential Adriatic Version of the European Prehistoric Pile-Dwelling Phenomenon

A typological analysis of the Zambratija pottery was culturally ascribed to local cultural groups and styles from the Late Neolithic to the Early Bronze Age (Koncani Uhač and Čuka 2015). This relative chronology was further supported with one radiocarbon date from a wooden pile confirming that the site was contemporary with the Eastern Adriatic Late Neolithic cultural complexes (Forenbaher et al. 2013, 592). The presence of wooden piles driven into the organic, submerged freshwater sediments however, indicated that Zambratija was a prehistoric pile-dwelling similar to those around the Alpine lakes, an architectural innovation which started in the Early Neolithic and lasted for around 3500 years (Menotti 2015a). The nearest Alpine pile-dwellings to the Eastern Adriatic are situated in Austria (Ruttkay et al. 2004), Italy (Marzatico 2004) and Slovenia (Velušček 2004) with the earliest known settlements there aligning with the one preliminary date from Zambratija.

In recent times, the 2011 UNESCO Heritage listing revoked the interest in prehistoric pile-dwellings across European archaeological circles. This interest is focused on contextualising the past and present with multidisciplinary research in Eastern Europe where the number of known pile-dwelling sites is regularly being updated with new, significant discoveries changing our understanding of the prehistoric pile-dwelling phenomenon (Hafner et al. 2021; Karkanas et al. 2011; Maczkowski et al. 2021; Mazurkevich and Dolbunova 2011; Mazurkevich et al. 2011; Reich et al. 2021). Croatia is no exception to this trend, where a few more prehistoric sites other than the one in Zambratija Bay show architectural, environmental and archaeological indicators of being prehistoric pile-dwelling sites. Three of the sites —Janice (Bekić et al. 2015), Šimuni (Bekić 2017) and Ričul (Čelhar et al. 2017)—are recent discoveries, whilst all others were late-nineteenth to mid-20th century findings that have resurfaced to the Croatian prehistoric pile-dwelling discussion as a result of desktop assessments for the author's Ph.D. project (Brunšmid 1900; Ljubić 1885, 1887; Majnarić-Pandžić 1993; Marović 2002; Milošević 1992, 1999). Together with those sites, and possibly even more, Zambratija Bay represents a knowledge gap between the Alps and the Balkans that could potentially connect the currently known geographical prehistoric pile-dwelling clusters and answer questions regarding the expansion of the prehistoric pile-dwelling lifestyle within Europe.

5.4 Remarks on Further Developments

As mentioned, the 2017 fieldwork campaign was composed of two components—seabed coring and waterlogged wood sample collection. Coring is a common minimal impact method used by researchers to investigate submerged palaeolandscapes, sea-level studies and wetland studies (Faught 2004, 278; Gifford 1983; Lambeck et al. 2004, 1569), which represented an ideal method for the project due to budget, travel and timeframe limitations. Since there were more than one hundred recorded wooden piles at the time, a wooden sample collection was also chosen as an assessment method to review the dendrochronological potential of the site. The results of both of these methods are currently underway to being published together with all the external collaborators that participated in the Ph.D. project.

It is evident that Zambratija represents a depository of archaeological and environmental data that can add value to a variety of archaeological and interdisciplinary fields. The main fields mentioned in this article include submerged prehistory and palaeolandscapes, Wetland Archaeology and Alpine pile-dwelling research and dendrochronology. There are however many other aspects of and cultural implications to Zambratija Bay, such as being a resource for human adaptations to climate change and adding archaeology to current climate change debates by using the Climate Change Archaeology model (Van de Noort 2011, 2013), and even serving as cultural evidence and background for applications in local tourism and sustainable development (Iveša et al. 2017; Koncani Uhač et al. 2017b).

References

Bekić L (2017) Šimuni, a new Bronze Age underwater site in Zadar County. Diadora 31:41–50

Bekić L, Pešić M, Scholz R, Meštrov M (2015) Underwater archaeological research at the Prehistoric site of Pakoštane-Janice. Diadora 29:7–22

Benjamin J, Bekić L, Komšo D, Koncani Uhač I, Bonsall C (2011) Investigating the submerged prehistory of the eastern adriatic: progress and prospects. In: Benjamin J, Bonsall C, Pickard C, Fischer A (eds) Submerged prehistory. Oxbow Books, Oxford, pp 193–206

Boetto G, Koncani Uhač I, Uhač M (2015) Sewn Ships from Istria (Croatia): the Shipwrecks of Zambratija and Pula. In: 14th international symposium on boat and ship archaeology: Baltic and beyond. change and continuity in shipbuilding. National Maritime Museum, Gdansk

Brunšmid J (1900) Naselbina bronsanoga doba kod Novoga grada na Savi, Kotar Brod na Savi. Vjesn Hrvatskoga Arkeološkoga Društva 4:44–45

Čelhar M, Parica M, Ilkić M, Vujević D (2017) A bronze age underwater site near the islet of Ričul in northern Dalmatia (Croatia). Skyllis 17(1):21–34

Corti P, Martinelli N, Rottoli M, Tinazzi O, Vitri S (1997) New data on the wooden structures from the pile-dwelling of Palù di Livenza. Preistoria Alpina 33: 73–80

Čufar K, Tegel W, Merela M, Kromer B, Velušček A (2015) Eneolithic pile dwellings south of the Alps precisely dated with tree-ring chronologies from the north. Dendrochronologia 35:91–98

Čufar K, Martinelli N (2004) Teleconnection of chronologies from Hočevarica and Palù di Livenza, Italy. In: Velušček A (ed) Hočevarica. An Eneolithic Pile-Dwelling in the Ljubljansko Barje. Institute of Archaeology at ZRC SAZU in association with ZRC Publishing, Ljubljana, pp 286–289

Čufar K, Velušček A (2004) Dendrochronological research of the Hočevarica pile dwelling settlement. In: Velušček A (ed) Hočevarica. An eneolithic pile-dwelling in the Ljubljansko Barje. Institute of Archaeology at ZRC SAZU in association with ZRC Publishing, Ljubljana, pp 274–280

Faught MK (2004) The underwater archaeology of Paleolandscapes, Apalachee Bay, Florida. Am Antiq 69(2):275–289

Felja I, Fontana A, Furlani S, Bajraktarević Z, Paradžik A, Topalović E, Rossato S, Ćosović V, Juračić M (2015) Environmental changes in the lower Mirna River valley (Istria, Croatia) during the Middle and Late Holocene. Geologia Croat 68(3):209–224

Flemming NC, Bailey GN, Courtillot V, King G, Lambeck K, Ryerson F, Vita-Finzi C (2003) Coastal and marine palaeo-environments and human dispersal points across the Africa-Eurasia boundary. In: Brebbia CA, Gambin T (eds) Maritime Heritage. Wessex Institute of Technology & The University of Malta, pp 61–73

Forenbaher S, Miracle PT (2005) The spread of farming in the Eastern Adriatic. Antiquity 79:514–528

Forenbaher S, Kaiser T, Miracle PT (2013) Dating the east adriatic neolithic. Eur J Archaeol 16(4):589–609

Forenbaher S (1999–2000) "Nakovana culture"-state of research. Opuscula Archaeologica 23–24:373–385

Gearey BR, Hopla E-J, Boomer I, Smith D, Marshall P, Fitch S, Griffiths S, Tappin DR (2017) Multi-proxy palaeoecological approaches to submerged landscapes: a case study from 'Doggerland', in the southern North Sea. In: Williams M, Hill T, Boomer I, Wilkinson IP (eds) The archaeological and forensic applications of microfossils: a deeper understanding of human history. The Micropalaeontological Society, Special Publications. Geological Society, London, pp 35–53

Gifford J (1983) Core sampling of a holocene marine sedimantary sequence and underlying neolithic cultural material off Franchthi Cave, Greece. In: Masters PM, Flemming NC (eds) Quaternary coastlines and marine archaeology: towards the prehistory of land bridges and continental shelves. Academic Press, London, pp 269–281

Hafner A, Reich J, Ballmer A, Bolliger M, Antolín F, Charles M, Emmenegger L, Fandré J, Francuz J, Gobet E, Hostettler M, Lotter AF, Maczkowski A, Morales-Molino C, Naumov G, Stäheli C, Szidat S, Taneski B, Todorovska V, Bogaad A, Kotsakis K, Tinner W (2021) First absolute chronologies of neolithic and bronze age settlements at lake Ohrid based on dendrochronology and radiocarbon dating. J Archaeol Sci Rep 38:1–17

Harff J, Bailey GN, Lüth F (eds) (2016a) Geology and archaeology: submerged landscapes of the continental shelf. Geological Society Special Publications, Geological Society, London

Harff J, Bailey GN, Lüth F (2016b) Geology and archaeology: submerged landscapes of the continental shelf: an introduction. In: Harff J, Bailey GN, Lüth F (eds) Geology and archaeology: submerged landscapes of the continental shelf. Geological Society, London, pp 1–8

Iveša N, Knežević I, Koncani Uhač I, Žužić A (2017) Zona Posebnog Upravljanja U Uvali Zambratija (Umag). Lokalna akcijska grupa u ribarstvu "Pinna nobilis", Novigrad

Karkanas P, Pavlopoulos K, Kouli K, Ntinou M, Tsartsidou G, Facorellis Y, Tsouru T (2011) Palaeoenvironments and site formation processes at the neolithic Lakeside settlement in Dispilo, Kastoria Northern Greece. Geoarchaeology Int J 26(1):83–117

Koncani Uhač I (2009) Podvodna arheološka istraživanja u uvali Zambratija. Histria Antiqua 17:263–268

Koncani Uhač I, Čuka M (2015) A contribution to a better understanding of the underwater Eneolithic site at the Zambratija cove. Histria Archaeologica 46:25–73

Koncani Uhač I, Iveša N, Mioković D, Žužić A (2017b) Opportunities and challeges of tourist valorization of Zambratija cove (Umag). Stud Univ Hereditati Znanstvena Revija Za Raziskave Teorijo Kulturne Dediščine 5(1):83–95

Koncani Uhač I, Čuka M (2018) Sito preistorico sommerso nelle acque della baia di Zambrattia (Umago, Croazia). In: Borgna E, Càssola Guida P, Corazza S (eds) Preistoria e Protostoria del Caput Adriae. Studi di Preistoria e Protostoria, vol 5. Istituto Italiano di Preistoria e Protostoria, Firenze, pp 491–506

Koncani Uhač I, Uhač M (2012) Prapovijesni brod iz uvale Zambratija—prva kampanja istraživanja Histria Antiqua 21:533–538

Koncani Uhač I, Boetto G, Uhač M (2017a) Zambratija. Prehistoric sewn boat. Archaeological Museum of Istria, Pula

Koncani Uhač I (2008) Zambratija—uvala, Croatian Archaeological Yearbook 5, Ministry of Culture, Directorate for Archives and Archaeological Heritage, Zagreb, pp 396–398

Lambeck K, Antonioli F, Purcell A, Silenzi S (2004) Sea-level change along the Italian coast for the past 10,000 yr. Quatern Sci Rev 23:1567–1598

Lewis Johnson L, Stright M (eds) (1992) Palaeoshorelines and prehistory: an investigation of method. CRC Press, Boca Raton

Ljubić Š (1885) Terramara u Hrvatskoj. Viestnik Hrvatskog Arkeologičkoga Društva 2:97

Ljubić Š (1887) Terramara u Hrvatskoj. Viestnik Hrvatskog Arkeologičkoga Društva 9:97

Maczkowski A, Bolliger M, Ballmer A, Gori M, Lera P, Oberweiler C, Szidat S, Touchais G, Hafner A (2021) The early bronze age dendrochronology of Sovjan (Albania): a first tree-ring sequence of the 24th–22nd c. BC for the Southwestern Balkans. Dendrochronologia 66:1–12

Majnarić-Pandžić N (1993) Srednje brončano doba u Slavoniji. Arheološka istraživanja u Slavonskom Brodu i Brodskom Posavlju. Izdanja Hrvatskog arheološkog društva. Zagreb

Marović I (2002) Sojeničko naselje na Dugišu kod Otoka (Sinj). Vjesnik za arheologiju i historiju dalmatinsku 94(1):217–296

Marzatico F (2004) 150 Years of lake-dwelling research in Northern Italy. In: Menotti F (ed) Living on the Lake in Prehistoric Europe. 150 Years of Lake-Dwelling Research. Routledge, London, pp 83–97

Masters PM, Flemming NC (eds) (1983) Quaternary coastlines and marine archaeology: towards the prehistory of land bridges and continental Shelves. Academic Press, London

Mazurkevich A, Dolbunova E (2011) Underwater Investigations in Northwest Russia: lacustrine archaeology of Neolithic pile dwellings. In: Benjamin J, Bonsall C, Pickard C, Fischer A (eds) Submerged prehistory. Oxbow Books, Oxford, pp 158–172

Mazurkevich A, Dolbunova E, Maigrot Y, Hookk D (2011) The results of underwater excavations at Serteya II, and research into pile-dwellings in Northwest Russia. Archaeologia Baltica 14:47–64

McClure SB, Podrug E, Moore AM, Culleton BJ, Kennett DJ (2014) AMS ^{14}C chronology and ceramic sequences of early farmers in the Eastern Adriat. Radiocarbon 56(3):1019–1038

Menotti F (2015) The 3500-year-long lake-dwelling tradition comes to an end: what is to blame? In: Menotti F (ed) The end of the lake-dwellings in the circum-alpine region. Oxbow Books, Oxford, pp 287–302

Menotti F (2004) The lake-dwelling phenomenon and Wetland archaeology. In: Menotti F (ed) Living on the Lake in Prehistoric Europe. 150 Years of Lake-Dwelling Research. Routledge, London, pp 1–6

Milošević A (1992) Arheološki nalazi u koritu rijeke Cetine u Sinjskom polju. Obavijesti Hrvatskog Arheološkog Društva XXIV/92(2):45–48

Milošević A (1999) archäologische probeuntersuchungen im flussbett der Cetina (Kroatien) Zwischen 1990 und 1994. Archäologisches Korrespondenzblatt 29(2):203–210

Reich J, Steiner P, Ballmer A, Emmenegger L, Hostettler M, Stäheli C, Naumov G, Taneski B, Todorovska V, Schindler K, Hafner A (2021) A novel structure from motion-based approach to underwater pile field documentation. J Archaeol Sci Rep 39:1–14

Reimer P, Austin W, Bard E, Bayliss A, Blackwell P, Bronk Ramsey C, Butzin M, Cheng H, Edwards R, Friedrich M, Grootes P, Guilderson T, Hajdas I, Heaton T, Hogg A, Hughen K, Kromer B, Manning S, Muscheler R, Palmer J, Pearson C, van der Plicht J, Reimer R, Richards D, Scott E, Southon J, Turney C, Wacker L, Adolphi F, Büntgen U, Capano M, Fahrni S, Fogtmann-Schulz A, Friedrich R, Köhler P, Kudsk S, Miyake F, Olsen J, Reinig F, Sakamoto M, Sookdeo A, Talamo S (2020) The IntCal20 Northern Hemisphere radiocarbon age calibration curve (0–55 cal kBP). Radiocarbon 62:1–33

Ruttkay E, Cichocki O, Pernicka E, Pucher E (2004) Prehistoric Lacustrine Villages on the Austrian Lakes. Past and present research developments. In: Menotti F (ed) Living on the Lake in Prehistoric Europe. 150 years of lake-dwelling research. Routledge, London, pp 50–68

Šegota T, Filipčić A (1991) Arheološki i geološki pokazatelji holocenskog položaja razine mora na istočnoj obali Jadranskog mora. Rad Hrvatske Akademije Znanosti i Umjetnosti 458:148–172

Smith DE, Harrison S, Firth CR, Jordan JT (2011) The early Holocene sea level rise. Quatern Sci Rev 30:1846–1860

Stocchi P, Spada G, Cianetti S (2005) Isostatic rebound following the Alpine deglaciation: impact on the sea level variations and vertical movements in the Mediterranean region. Geophys J Int 162:137–147

Tasca G, Bassetti M, Degasperi N, Salvador S, Miheli R (2019) Piastra di cottura dal sito palafitticolo neolitico di Palù di Livenza. IpoTESI Di Preistoria 12:17–26

Turney CSM, Brown H (2007) Catastrophic early Holocene sea level rise, human migration and the Neolithic transition in Europe. Quatern Sci Rev 26:2036–2041

UNESCO (2017–2018) Prehistoric pile dwellings around the alps. https://www.palafittes.org

Van de Noort R (2011) Conceptualising climate change archaeology. Antiquity 85:1039–1048

Van de Noort R (2013) Climate change archaeology: building resilience from research in the World's coastal Wetlands. Oxford University Press, Oxford

Velušček A (2004) Past and present lake-dwelling studies in Slovenia. Ljubljansko Barje (The Ljubljana Marsh). In: Menotti F (ed) Living on the lake in Prehistoric Europe. 150 Years of Lake-Dwelling Research. Routledge, London, pp 69–82

Neolithic and Bronze Age Pile-Dwellings at Lake Ohrid: Underwater Excavations at Ploča Mičov Grad (North Macedonia)

6

Johannes Reich, Marco Hostettler,
Ariane Ballmer, and Albert Hafner

Abstract

One of the regions of Europe where archaeological lakeshore sites or so-called pile-dwellings are found is the tripoint of Albania, Greece and North Macedonia, which includes dozens of known lakeshore sites on and in different lakes. One of the aims of the ERC Synergy Project EXPLO is the systematic underwater archaeological investigation of key pile-dwelling sites in this region. Currently, more than a dozen prehistoric lake- or river shore sites from the Neolithic and the Bronze and Iron Ages are known in the area surrounding the lakes of Ohrid, Prespa and Maliq (Fig. 5.1) (Naumov in Plattform 23:10–20, 2016; Andoni et al. in New Archaeological Discoveries in the Albanian Regions.

Procedings of the International Conference, Tirana 30–31 January 2017. Botimet Albanologjike, Tirana, pp 123–140, 2017; Oberweiler et al. in Bulletin Archéologique Des Écoles Françaises à L'étranger 2020; Lera et al. in Bulletin archéologique des Écoles françaises à l'étranger 2020). Several of these sites still show waterlogged archaeological layers, in which organic matter and wooden construction elements are well preserved. However, only a few are systematically investigated and the number of absolute dates for most of the sites is still rather low. This paper summarises the recent archaeological fieldwork conducted by the University of Bern in partnership with the Museum of Ohrid and the Center for Prehistoric Research in Skopje on the site Ploča Mičov Grad in North Macedonia based on the publications of Hafner et al. (J Archaeol Sci Rep 38, 2021) and Reich et al. (J Archaeol Sci Rep 39, 2021) as well as new preliminary results of the ongoing research. We focus on ongoing research that has provided the first absolute chronologies of Neolithic and Bronze Age settlements at Lake Ohrid (Hafner et al. J Archaeol Sci Rep 38, 2021). The main methodology relies on both dendrochronological and radiocarbon datings. This high precision dating, in combination with studies of the material culture, contributes to a better chronological understanding of the prehistory of the surrounding basins and the southwestern Balkans in general.

J. Reich (✉) · M. Hostettler · A. Ballmer · A. Hafner
Institute of Archaeological Sciences and Oeschger Centre for Climate Change Research (OCCR), University of Bern, Bern, Switzerland
e-mail: johannes.reich@unibe.ch

M. Hostettler
e-mail: marco.hostettler@unibe.ch

A. Hafner
e-mail: albert.hafner@unibe.ch

A. Ballmer
Independent Researcher, Bern, Switzerland
e-mail: mail@arianeballmer.com

A. Ballmer et al. (eds.), *Prehistoric Wetland Sites of Southern Europe*,
Natural Science in Archaeology, https://doi.org/10.1007/978-3-031-52780-7_6

Keywords

Lake Ohrid · Underwater archaeology ·
Lakeshore settlements · Pile-dwellings ·
Neolithic and Bronze Age Balkans · EXPLO

6.1 Natural Setting

The research area covers a very old, mountainous lake area with three partially interlinked large lakes: Ohrid, Great Prespa and Small Prespa (Fig. 6.1). Lake Ohrid has a surface area of 358 square kilometres and a maximal depth of 289 m. The Great Prespa lake has a surface area of 259 square kilometres and a maximal depth of 54 m (GIZ 2015). Deep lake sediment drillings

in Lake Ohrid recently revealed that the lake was established around 1.36 million years ago (Wagner et al. 2019; Wilke et al. 2020). South of Lake Ohrid the elevated plain of Korçë extends, where former Lake Maliq was located until its drainage in the middle of the 20th century (Fouache et al. 2010).

These lake basins are surrounded by several North–South-oriented mountain ranges with peaks reaching altitudes of up to 2300 m. The mountain ranges of Galičica and Mali i Thatë separate the two basins of Lake Ohrid and Lake Prespa. Great Prespa has no river outlet but is connected to Lake Ohrid by an underground karst aquifer system. The water flows from Lake Prespa, c. 150 m higher up, through the karst into Lake Ohrid, where it emerges

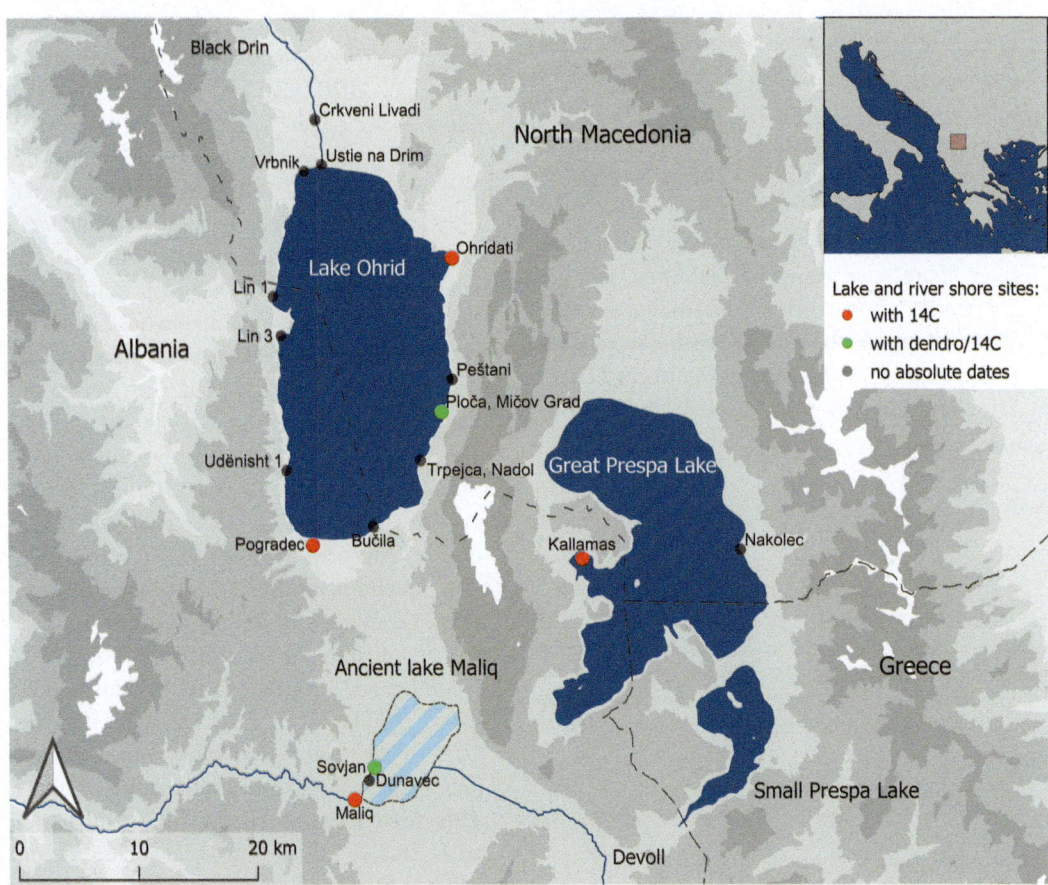

Fig. 6.1 Prehistoric lake and rivershore settlements around the tripoint of Albania, Greece and North Macedonia from the Neolithic and the Bronze Age in the three lacustrine sub-basins of Maliq, Prespa and Ohrid (J. Reich, EXPLO/UBern)

again in several spring areas (Matzinger et al. 2006; Amataj et al. 2007; Vogel et al. 2010; Hauffe et al. 2011; Jordanoska et al. 2013). Lake Ohrid feeds the Black Drin river which flows northwards and discharges into the Adriatic Sea. The Korçë basin, with former Lake Maliq, is not connected to this karst aquifer system and is drained by the river Devoll, discharging to the Adriatic Sea as well (Fouache et al. 2010).

6.2 Ploča Mičov Grad: Description and Research History

To date, the locations of eleven prehistoric sites are known from the shores of Lake Ohrid. Their chronological range spans from the Middle Neolithic until the Late Bronze Age and Early Iron Age. Apart from single examples where radiocarbon dating had been performed earlier, the chronology of the sites and the surrounding region relied entirely on typochronological analyses until recently.

The archaeological site of Ploča Mičov Grad, also known as the Bay of Bones, is situated in a bay south of the Gradište peninsula on the eastern shore of Lake Ohrid. Today the site lies in three to five metres deep water (Fig. 6.2). Towards the lake, the lakebed outside the bay drops rather steeply. Landwards in the shallow water, the lakebed is formed by bedrock. The first prehistoric ceramic finds from the area around the peninsula of Gradište were already made in the 1970s by local fishermen (Kuzman 2013). After the site's discovery by professional diver Milutin 'Mičo' Sekuloski in 1997, the first underwater archaeological investigations led by archaeologist Pasko Kuzman were started. In several campaigns up to 2005, more than 6000 piles were documented, and the pile field extension could be estimated at up to 8000 m^2. Details on the methodological approach and its results have remained mostly unpublished until now. Preliminary reports mention the presence of an archaeological layer of 150 cm thickness underneath the covering layer of the present lake bottom (Kuzman 2013).

Fig. 6.2 Ploča Mičov Grad. The bay of Ploča Mičov Grad south of the Gradište peninsula at the foot of the Galičica mountains on the east shore of Lake Ohrid. The open-air reconstruction of a prehistoric lakeshore settlement marks part of the estimated site extension (G. Milevski)

The large number of finds and piles underwater were interpreted to originate from a wooden platform on which houses of a single habitation phase were built (Kuzman 2013; Naumov 2016). Based on typochronological comparisons, this phase was dated to the Late Bronze and Early Iron Ages. This interpretation was directly applied to a life-size reconstruction of a prehistoric village on site. The platform carrying houses was built as an open-air museum ('Museum on Water') which currently is one of the most visited places by tourists around Lake Ohrid.

6.3   Ploča Mičov Grad: Current Investigation

Research on the site resumed in 2018. After a pilot study initiated by the Institute of Archaeological Sciences of the University of Bern, in partnership with the Center for Prehistoric Research in Skopje and the Museum of Ohrid, the investigations were continued in 2019 by the same institutions but in the framework of the ERC Synergy Project EXPLO. One major aim of the field campaigns, taking place over several weeks in two summers, was to obtain the first dendrochronological data from the thousands of wooden piles. In combination with an extensive application of radiocarbon dating the first reliable absolute chronological data of a prehistoric lakeshore settlement site at Lake Ohrid could be achieved.

During the first campaign, an area of 40 m^2 in the centre of the known extension of the pile field was documented and sampled (Fig. 6.3, Field 1a). In water depths of up to five metres, the lakebed was cleaned of macrophytes and a 10–20 cm thick sandy surface layer with stones was removed. The surface layer contained many ceramic fragments, stone tools and animal bones. The piles were mostly covered by this surface layer and had flatly eroded pile heads. In this field campaign, a total of 265 vertical piles were documented and a slice of 10 cm thickness was sawn from each pile as a sample for dendrochronological analysis.

In the second year, the pile field documentation and the dendrochronological sampling were continued resulting in a total area of 10 m × 9 m (Fig. 6.3, Field 1). Additionally, an area of 2 m 3 m was excavated into a depth of 50 cm into the sediments (Fig. 6.3, Field 2). The excavated area was laid out adjoining the NW corner of Field 1. The second field campaign added 516 documented and mostly sampled vertical piles. During these two field campaigns, a total number of 781 vertical piles in an area of 96 m^2 were documented resulting in a density of 8.1 piles/m^2. The dendrochronological analysis of the wooden samples from 2018 is completed (Hafner et al. 2021), whereas the samples of 2019 are currently under investigation (see Chap. 14 in this book).

Additionally, in 2019 the safely accessible part of the site was drill cored in two transects with an UWITEC drilling platform by a team of the Institute of Plant Sciences of the University of Bern, also one of the institutions carrying the ERC Synergy Project EXPLO (Figs. 6.4 and 6.5).

6.4   New Underwater Documentation Approach

For the further analysis of a pile field based on the dendrochronological information, it is crucial to ensure the correlation between the in situ location of the pile and its sample during and after fieldwork. Only when each pile is accurately mapped, can the full potential of the high-resolution absolute chronological data be exploited and leads to insights into the spatial and temporal relationships within a site.

To meet this criterion, a workflow had to be developed that guarantees high accuracy and flexibility and that is at the same time fast and robust. For the scope of such short field campaigns of only a few weeks, the common methods used to document pile fields were either too slow, too expensive or logistically too complex. In consequence, a novel methodology and workflow for the underwater documentation of pile fields in archaeological lakeside settlement sites were developed, adapting approaches using Structure from Motion (SfM) from other fields

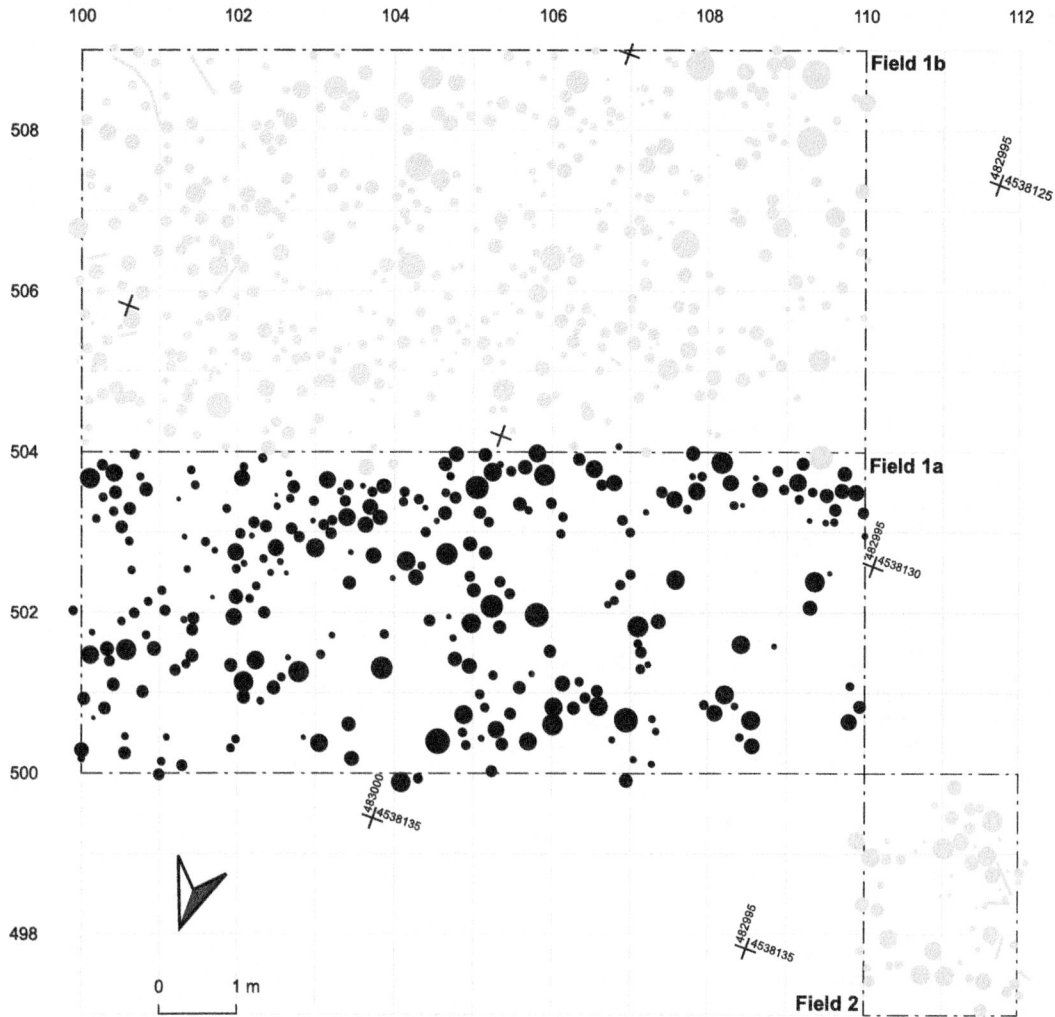

Fig. 6.3 Ploča Mičov Grad. Vectorised plan of the recovered wood samples of the excavation campaigns 2018 (black) and 2019 (grey). The circles mark the location of each vertical pile, indicating its surface area as diameter (CRS: local and EPSG:32634) (J. Reich, EXPLO/UBern)

concerned with underwater documentation such as maritime archaeology (Fig. 6.6) (cf. McCarthy et al. 2019). As the reproducibility, effectiveness and high suitability of this approach have been proven in 2018 (Reich et al. 2019), it was further used in the field campaign of 2019 (Reich et al. 2021).

For the documentation of the pile field, the investigated area is structured in a local excavation grid. In the case of Ploča Mičov Grad, a grid of 10 m by 10 m was set up (cf. Figure 6.3, Field 1). The corner points were marked with wooden

measuring posts driven c. 1 m deep into the lake bottom. For georeferencing, the corner points were measured with an RTK-GNSS receiver mounted on a buoy, which was positioned over the measuring posts with a rope by a diver. An alternative measurement concept was tested in 2019 in collaboration with the Institute of Geodesy and Photogrammetry of the Swiss Federal Institute of Technology (ETH) in Zurich (Fandré 2020).

After the grid was set up, the plants, stones and sand were removed on a strip of 10 m by

Fig. 6.4 Ploča Mičov Grad. Drill coring with the UWITEC platform and simultaneous pile field documentation and sampling under water (M. Hostettler, EXPLO/UBern)

1 m until the surface of the first archaeological layer was reached (Fig. 6.7). After exposure, all piles were tagged with a number. For the photogrammetrical documentation, ground control points were set according to the excavation grid using coded targets. To get the three-dimensional coordinates of each ground control point, only the water depth above each point had to be measured, as the x- and y-coordinates were given by the excavation grid. In the subsequent photographic recording, the strip was continuously photographed from two sides 1–1.5 m above the lake bottom. Special attention was paid to an overlap of at least 60% between the successive photos. In this way, between 60 and 110 photos were taken for a strip of 10 m^2 (Fig. 6.8). From those photos, the strip could be reconstructed three-dimensionally and a georeferenced orthophoto could be generated (Fig. 6.9). The

orthophotos of several strips were combined in a GIS and the location of each pile was vectorised (cf. Figure 6.3).

After an evaluation of the recorded photos and their suitability for a three-dimensional reconstruction using SfM software, each pile was sampled with a handsaw for the dendrochronological analysis. This procedure was repeated for each strip of 10 m^2.

During the two archaeological diving campaigns at Ploča Mičov Grad, a total of 90 m^2 (Field 1) were documented following the described workflow. In the additional 6 square metres of Field 2, the surface of each artificial spit going into the archaeological layer was as well documented by means of SfM.

6.5 Archaeological Layer

The excavation in Field 2 was opened on an area of about 5 m^2. The excavation penetrated approximately 40 cm into the layer, without reaching the lower bottom. It was not possible to distinguish different sub-layers within the organic layer. Part of the small trench was excavated in four artificial spits, while less than half was only excavated in two artificial spits. Nevertheless, a total amount of 58.1 kg pottery, 3.1 kg stone tools and 1.8 kg bone and antler were recovered from the trench. The abundant presence of organic material and non-organic artefacts allows for a characterisation of the organic layer as an anthropogenic layer, originating from settling activities on the site.

Additional information on the nature of this organic layer could be retrieved from drill cores extracted on two crossed transects (cf. Figure 6.5). The 14 drill cores showed the presence of the same organic layer over the entire site and its thickness of up to 1.7 m (Fig. 6.10). Until now the layer appears not to be subdivided by any sterile layers and must be considered as the result of a continuous accumulation process. Below the organic layer, sterile lake marl or layers with a mixture of lake marl and organic material were found. The content of these drill cores is currently under investigation by

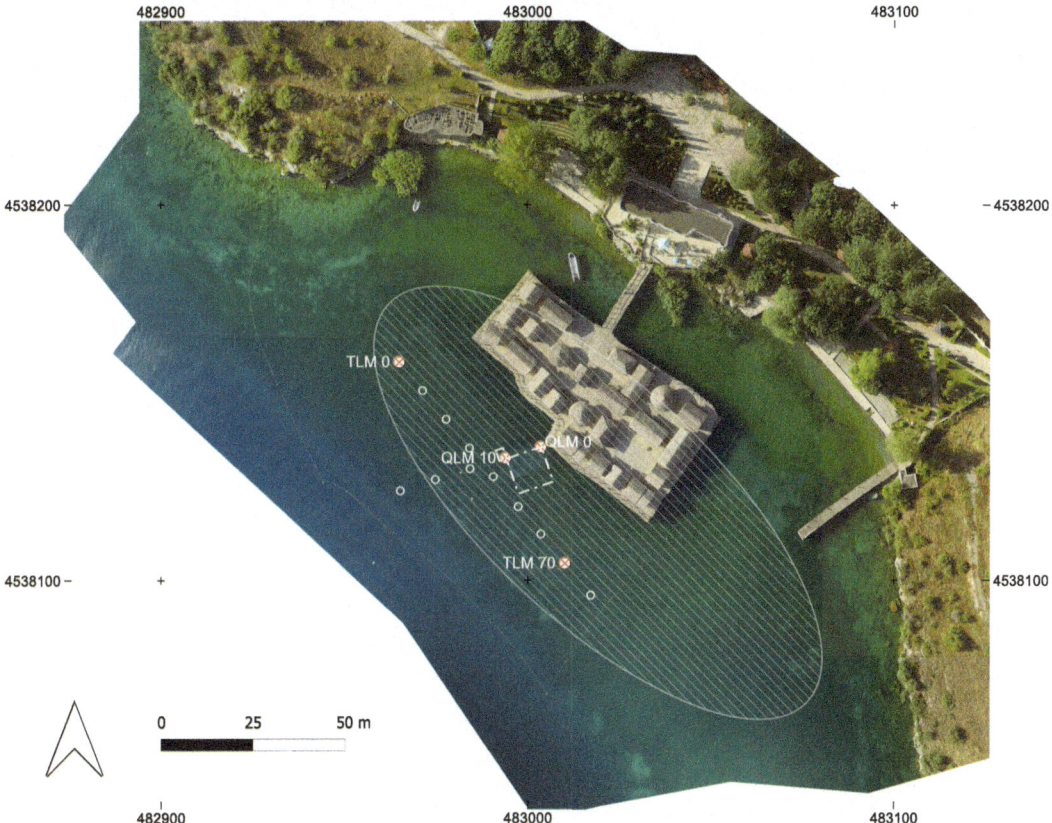

Fig. 6.5 Ploča Mičov Grad. Orthophoto of the archaeological site underwater with the sampled area (square) in front of the open-air 'Museum on Water'. The hatched area marks the estimated extension of the pile field based on the archaeological interventions between 1997 and 2005. Circles = location of core drillings from 2019; circles with red cross = core drillings with radiocarbon dated samples (CRS: EPSG:32634) (J. Reich, M. Hostettler, G. Milevski, EXPLO/UBern)

specialists of the School of Archaeology at the University of Oxford (see Chap. 17 in this book).

6.6 Absolute Chronological Data

Until now the dendrochronological analysis of 268 wooden elements (265 vertically standing piles and three horizontally oriented elements) from the 40 m² excavated in 2018 is completed (Hafner et al. 2021). More detailed information on the dendrochronological analysis and its current state is presented in Chap. 14.

Ninety-three % of the 268 samples are either oak or coniferous woods, and half of them have more than 50 annual rings and are therefore most suitable for dendrochronological analyses. One hundred and seventeen samples (114 oaks and three junipers) were dendrochronologically measured and analysed. From these measured wood samples, a series of 36 samples were taken for radiocarbon dating. The radiocarbon data was combined with the dendrochronological results and a first chronological classification was achieved through wiggle-matching (Bronk Ramsey et al. 2001; Galimberti et al. 2004).

The results of this combined dendrochronological and radiocarbon dating approach show the existence of up to four main settlement phases both in the 5th and 2nd millennium cal BC. The construction timbers, which indicate settlement activities from the 5th millennium cal BC,

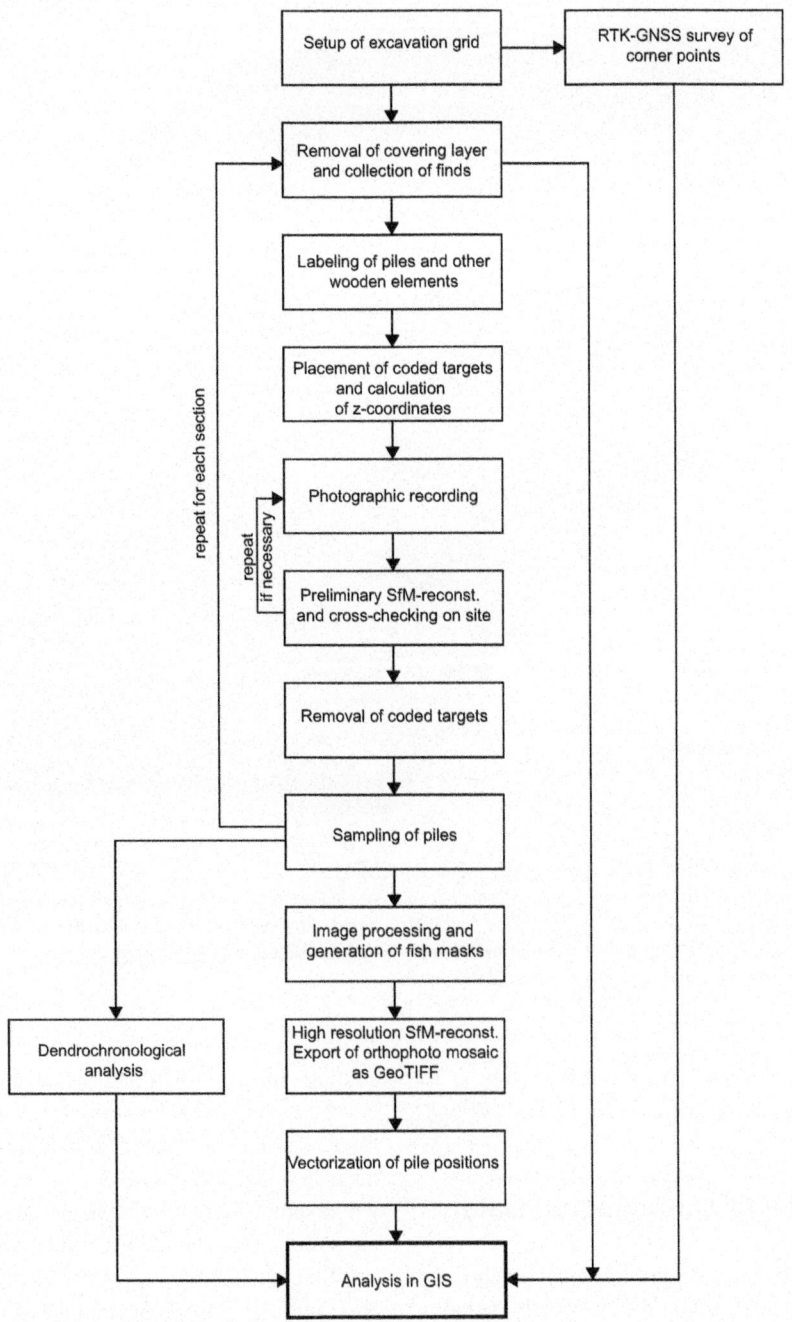

Fig. 6.6 Working steps of the proposed novel Structure from Motion-based workflow for underwater pile field documentation and pile field analysis (Reich et al. 2021)

all date consistently to the middle of the millennium. In the 2nd millennium cal BC, settlement activities are evidenced around 1800, 1400 and 1300 cal BC (Fig. 6.11).

To clarify the relationship between the anthropogenic layer and the construction phases, samples from the organic layer consisting of short-lived plant remains were selected and

Fig. 6.7 Ploča Mičov Grad. A diver is removing vegetation (macrophytes) and the covering layer on the lake bottom, exposing more wooden piles as seen on the left in the already exposed part (M. Hostettler, EXPLO/UBern)

Fig. 6.8 Ploča Mičov Grad. **a** Diver taking pictures of a prepared strip of the pile field with all wooden elements marked by a number tag. **b** Single documentation picture used for 3D-reconstruction by means of Structure from Motion. On the left and the right sides, a coded target is visible, and these are automatically detected in the software Agisoft Metashape and used for georeferencing. **c** A roughly 5.5 m long and 1 m deep strip of the pile field visualised in an oblique view on a textured mesh. The blue squares represent camera positions used for the 3D-reconstruction. The red square marks the position of the camera while taking the photo shown in Fig. 6.8b (**a** A. Ulisch, EXPLO/UBern; **b** and **c** J. Reich, EXPLO/UBern)

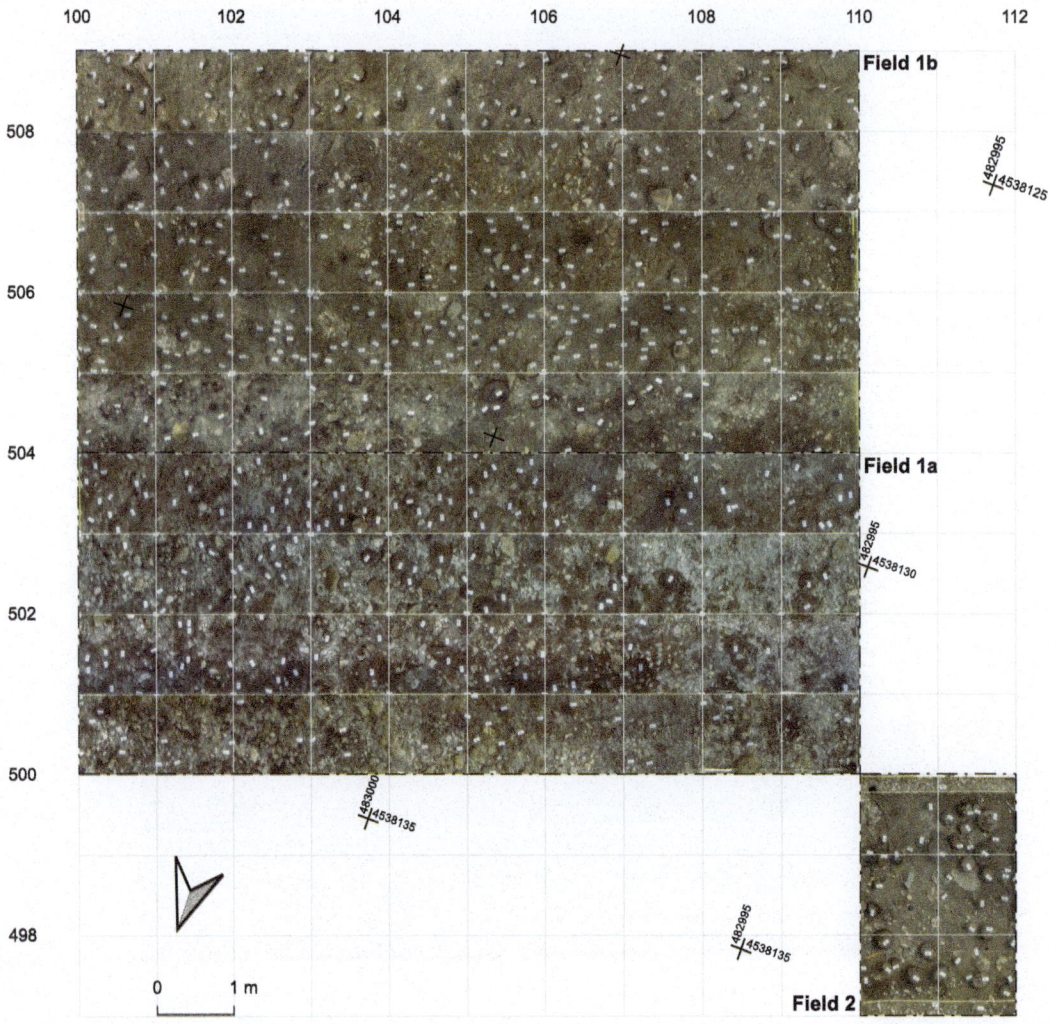

Fig. 6.9 Ploča Mičov Grad. Joined orthophoto mosaics of the documented and sampled area during the excavation campaigns 2018 and 2019. Field 1a (2018): axis 500 to 504; Field 1b (2019): axis 504 to 509; Field 2: axis 497 to 500 (CRS: local and EPSG:32634) (J. Reich, EXPLO/UBern)

radiocarbon dated. A series of eight cereal chaff samples were taken from four drill cores. From each of these four drill cores, a cereal chaff sample was taken both from the upper and the lower parts of the main organic layer. Additionally, two macrofossil samples taken in 2018 from the top part of the layer were dated.

A bi-phased sequence-phase model was used for the calibration of the radiocarbon ages to calendar ages with the IntCal20 calibration curve in OxCal v.4.4.2 (Bronk Ramsey 2009a, b; Reimer et al. 2020). The sequence was modelled by taking into account the stratigraphic position of the samples within the main organic layer and by assuming that the six samples from the upper part represent a later phase than the one captured by the samples from the lower part.

To visualise the two phases, Kernel Density plots (KDE plots) were used (Bronk Ramsey 2017). The sequence-phase model shows that the time of accumulation of the main organic layer can be estimated between the 45th and the 44th century cal BC (Fig. 6.12).

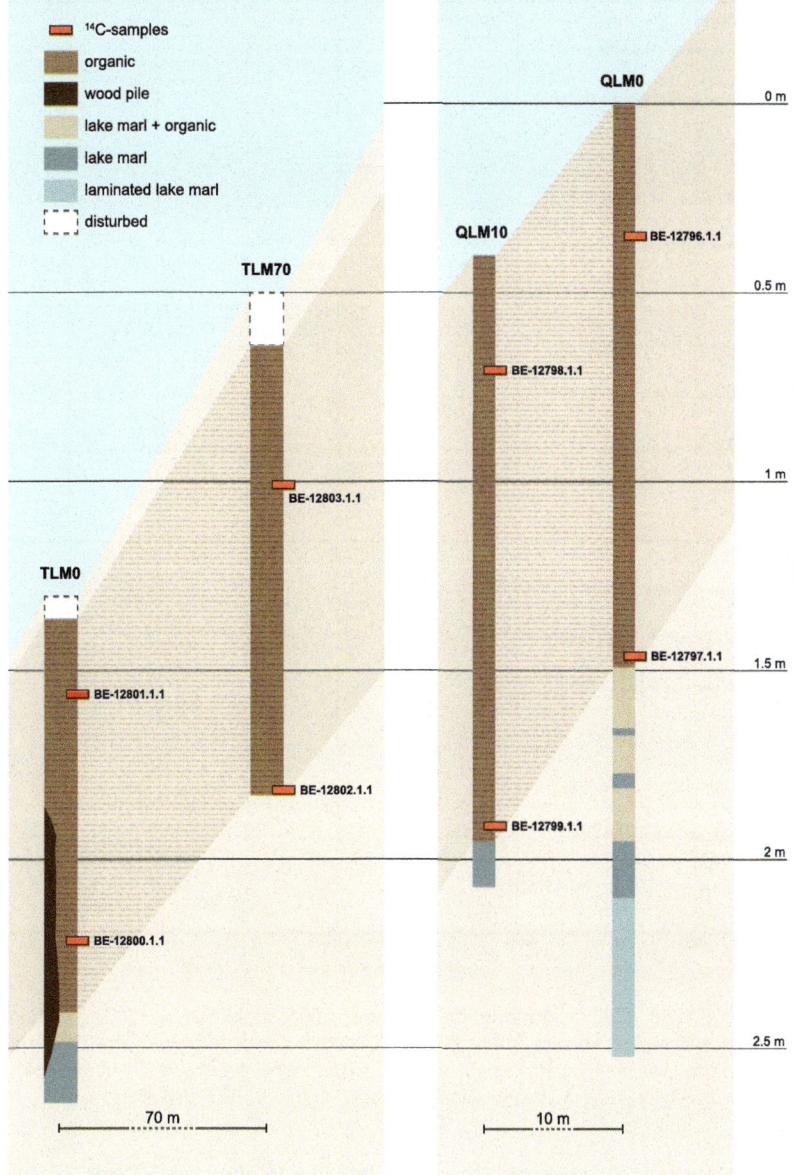

Fig. 6.10 Ploča Mičov Grad. Schematic illustration of drill cores showing the massive organic layer and the location of the radiocarbon dated samples (A. Bieri, UBern, J. Reich and A. Ballmer EXPLO/UBern)

In combination with the already completed dendrochronological analysis of the wood samples, this result is even clearer, as the mean curves including sapwood (MK 5, MK 4 and MK 10) all point to the forty-fifth and the forty-fourth century cal BC. The current results suggest that the whole preserved organic layer in Ploča Mičov Grad dates between the forty-fifth and the forty-fourth century cal BC and was probably accumulated in around 100 years of settlement activity.

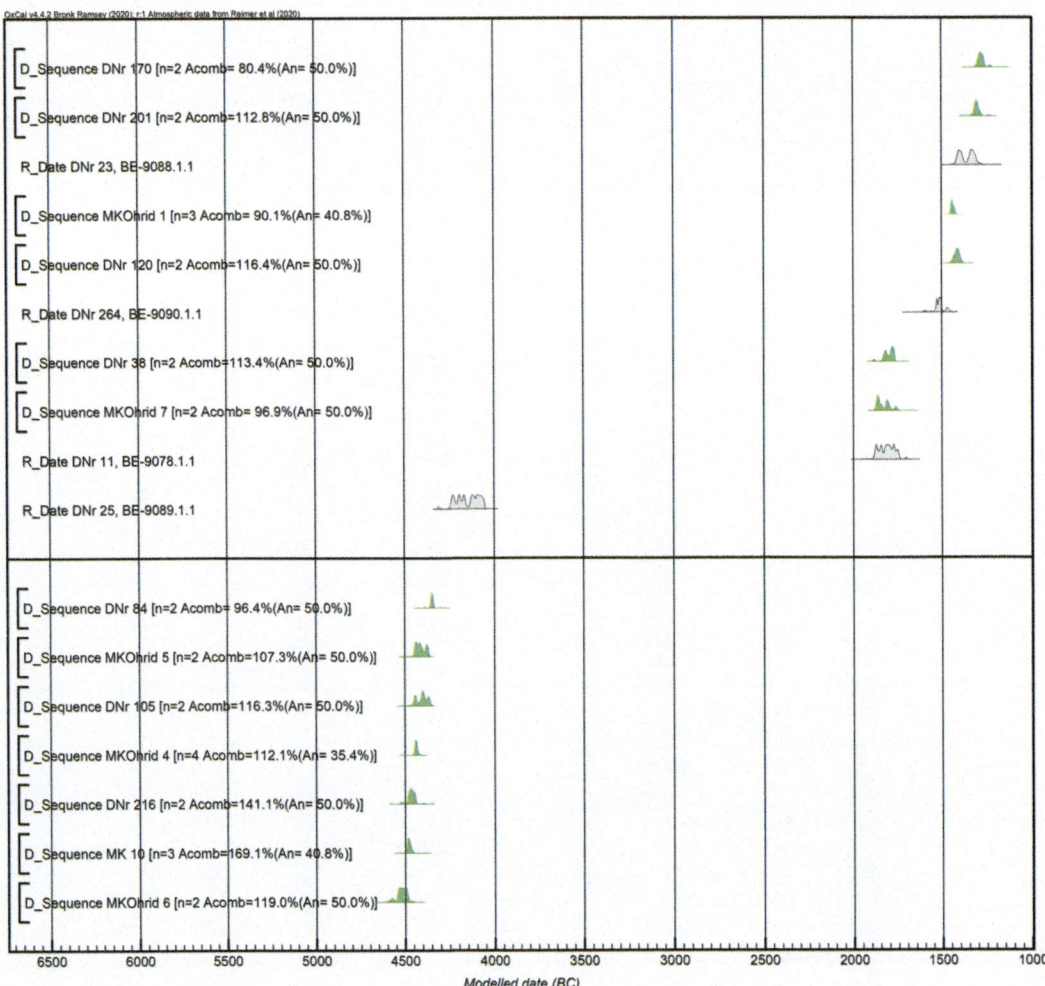

Fig. 6.11 Ploča Mičov Grad. Calibrated results of the radiocarbon dates of the wood samples from 2018. Green = results of wiggle matching; grey = single calibrations. Modelled with OxCal v4.4.2 (Bronk Ramsey et al. 2001; Bronk Ramsey 2009a, b; Reimer et al. 2020). Underlying data: https://doi.org/10.5281/zenodo.4613033. Code: https://doi.org/10.5281/zenodo.4560125 (M. Bolliger, J. Reich, EXPLO/UBern)

6.7 Material Culture: First Chronological Conclusions

This newly gained absolute chronological data allows making further conclusions concerning the material culture found in the surface layer and the organic layer. Firstly, the preserved organic layer and all the included material culture can be attributed accurately to the middle of the 5th millennium cal BC. Secondly, all archaeological remains from the settlement phases dated to the 2nd millennium cal BC are situated above the organic layer and form part of today's surface layer and lake bottom.

The fragmentary information available from the excavations between 1997 and 2005 indicates that most of the published find material originates from the surface layer (Kuzman 2013). Thus, the finds of the previous excavations can be related to the younger phases around 1800, 1400 and 1300 cal BC. Artefacts typochronologically pointing to the Middle Bronze Age around 1800 cal BC are still to be identified. At

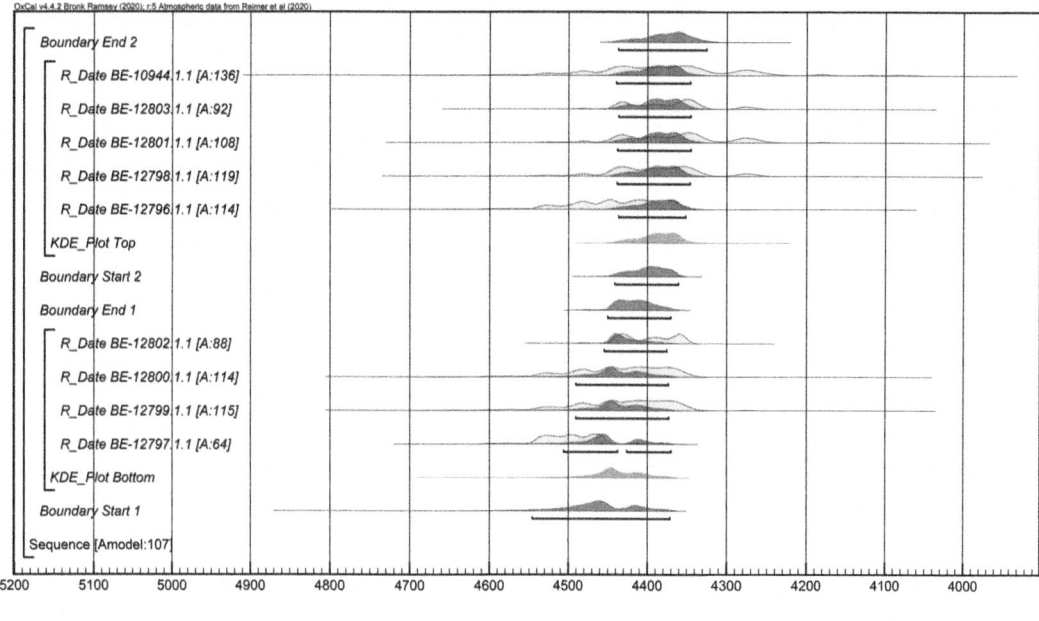

Fig. 6.12 Ploča Mičov Grad. Sequence-phase model of archaeobotanical samples representing the 'top' and 'bottom' phase of the organic layer; brown = modelled Kernel Density estimation plots; grey = single dates. Modelled with OxCal v4.4.2 (Bronk Ramsey et al. 2001; Bronk Ramsey 2009a, b; Reimer et al. 2020). Underlying data: https://doi.org/10.5281/zenodo.4612910; Code: https://doi.org/10.5281/zenodo.4560161 (J. Reich, EXPLO/UBern)

least the two younger phases are in agreement with the attribution of the so far known material to the Late Bronze Age. However, the also postulated Early Iron Age was not detected by means of dendrochronology nor radiocarbon dating of organic remains.

The pottery found in the organic layer is in a very well-preserved condition. Most of the recorded sherds have an intact surface and are decorated in some cases with linear incisions or knobs or painting. Typochronologically, these elements can be attributed to the Final Neolithic of the southwestern Balkans. The conclusions from the scientific dating allow placing the finds on the absolute timescale between the forty-fifth and the 44th century cal BC.

In contrast, the preservation of the finds from the surface layer shows mostly eroded surfaces and rounded edges. The finds include decoration with knobs or ledges on the surface. Handles seem to be a frequent feature although a systematic evaluation of the material has not taken place yet. Currently, the main features support a rough classification to the Late Bronze Age.

6.8 Material Culture: Regional Contextualisation

For a typochronological comparison with other sites in the region, long stratigraphic sequences are particularly suitable. In the region, two sites have been extensively excavated and show stratigraphies ranging from the Neolithic period to the Iron Age. These are the two sites of Sovjan and Maliq, both located in the Korçë basin at the western shore of ancient Lake Maliq (Prendi 2018; Oberweiler et al. 2020).

Maliq is one of the best-known prehistoric sites of Albania. The site is located on the right side of the present course of the river Devoll. It was discovered in the middle of the 20th century

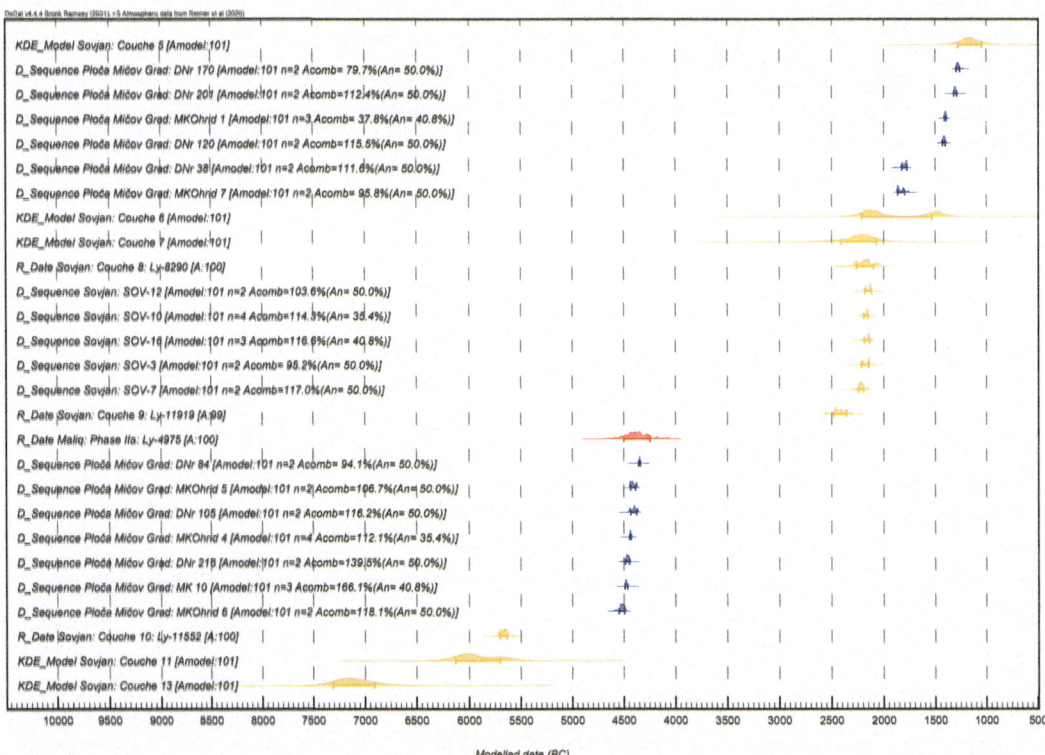

Fig. 6.13 All published radiocarbon data of the lake-shore settlement sites of Maliq (red), Sovjan (orange) and Ploča Mičov Grad (blue). Modelled with OxCal v4.4.2 (Bronk Ramsey et al. 2001; Bronk Ramsey 2009a, b; Reimer et al. 2020). Underlying data: (Guilaine and Prendi 1991; Lera et al. 1996; Lera and Touchais 2002, 2003; Gori 2015; Oberweiler et al. 2020; Maczkowski et al. 2021); https://doi.org/10.5281/zenodo.4613033 (J. Reich, EXPLO/UBern)

during drainage works in the Korçë plain and was subsequently excavated. The excavated areas are grouped in sectors A–C, which are separated by several hundred metres (Hasa 2018; Prendi 2018). The stratigraphic sequence of Maliq consists of four main phases (Maliq I–IV) which are roughly attributed to the Late Neolithic (Maliq I); the Final Neolithic, Eneolithic or Chalcolithic, depending on the nomenclature (Maliq II); the Bronze Age (Maliq III) and the Iron Age (Maliq IV). The individual phases are again subdivided based on stratigraphical and typochronological attributions. Assignment of the published find material to the excavated sectors and layers is not possible based on the published information. This has repeatedly led to doubts about the validity of the phase classification and its dating, especially for the sub-phases of Maliq III (e.g. Gori 2015; Maran 1998).

Until now only one radiocarbon date from Maliq has been published (Fig. 6.13) (Guilaine and Prendi 1993). It dates Maliq phase IIa to the middle and second half of the 5th millennium cal BC. In comparison to the available dates from Ploča Mičov Grad at Lake Ohrid, a contemporaneity between the settlement activities in connection to the organic layer with Maliq IIa can be assumed. At least for the early phases, water-logged wooden construction remains were found providing potential for future dendrochronological analysis. An initial assessment of the material culture at both sites, using the example of ceramic bowls, shows that the similarities are striking (Fig. 6.14) (Prendi 2018, 353–356). The profiles of the vessels, the described surface treatment, and the decorations mainly through linear and simple geometrical incisions or white paint on a black polished surface are highly comparable.

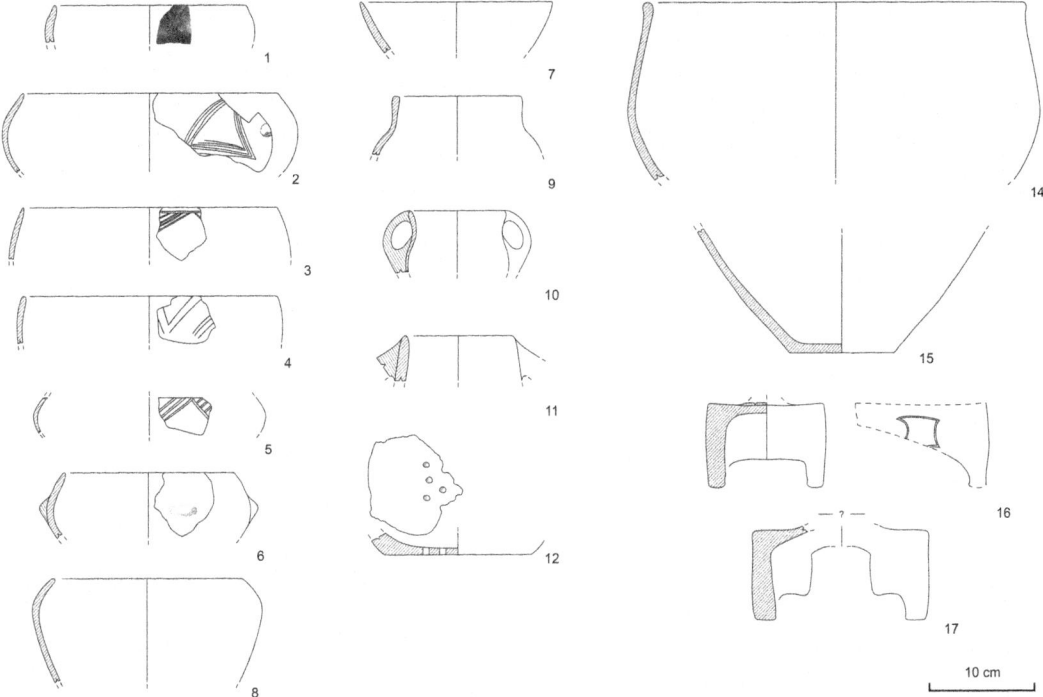

Fig. 6.14 Ploča Mičov Grad. A selection of pottery from spits 2–4 in Field 2 excavated in 2019 (Drawings G. Naumov; Re-drawing: J. Reich, EXPLO/UBern)

The site of Sovjan is located about 5 km north of Maliq. It was discovered in 1988 after agricultural activities and intensively excavated between 1993 and 2006 by the Franco-Albanian Korçë Basin Archaeological Expedition. The site has a stratigraphy with 13 layers, with the oldest layers dating to the 7th millennium cal BC. After repeated settlement activities through the Neolithic and the Bronze Age, the site was finally abandoned around the 8th or 7th century cal BC when rising water levels led to an inundation of the site (Fouache et al. 2010). The deeper layers of the site, especially layer 8, showed very good organic preservation including construction timbers. In comparison to Maliq, the stratigraphic sequence from Sovjan is much better dated with scientific dating methods. In total there are 40 published radiocarbon dates (cf. Figure 6.13) (Lera et al. 1996; Lera and Touchais 2002, 2003; Gori 2015; Oberweiler et al. 2020; Maczkowski et al. 2021). Some of them were only recently sampled, analysed and published within the EXPLO project. A series of samples from wooden construction elements connected to Level 8 was dendrochronologically measured and dated by means of wiggle-matching (Maczkowski et al. 2021).

The radiocarbon dates from Sovjan Level 5c show, around 1300 cal BC, a possible contemporaneous phase with Ploča Mičov Grad. Level 5c is also typochronologically firmly attributed to the Late Bronze Age and further subdivided into three sub-levels (Gori and Krapf 2016). The published selection of pottery from the youngest Sub-level 5c1 is very well comparable with the pottery found in the surface layer in Ploča Mičov Grad (Gori and Krapf 2016, 114, 116). The vessel shapes as well as the shape and position of the handles can be directly compared with the surface finds from Ploča Mičov Grad and show a high degree of resemblance (Kuzman 2013, XIV–XXVI). Taking into account the published material from Maliq phase IIId (Prendi 2018, 452–461), the same observations can be made. The shape of the vessels, but also the shape of the handles as well as their position point to some degree of chronological relation between the three sites. However, more precise conclusions about the exact relative and absolute chronology between the three different phases of the Late Bronze Age must be taken with caution at this point.

6.9 Outlook

Recent research on the prehistoric underwater lakeshore settlement of Ploča Mičov Grad has extraordinary potential to produce new insights into the prehistory of the Albania—Greece—North Macedonia tripoint region. With the new data and results presented here, and ongoing research into other lakeside settlements, the basis for a new local reference chronology around Lake Ohrid is set.

Acknowledgements The pilot study in 2018 was funded by the Institute of Archaeological Sciences, University of Bern, the Association of Swiss Underwater Archaeology and the Johanna Dürmüller-Bol Foundation, Muri b. Bern. From 2019 on this project has received funding from the European Research Council (ERC) under the European Union's Horizon 2020 research and innovation programme grant agreement No 810586 (project EXPLO). For their technical, personnel and administrative support, we thank the following institutions: Association of Swiss Underwater Archaeology; Archaeological Service Canton of Bern, Underwater Archaeology and Dendrochronology; Center for Prehistoric Research, Skopje; Department of Underwater Archaeology and Dendroarchaeology, Office for Urbanism Zurich; Diving Center Amfora, Ohrid; Embassy of Switzerland in North Macedonia, Skopje; Institute of Geodesy and Photogrammetry of the Swiss Federal Institute of Technology Zurich; Institute for Protection of Monuments and Museum, Ohrid; Space Research and Planetary Sciences, Physics Institute, University of Bern. Special thanks go to all those involved in the 2018 and 2019 field campaigns, whose personal commitment contributed to the successful outcome.

References

Amataj S, Anovski T, Benischke R, Eftimi R, Gourcy LL, Kola L, Leontiadis I, Micevski E, Stamos A, Zoto J (2007) Tracer methods used to verify the hypothesis of Cvijić about the underground connection between Prespa and Ohrid Lake. Environ Geol 51:749–753. https://doi.org/10.1007/s00254-006-0388-9

Andoni E, Hasa E, Gjipali I (2017) Neolithic settlements on the western bank of Lake Ohrid: Pogradec and Lin 3. In: New archaeological discoveries in the Albanian Regions. proceedings of the international conference, Tirana 30–31 January 2017. Botimet Albanologjike, Tirana, pp 123–140

Bronk Ramsey C (2009a) Bayesian analysis of radiocarbon dates. Radiocarbon 51:337–360. https://doi.org/10.1017/S0033822200033865

Bronk Ramsey C (2009b) Dealing with outliers and offsets in radiocarbon dating. Radiocarbon 51:1023–1045. https://doi.org/10.1017/S0033822200034093

Bronk Ramsey C (2017) Methods for summarizing radiocarbon datasets. Radiocarbon 59:1809–1833. https://doi.org/10.1017/RDC.2017.108

Bronk Ramsey C, van der Plicht J, Weninger B (2001) 'Wiggle matching' radiocarbon dates. Radiocarbon 43:381–389. https://doi.org/10.1017/S0033822200038248

Fandré M-J (2020) Koordinatentransformationen und Messsysteme für Unterwasser-Archäologische Projekte in Südosteuropa. Bachelor thesis, Institute of Geodesy and Photogrammetry, ETH Zürich

Fouache E, Desruelles S, Magny M, Bordon A, Oberweiler C, Coussot C, Touchais G, Lera P, Lézine A-M, Fadin L, Roger R (2010) Palaeogeographical reconstructions of Lake Maliq (Korça Basin, Albania) between 14,000 BP and 2000 BP. J Archaeol Sci 37:525–535. https://doi.org/10.1016/j.jas.2009.10.017

Galimberti M, Ramsey CB, Manning SW (2004) Wiggle-match dating of tree-ring sequences. Radiocarbon 46:917–924. https://doi.org/10.1017/S0033822200035967

GIZ (2015) Initial characterisation of lakes Prespa, Ohrid and Shkodra/Skadar. Implementing the EU Water Framework Directive in South-Eastern Europe. Deutsche Gesellschaft für Internationale Zusammenarbeit (GIZ) GmbH, Bonn, Eschborn

Gori M, Krapf T (2016) The bronze and iron age pottery from Sovjan. Iliria 39:91–135

Gori M (2015) Along the rivers and through the mountains: a revised chrono-cultural framework for the south-western Balkans during the late 3rd and early 2nd millennium BCE. Habelt, Bonn

Guilaine J, Prendi F (1991) Dating the copper age in Albania. Antiquity 65:574–578. https://doi.org/10.1017/S0003598X00080200

Guilaine J, Prendi F (1993) Një datim për epokën e bakrit në Shqipëri/Une datation pour l'âge du cuivre de l'Albanie. Iliria 23:95–100. https://doi.org/10.3406/iliri.1993.1618

Hafner A, Reich J, Ballmer A, Bolliger M, Antolín F, Charles M, Emmenegger L, Fandré J, Francuz J, Gobet E, Hostettler M, Lotter AF, Maczkowski A, Morales-Molino C, Naumov G, Stäheli C, Szidat S, Taneski B, Todoroska V, Bogaard A, Kotsakis K, Tinner W (2021) First absolute chronologies of neolithic and bronze age settlements at Lake Ohrid based on dendrochronology and radiocarbon dating.

J Archaeol Sci Rep 38:103107. https://doi.org/10.1016/j.jasrep.2021.103107

Hasa E (2018) Sondazh arkeologjik në vendbanimin prehistorik të Maliqit. Candavia 7:417–432

Hauffe T, Albrecht C, Schreiber K, Birkhofer K, Trajanovski S, Wilke T (2011) Spatially explicit analysis of gastropod biodiversity in ancient Lake Ohrid. Biogeosciences 8:175–188. https://doi.org/10.5194/bg-8-175-2011

Jordanoska B, Stafilov T, Wuest A (2013) Assessment of ecological importance and anthropogenic change of subaquatic springs in ancient lake ohrid. Water Res Manage 3:9–17

Kuzman P (2013) Praistoriski palafitni naselbi vo Makedonija. In: Dimitrova E, Donev J (eds) Makedonija: mileniumski kulturno-istoriski fakti. Skopje, pp 297–430

Lera P, Touchais G (2002) Sovjan (Albanie). Bullet Correspondance Hellén 126:627–645. https://doi.org/10.3406/bch.2002.7114

Lera P, Touchais G (2003) Sovjan (Albanie). Bull Correspondance Hellén 127:578–609. https://doi.org/10.3406/bch.2003.7337

Lera P, Prendi F, Touchais G (1996) Sovjan (Albanie). Bull Correspondance Hellén 120:995–1026. https://doi.org/10.3406/bch.1996.7038

Lera P, Touchais G, Oberweiler C, Aslaksen OC, Blein C, Boleti, A, Elezi G, Fadin L, Gori M, Krapf, T Maniatis Y, Odie C (2020) Bassin de Korçë, Kallamas. Bull Archéologique Écoles Françaises à l'étranger

Maczkowski A, Bolliger M, Ballmer A, Gori M, Lera P, Oberweiler C, Szidat S, Touchais G, Hafner A (2021) The early bronze age dendrochronology of Sovjan (Albania): A first tree-ring sequence of the 24th–22nd c. BC for the southwestern Balkans. Dendrochronologia 66. https://doi.org/10.1016/j.dendro.2021.125811

Matzinger A, Spirkovski Z, Patceva S, Wüest A (2006) Sensitivity of ancient lake ohrid to local anthropogenic impacts and global warming. J Great Lakes Res 32:158–179. https://doi.org/10.3394/0380-1330(2006)32[158:SOALOT]2.0.CO;2

Maran J (1998) Kulturwandel auf dem griechischen Festland und den Kykladen im späten 3. Jahrtausend v. Chr. Studien zu den kulturellen Verhältnissen in Südosteuropa und dem zentralen sowie östlichen Mittelmeerraum in der späten Kupfer- und frühen Bronzezeit. Habelt, Bonn ISBN: 978-3-7749-2870-1

McCarthy JK, Benjamin J, Winton T, van Duivenvoorde W (eds) (2019) 3D recording and interpretation for maritime archaeology. Springer, Cham

Naumov G (2016) Prähistorische Pfahlbauten im Ohrid-See, Republik Mazedonien. Plattform 23(24):10–20

Oberweiler C, Lera P, Kurti R, Touchais G, Aslaksen OC, Blein C, Elezi G, Gori M, Krapf T, Maniatis Y, Wagner S (2020) Mission archéologique franco-albanaise du bassin de Korçë. Bull Archéologique Écoles Françaises à L'étranger. https://doi.org/10.4000/baefe.1660

Prendi F (2018) Vendbanimi prehistorik i Maliqit: The prehistoric settlement of Maliq. Botimet M&B, Tiranë

Reich J, Steiner P, Ballmer A, Emmenegger L, Hostettler M, Stäheli C, Naumov G, Taneski B, Todoroska V, Schindler K, Hafner A (2021) A novel Structure from Motion-based approach to underwater pile field documentation. J Archaeol Sci Rep 39:103120. https://doi.org/10.1016/j.jasrep.2021.103120

Reich J, Emmenegger L, Hostettler M, Stäheli C, Naumov G, Hafner A (2019) A new approach for structure from motion underwater pile-field documentation. https://doi.org/10.5281/zenodo.2552068

Reimer PJ, Austin WEN, Bard E, Bayliss A, Blackwell PG, Bronk Ramsey C, Butzin M, Cheng H, Edwards RL, Friedrich M, Grootes PM, Guilderson TP, Hajdas I, Heaton TJ, Hogg AG, Hughen KA, Kromer B, Manning SW, Muscheler R, Palmer JG, Pearson C, van der Plicht J, Reimer RW, Richards DA, Scott EM, Southon JR, Turney CSM, Wacker L, Adolphi F, Büntgen U, Capano M, Fahrni SM, Fogtmann-Schulz A, Friedrich R, Köhler P, Kudsk S, Miyake F, Olsen J, Reinig F, Sakamoto M, Sookdeo A, Talamo A (2020) The IntCal20 Northern hemisphere radiocarbon age calibration curve (0–55 cal kBP). Radiocarbon 62:725–757. https://doi.org/10.1017/RDC.2020.41

Vogel H, Wagner B, Zanchetta G, Sulpizio R, Rosén P (2010) A paleoclimate record with tephrochronological age control for the last glacial-interglacial cycle from Lake Ohrid, Albania and Macedonia. J Paleolimnol 44:295–310. https://doi.org/10.1007/s10933-009-9404-x

Wagner B, Vogel H, Francke A, Friedrich T, Donders T, Lacey JH, Leng MJ, Regattieri E, Sadori L, Wilke T, Zanchetta G, Albrecht C, Bertini A, Combourieu-Nebout N, Cvetkoska A, Giaccio B, Grazhdani A, Hauffe T, Holtvoeth J, Joannin S, Jovanovska E, Just J, Kouli K, Kousis I, Koutsodendris A, Krastel S, Lagos M, Leicher N, Levkov Z, Lindhorst K, Masi A, Melles M, Mercuri AM, Nomade S, Nowaczyk N, Panagiotopoulos K, Peyron O, Reed JM, Sagnotti L, Sinopoli G, Stelbrink B, Sulpizio R, Timmermann A, Tofilovska S, Torri P, Wagner-Cremer F, Wonik T, Zhang X (2019) Mediterranean winter rainfall in phase with African monsoons during the past 1.36 million years. Nature 573:256–260. https://doi.org/10.1038/s41586-019-1529-0

Wilke T, Hauffe T, Jovanovska E, Cvetkoska A, Donders T, Ekschmitt K, Francke A, Lacey JH, Levkov Z, Marshall CR, Neubauer TA, Silvestro D, Stelbrink B, Vogel H, Albrecht C, Holtvoeth J, Krastel S, Leicher N, Leng, MJ Lindhorst K, Masi A, Ognjanova-Rumenova N, Panagiotopoulos K, Reed JM, Sadori L, Tofilovska S, Van Bocxlaer B, Wagner-Cremer F, Wesselingh FP, Wolters V, Zanchetta G, Zhangand X, Wagner B (2020) Deep drilling reveals massive shifts in evolutionary dynamics after formation of ancient ecosystem. Sci Adv 6: eabb2943. https://doi.org/10.1126/sciadv.abb2943

The Lakeside Settlement of Sovjan (Southeastern Albania) During the Bronze Age in the Light of New Chronological Data

Gilles Touchais, Cécile Oberweiler, and Petrika Lera

Abstract

While preparing the final publication of the excavations carried out on the prehistoric lakeside settlement of Sovjan (1993–2006), during the last years, we have undertaken (1) the complete re-examination of the stratigraphic data, (2) the detailed study of the pottery of the Bronze and Iron Age levels, and (3) the realisation of new absolute dates, by radiocarbon and even more so by dendrochronology. The latter is based on the analysis of a series of wood samples taken during the partial reopening of the excavation trench in 2018. From all these data, the relative and absolute chronological sequence of Sovjan could be completed, corrected, and updated, and the successive phases of human occupation synchronised with environmental change. As the most significant updates concern the Bronze Age, the present paper focuses on the definitive chronological sequence of this period and summarises the main related data.

Keywords

Albania · Lake Maliq · Lakeside settlement · Bronze Age · Dendrochronology

7.1 Introduction

The prehistoric site of Sovjan, discovered by chance during the digging of a drainage canal at the end of the 1980ies, is located in the Korça plain, one of the closed little basins typical of the Balkans, at an altitude of about 800 m a.s.l. It is situated very close to the well-known prehistoric settlement of Maliq, which was explored in the early 1960ies by Frano Prendi, one of the pioneers of Albanian archaeology (Prendi 1966, 2018). These two sites, like several others, were established on the western edge of the shallow marshy Lake Maliq, now dried up, which was located in the northwestern part of the plain.

At Sovjan, some trenches were opened in 1990–1991 by the Albanian Institute of Archaeology in Tirana (Korkuti and Petruso 1993, 715), and a systematic excavation was conducted from 1993 to 2006 by the French-Albanian mission in the Korça basin, the fruit of a cooperative agreement between the Albanian Institute of Archaeology at Tirana, the University of Paris 1

G. Touchais (✉) · C. Oberweiler
CNRS, UMR 7041 ArScAn, Université Paris I Panthéon Sorbonne, Université Paris Nanterre, Paris, France
e-mail: gilles.touchais@univ-paris1.fr

C. Oberweiler
e-mail: cecile.oberweiler@protonmail.com

P. Lera
Institute of Archaeology, Academy of Albanological Studies, Tirana, Albania
e-mail: petrikaa_lera@yahoo.it

A. Ballmer et al. (eds.), *Prehistoric Wetland Sites of Southern Europe*, Natural Science in Archaeology, https://doi.org/10.1007/978-3-031-52780-7_7

Panthéon-Sorbonne, and the French School in Athens.[1] One of the main objectives of the joint French-Albanian project was to carry out in parallel the archaeological exploration of the prehistoric settlement and the systematic study of the palaeoenvironment at a local and regional level, in order to: (1) define more precisely the interaction between the inhabitants of the prehistoric settlement and both their immediate and more distant environment in terms of the exploitation of minerals, as well as vegetal and faunal resources; and (2) put the settlement of Sovjan in its natural geographical setting, by studying for instance the impact of the lake level fluctuations during the occupation of the village (Touchais and Fouache 2007; Fouache et al. 2010). To meet this last objective, several series of core drillings were carried out in the Korça basin, including a ten-core transect beginning on the summit of the tell of Sovjan and extending eastwards towards the lake, which provided a comprehensive stratigraphic section of the littoral sediment sequence (Lera et al. 2008, 885–887).

The site of Sovjan appears as a very low tell, oval in shape and of moderate extension (c. 1.5 ha); it has been cut by the canal from north to south (Fig. 7.1). The excavation made it possible to observe a stratigraphic sequence of fourteen strata corresponding to several occupation periods which extend from the earliest Neolithic (around 7000 BC) to the Early Iron Age (around 700 BC), when the level of the lake rose, and the site was flooded. However, there seems to be a gap in the occupation between the Late Neolithic and the Early Bronze Ages.

This sequence has been presented several times in the past, in a more or less complete way (Touchais et al. 2005; Touchais and Lera 2007; Touchais 2008; Lera et al. 2010; Oberweiler et al. 2014). However, in view of the final publication, we recently conducted a systematic review of all the stratigraphic data from the excavations. This re-examination, combined with a detailed study of the archaeological material—especially ceramics —and new absolute chronological data, made it

possible to clearly distinguish seven main occupation phases, for the first time, dating to the Neolithic (Sovjan I–II = levels 13, 12, 11),[2] the Bronze Age (Sovjan VI, V, IV, III = levels 9, 8, 7, 6, 5c), and the Early Iron Age (Sovjan VII = levels 5a, 5b),[3] and secondly, to correct and clarify several points in the chronological and architectural sequence of the settlement during the Bronze Age.

Thanks to the many parallels observed between the ceramics of levels 9 to 5c and those from corresponding levels of nearby sites—such as Maliq—or more distant ones—from Macedonia (e.g. Kastanas, Archontiko) and the northern Balkans to Central and Southern Greece —and with the support of absolute dating, it has been possible not only to date the four Bronze Age phases of Sovjan more accurately, but also to establish reliable synchronisms with the phases of other sites and/or cultural areas of the southwestern Balkans (Fig. 7.2).

These latest updates are summarised here, as a preview of the final publication of our research at Sovjan (Touchais et al. 2023) in which all the data are presented in detail. We will focus on Phases VI to III, with particular emphasis on the latter which features the newest absolute chronological data, mainly from dendrochronology.

7.2 Sovjan VI

During the systematic re-examination of the stratigraphy and the finds, layer 5c was subdivided into three strata: 5c1, 5c2 and 5c3 (Fig. 7.3) (Lera et al. 2010). However, the detailed study of 5c1 showed that the upper stratigraphic units of this stratum contain pottery which belongs to a transitional phase between the Late Bronze Age (LBA) and the Early Iron Age (EIA), while the pottery of the two underlying strata shows exclusively LBA-characteristics. This transitional phase, corresponding with Kastanas V (see Fig. 7.2), has been defined as Sovjan VI.

---

[1] See preliminary reports in the *Bulletin de Correspondance Hellénique* (*BCH*) 118 (1994) to 131 (2007).

[2] Level 14 corresponds to the natural soil, and level 10 to a gap in the occupation of the settlement (see above).

[3] Levels 4 to 1 follow the flooding of the site. They are natural sediments and do not show traces of human activity.

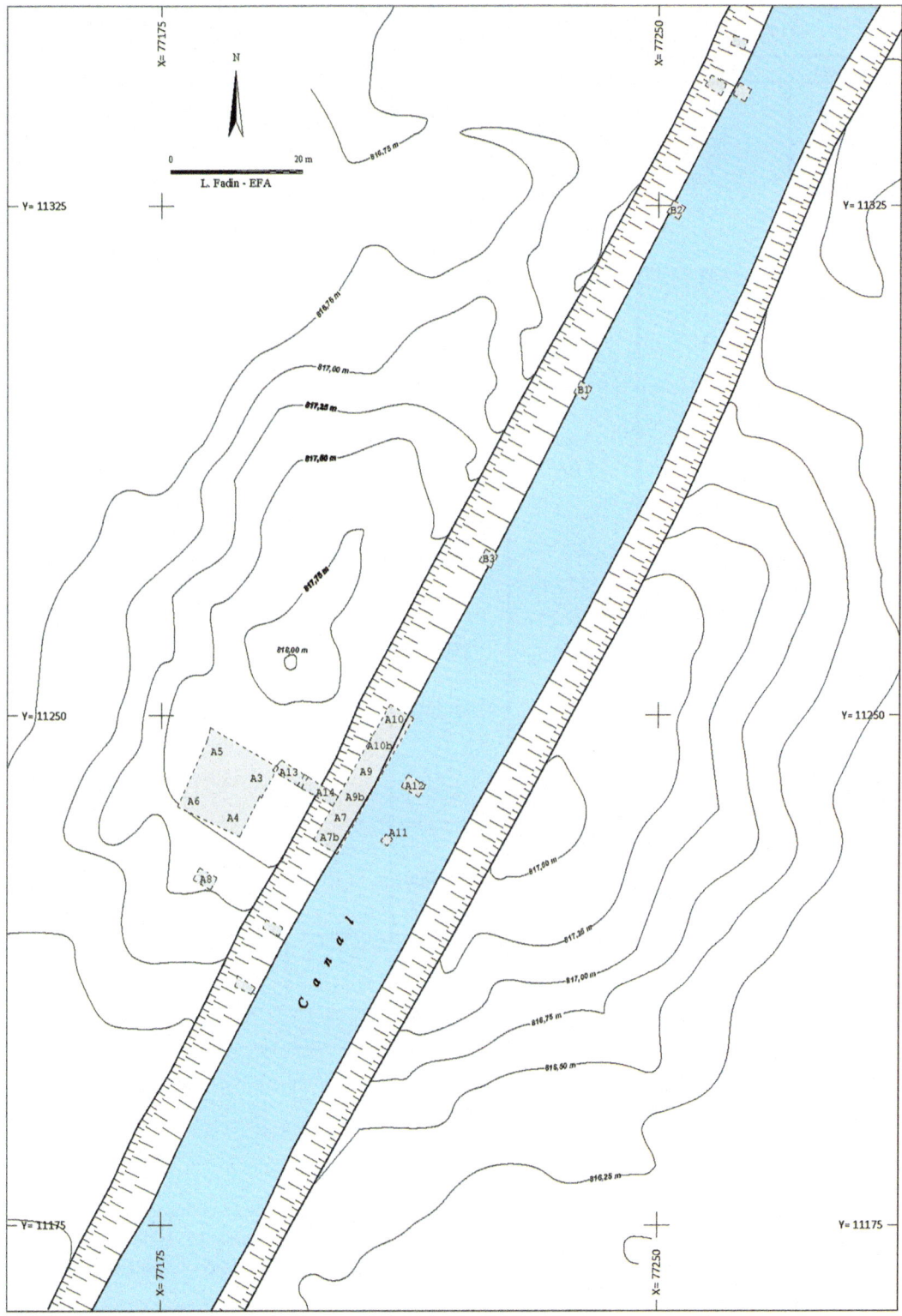

Fig. 7.1 Topographical plan of Sovjan with location of the excavation trenches (L. Fadin, EFA)

SOVJAN		Periods (Albanian chronology)	Absolute dates BC	Maliq phases	Kastanas phases	Periods (Aegean chronology)
Levels	Phases					
1	/					
2	/					
3	/					
4	/					
5a [*S1*]	**VII**	EIA	900–700	IVb IVa	VI-VII (c. 5–10)	[Protocorinthian] Geometric
5b						
5c1	**VI**	Transition LBA/EIA	1100–900	IIId2	V (c. 11–13)	Protogeometric/ Submycenaean
5c1 [*S2*]			1200		IV (c. 14–17)	LH IIIB–C LH IIIA
	V	LBA		IIId 1		
5c2 [*S3*]			1400			LH II LH I
5c3			1600–1400		III (c. 18–19)	
gap					*gap* (c. 20)	MH III MH II
				IIIc		
6 [*S4*]	**IV**	MBA	1900		II (c. 21–22a)	MH I
7				IIIb		
8			2200–2000	IIIa	I (c. 22b–28)	EH III
	III	EBA III				
9			2500–2300			
10						
11	**II**	LN I MN	5200–5000 6000–5600			LN I MN
12						
13	**I**	EN/Mesolithic?	7300–6700			Mesolithic
14		*Virgin soil*				

Fig. 7.2 Chrono-stratigraphic sequence of Sovjan with the main synchronisms. *EN* Early Neolithic; *LN* Late Neolithic; *EBA* Early Bronze Age; *MBA* Middle Bronze Age; *LBA* Late Bronze Age; *EIA* Early Iron Age; *EH* Early Helladic; *MH* Middle Helladic; *LH* Late Helladic (C. Oberweiler, T. Krapf, G. Touchais)

Only very few architectural features could be associated with this phase: a small number of heaps of burnt or decomposed clay from collapsed walls, several post holes, and one single oak stake. Three radiocarbon dates are available for this phase, ranging between the 14th and 9th centuries BC, whereas the dendrochronological analysis of the oak stake points to 1008 BC.[4] Sovjan VI might therefore be considered roughly

---

[4] Analysis performed at Cornell University laboratory, 1995.

Fig. 7.3 Large pit formed by level 6 in trenches A9–A10b (G. Touchais, EFA)

contemporary with the (Aegean) Submycenaean/
Protogeometric periods.

7.3 Sovjan V

Phase V includes three habitation levels, all
dating to the LBA: the latest one, to which the
floor S2 belongs,[5] corresponds to the lower
stratigraphic units of level 5c1 (see above);
underneath comes level 5c2 with its floor S3, and
then level 5c3. All these levels consist of a
similar soil, brownish grey to greenish in colour.
They have a sandy-clay texture and contain
varied amounts of micro-charcoal and reddish
clay nodules, resulting from the decomposition
of architectural structures.

7.3.1 Level 5c1

Floor S2, found in all sectors, is very well
characterised, both by associated remains of
architectural structures—which unfortunately do

not provide any information of the plan, neither
of the dimensions of the houses, but which
nevertheless show a general orientation of the
buildings along two axes NS and EW—and by a
large amount of pottery (more than 11,000
sherds) and other artefacts (more than 160).
There is evidence to suggest that S2 was
destroyed (accidentally?) by fire: by baking the
loam constructions, the fire partially conserved
the buildings in situ. Many pottery fragments
were baked to a degree of vitrification.

The architectural traces consist mainly of
collapsed loam walls, fragments of floors made
of earth and loam, of wattle and daub walls or
roofs, about sixty wooden piles, many post holes,
and at least nine domestic hearths or ovens. We
also note the presence of many fragments
belonging to one (or more) perforated kiln bottoms, finding parallels in Greece—and even in
Macedonia—during the LBA, and which prove
the presence of one (or more) potter(s) at Sovjan
during this period. Among the rich set of artefacts associated with floor S2, there are mill-
stones and other macrolithic tools, bone and
antler tools, bone pins with carved heads (Touchais and Lera 2007, Fig. 12), clay spindle
whorls, loom weights, and small perforated discs

---

[5] Floors («sols») are numbered continuously from top to
bottom; floor S1 belongs to layer 5a (Sovjan VII: EIA).

—which could have functioned as textile implements (Cheval 2012), one bronze knife of the type Sandars 1a (Touchais and Lera 2007, Fig. 9), and one spearhead of the 'willow leaf' type that points to contacts with the central Balkan area (Vasic 2015, 34–100). We would like to highlight the presence of a clay bellow's tip with traces of burning, testifying to local metallurgical activity, as is also confirmed by two stone moulds for bronze axes (Touchais and Lera 2007, Fig. 11) and one single-use clay mould from level 5c2 (see below). If we add the fact that at least some of the bone pins were made on site, as evidenced by the many blanks and preforms (Christidou 2007), we conclude that Sovjan V was not just a rural village subsisting on agriculture and livestock, but also a place of active artisan production and craftsmanship.

Beside the local pottery, which presents all the characteristic features of the LBA-ceramics in the Korça basin and the neighbouring areas (Gori and Krapf 2015, 115–117), level 5c1 yielded some Mycenaean sherds which can be attributed to the phases LH III (A–) B and LH IIIC (Touchais and Lera 2007, Figs. 6–8). Furthermore, two bronze objects, as well as bone pins with ornate heads, belong to the objects circulating in the Balkans and the Aegean during the LBA. The relative chronology based on the find material corresponds to the very coherent series of eleven radiocarbon dates available for the stratigraphic units of level 5c1. These can be attributed to phase V since all of them fall into the second half of the 2nd millennium BC (between 1501 and 1013 cal BC).

7.3.2 Level 5c2

The lower part of level 5c2 corresponds to floor S3, which has been identified over a smaller surface than floor S2. It is characterised by a large amount of reddish clay nodules and large charcoal fragments, but also by several concentrations of ash. The identifiable architectural remains related to this floor are limited to five wooden piles and four poorly preserved ovens or hearths.

The pottery is very fragmented and only a few vases could be partly reconstructed. Nevertheless, the pottery shapes are similar to those of level 5c1. However, compared to the underlying level 5c3, we note several innovations that are typical of the LBA, such as incised decorations or the pattern of three perforations or triangle prints on the upper attachment of *kantharoi* handles (Gori and Krapf 2015, 111–113). The forty-nine lithic, bone, and clay artefacts hardly differ from those of level 5c1, except a fragment of a single-use clay mould which indicates the mastering of the lost-wax casting technique.

The five radiocarbon dates available from this level range between 1450 and 1229 cal BC. Hence, they are very close to those of 5c1, without any of them dating later than the middle of the 12th century. This suggests a continuous occupation but also a relatively short duration for level 5c2.

7.3.3 Level 5c3

The underlying level 5c3, which is relatively thin, constitutes a 'buffer layer' between the floors S3 and S4, the latter belonging to level 6 (see below). It has been interpreted as an abandonment level, at least based on the areas that have been excavated: in fact, no architectural remains have been discovered. However, the pottery from this layer, although quite scarce and very fragmented, fits coherently into the typological development from level 6 to level 5c2. It shows not only the first appearance of types that are characteristic of the LBA, such as the *kantharos* and the *pyraunos* (cooking pot with integrated support), but also the introduction of the oxidation-firing process, which produces light-surfaced vases. This pottery assemblage could potentially evidence an occupation of this period in another area of the settlement. Such ceramics are not represented in the upper or the lower layers. Furthermore, the two radiocarbon dates obtained for level 5c3 (between 1628 and 1306 cal BC) are a little older than those of 5c2 but in perfect continuity. This suggests that the village of Sovjan did not experience a real period of abandonment like the one observed a few centuries earlier (see below), but rather that the dwellings of this phase are located somewhere outside of the explored area.

7.4 Sovjan IV

Sovjan IV corresponds to level 6, which is easily distinguishable from the overlying levels by its yellow colour and its sandy texture. This level is associated with a floor, S4, which can be subdivided into two successive horizons, h1 and h2.[6]

In the main sector at the edge of the canal, level 6 formed a large pit, oriented NW–SE and roughly corresponding to the trenches A9b–A9–A10b–A10 (Fig. 7.3). This pit was about 3 m wide, over 15 m long, and 0.70 m deep; its bottom was lined with a thin black layer, which contained many fragments of charred wood. In the upper part of the pit, floor S4 was identified. It was delimited by differently oriented wall segments, clearly corresponding to two successive horizons (Fig. 7.4). The lower part of the pit, filled with the same sandy yellow soil, did not contain any archaeological material. As the pit lies directly above an older floor (S5) which belongs to level 7 (see below), it is assumed that it is a backfill pit, containing material accumulated while cleaning up the area before installing new dwelling structures: the inhabitants of level S4 would therefore have first burned the previous dwelling level in order to make space—which would explain the presence of the thin carbon layer at the bottom of the pit—and then they would have backfilled the area where they intended to settle with sand, in order to level it out and to protect it from moisture.

The wall remains corresponding to the upper horizon h1, one of which (*locus* 917) had a narrow foundation trench in which were stuck, at regular intervals, plank-shaped wooden beams (Fig. 7.4), allow us to assume a long, rectangular building (Building A) with dimensions of c. 8 × 2 m and oriented NE–SW. Inside, a large

Fig. 7.4 Remains of walls of Building A (917) and Building B (918) (G. Touchais, EFA)

storage jar and at least one oven or hearth were found. Most of the structures associated with this horizon were in poor condition, as was the pottery, suggesting significant post-depositional natural and/or anthropogenic formation processes. Despite this, at least two complete vases were recorded, including a dark burnished *kantharos* with raised handles of the 'pseudo-Minyan' type (Touchais and Lera 2007, Fig. 4), which clearly imitates Aegean prototypes of the early Middle Bronze Age (MBA).

The lower horizon h2 was better preserved and richer in archaeological findings. Remains of a house (Building B) were recovered, which had at least one wall built in a different technique (*locus* 918), i.e. a wooden frame supported by a double alignment of cylindrical piles (Fig. 7.4). This building was over 7 m long and over 2 m wide (possibly c. 15 × 4 m) and its orientation (N–S) differed from that of Building A. Several

---

[6] Like the floors, the horizons are numbered continuously from top to bottom.

hearths or ovens, as well as an oval pit containing charred seeds (barley, einkorn, wheat, spelt, bread wheat, bitter vetch: Allen 2005) suggesting that it was probably a buried container made of perishable material, also belong to this horizon— perhaps even to this building. The pottery of horizon h1 includes several almost complete, light-surfaced vases which all belong to types that were already present (in dark wares) in level 7 (or even 8). Hence, they testify both to the persistence of the traditions of the previous phase with regard to the repertoire, but also to a technological innovation which will be generalised in the subsequent phase, i.e. the oxidation-firing process.

The exact chronology of level 6 is difficult to determine. Based on the incomplete data, it was initially dated to the beginning of the LBA (Touchais and Lera 2007; Touchais 2008). However, the detailed study of the ceramic typology (Gori and Krapf 2015, 107–109) and a series of radiocarbon dates ranging between 2567 and 1931 cal BC now clearly prove that this level does not date later than the early MBA in the Aegean but might even date a little earlier. Consequently, level 6 seems to represent the transition from the Early to the Middle Bronze Age at the turn of the 3rd millennium BC. This also implies a gap of three or four centuries in the occupation of the settlement, namely between the end of phase IV and the beginning of phase V, which could possibly explain the significant post-depositional transformations observed in horizon h1.

7.5 Sovjan III

Phase III encompasses three distinct stratified levels that are easy to distinguish from each other: level 7 with floors S5 and S6, level 8 with several successive layers of wooden constructions, and layer 9 which did not yield any architectural remains and contained only a very small amount of pottery. These three levels are characterised by a monochrome dark burnished ware, the same as in Maliq IIIc (Prendi 2018, Pls. CXXIV–CXXXI). The levels also all lie below the water level, leading to permanent waterlogged, which makes their exploration more delicate and requires the ground water to be constantly pumped out during the excavation process.

7.5.1 Level 7

Level 7 is a very clayey layer of bluish-grey colour, containing many small charcoal fragments. According to sedimentological analysis, its formation is linked to a rise in the level of the lake (see below). In A7, A7b, and A9, this level included floor S5, split into two successive horizons, h3 and h4 (Fig. 7.5). This floor carried at least four ovens or hearths, as well as a pebbled area referring to heated stone hearths as known from the Neolithic lakeside settlements in the circum-alpine area. Since floor S5 was lying one level below the top of the wooden planks which supported the wattle walls of the previously built 'Maison du Canal' (see below, layer 8), floor S5 is assumed to represent the last state of this house. The floor deposit was characterised by a high density and variety of material: crushed vases, numerous artefacts (chipped-stone, ground stone and bone tools, clay tools), plant remains, and fauna debris, including a significant concentration of fish bones.

In A10b, another floor (S6), delimited to the west by an alignment of post holes which corresponded to a wall that runs parallel to the east wall of the previous house at c. 2 m distance: this floor has been attributed to the latest stage of the underlying Building C, the entire western part of which had been cut off by the canal (see below, layer 8). The floor deposit, which included six clay loom weights, bore traces of a violent fire; here, the only anthropomorphic figurine found in Sovjan was discovered (Fig. 7.6). Its style recalls that of a late Early Bronze Age (EBA) figurine from Gradište Pelince in North Macedonia (Gori 2015, Fig. 77a).

On the basis of clear parallels between the pottery of level 7 and that of Maliq IIIc—considered as typical of the MBA according to the traditional Albanian chronology—level 7 was from the start assigned to this period, i.e. to the first half of the 2nd millennium BC (Touchais

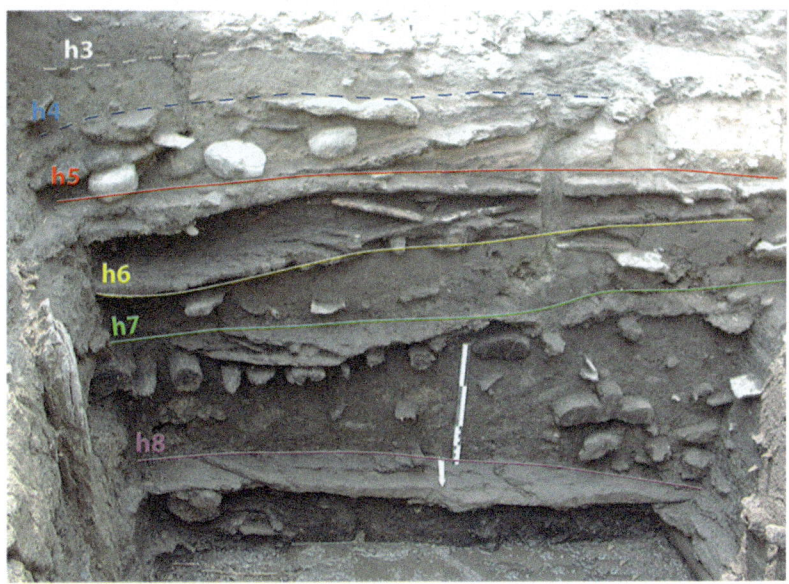

Fig. 7.5 South section of trench A9, with indication of the stratigraphic horizons of levels 7 and 8 (G. Touchais, C. Oberweiler, EFA)

Fig. 7.6 Clay figurine from level 7 (Sv 02/186.5) (G. Touchais, EFA)

2008). However, the in-depth study of pottery (Gori 2015), supported by six radiocarbon dates close to those of level 6 (between 2452 and 1954 cal BC)—has obliged us to question this former dating. In fact, the presence of ceramic types, such as the two-handled globular bowl with flaring rim or the so-called 'smoking pot' (Gori 2015, 82–84, 117–120), which are attested in well dated EB III contexts in Greece and the Balkans, argues that Sovjan 7—as well as Maliq IIIc—should be synchronised with the latest phase of the EBA, i.e. the last centuries of the 3rd millennium BC. Sovjan 7 therefore appears to be roughly contemporary with phase B at Archontiko in western Macedonia, whose absolute dates also fall between 2300 and 1900 BC (Gori and Krapf 2015, 94).

7.5.2 Level 8

Level 8 is mainly characterised by the presence of numerous remains of well-preserved wooden constructions, most of them in situ, which gives Sovjan a special status among the known Bronze Age settlements of the Balkans.

Level 8 is 0.70 m thick on average and characterised by a dark brown to black clay sediment, rich in rather decomposed organic matter (peat). The careful examination of the stratigraphy made it possible to distinguish at least four successive horizons within this level (h5, h6, h7, h8), related to various timber structures: floors of logs, wattle walls, piles, fallen beams, etc. Most of these wooden structures are of oak, except certain floors which sometimes combine wood from several different tree species. The structures essentially belong to four different units: the 'Maison du canal', the Building C, the 'Chemin de rondins' (Log Path), and the 'Maison du Pêcheur' (Fisherman's House).

7.5.2.1 The Timber Structures

The most spectacular find is a large apsidal house measuring more than 15 m in length and over 4 m in width, which has been called the 'Maison du Canal'. It is built of wattle and daub and preserved to a height of about 0.50 m. Orientated

N–S, it consists of a large space separated from a small apsidal room to the north by a partition wall (Fig. 7.7). A rather narrow opening (0.80 m wide) in the eastern exterior wall (towards the lake side), interpreted as door, was preserved, as well as remains of a bark layer in the northern part of the large space, which probably served as floor insulation (Fig. 7.8). The exact shape of the southern border of the house is not known because it was destroyed. However, one can imagine that it had a wide opening, like the huts of the culture of Palma Campania (first half of the 2nd millennium BC) discovered near Nola, in Campania (Albore Livadie 2002), which constitute the closest formal parallels to Sovjan's 'Maison du canal'. Another common feature is the overhanging roof that extends down to the ground, which creates a sort of covered gallery all around the house where food could be stored.

In addition to the wattle and daub walls and the layer of bark, further architectural elements of the 'Maison du canal' have been preserved:

Fig. 7.7 'Maison du Canal' from south (G. Touchais, EFA)

Fig. 7.8 Layer of bark in the 'Maison du Canal' (G. Touchais, EFA)

several areas of scattered plants or compacted branches, one of which (L1) covered most of the interior surface and was made up of wood and bark splinters, probably resulting from on-site timber cutting for the construction of the house; four wooden floors located at different levels, two of which were visible in A9 (*loci* 870 and 969) and two others in A7 (*loci* 796 and 933) (Fig. 7.9); a threshold related to a previous stage of the eastern exterior wall, slightly shifted to the west (Fig. 7.9); several large pieces of wood fallen to the ground and collapsed sections of wattle walls; as well as some elements that may belong to the roof. A single hearth, surrounded by several vases and tools, could be connected to the house. It belongs to the late floor, which corresponds to horizon h5 (see below).

Based on the study of all these elements, it is possible to distinguish four architectural stages of the 'Maison du canal', which largely correspond to the four horizons (h5–h8) mentioned above. For various reasons, mainly due to the difficult excavation conditions, it has not been possible to relate each of these stages to a specific floor. However, each of them could be precisely dated by dendrochronology (see below).

Fig. 7.9 Wooden floors 796 and 933, and the sill of the eastern door belonging to the 3rd stage of the 'Maison du Canal'; in the background, the eastern exterior wattle wall of the 4th stage (G. Touchais, EFA)

Immediately east of the 'Maison du canal' stood Building C, whose latest stage was already visible in level 7 (see above). The traces of its western wall consisted of a double row of post holes that could be traced over a 3 m distance. Further, a ridge post with a mortise at one of its two ends was found in the context of this building, as well as the remains of a wooden floor and a section of collapsed wattle wall.

Very close to the 'Maison du canal' and Building C, to the north, was the 'Chemin de rondins', a horizontal timber structure of 2 m width and more than 4 m length, oriented E–W. It seems to have been built of about twenty oak half-logs resting on four large sleepers (Fig. 7.10). This structure is interpreted as a path built directly on the ground, allowing people to move from one dwelling to another while keeping their feet dry. This is where the complete antler hammer-axe with a beech wood handle was discovered, lying between two sleepers (Touchais et al. 2005, Fig. 2).

On the north side of this path the access to another house was found, of which only the entrance area and a small part of wall were excavated. The door consisted of an oak sill of 1.20 m length with two vertical posts on each side, whereas the wall was built in the same wattle technique as the one applied in the 'Maison du canal' (Fig. 7.10). This house was called 'Maison du Pêcheur' because of the large amount of fish scales found inside.

7.5.2.2 The Relative Chronology of Level 8

The pottery of level 8, of which over a third comes from the 'Maison du Canal' (as is the case for all the material categories), is very similar to that of level 7. From a technological point of view, it shows the same predominance of dark monochrome vases betraying a clear preference for reduced firing conditions—which will change in LBA (see above), and the same exclusive use of plastic decoration (especially, fingered bands). From a typological point of view, the repertoire includes many shapes also occurring in level 7 (globular bowls, conical cups, 'smoking pot'), although few types, like the tankard, are found almost only in level 8 (Gori 2015, 84–87). All these features find their closest parallels in the sites of

Fig. 7.10 'Chemin de rondins' from west; in the upper left corner, the 'Maison du Pêcheur' (G. Touchais, EFA)

Kastanas layer 27 and Sitagroi Vb (Macedonia). Comparable pottery types from more distant regions (Serbia, Peloponnese) also fall into the last phase of the EBA (Gori and Krapf 2015, 103–104).

7.5.2.3 The Absolute Chronology of Level 8

The lack of a significant chronological difference between the levels 7 and 8 and their common attribution to the last quarter of the 3rd millennium BC are confirmed by the single radiocarbon date available for level 8 (2336–2033 cal BC). This date was provided by a charcoal sample collected on the oldest wooden floor of the 'Maison du Canal' (*locus* 796).

New absolute data have been provided by dendrochronology. Although many wood samples were taken during the excavations, they could not be properly analysed for various reasons. In 2018, we decided to reopen the main trench (A7b–A10) in order to sample further wood. Its dendrochronological analysis was carried out by specialists at the Institute of Archaeological Sciences at the University of Bern, under the supervision of Prof. A. Hafner. From the new 34 samples—including ten from the 'Maison du Canal' alone—six 'local' curves (SOV-3, SOV-4, SOV-7, SOV-10, SOV-12, SOV-16) could be established for all the structures of level 8, from which an average curve was calculated (SOV-18) (Fig. 7.11). This last curve, which covers 269 years, represents the relative chronology of level 8 (Maczkowski et al. 2021).

This relative sequence of 269 years could be converted into absolute years thanks to radiocarbon dating carried out on the same wood samples. The modelling of the combined radiocarbon and dendrochronological data made it possible to obtain a relatively precise date for the felling of the most recent timber, which was used in the last (fourth) architectural stage of the 'Maison du Canal'. This date, between 2158 and 2142 cal BC, marks the end of level 8 and therefore provides a *terminus post quem* for the beginning of level 7.

Thus, it is possible to assign calendar dates to the four successive building phases (Fig. 7.12) distinguished in layer 8 (from top to bottom):

Phase 4. In the latest phase (between 2158 and 2142 cal BC), the dendrochronological data allows to interrelate three distinct structures with the same architectural programme: 1) the fourth stage of the 'Maison du Canal', 2) the 'Chemin de rondins', and 3) the 'Maison du Pêcheur'. The fourth stage of the 'Maison du Canal' corresponds to horizon h5, to which the bark layer and the refuse or 'litter' L1 seem to belong (see above). It also incorporates some re-usage of older planks, two of which come from the eastern wall of the previous stage (see below). On the other hand, the curves SOV-3 and SOV-4 show that the other two structures had undergone repair work shortly after phase 4.

Phase 3. This building phase corresponds to the third stage of the 'Maison du Canal' (horizon h6), in which the eastern exterior wall and its door were located 0.50 m further west than at the final stage (see above). The dendrochronological data show that the timber used in this phase was felled ten to twenty-four years earlier than the one of phase 4, i.e. between 2168/2152 and 2182/2166 cal BC. If we assume, for the sake of convenience, an interval of twenty years—which roughly corresponds to a human generation—phase 3 can be dated between 2178 and 2162 cal BC. The presence of some older piles may be explained either by the re-use of older timbers, or by the continuous use of structures built during the previous phase. On the other hand, we note that the timber used during this phase is almost exclusively oak and comes from quite old trees, i.e. between 170 and 190 years old at the moment of felling. The material is therefore very different from that which was used during the two previous phases (see below).

Phase 2. This phase is synchronous with the second stage of the 'Maison du Canal' (horizon h7), which is mainly represented by the wooden floor 969. Its structure indicates that it was not resting directly on the ground but was slightly raised. Based on combined dendrochronology and radiocarbon data, this phase is dated between 2218 and 2202 cal BC. It therefore predates phase 3 by around forty years. The logs used in the construction of the floors in this phase, like those in phase 1, belong to different species (oak,

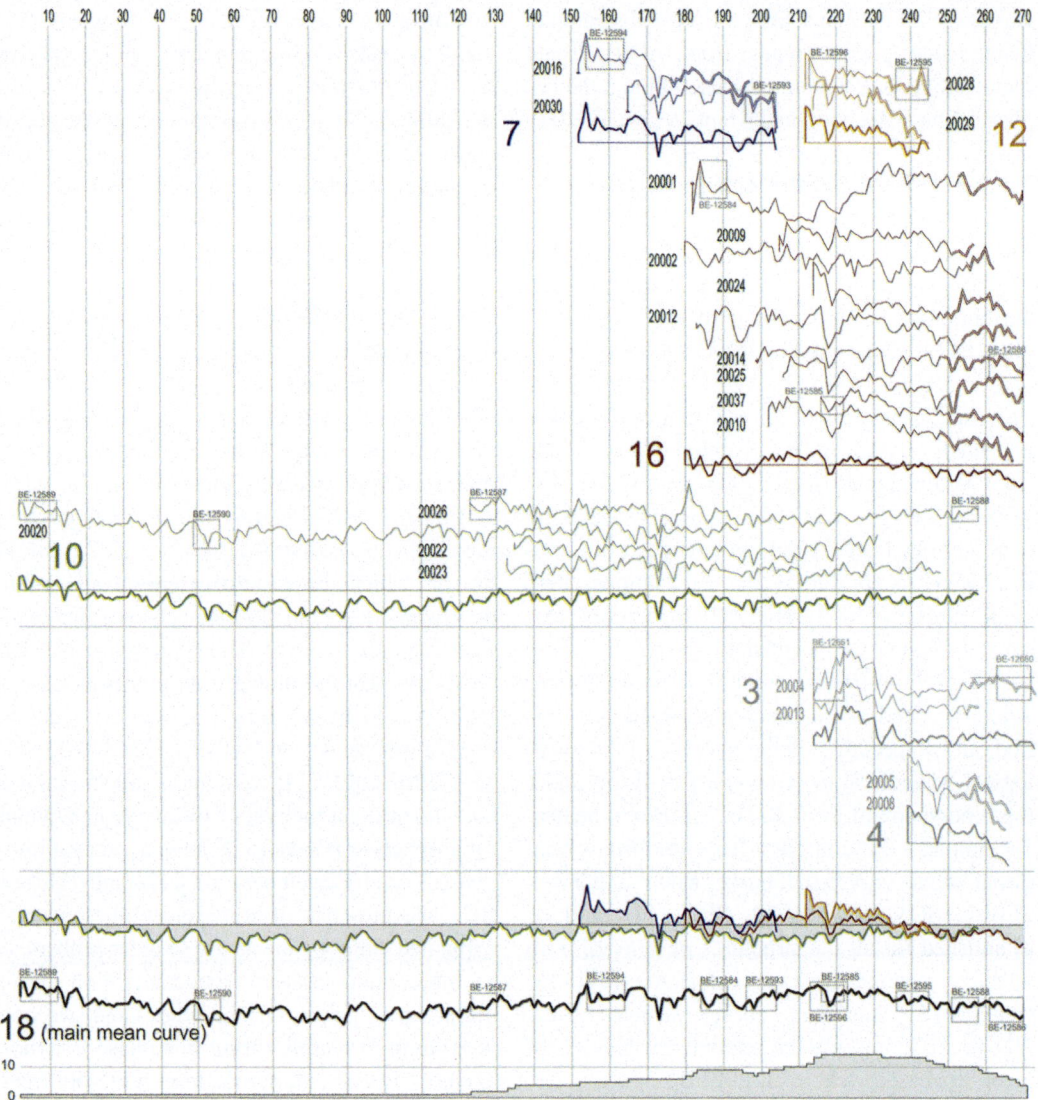

Fig. 7.11 Plotted tree-ring widths (semi-logarithmic scale) of the individual wood samples and the constructed mean curves discussed in the text. Numbers above the vertical straight lines correspond to the relative ring number. The histogram at the bottom of the image represents the sample replication in the mean curve SOV-18. The small black rectangles with 'BE-#' labels represent the positions of samples taken for radiocarbon dating (Matthias Bolliger, EXPLO/UBern)

willow, elm, beech, ash, fir, pome fruit tree) and were felled at a young age (10–15 years).

Phase 1. There are no dendrochronological data indicating a building phase prior to phase 2, yet an older phase has left traces in the stratigraphy, i.e. horizon h8, which corresponds to the first stage of the 'Maison du Canal'. The two lowest wooden floors (796 and 870), the first of which is radiocarbon dated to 2336–2033 cal BC (see above), belong to this initial stage. As this first floor constitutes the stratigraphic limit between levels 8 and 9, it provides a *terminus ante quem* for the beginning of level 8 and therefore allows to date the first building phase

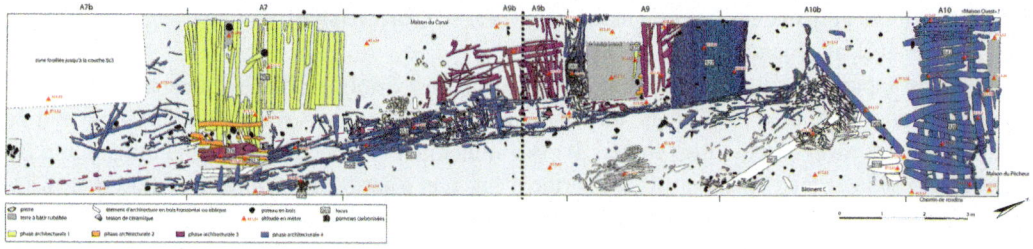

Fig. 7.12 Four architectural phases of the wooden structures of level 8 in the main trenches A7b–A10 (L. Fadin, C. Oberweiler, EFA)

between 2336 and 2202 cal BC. Floor 870, on the other hand, testifies to a first raised floor level of the house, which suggests that the inhabitants were seeking to insulate the interior of the buildings from humidity.

7.5.3 Level 9

Level 9 is a compact clay-peat layer varying in thickness between c. 1.00 and 1.80 m. It is only evidenced stratigraphically in the trenches A7 and A9, where it appears immediately underneath level 8, while its stratigraphic position is a little less certain in A12. In A9, the layer lacks any archaeological material, which explains why it was initially interpreted as virgin soil. This was the case until the excavation could advance deeper in 2001 and reach the Neolithic layers. In A12, the small amount of pottery found in the upper part of level 9 hardly differs from that of level 8.

The only architectural remains found in this level (A12) are nearly thirty large oak stakes, the pointed ends of which were planted more than two metres into the underlying layers. Some of them had a fork at the top and supported the remains of a raised wooden floor, similar to those of the first two building phases of level 8 (see above).

Therefore, everything seems to indicate that level 9 is an archaeologically sterile level and that the structures, as well as the rare artefacts found in its upper part, actually belong to level 8. The only radiocarbon date available for this layer (2563–2308 cal BC) are barely two centuries earlier than that of level 8 (see above).

7.6 Human Settlement and Variations of the Lake Level

If we attempt to combine the results of the archaeological excavations with those of the palaeoenvironmental study carried out in parallel (see above), the impact of the lake level variations on the settlement occupation during the Bronze Age becomes quite clear.

The thick layer of sterile peat that precedes the settlement of level 8 is easily identifiable in the ten sediment cores taken from the transect mentioned above. It reflects a period of decreasing lake level, with a resumption of peatland formation that begins around 4000 BC (Lera et al. 2008, 885–887). Level 9, as it was recognised in the archaeological trenches, therefore corresponds to the end of this period and is evidence of a moment before the water rise level attested by level 7 (see below).

The first identifiable dwelling of Sovjan III (= 1st stage of the 'Maison du Canal') is built on the shore of the lake, close enough that its foundations got wet during periods of lacustrine transgression. A generation later, the inhabitants rebuilt this first dwelling (= 2nd stage) with a raised floor, which indicates the lake level's overall rising trend. The same phenomenon recurred some forty years later, with the 3rd stage of the 'Maison du canal', featuring another raised wooden floor. Finally, some twenty years later, the construction of a wooden path to move between the houses suggests that the dwellings are in contact with surface water, which is confirmed by the grey clay in layer 7, indicating a high lake level at around 2100/2000 cal BC. This episode apparently forced the inhabitants to temporarily abandon their dwellings and presumably move away from the shore. It is tempting to relate this episode of lacustrine transgression to the brief cold and humid climatic event ('4.2 ka event') which occurs at about the same time, i.e. between 4300 and 3800 cal BP, causing a phase of high water levels in the lakes of western-central Europe (Magny et al. 2009). However, this hypothesis still lacks evidence.

Taking advantage of a period of low water, the inhabitants of Sovjan phase IV resettled on top of the ruins of the 'Maison du canal' after having cleaned and filled the entire area with sand (see above). They put up a first building (B) that has the same orientation as the 'Maison du Canal', which shows that its memory had not been lost. Later, still during phase IV, a further building (A) is built on top of the previous one. This time, its orientation is different. In fact, the LBA settlement will refer to it. Phase IV ends with a gap of about three to four centuries, i.e. based on radiocarbon dating between 1900 and 1600 cal BC. This temporary abandonment of the settlement does not seem linked to a lake level rise, as according to sedimentological analyses, the beginning of the 2nd millennium is a period of lake regression which will last for the entire millennium.

The period which spans the entire second half of the 2nd millennium BC is marked both by a low lake level and by an elevation of the tell itself. Consequently, the inhabitants of the phases V and VI are installed on dry land and further away from the shore than their predecessors. Several successive floors are related to these phases, which prove that the settlement is continuously inhabited. The finds show that it reached its peak towards the end of the LBA, i.e. between 1300 and 1100 BC (level 5c1, floor S2). However, even if various—and sometimes sophisticated—craft productions of this period are well documented, the dwellings are less known. Rather, they are most often reduced to shapeless clay masses, while the rare indications at best hint at the general structure and orientation of the settlement.

7.7 Conclusion

It is now firmly established that the Bronze Age levels explored at Sovjan range from the mid-3rd millennium to the end of the 2nd millennium BC, i.e. over a period of 1500 years—however, with a gap covering almost the entire first half of the 2nd millennium (MBA). Whereas the lake dynamics had an undeniable impact on the evolution of the settlement, they cannot exclusively explain all the changes observed during this long period and in particular not the abandonment phases. Some of these changes are surely related to human factors or choices, which however cannot be directly revealed by archaeology. Only a systematic exploration carried out at the scale of the entire Korça basin, by excavating the synchronous settlements identified during the archaeological survey of the PALM Project (2007–2013),[7] could possibly allow to observe common trends and suggest explanatory hypotheses.

Acknowledgements The 'Mission archéologique franco-albanaise du bassin de Korçë', which is currently under the co-direction of A. Gardeisen and R. Kurti, has carried out several archaeological projects in the Korça basin since 1993, with the constant support of the École française d'Athènes, the French Ministry of Foreign Affairs and the Albanian Institute of Archaeology at Tirana, to all of whom we would like to express our warm gratitude.

References

Albore Livadie C (2002) A first Pompeii: the Early Bronze Age village of Nola-Croche del Papa (Palma Campania phase). Antiquity 76:941–942

Allen SE (2005) A living landscape: the Palaeoethnobotany of Sovjan, Albania. Ph.D. dissertation, Boston University, Ann Arbor

Cheval C (2012) Une utilisation des disques perforés en terre cuite: la série de Sovjan en Albanie méridionale. BSPF 109:157–160

Christidou R (2007) Aperçu des industries osseuses de l'habitat protohistorique lacustre de Sovjan (bassin de Korçë, Albanie sud-orientale). BCH 131:755–803

Fouache E, Desruelles S, Magny M, Bordon A, Oberweiler C, Coussot C, Touchais G, Lera P, Lézine A-M, Fadin L, Roger R (2010) Palaeogeographical

---

[7] See preliminary reports in BCH 131 (2007) to 138 (2013).

reconstructions of Lake Maliq (Korça Basin, Albania) between 14000 BP and 2000 BP. JAS 37:525–535

Gori M (2015) Along the rivers and through the mountains. A revised chrono-cultural framework for the south-western Balkans during the late 3rd and early 2nd millennium BCE. Rudolf Habelt, Bonn

Gori M, Krapf T (2015) The Bronze and Iron Age pottery from Sovjan. Iliria 39:91–136. https://doi.org/10.3406/iliri.2015.2500

Korkuti M, Petruso K (1993) Archaeology in Albania. AJA 97:703–743

Lera P, Touchais G et al (2008) Rapport sur les travaux de l'École française d'Athènes en en 2007. Sovjan. Étude Et Prospection. BCH 132:875–903

Lera P, Oberweiler C, Touchais G (2010) Le passage du Bronze Récent au Fer Ancien sur le site de Sovjan (bassin de Korçë, Albanie): nouvelles données chronologiques. In: Lamboley J-L, Castiglioni MP (eds) L'Illyrie méridionale et l'Épire dans l'Antiquité V, Actes du Ve colloque international de Grenoble, 8–11 octobre 2008. De Boccard, Paris, pp 41–52

Maczkowski A, Bolliger M, Ballmer A, Gori M, Lera P, Oberweiler C, Szidat S, Touchais G, Hafner A (2021) The Early Bronze Age dendrochronology of Sovjan (Albania): a first tree-ring sequence of the 24th–22nd c. BC for the Southwestern Balkans. Dendrochronologia 66:1–12. https://doi.org/10.1016/j.dendro.2021.125811

Magny M, Vannière B, Zanchetta G, Fouache E, Touchais G, Lera P, Coussot C, Walter-Simonnet A-V, Arnaud F (2009) Possible complexity of the climatic event around 4300–3800 cal. BP in the central and western Mediterranean. The Holocene 19:823–833

Oberweiler C, Touchais G, Lera P (2014) Les recherches franco-albanaises dans la région de Korçë: nouvelles données sur la chronologie absolue de la préhistoire albanaise. In: Përzhita L, Gjipali I, Hoxha G, Muka B (eds) Proceedings of the international congress of Albanian archaeological studies, 65th anniversary of Albanian archaeology (21–22 November, Tirana 2013). Botimet Albanologjike, Tirana, pp 83–92

Prendi F (1966) La civilisation préhistorique de Maliq. Studia Albanica 3(1):255–280

Prendi Fr (2018) Vendbanimi Prehistorik i Maliqit. The Prehistoric Settlement of Maliq. Posthumous edn by Bunguri A, Gashi S. Botimet M&B, Tirana

Touchais G (2008) Sovjan et l'Âge du Bronze en Albanie. In: Guilaine J (ed) Villes, villages, campagnes de l'Âge du Bronze. Séminaire du Collège de France, Paris, pp 108–123

Touchais G, Fouache E (2007) La dynamique des occupations de bord de lac dans le Sud-Ouest des Balkans: l'exemple de Sovjan, bassin de Korçë (Albanie) ». In: Richard H, Magny M, Mordant C (eds) Environnements et cultures à l'Âge du Bronze en Europe occidentale. Actes du 129e congrès national des sociétés historiques et scientifiques (CTHS), Besançon, 19–21 avril 2004. Paris, pp 375–386

Touchais G, Lera P (2007) L'Albanie méridionale et le monde égéen à l'âge du Bronze: problèmes chronologiques et rapports culturels. In: Galanaki I, Tomas H, Galanakis Y, Laffineur R (eds) Between the Aegean and Baltic Seas: prehistory across borders. Proceedings of the international conference, Zagreb, 11–14 April 2005. Aegaeum, vol 27. Liège, pp 141–147

Touchais G, Lera P, Oberweiler C (2005). L'habitat préhistorique lacustre de Sovjan (Albanie): dix ans de recherches franco-albanaises (1993–2003). In: Della Casa Ph, Trachsel M (eds) WES '04. Wetland economies and societies. Proceedings of the international conference, Zurich, 10–13 march 2004. Collectio Archælogica, vol 3. Chronos, Zurich, pp 255–258

Touchais G, Lera P, Oberweiler C (eds) (2023) Sovjan, village préhistorique lacustre d'Albanie sud-orientale. 1: Le site dans son environnement, Recherches Archéologiques Franco-Albanaises, vol 5. Athènes

Vasic R (2015) Die Lanzen- und Pfeilspitzen im Zentralbalkan. Prähistorische Bronzefunde, Abteilung V, Band 8. Franz Steiner Verlag, Stuttgart

Neolithic Lake Settlements in Western Macedonia, Greece: New Evidence from Dispilio and Amindeon Basin

8

Kostas Kotsakis and Tryfon Giagkoulis

Abstract

The discovery and investigation of Neolithic habitations established in marginal zones of West Macedonia lakes and marshes constituted an exceptional occasion for Greek prehistoric archaeology. The distinctive site-formation processes and the uniqueness of the preserved organic materials posed unprecedented practical, methodological, and interpretive challenges, offering at the same time new potentials for the study of the communities' diachronic development. The present paper summarises the research results derived from the Rescue Excavations Project in the Four Lakes Region (Amindeon Basin) and the latest data from the ongoing study by the Aristotle University of Thessaloniki (AUTH) of the Dispilio lake settlement in the framework of the ERC-funded EXPLO project. The aim is to codify and evaluate the available information about the chronology of the habitations, their location in specific environmental settings, and the evidence regarding the construction and organisation of their built space. Subsequently, these characteristics are juxtaposed with documented parallels from the neighbouring areas to contextualise the specific attributes of the wetlands into the regional Neolithic. This comparative approach allows us to consider to what degree the wetland habitations constitute a unique phenomenon, or they are simply one of the diverse manifestations of the Western Macedonia Neolithic developed for specific—so far not detectable—reasons.

Keywords

Western Macedonia (Greece) · Neolithic communities · Wetland habitations · Environmental setting · Chronology · Regional contextualisation · EXPLO

8.1 Introduction

The lakeside site of Dispilio represents an advanced stage of the Neolithic settlement in the broader region of Northern Greece and Thessaly. Before Dispilio, a long string of dry-land sites goes back at least one thousand years before the local neolithic people built their dwellings on waterlogged ground. So why was this choice preferred? It certainly looks out of place in Thessaly, even in Western and Central Macedonia, where recent research excavated some of the pioneer sites of the earliest Neolithic (Kotsakis 2014; Maniatis 2014; Maniatis et al. 2015; Karamitrou-Mentesidi et al. 2015).

K. Kotsakis (✉) · T. Giagkoulis
Department of Archaeology, School of History and Archaeology, Aristotle University of Thessaloniki, Thessaloniki, Greece
e-mail: kotsakis@hist.auth.gr

T. Giagkoulis
e-mail: tgiagkou@hist.auth.gr

A. Ballmer et al. (eds.), *Prehistoric Wetland Sites of Southern Europe*,
Natural Science in Archaeology, https://doi.org/10.1007/978-3-031-52780-7_8

The contrast with the long tradition of dry-land settlements is stark and calls for some explanation, especially since there is plenty of dry ground around the immediate vicinity of the Dispilio site. Broadly contemporary with Dispilio dry sites have been excavated in the vicinity, on the hillslopes of Avgi (Stratouli et al. 2020) and the banks of Aliakmon River (Kolokinthou, Stratouli et al. 2019 and Trita Koromilias, Tsouggaris et al. 2004). Environmental considerations come to mind, e.g. a change in climate parameters acting as a pull factor is a possibility—but cultural, social, and economic factors should also play an essential part in this innovative choice. The abundance of palaeoenvironmental evidence preserved in the waterlogged deposits of Dispilio makes it an ideal case to shed light on the interaction of environmental and cultural factors.[1]

Since the beginning of the excavations at Dispilio, the ongoing research added more waterlogged sites to the Neolithic habitation of the region. Some of them were excavated, and we have a good idea of their material culture. With all the affinities, relations, and influences, they form an exciting picture of the Neolithic culture in this region which will be briefly presented further down before focusing on the Dispilio site itself.

8.2 Waterlogged Sites of Amindeon Basin

To examine the new waterlogged sites of Western Macedonia, we have to move further north to the region of Amindeon, which according to its geomorphology, is a typical basin situated at

an altitude of approximately 600 m a.s.l. defined by relatively high mountainous ranges. An extensive shallow-water lake covered the basin, formed in the late Miocene to the early Pliocene periods. The four surviving up to this day lakes are the remains of this original palaeolake Eordaea dated in the late Miocene to the early Pliocene periods (Kloosterboer-van Hoeve et al. 2001) (Fig. 8.1).

Former and recent palynological and anthracological investigations document a predominance of mixed oak-hornbeam forests in Lakes Chimaditis and Zazari during almost all of the Early, Middle, and Late Neolithic, with one maximum expansion between the mid-6th and mid-5th millennium BCE. Moreover, coastal halophytic and alluvial hardwood forests covered these wetlands (Bottema 1974, 1982; Gassner et al. 2020; Gerasimidis and Athanasiadis 1995; Marinova and Ntinou 2018; Ntinou 2014).

This region had attracted little attention from archaeological research, and consequently, the available information regarding the development of the prehistoric habitation was derived from surveys and collection of surface material (Kokkinidou and Trantalidou 1991; Trantalidou 1989). However, this situation changed during the last two decades due to the intensification of the lignite-mining activity that necessitated a large-scale prospection and rescue excavations project by the Florina Ephorate of Antiquities. According to preliminary reports, fifty-four new sites were identified, dating from prehistoric to later historic periods (Chrysostomou and Giagkoulis 2016; Chrysostomou et al. 2015). Of particular interest is the presence of thirteen settlements dating to the late-7th millennium BCE; yet, except from one short reference to potential Early Neolithic pit-houses in the southeastern part of the dry-land habitation Anarghiri XI (Chrysostomou and Giagkoulis 2018), no further information on these early settlements and no ^{14}C dates are published so far.

It seems plausible that the local farming communities expanded their activities in zones closer to the local waterscapes about the mid-6th millennium BCE. Indications are that nineteen Neolithic 'lakeshore habitations' were established in

[1] An interdisciplinary team from the universities of Bern, Oxford, and Thessaloniki was awarded an ERC Synergy Grant in 2018 for the project entitled 'Exploring the dynamics and causes of prehistoric land use change in the cradle of European farming' (EXPLO). EXPLO (2019–2025) seeks evidence for the human–environment interface. Dispilio is one of the focuses of the project. The principal investigators (PI) of the project are Prof. Albert Hafner and Willy Tinner (University of Bern), Prof. Amy Bogaard (University of Oxford), and Kostas Kotsakis (Aristotle University of Thessaloniki).

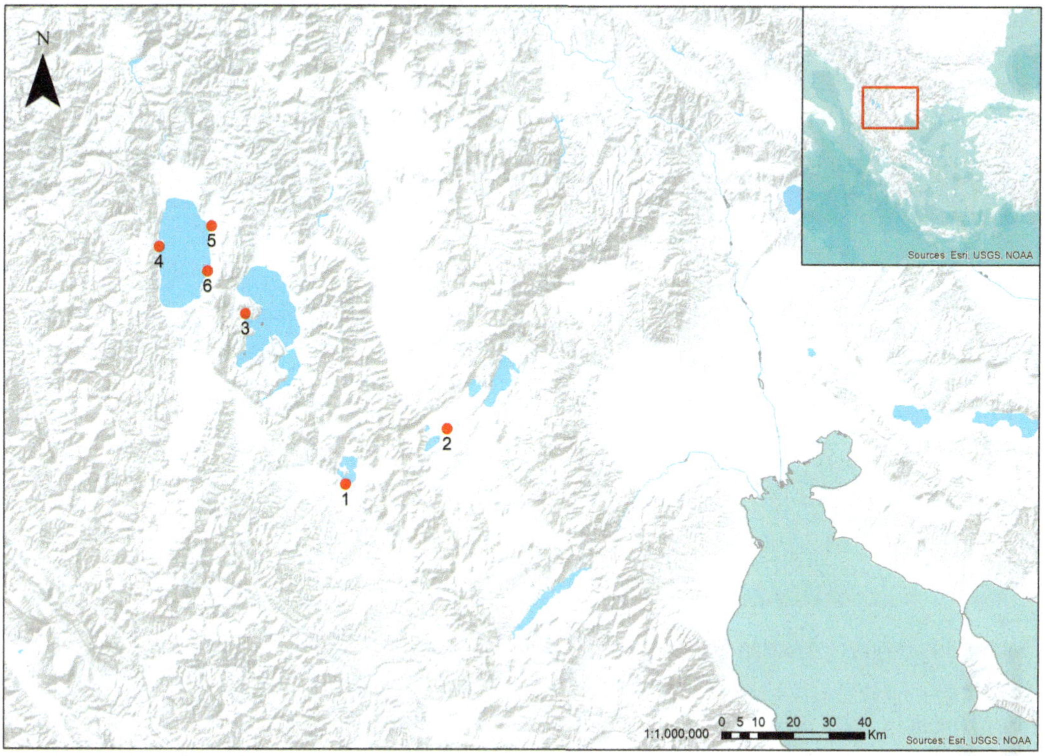

Fig. 8.1 Wetland habitations in Greek Macedonia, Albania and Northern Macedonia: **1.** Dispilio; **2.** Amindeon Basin waterlogged habitations (Anarghiri III, IXa and IXb, Limnochori II and III); **3.** Kallamas; **4.** Lin II; **5.** Ohridati; **6.** Ploca Micov Grad (Digital map by Filippos Stefanou)

the surroundings of the prehistoric wetlands, influenced even periodically by water fluctuations. Another eight occupations on the northern shore of Lake Chimaditis are described as 'typical lakeside pile-dwellings'. The findings of Limnochori II and Anarghiri III represent this last type of habitation and are regarded as the earliest attempts of the local communities to settle in close relationship to water during the mid-6th millennium BCE. In both settlements, the vertical posts, the horizontal wooden elements, and destruction layers are presumed to belong to stilted houses built in the Lake Chimaditis littoral zone (Chrysostomou and Giagkoulis 2016, 7). The excavator interpreted likewise the posts and horizontal wood found in the earliest habitation phase of Anarghiri IXa. The remains in the centre of the excavation are attributed to another two-storey stilted house, containing several clay structures and numerous household artefacts which belong to the early-4th

millennium Final Neolithic of Northern Greece (Chrysostomou and Giagkoulis 2018, 208–215).

The scale of the rescue excavation on the edge of the lignite-mining zone, the preservation of finds, and studies' progress (Arampatzis 2019; Giagkoulis 2019, 2020; Papadopoulou 2020) make Anarghiri IXb the most extensively investigated and intriguing site in the Amindeon Basin. The habitation was located at Lake Chimaditis' northeastern edges, in an area covered until the 1960s by shallow-water marshes. The excavation down to the natural soil of approximately 12,000 m^2 on the periphery of the settlement of 28,000 m^2 of the occupation's total area produced 80 ^{14}C dates of structural wood and other carbonised organic materials (Fig. 8.2). The Laboratory for the Analysis of Radiocarbon with AMS of the University of Bern analysed the samples and established the earliest habitation of the site in the Late Neolithic I period

Fig. 8.2 Excavation of Anarghiri IXb (2015 campaign) on the edge of the lignite-mining zone (Giagkoulis 2019)

(approximately 5400 cal BCE), with continuous human presence until the beginning of Final Neolithic period (approximately 4200 cal BCE) (Giagkoulis 2019, Vol. III Plan 3).

The exposed pile field of Anarghiri IXb consisted of more than 3600 vertical and horizontal wooden elements of various sizes and types (Fig. 8.3). Despite their number, their stratigraphic and spatial distribution, together with the dating of specific vertical elements, permitted the identification of enclosing structures and trackways on the eastern periphery of the habitation (Fig. 8.4). The earliest structural complex comprising a double fence and a trackway was established and used from 5300 to 5000 cal BCE. The relocation of the trackways and the fences after 5000 BCE correlates with the rearrangement of the main habitation area, probably imposed by the elevation of the lake water. The only documented evidence of the use of the site after the abandonment of the habitation around 4200 BCE is the bridge-like structure connecting the eastern edge of the settlement dated in EBA, namely the mid-3rd mil. BCE (Giagkoulis 2019, Vol. I 176–179, Vol. III Plans 24–29).

8.3 Dispilio Lake Settlement

The decision of Kastoria local authorities in 1935 to lower the Orestias Lake's water level to create arable land at its southern shore resulted in the exposure of several posts and artefacts at Dispilio. Three years later, Prof. Antonios Keramopoullos conducted the first trial excavation, and in his reports, he concluded that the wooden elements and the movable finds were the remains of a 'Pre-Hellenic' lakeside dwelling (Keramopoullos 1939, 1941). Nevertheless, the systematic excavations at Dispilio started in 1992 by Prof. George Chourmouziadis from the Aristotle University of Thessaloniki and lasted until his passing in 2013 (Chourmouziadis 2002). Thus, from its very beginning, the excavation operates as an institutional centre for the education of students and young researchers in prehistoric wetland archaeology, while Dispilio Open Air Museum—namely a representation in the actual scale of a lakeside Neolithic settlement at the shore of Kastoria Lake—is one of the most visited places in Greece related to prehistory.

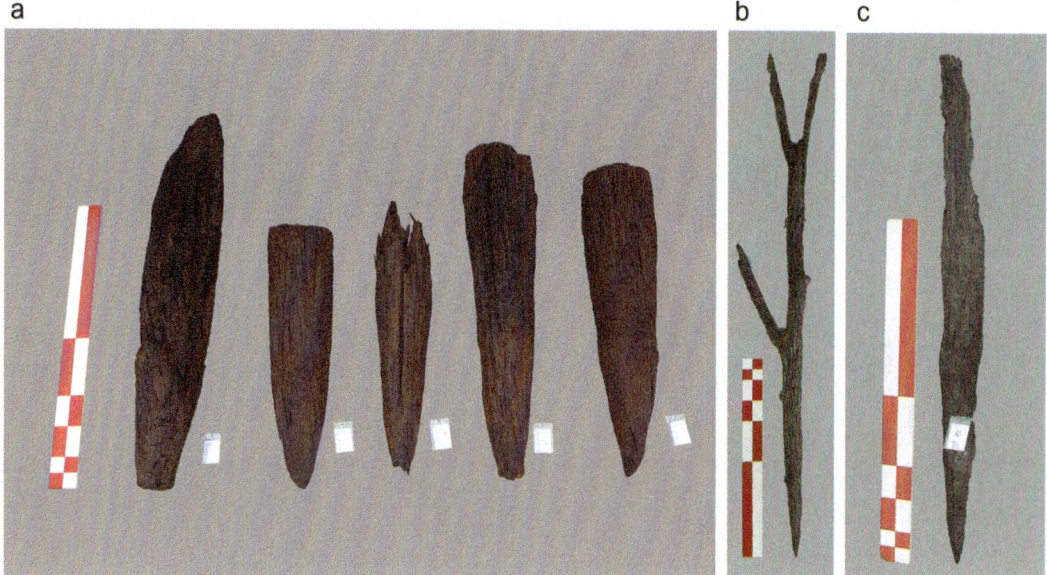

Fig. 8.3 Vertical posts extracted from the Late Neolithic layers of Anarghiri IXb (Giagkoulis 2019)

The settlement of Dispilio situated at a location called Nissi ('the island') was surrounded by water until the mid-1970s. The anthropogenic layers cover about 17,000 m^2, but since the shallow lake water north of the site remains uninvestigated, there is no secure estimation of the actual extent of the habitation (Fig. 8.5). The excavation was conducted mainly in three sectors and covered an area of approximately 1430 m^2. Regarding the settlement's palaeoenvironmental setting, palynological investigations document the occurrence of dense, mixed deciduous oak forests at intermediate and low altitudes, and coniferous and beech forests in mountainous areas. Pines and oaks are the dominant arboreal taxa, while the presence of Junipers should also be noted (Kouli 2015; Kouli and Dermitzakis 2008). The palaeobotanical results are confirmed by the anthracological analysis, according to which, despite the continuous use of the adjacent mixed forest for firewood and timber by Dispilio settlers, the local woodland experienced little change, perhaps because of the small-scale farming activities combined with the high potential for vegetation regrowth in this hinterland area (Ntinou 2010).

A preliminary interpretation of stratigraphy led the excavator to distinguish three phases. Backed by a geomorphological examination (Karkanas et al. 2011, Fig. 12), G. Chourmouziadis proposed a preliminary model for the site's development from a settlement built over the water to an almost dry-land habitation on the low mound formed by anthropogenic activities. A series of ^{14}C dates supported this conclusion, setting the earliest occupation at the end of the Greek Middle Neolithic (mid-6th millennium BCE) and the last at the Final Neolithic (4th millennium BCE), a span of more than 2000 years (Facorellis et al. 2014).

However, ascribing over a thousand piles to the three-phase model proved harder than anticipated. Of equal difficulty was the connection of piles to specific architectural units. Consequently, the early studies of the preserved wooden elements in the lowest excavation layers led only to general remarks about habitation structures (Chatzitoulousis 2006; Chourmouziadi and Giagkoulis 2002, 2004), but the overall architectural picture of the settlement together with its evolution through time remained little understood.

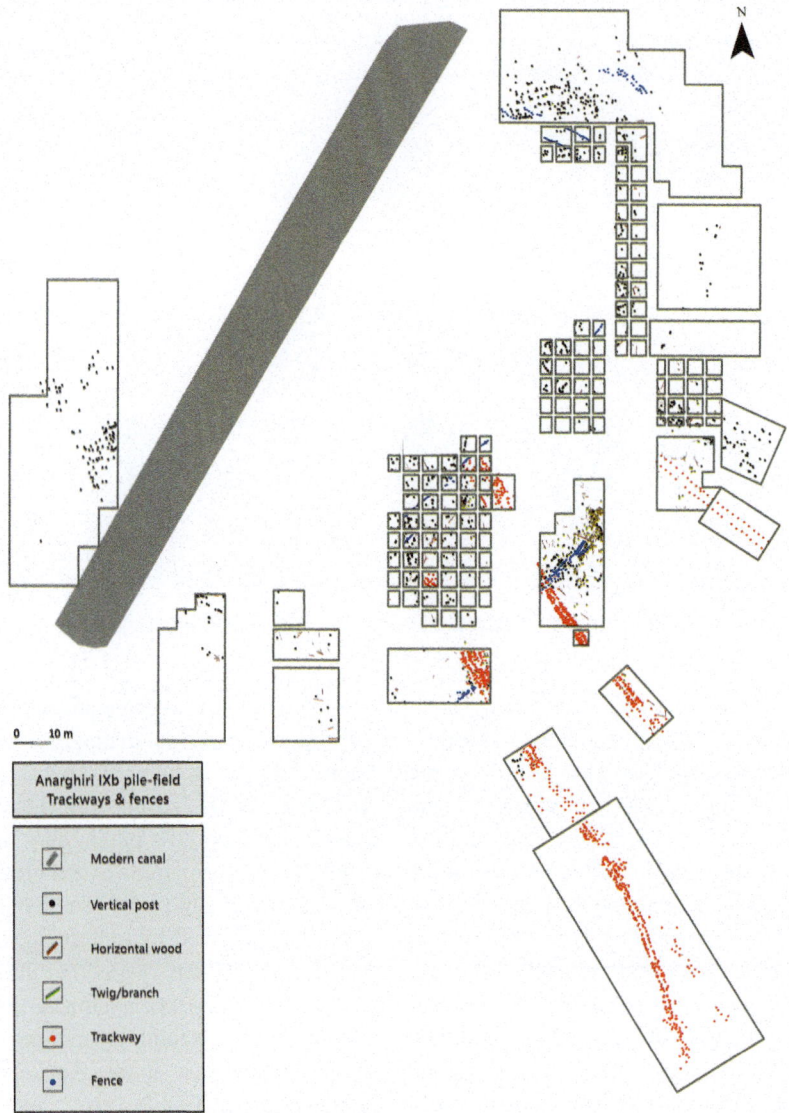

Fig. 8.4 Anarghiri IXb pile field and the accessing and enclosing wooden structures (Giagkoulis 2019)

The new period of research of the Aristotle University of Thessaloniki started in 2014 and aimed to clarify the issues and complete the research previously accomplished. Primarily, the focus was on re-examining the stratigraphy preserved in the excavated trenches. However, it was soon realised that sorting out the palimpsest of the piles required a thorough dendrochronological approach. Thus, the participation of the Aristotle University of Thessaloniki in the EXPLO project came as the natural next step. Within the EXPLO framework, the plan is to complete the dendrochronological analysis of the Dispilio pile field, define the habitation's architectural character, and collect new, stratigraphically sound, bioarchaeological, palaeoenvironmental, and geoarchaeological stratigraphic evidence (Fig. 8.6). Furthermore, the underwater exploration of the settlement's waterfront and the definition of the extent of the pile field there is a supplementary objective, which would allow comparisons with other lakeside settlements

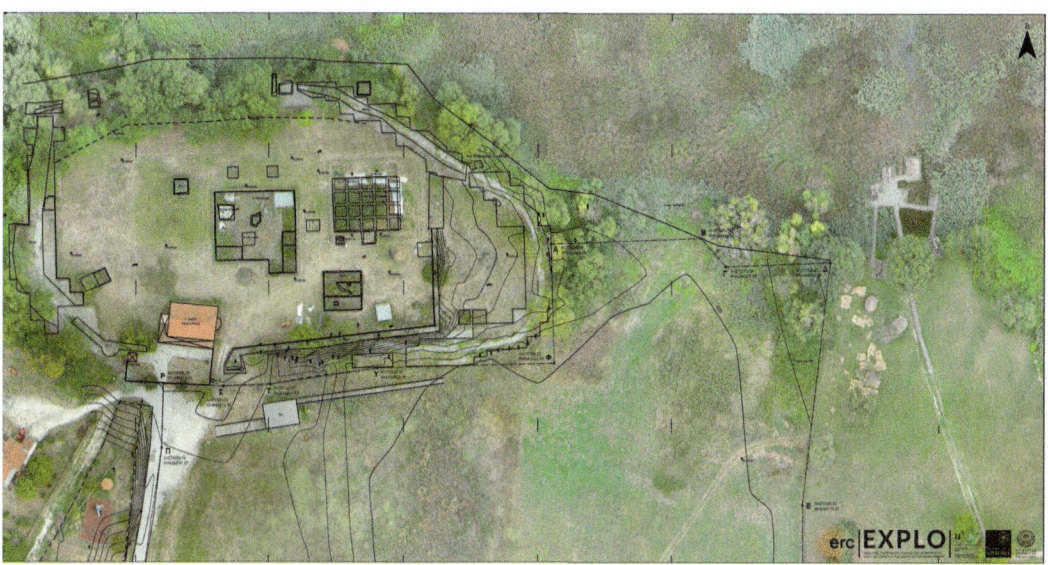

Fig. 8.5 Aerial view of Dispilio 'Nissi' and the excavated sectors (Survey: Filippos Stefanou)

investigated by EXPLO in North Macedonia and Albania. All the above goals, together with the incorporation of detailed expert studies of the old archaeological material, will sum up the complete publication of the site.

In the first two years of EXPLO's implementation, the AUTH team realised several research tasks: the detailed re-examination and evaluation of the rich excavation archive, recording and studying various classes of archaeological material, and significant fieldwork orientated to the accomplishment of the objectives mentioned above.

In 2019 the team carried out the 3D photogrammetric surveying of posts, which resulted in the detailed documentation of the pile field and the creation of a multivariate digital plan (Fig. 8.7). Following this, the systematic sampling of structural wood provided 787 samples, analysed by the dendrochronologists at the University of Bern. According to the on-site visual wood-sorting, the dominance of juniper timbers (*Juniperus* sp., 506 samples, i.e. 64% of total) and the almost equal occurrence of conifers (most probably *Pinus* sp., 116 samples, i.e. 15% of total) and oaks (*Quercus* sp., 165 samples, i.e. 21% of total) are attested (Fig. 8.8).

The first results were derived from the analysis of the oak samples combined with radiocarbon data.[2] Most of these samples are categorised in four reliable wiggle-matched mean curves. The end felling date (5727–5661 cal BCE) of the horizontal wood Disp10547 is the earliest indication of human activity in the excavated area of the East Sector, but it does not represent a settlement phase so far, and the same is for the horizontal wood Disp10930 (5471–5404 cal BCE). The earliest mean curve is MCDisp6005 (end felling dates 5612–5594 cal BCE), with posts situated in the western part of the excavation. Mean curve MCDisp6001(end felling dates 5322–5293 cal BCE) is built by 58 comprehensively wiggle-matched posts covering about 50 felling years that indicate a more or less continuous building activity, with a possible 10-year gap. This is the densest use of oak trees as foundation posts, with a balanced distribution all over the excavated area. MCDisp6003 with only

---

[2] In this paper we present the preliminary results of the dendrochronological analysis realised by the EXPLO dendro team (M. Bolliger, A. Mackzowski, J. Francuz and S. Szidat). The analytic publication of this work and its correlation to the archaeological data is in preparation.

Fig. 8.6 View of the excavation's west profile with vertical posts and structural elements of the lowest excavation layers (Dispilio Excavations archive)

Fig. 8.7 Three-dimensional photorealistic view of Dispilio pile field in 2019 (Dispilio Excavations archive)

a few samples (end felling dates 5291–5209 cal BCE) and MCDisp86 (end felling dates 5278–5200 cal BCE) may represent two hardly detectable building episodes.

Based on these results, three settlement phases can be distinguished: one around 5600 cal BCE, a second around 5300 cal BCE, and possibly a third around 5250 BCE. The possibility that these two last phases are continuous or the open question of the 300-year gap between the earliest

and the second phases will hopefully be clarified by the analysis of juniper and other conifer samples.

In parallel to the dendrochronological study, the systematic examination of the excavation records, plans, and photos led to the detection of seven distinctive stratigraphic horizons, comprising concentrations of wooden and clay building elements, architectural interventions of various types, inorganic sediments, and movable

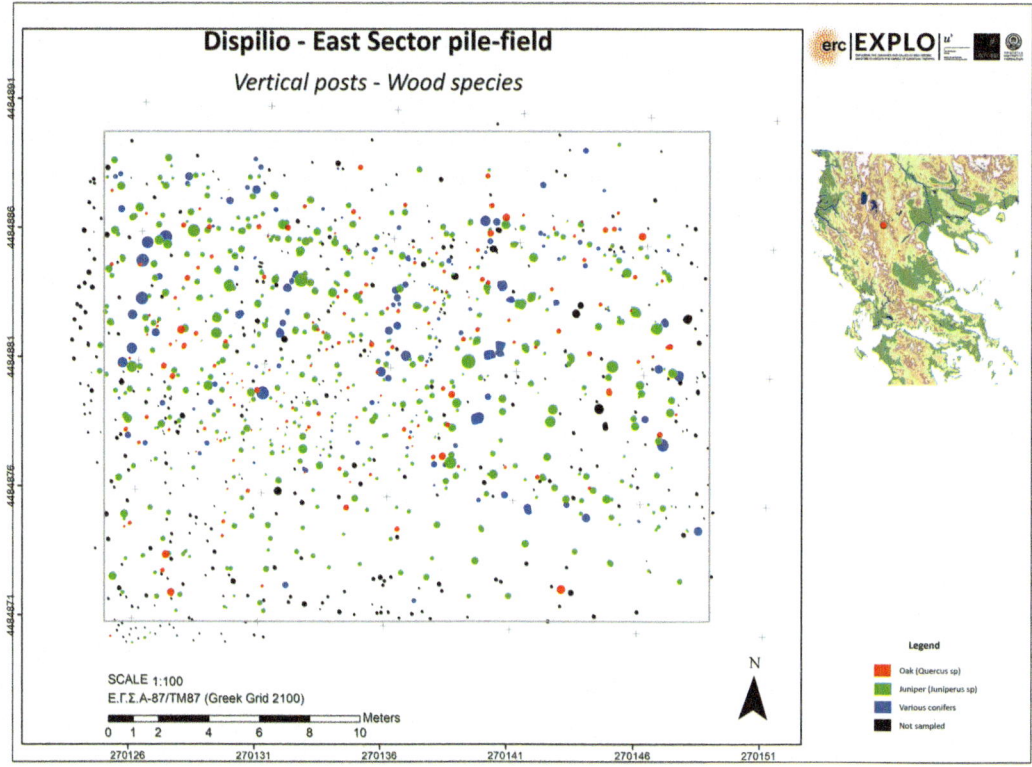

Fig. 8.8 GIS map of Dispilio East Sector pile field and the wood species of sampled vertical posts (Digital map by Tryfon Giagkoulis and Filippos Stefanou)

finds. The earliest horizons of Dispilio I–III are rich in organic materials and have an abundance of pottery and artefacts, preserving a particularly revealing spatial distribution. The subsequent horizons Dispilio IV–VII represent the later phases of the settlement's development, with an evident alteration in the state of organic preservation and the density and form of the architectural traces that correspond to a dry-land occupation.

As usual in Neolithic Greece, pottery is noteworthy for its quantity. For some reason, Dispilio potters have developed superior expertise that other dry-land Neolithic sites of the region do not easily match. The quality of execution is impressive; nevertheless, this phenomenon does not inhibit pottery styles and fashions, indicating influences from the surrounding communities. Although relations between communities tend, as a rule, to be

opportunistic, the individual and unique daily life of the lake settlements would encourage and reinforce social interregional alliances and bonds with communities of other similar settlements further north. Within these communities of regular everyday practices and landscapes, the bonds created are imprinted in material culture. Ceramics point strongly to networks of such 'communities of practice'. Pots like arcade barbotine, or bitumen decoration, are virtually unknown in Thessaly or even southern parts of Macedonia, Greece, but much at home in south Albania or North Macedonia. The same also holds for the rest of the material culture, especially figurines, which present some unique forms, among which one 'statue' of a seated woman more than 40 cm in height. The symbolic repertoire of this exquisite culture is completed by fine bone tools and ornaments, white marble bracelets, and spondylus jewellery. Some indications for specialisation in macrolithic stone tool production exploiting a local source are currently under investigation.

8.4 Discussion and Outlook

Although wetland archaeology in the cross-border area of Greece, North Macedonia, and Albania is a relatively young research field and the available information is still under processing (Naumov 2020), similar geomorphological as well as environmental features—especially the presence of several lakes—seem to be of importance for the development of the farming communities during the Neolithic (Gkouma and Karkanas 2018). It is still too early to connect this securely with specific environmental conditions that EXPLO is investigating. However, with the evidence available, we can conclude that the mid-6th millennium marks a wave of expansion towards lakes and waterlogged areas in SE Europe. The earliest reliable indication for habitation on a lakeshore, for the time being, is the Mean Curve 6005 from Dispilio, dated around 5600 cal BCE, within the Greek Middle Neolithic phase. A single dated post from the rescue excavation of the settlement 'Ohridati' is the only comparable information on habitation on the northern shore of Lake Ohrid in the mid-6th mil BCE (Westphal et al. 2010). The long series of ^{14}C dates from Anarghiri IXb, the well-documented Mean Curves from Dispilio, and some recent dates from Kallamas on the Albanian shore of Great Prespa Lake certify the existence and diachronic development of farming communities around the lakes securely during the Late Neolithic I and II (Lera et al. 2020).

Moreover, some other known prehistoric wetlands of the region, for example, Dunavec, Maliq, and Sovjan in Korça Basin or Ustje na Drim in Ohrid Lake, are dated in various Neolithic phases. Yet, the lack of ^{14}C dates and the problems in the terminology of 'early, middle, and late' Neolithic in correlation to specific pottery categories do not facilitate the secure integration of these settlements into an established chronological framework. The preliminary results of the ongoing dendrochronological research of the samples from Ploca Mičov Grad promise to create a robust chronological framework for the 5th millennium BC lake habitations of the region (Hafner et al. 2021), combined with those that will hopefully emerge from the recent research at Lin in Ohrid Lake.

The contribution of EXPLO to sorting out Dispilio was significant. We hope that the future results of the project, in combination with those of Dispilio, will shed more light on the choice for wetland habitation in SE Europe.

References

Arampatzis C (2019) Antler artifacts from the Neolithic lakeside settlement Anarghiri IXb, Western Macedonia, Greece. Ph.D. thesis. Faculty of Humanities, University of Bern. https://www.swissbib.ch/Record/575617438. Accessed 17 May 2021

Bottema S (1974) Late quaternary vegetation history of Northwestern Greece. Ph.D. thesis. Rijksuniversiteit Groningen

Bottema S (1982) Palynological investigations in Greece with special reference to pollen as an indicator of human activity. Palaeohistoria 24:257–288

Chatzitoulousis S (2006) To xilo os archaeologiko yliko stin proistoria: to paradeigma tou neolithikou limnaiou oikimsou sto Dispilio Kastorias. Ph.D. thesis. Aristotle University of Thessaloniki

Chourmouziadi A, Giagkoulis T (2002) Provlimata kai methodoi prosegisis tou chorou. In: Chourmouziadis G (ed) Dispilio 7500 chronia meta. University Studio Press, Thessaloniki, pp 37–74

Chourmouziadi A, Giagkoulis T (2004) Dispilio 2002: Eikones enos (apo)domimenou chorou. In: Ministry of Culture and Sports, Aristotle University of Thessaloniki (eds) To Archaeologiko Ergo sti Makedonia kai Thraki to 2002 (AEMTH 16). Altintzis, Thessaloniki, pp 641–648

Chourmouziadis G (ed) (2002) Dispilio 7500 chronia meta. University Studio Press, Thessaloniki

Chrysostomou P, Giagkoulis T (2016) Land und Seeufersiedlungen der „Kultur der Vier Seen", Griechenland. Neue archäologische Einblicke in den prähistorischen Siedlungen im Amindeon-Becken. Plattform 23/24:4–9

Chrysostomou P, Giagkoulis T (2018) Entos kai ektos orion: opseis tis chororganosis stous proistorikous oikismous Anarghiri IXa and Anarghiri XI sto lekanopedio Amyntaiou. In: Karamitrou-Mentesidou G (ed) To Archaeologiko Ergo stin Ano Makedonia (3rd scientific meeting proceedings, archaeological Museum of Aiani, November 2013), vol A. Archaeological Museum, Aiani, pp 207–228

Chrysostomou P, Giagkoulis T, Mäder A (2015) The "culture of four lakes". Prehistoric lakeside settlements (6th–2nd mil. BC) in the Amindeon Basin, Western Macedonia, Greece. Archäologie Schweiz 38(3):24–32

Facorellis Y, Sofronidou M, Chourmouziadis G (2014) Radiocarbon dating of the Neolithic lakeside settlement of Dispilio, Kastoria Northern Greece. Radiocarbon 56(2):511–528. https://doi.org/10.2458/56.17456

Gassner S, Gobet E, Schwoerer C, van Leeuwen J, Vogel H, Giagkoulis T, Makri S, Grosjean M, Panajiotidis S, Hafner A, Tinner W (2020) 20,000 years of interactions between climate, vegetation and land use in Northern Greece. Veg Hist Archaeobotany 29:75–90. https://doi.org/10.1007/s00334-019-00734-5

Gerasimidis A, Athanasiadis N (1995) Woodland history of northern Greece from the mid-Holocene to recent time based on evidence from peat pollen profiles. Veg Hist Archaeobotany 4:109–116

Giagkoulis T (2019) The pile-field and the wooden structures of the Neolithic lakeside settlement Anarghiri IXb Western Macedonia, Greece, vols I–III. Ph.D. thesis. Faculty of Humanities, University of Bern. https://boristheses.unibe.ch/1123/. Accessed 17 May 2021

Giagkoulis T (2020) On the edge: the pile-field of the Neolithic Lakeside Settlement Anarghiri IXb (Amindeon, Western Macedonia, Greece) and the Non-Residential Wooden Structures on the Periphery of the Habitation. In: Hafner A, Dolbunova E, Mazurkevich A, Pranckenaite E, Hinz M (eds) Settling waterscapes in Europe. The archaeology of Neolithic and Bronze Age Pile-Dwellings. OSPA—open series in prehistoric archaeology, vol 1. Propylaeum, Bern, Heidelberg, pp 137–155

Gkouma M, Karkanas P (2018) The physical environment in Northern Greece at the advent of the Neolithic. Quatern Int 496:14–23

Hafner A, Reich J, Ballmer A, Bolliger M, Antolín F, Charles M, Emmenegger L, Fandré J, Francuz J, Gobet E, Hostettler M, Lotter A, Maczkowski A, Morales-Molino C, Naumov G, Stäheli C, Szidat S, Taneski B, Todoroska V, Bogaard A, Kotsakis K, Tinner W (2021) First absolute chronologies of neolithic and bronze age settlements at Lake Ohrid based on dendrochronology and radiocarbon dating. J Archaeol Sci Rep 38:103107. https://www.sciencedirect.com/science/article/pii/S2352409X21003199?via%3Dihub. Accessed 15 Nov 2021

Karamitrou-Mentesidi G, Efstratiou N, Kaczanowska M, Kozłowski JK (2015) Early Neolithic settlement of Mavropigi in western Greek Macedonia. Eurasian Prehistory 12(1–2):47–116

Karkanas P, Pavlopoulos K, Kouli K, Ntinou M, Tsartsidou G, Facorellis Y, Tsourou T (2011) Palaeoenvironments and Site Formation Processes at the Neolithic Lakeside Settlement of Dispilio, Kastoria, Northern Greece. Geoarchaeology Int J 26(1):83–117

Keramopoullos A (1939) Anaskafai kai erevnai en ti Dytiki Makedonia ypo Ant. D. Keramopoullo.

Praktika tis en Athinais Archaeologikis Etaireias tou etous 1938, pp 53–66

Keramopoullos A (1941) Anaskafi en Kastoria ypo Ant. D. Keramopoullo. In: Praktika tis en Athinais Archaeologikis Etaireias tou etous 1940, pp 22–23

Kloosterboer-van Hoeve M, Steenbrink J, Brinkhuis H (2001) A short-term cooling event, 4025 million years ago, in the Ptolemais Basin, northern Greece. Palaeogeogr Palaeoclimatol Palaeoecol 173:61–73

Kokkinidou D, Trantalidou K (1991) Neolithic and Bronze age settlement in western Macedonia. Annu Br Sch Archaeol Athens 86:93–106

Kotsakis K (2014) Domesticating the periphery: new research into the Neolithic of Greece. Pharos 20:41–73

Kouli K (2015) Plant landscape and land use at the Neolithic lake settlement of Dispilio (Macedonia, northern Greece). Plant Biosyst Int J Deal Aspects Plant Biol Off J Soc Bot Ital 149(1):195–204. https://doi.org/10.1080/11263504.2014.992998

Kouli K, Dermitzakis M (2008) Natural and cultural landscape of the Neolithic settlement of Dispilio: Palynological results. Hellenic J Geosci 43:29–39

Lera P, Touchais G, Oberweiler C, Aslaksen OC, Blein C, Boleti A, Elezi G, Fadin L, Gori M, Krapf T, Maniatis Y, Odie O (2020) Bassin de Korçë, Kallamas. Bulletin archéologique des Écoles françaises à l'étranger, Balkans. http://journals.openedition.org/baefe/1362

Maniatis Y (2014) Hronologisi me ^{14}C ton megalon politismion allagon stin proistoriki Makedonia: prosfates exelixeis. In: Stefani E, Merousis N, Dimoula A (eds), A century of research in prehistoric Macedonia 1912–2012. International conference proceedings, Archaeological Museum of Thessaloniki, 22–24 Nov 2012. Zitis, Thessaloniki, pp 205–222

Maniatis Y, Kotsakis K, Halstead P (2015) Paliampela Kolindrou. Nees hronologies tis arxaioteris neolithikis. Παλιάμπελα Κολινδρού. In: Ministry of culture and sports, Aristotle University of Thessaloniki (eds) To Archaeologiko Ergo sti Makedonia kai Thraki to 2011 (AEMTH 25). Zitis, Thessaloniki, pp 149–156

Marinova E, Ntinou M (2018) Neolithic woodland management and land-use in southeastern Europe: the anthracological evidence from Northern Greece and Bulgaria. Quatern Int 496:51–67. https://doi.org/10.1016/j.quaint.2017.04.004

Naumov G (2020) Neolithic wetland and lakeside settlements in the Balkans. In: Hafner A, Dolbunova E, Mazurkevich A, Pranckenaite E, Hinz M (eds) Settling waterscapes in Europe. The Archaeology of Neolithic and Bronze Age Pile-Dwellings. OSPA—Open Series in Prehistoric Archaeology, vol 1. Propylaeum, Bern, Heidelberg, pp 111–135

Ntinou M (2010) Palaeoperivallon kai anthropines drastiroities. I anthrakologia sto limnaio neolithiko oikismo sto Dispilio Kastorias. Anaskamma 4:45–60

Ntinou M (2014) I physiki vlastisi kai i proistorikes koinotites tis Makedonias. Mia synthesi ton pliroforion tis anthrakologikis erevnas. In: Stefani E, Merousis N, Dimoula A (eds) A century of research in prehistoric Macedonia 1912–2012. International conference proceedings, archaeological Museum of Thessaloniki, 22–24 Nov 2012. Zitis, Thessaloniki, pp 409–417

Papadopoulou S (2020) Chipped stone industries from Western Macedonia, Greece. The case of the Neolithic lakeside settlement Anarghiri IXb. Ph.D. thesis, vols. I–III. Faculty of Humanities, University of Bern. https://www.swissbib.ch/Record/604544367. Accessed 17 May 2021

Stratouli G, Katsikaridis N, Bekiaris T, Maousidou E, Petaslnikos I, Tsiola E (2019) Trita Koromilias. Mia parapotamia neolithiki thesis tin Periferiaki Enotita Kastorias. In: Ministry of Culture and Sports, Aristotle University of Thessaloniki (eds) To Archaeologiko Ergo sti Makedonia kai Thraki to 2014 (AEMTH 28). Zitis, Thessaloniki, pp 1–8

Stratouli G, Katsikaridis N, Bekiaris T, Kloukinas D, Koromila G, Kyrillidou S (2020) New excavations in Northwestern Greece: the Neolithic settlement of Avgi, Kastoria. J Greek Archaeol 5:63–134

Trantalidou K (1989) Proistoriki oikismoi stis lekanes tis Florinas kai tou Amyntaiou (Ditiki Makedonia). In: Institute for Balkan Studies (ed) Ancient Macedonia V. Institute for Balkan Studies, Thessaloniki, pp 1593–1622

Tsouggaris C, Salonidis T, Douma A, Sariggianidou C (2004) Kolokinthou. Enas neos parapotamios neolithikos oikismos tou N. Kastorias. In: Ministry of Culture and Sports, Aristotle University of Thessaloniki (eds) To Archaeologiko Ergo sti Makedonia kai Thraki to 2002 (AEMTH 16). Altintzis, Thessaloniki, pp 625–640

Westphal T, Tegel W, Heußner K-U, Lera P, Rittershofer K-F (2010) Erste dendrochronologische Datierungen historischer Hölzer in Albanien. Archäologischer Anzeiger 2:75–95

Prehistoric Wetland Settlements of the Bulgarian Black Sea Coast

Ariane Ballmer, Kalin Dimitrov, Nayden Prahov, and Pavel Georgiev

Abstract

Along Bulgaria's Black Sea coast, a series of settlement remains from the Eneolithic and Early Bronze Age have been preserved underwater and in marshy environments. A few of them seem originally to have been terrestrial settlements which were flooded and abandoned when the level of the Black Sea rose, while others were actual wetland settlements. Specific remains of 'pile-dwelling' architecture have been found, for example, at sites in the Varna Lakes area, in Sozopol harbour, and at Ropotamo and Urdoviza. As well as drawing parallels with the pan-European phenomenon of pile-dwellings, the paper examines the broader significance of the Bulgarian wetland settlements, including their complex interrelationship with the water system and their role in the overall cultural landscape at the time of their occupation.

Keywords

Bulgaria · Black Sea coast · Wetland settlements · Pile-dwellings · Eneolithic · Early Bronze Age

A. Ballmer
Independent Researcher, Bern, Switzerland
e-mail: mail@arianeballmer.com

Institute of Archaeological Sciences and Oeschger Centre for Climate Change Research (OCCR), University of Bern, Bern, Switzerland

K. Dimitrov (✉) · N. Prahov
National Archaeological Institute with Museum, Bulgarian Academy of Sciences, Sofia, Bulgaria
e-mail: kalin.d@abv.bg

K. Dimitrov · N. Prahov · P. Georgiev
Centre for Underwater Archaeology,
Ministry of Culture of the Republic of Bulgaria, Sozopol, Bulgaria
e-mail: p.y.georgiev@soton.ac.uk

P. Georgiev
Centre for Maritime Archaeology,
University of Southampton, Southampton, UK

9.1 Introduction

At present, only some of the Bulgarian underwater sites can be securely addressed as actual wetland settlements. It will be a matter for future research to determine which of the known underwater remains were originally built on humid ground and featured specific architectural adaptations, such as pile substructions. What is so far known about the Bulgarian wetland settlements, however, integrates well with the discoveries in other regions of the Balkans (and beyond). All the known wetland sites in Bulgaria are located along the Black Sea coast (Peev et al. 2020). So far, none are known inland. However, this picture is possibly only a reflection of the current state of research.

The Black Sea coast has been a focus of underwater archaeology since early on, with ancient port facilities and shipwrecks being of particular interest.

© The Author(s) 2025
A. Ballmer et al. (eds.), *Prehistoric Wetland Sites of Southern Europe*,
Natural Science in Archaeology, https://doi.org/10.1007/978-3-031-52780-7_9

Inundated prehistoric settlements first became a focus of attention in the second decade of the 20th century and have been archaeologically explored since the 1980s in campaigns involving expert scientific divers and interdisciplinary research teams (for an overview of the history of research along the Bulgarian Black Sea coast, see Stanimirov (2003); Krasteva and Hristov 2014; Dimitrov et al. 2020). Today, the Centre for Underwater Archaeology (= **CUA**) in Sozopol is systematically excavating and evaluating these remains.

In Bulgaria, the earliest Neolithic settlements appeared shortly before 6000 BC (for the absolute chronology of Bulgarian prehistory, see Boyadziev 1995; Bojadziev 1998; Görsdorf and Bojadžiev 1996). The wetland settlements known to date emerged only later, in the Eneolithic (= **ENE**), or rather, the late ENE (Kodzhadermen–Gumelniţa–Karanovo (= **KGK**) VI to Varna III, i.e. c. 4450–3950 BC). A second phase of wetland settlement is dated to the Early Bronze Age (= **EBA**), i.e. c. 3175–2525 BC. In between lies the so-called Transitional Period (c. 3800–3175 BC), generally characterised by a gap in the settlement evidence (Вайсов 1992; Todorova 1995, 2003; Tsirtsoni 2016) and in the coastal region, specifically, by higher water levels (Filipova-Marinova 2007) and indications of forest recovery (Filipova-Marinova et al. 2011). Finally, finds from the Varna Lakes indicate a third, Late Bronze Age (= **LBA**) phase of wetland settlement in the second half of the second millennium BC (Тончева 1972; Toncheva 1981).

Given its favourable environment, with rich and diverse resources, it must be assumed that the western Black Sea coastline was an attractive habitat from the Mesolithic onwards. In recent years, evidence of Late Neolithic coastal sites from the late 6th millennium BC has been discovered (Leshtakov 2010; Leshtakov et al. 2020a, b). These were terrestrial sites. It is unclear to what extent the absence of settlement evidence from before the Late Neolithic is related to changes in the coastline during the late 7th millennium BC (see *infra*). These could have resulted in the archaeological remains being buried under thick packages of marine sediment, below today's sea level.

It is crucial that past activities along the western Black Sea coast (as well as the preservation of their traces) are understood against the background of water level fluctuations, the associated development of the coastline, and, last but not least, variations in water salinity (see Sect. 9.3.2). While the scientific debate on the evolution of the Black Sea cannot be rehearsed here, the aspects that are of immediate importance to the present subject will be briefly addressed. Thus, in the course of the late Quaternary, the Black Sea basin underwent various changes, some of them dramatic (Yanko-Hombach et al. 2007a, b; Kislov and Toporov 2011; Lericolais et al. 2011; Nicholas et al. 2011; Benjamin et al. 2017; Kislov 2018). During the period of most extreme low water, the Black Sea was an autonomous freshwater body, not connected to the global ocean system. While the rhythm and speed of rise and fall in Black Sea water levels are debated, what is significant for the present discussion is that repeated water level fluctuations, major and minor, are proven to have occurred during the Mid- and the first part of the Late Holocene. These had decisive effects on the coastline and thus on areas of coastal prehistoric settlement and economic activity. During the periods discussed in this article (especially 4450–2525 BC), the water level may have been about 6–8 m lower than today, at least during the period of occupation of the settlements. These sites were later flooded during a complex series of relatively rapid, short-term sea level fluctuations —and eventually submerged by a significant water level rise.

9.2 Archaeological Remains of Wetland Settlements: An Overview

In Bulgaria, two main concentrations of prehistoric wetland sites can be found: one in the region of Varna on the northern Black Sea coast, and the other south of Burgas (Fig. 9.1). In the following, the most important sites are presented, taking into account research data up to 2021. At the time of their occupation, the presented sites

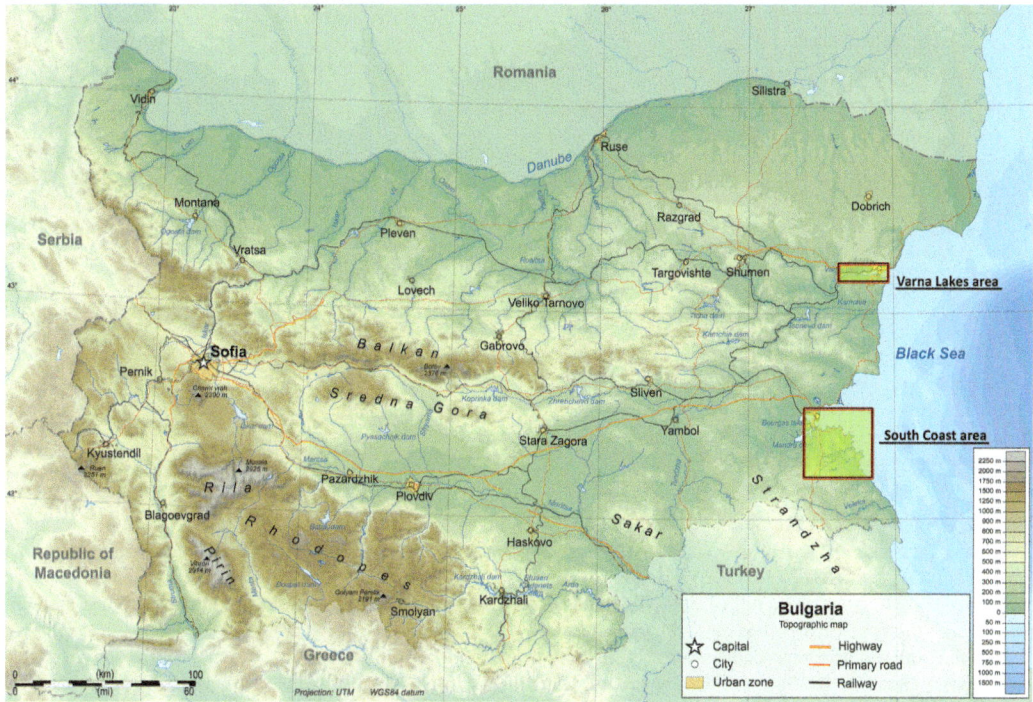

Fig. 9.1 Bulgarian Black Sea coast with concentrations of known prehistoric wetland sites (K. Dimitrov)

were presumably located on permanently marshy or periodically inundated ground, since they are characterised by an adapted architecture, with substructures elevating the floors above ground level (for instance, in the form of houses on wooden stilts, commonly known as 'pile-dwellings').

9.2.1 The Varna Lakes

Lake Varna and Lake Beloslav are located west of today's city of Varna. Before modern construction works, they were connected to each other by marshlands and independent of the sea (Fig. 9.2). In order to create a waterway trade route in the first third of the last century, the water levels of both lakes were considerably lowered and canals were installed between the lakes and the sea. It was during the construction of the canal between the two lakes in 1921 that the very first pile-dwelling in Bulgaria was discovered: Strashimirovo 1 (Шкорпил and

Шкорпил 1921). Since then, more than ten waterlogged sites of prehistoric wetland settlements have been recorded in Lakes Varna and Beloslav in the course of canal-, harbour-, and dike-building works (Table 9.1). As most of these sites have not been properly explored, their interpretation is complicated. However, some relevant conclusions can be drawn from published preliminary reports and the finds stored at the Varna Archaeological Museum. Most of the sites delivered material from two main phases: the late ENE (KGK VI, variant Varna II and Varna III) and the EBA (II and III). Two radio-carbon dates were obtained from the Ezerovo II site (Table 9.2) which correspond with the EBA find material. Interestingly, the finds from the Ezerovo I and Old Canal sites indicate an LBA wetland occupation phase at Lake Varna (Тончева 1972; Toncheva 1981; Прахов et al. 2019) (Fig. 9.2). It should also be mentioned that during the archaeological excavations of Arse-nala in 1986, 37 pieces of wood, mainly standing posts, were sampled for dendrochronological

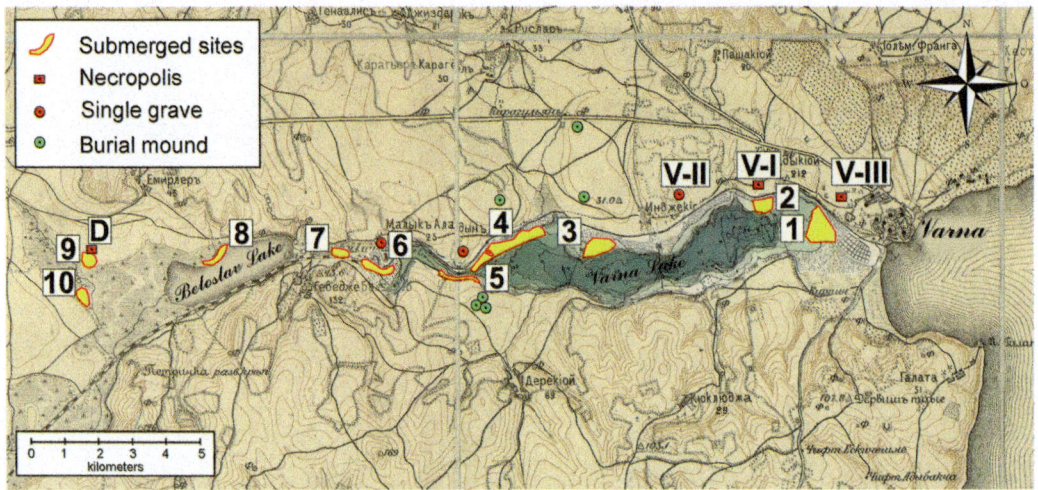

Fig. 9.2 Prehistoric wetland sites in the Varna and Beloslav Lakes area and the nearby (terrestrial) funerary sites (Map basis: Russian historical topographic map (scale 1:42 000) from the end of the 19th century, representing the landscape before the industrial interventions of the 20th century; blue shades: bathymetric map of Lake Varna (Божков 1936)). (**1**) **Outflow from Lake Varna into the sea**, EBA and LBA settlements; (**2**) **Morflot**, evidence of late ENE and EBA settlements; (**3**) **Topolite**, evidence of EBA settlement; (**4**) **Ezerovo**, late ENE and BA finds, areas with in situ preservation; (**5**) **Arsenala-Ladzhata**, Late ENE and EBA settlements, areas with in situ preservation; (**6**) **Strashimirovo**, late ENE and EBA settlements; (**7**) **Beloslav**, unclear evidence of EBA settlement; (**8**) **Povelyanovo**, late ENE and EBA settlements; (**9**) **Devnya**, late ENE settlement, evidence of waterlogged pile-dwelling and (**D**) necropolis (Тодорова & Симеонова 1971); (**10**) **Baltata**, EBA settlement. (**V–I**) **Varna** necropolis I, 329 burials with very rich finds from the late ENE; (**V–II**) **Varna** necropolis II, rich early ENE burials; (**V–III**) **Varna** necropolis III, rich late ENE burials (K. Dimitrov)

examination (Orcel and Orcel 1991). This preliminary study, which revealed two settlement phases, was to remain the only dendrochronological investigation of Bulgarian wetland settlements for a long time.

As well as common settlement finds, such as pottery, all the sites yielded construction timbers and artefacts made from organic materials. In some cases, the reports explicitly mention that the archaeological finds were embedded in layers containing seashells, while the wooden piles were vertically inserted into mud, clay, or marl layers with no finds. Furthermore, it was noticed at the time that the archaeological layers were covered by sediments saturated with freshwater molluscs, silt, and peat. These observations have led to the hypothesis of the presence of pile-dwellings, originally located on the shores of a deep, sheltered, semi-freshwater bay but eventually flooded due to a sea level rise (Маргос

1965, 1973; Иванов 1973; Божилова and Иванов 1985; Ivanov 1993).

Despite the extensive interference caused by construction activities, there is reason to assume that many of the relevant sites still have archaeological remains preserved in situ (Прахов et al. 2019). Action has been taken by the CUA for future site monitoring, research, and conservation.

9.2.2 The Black Sea Coast South of Burgas

The Bulgarian wetland sites at which most archaeological research has been carried out are concentrated along the southwestern Black Sea coast, south of Burgas. The relatively rugged shoreline is characterised by rocky outcrops and bays with lagoons (Fig. 9.3). Most of the

Table 9.1 Overview of the wetland settlements in the Varna region, i.e. in the area of Lakes Varna and Beloslav. They were all discovered during construction work in the last century. With two exceptions, all the sites have brought to light wooden posts, typical of prehistoric pile-dwellings. The sites' dating is based on pottery typology (K. Dimitrov and A. Ballmer)

No. on Fig. 9.2	Site name according to Ivanov (1993)	Site name *alias*	Date of discovery; archaeological interventions	Location and context with other sites	m bpsl	Finds	Dating	Bibliography
1	Rodopa	Asparuhovo	1964	On the S shore of Lake Varna, on the sandy strip between the lake and the sea	8.5–9.0	Wooden posts	Bronze Age?	Маргос (1965, 1969), Тончева (1972), Ivanov (1993), Лазаров (2009)
1	Old Canal	Settlement in front of the Electic power plant	1968, 2018	In the NE part of Lake Varna, close to the Varna III Late ENE necropolis	6.5	Wooden posts, pottery	LBA	Тончева (1972), Прахов et al. (2019)
1	Factory 'Hristo Botev'	Factory 'Vassil Kolarov', Varna I, Varna II	1937, 1960	In the NE part of Lake Varna, close to the Varna III Late ENE necropolis	7.0	Pottery and further archaeological material	EBA	Маргос (1961a; b), Тончева (1972), Ivanov (1993), Лазаров (2009)
2	Morflot (= Morflotte)	Varna I, Maxuda–Bozkov Chair, Factory 'Yanko Kostov', Factory 'Rodopa', Varna 2	1960, 2010	At the NE shore of Lake Varna, 150–200 m from the Varna I Late ENE necropolis	2.0–4.0 (?)	Wooden posts, pottery	Late ENE, EBA	Маргос (1961a; b), Тончева (1972), Ivanov (1993), Лазаров (2009)
3	Topolite	–	1987	At the N shore of the Varna lake	6	Pottery	EBA	Ivanov (1993), Лазаров (2009)
4	Ezerovo I	Ezerovo 'The railway station', Ezerovo IV, Copper age site Ezerovo	1966, 1967–1968	At the N shore of Lake Varna, close to the Varna power plant; EBA burial mound at a distance of 900 m	c. 7.5–8.5	Wooden posts, pottery and further archaeological material	Late ENE, LBA	Тончева (1968, 1972), Todorova and Tonceva (1975), Toncheva (1981), Змейкова (1991), Ivanov (1993), Лазаров (2009)

(continued)

Table 9.1 (continued)

No. on Fig. 9.2	Site name according to Ivanov (1993)	Site name *alias*	Date of discovery; archaeological interventions	Location and context with other sites	m bpsl	Finds	Dating	Bibliography
4	Ezerovo II	Ezerovo, Ezerovo3, Settlement at the Varna power plant, Ezerovo 'The railway station'	1966, 1967–1968, 2018	At the N shore of Lake Varna, along the piers of the Varna power plant; single ENE burial in the vicinity	6.5–8.5 (EBA)	Wooden posts, ceramics pottery and further archaeological material	Late ENE, EBA	Тончева (1968), Маргос (1969), Toncheva (1981), Zmeykova (1991), Ivanov (1993), Boyadziev (1995), Лазаров (2009), Прахов et al. (2019)
5	Arsenala	–	1976, 1978–1979; archaeological survey 1979; archaeological excavations 1984–1986	At the S shore of Lake Varna, E from the Ladzhata site, close to the canal connecting the Lakes Varna and Beloslav, at the outflow of the palaeo-Provadiiska river; EBA burial mound at a distance of 600 m	2.5–8.0	Wooden posts, pottery	Late ENE, EBA	Иванов (1980), Божилова and Иванов (1985), Иванов (1985), Иванов (1987), Orcel and Orcel (1991), Ivanov (1993), Лазаров (2009)
5	Ladzhata (= Ladjata)	Ezerovo-Boaza, Ezerovo I, Ezerovo II, Ezerovo III	1958	At the S shore of the Lake Varna, close to the canal connecting the Lakes Varna and Beloslav; EBA burial mound at a distance of 600 m	6.5–7.0	Wooden posts, ceramics, pottery and further archaeological material	EBA	Тончева and Маргос (1959), Маргос (1960, 1961a, b, 1965), Маргос and Тончева (1962), Тончева (1968, 1972), Toncheva (1981), Zmejkova (1991), Ivanov (1993), Лазаров (2009)
6	Strashimirovo 1	Strashimirovo-Est	1921, 1957, 1969–1970	In the most W part of the Lake Varna, at the canal connecting the Lakes Varna and Beloslav, at the palaeo-Provadiiska river; single ENE burial in the vicinity; EBA burial mound at a distance of 1600 m	3.5–4.5	Wooden posts, pottery and further archaeological material, remains of metal production	Late ENE, EBA	Шкорпил and Шкорпил (1921), Миков (1950), Маргос (1965, 1973), Тончева (1972), Ivanov (1993), Лазаров (2007)

(continued)

Table 9.1 (continued)

No. on Fig. 9.2	Site name according to Ivanov (1993)	Site name *alias*	Date of discovery; archaeological interventions	Location and context with other sites	m bpsl	Finds	Dating	Bibliography
6	Strashimirovo 2	Strashimirovo-West	1970	In the most W part of the Lake Varna, at the canal connecting the Lakes Varna and Beloslav, at the palaeo-Provadiiska river; single ENE burial in the vicinity; EBA burial mound at a distance of 1600 m	3.3	Wooden posts, pottery	EBA	Тончева (1972), Иванов (1973), Ivanov (1993), Лазаров (2009)
7	Beloslav	–	1970	In the middle of the canal connecting the Lakes Varna and Beloslav	?	Pottery	EBA	Ivanov (1993), Лазаров (2009)
8	Povelyanovo (= Poveljanovo)	–	1970, 1975–1976	In the most W part of the Lake Beloslav	2.0–6.5	Wooden posts, pottery	Late ENE, EBA	Тончева (1972), Ivanov (1993), Лазаров (2009), Parvanov (2019)
9, 10	Devnya	Povelyanovo	1971	In the NW part of Beloslav lake, probably related to the Late ENE Devnyanecropolis	?	Pottery, wooden posts?	Late ENE	Тодорова-Симеонова (1971)
11	Baltata	–	1970–1971	On a marshy area in the most W part of Lake Beloslav, close to the palaeo-bed of the Provadiiska river	4.2–4.4	Wooden posts, pottery	EBA	Ivanov (1993), Лазаров (2009)

Table 9.2 Radiocarbon dates from Bulgarian wetland sites mentioned in the article, sorted by site and lab number. Calibrated using OxCal 4.4 (Bronk Ramsey 2009) and the IntCal20 atmospheric calibration curve (Reimer et al. 2020). Obvious outliers are not included in the table. This also applies to radiocarbon dates specifically produced for dendrochronology and wigglematching (as at Sozopol harbour and Urdoviza (Kiten), see Kuniholm et al.1998, 2007). For Ropotamo, where the dendrochronological analyses are still ongoing, the 14C dates of the pile samples' outermost rings are provided in this table, as they are indicative of the dates when the tree were felled and thus of possible construction events (compilation of dates: A. Ballmer)

Site and intervention	Lab no.	Sample material	^{14}C Age (yr BP)	cal BC 2σ (95.4%)	Sample context and dated event	Publication
Ezerovo II	Bln-2391	Charcoal	4155 ± 60	2889–2576	Unknown	Boyadziev (1995)
Ezerovo II	Ki-89	Charcoal	4210 ± 60	2916–2584	Unknown	Boyadziev (1995)
Sozopol harbour Core Sz-I	OS-57882	Charcoal	5170 ± 30	4049–3820	Lower cultural layer; first occupation phase	Filipova-Marinova et al. (2011)
Sozopol harbour Core Sz-I	OS-58077	Charcoal	4560 ± 40	3491–3101	Upper cultural layer; second occupation phase	Filipova-Marinova et al. (2011)
Sozopol harbour Core Sz-D	VERA-1652	Wood piece	5310 ± 40	4315–3997	Lower cultural layer; first occupation phase	Filipova-Marinova et al. (2011)
Alepulagoon 2012 coring; core SOZ-7 ter(2)	Lyon-10429	Charcoal	4505 ± 30	3353–3098	Organic layer; 'occupation' (?) phase	Flaux et al. (2016)
Alepu lagoon 2012 coring; core SOZ-7 bis	Lyon-10430	Wooden stake	4550 ± 30	3371–3102	Sapwood; tree felling = set-up of wood structure	Flaux et al. (2016)
Alepulagoon 2012 coring; core SOZ-7 ter(1)	Lyon-10871	Charcoal	4475 ± 30	3341–3029	Organic layer; 'occupation' (?) phase	Flaux et al. (2016)
Alepu lagoon 2012 coring; core SOZ-7	Lyon-10873	Charcoal	4525 ± 30	3362–3102	Organic layer; 'occupation' (?) phase	Flaux et al. (2016)
Alepulagoon 2012 coring; core SOZ-7	Lyon-10875	Charcoal and organic macro-remains	4525 ± 30	3362–3102	Organic layer; 'occupation' (?) phase	Flaux et al. (2016)
Alepu lagoon 2012 coring; core SOZ-7	Lyon-10876	Charcoal	4570 ± 35	3494–3103	Lagoon muds; ante- 'occupation' (?)	Flaux et al. (2016)
Alepulagoon 2012 coring; core SOZ-7 ter(2)	Lyon-10877	Charcoal and organic macro-remains	4855 ± 30	3706–3531	Lagoon muds; ante- 'occupation' (?)	Flaux et al. (2016)

(continued)

Table 9.2 (continued)

Site and intervention	Lab no.	Sample material	^{14}C Age (yr BP)	cal BC 2σ (95.4%)	Sample context and dated event	Publication
Ropotamo 2017 excavation; square T2	SUERC-77016	Wooden pile (no. 0)	4474 ± 21	3336–3032	Outermost rings of sapwood; tree felling = construction event	Grant (2018), Вагалински et al. (2018)
Ropotamo 2018 excavation; square T3	SUERC-84589	Wooden pile (no. 3)	4390 ± 34	3264–2907	Outermost rings of sapwood; tree felling = construction event	Grant (2018)
Ropotamo 2018 excavation; square T3	SUERC-84596	Wooden pile (no. 7)	4386 ± 34	3261–2907	Outermost rings of sapwood; tree felling = construction event	Grant (2018)
Urdoviza (Kiten) 1989 excavation; layer III, square A3	Bln-4111	Wooden pile (no. 6)	4170 ± 50	2891–2584	Cultural layer; occupation phase	Görsdorf and Boyadžiev (1996)
Urdoviza (Kiten) 1989 excavation, Layer III, square A	Bln-4112	Wooden pile (no. 3)	4050 ± 50	2858–2466	Cultural layer; occupation phase	Görsdorf and Bojadžiev (1996)
Urdoviza (Kiten) 1989 excavation, Layer III, square A	Bln-4114	Wooden pile (no. 13)	3980 ± 60	2837–2291	Cultural layer; occupation phase	Görsdorf and Bojadžiev (1996)
Urdoviza (Kiten) 1989 excavation, Layer III, square A3	Bln-4115	Charred wooden pile (no number)	4040 ± 50	2857–2462	Cultural layer; occupation phase	Görsdorf and Bojadžiev (1996)
Urdoviza (Kiten) 1989 excavation, Layer III, square B	Bln-4117	Charred wooden pile (no. 24)	4060 ± 50	2861–2468	Cultural layer; occupation phase	Görsdorf and Bojadžiev (1996)
Urdoviza (Kiten) 1989 excavation, Layer III, square B	Bln-4118	Charred wooden pile (no. 64)	4070 ± 60	2868–2470	Cultural layer; occupation phase	Görsdorf and Bojadžiev (1996)
Urdoviza (Kiten) 1989 excavation, Layer III, square B	Bln-4119	Charred wooden pile (no. 16)	4160 ± 50	2886–2582	Cultural layer; occupation phase	Görsdorf and Bojadžiev (1996)

Fig. 9.3 Southern Bulgarian Black Sea coast with prehistoric wetland sites (K. Dimitrov)

ubmerged remains were discovered by chance during harbour construction works.

9.2.2.1 Sozopol Harbour

The historical town of Sozopol, ancient Apollonia Pontica, is located on a rocky peninsula on the south side of Burgas Bay, connected with the mainland by a narrow sandy isthmus (Fig. 9.3).

During the construction of a new harbour in 1927, dredging was carried out in the historical harbour of Sozopol. Rich archaeological remains were recovered from the sea bottom. When the

harbour was extended at the end of the 1980s, settlement remains from the ENE and the EBA were discovered. Between 1990 and 2021, systematic underwater excavations were carried out, bringing to light further ENE and EBA settlement relics (Fig. 9.4) (Ангелова et al. 1994, 1996; Draganov 1995, 1998; Angelova and Draganov 2003; Filipova-Marinova et al. 2011; Dimitrov et al. 2020; Димитров et al. 2021a, in press). Repeated dredging activities since the 1920s have severely affected the archaeological remains lying at less than 4.5 m bpsl. In the

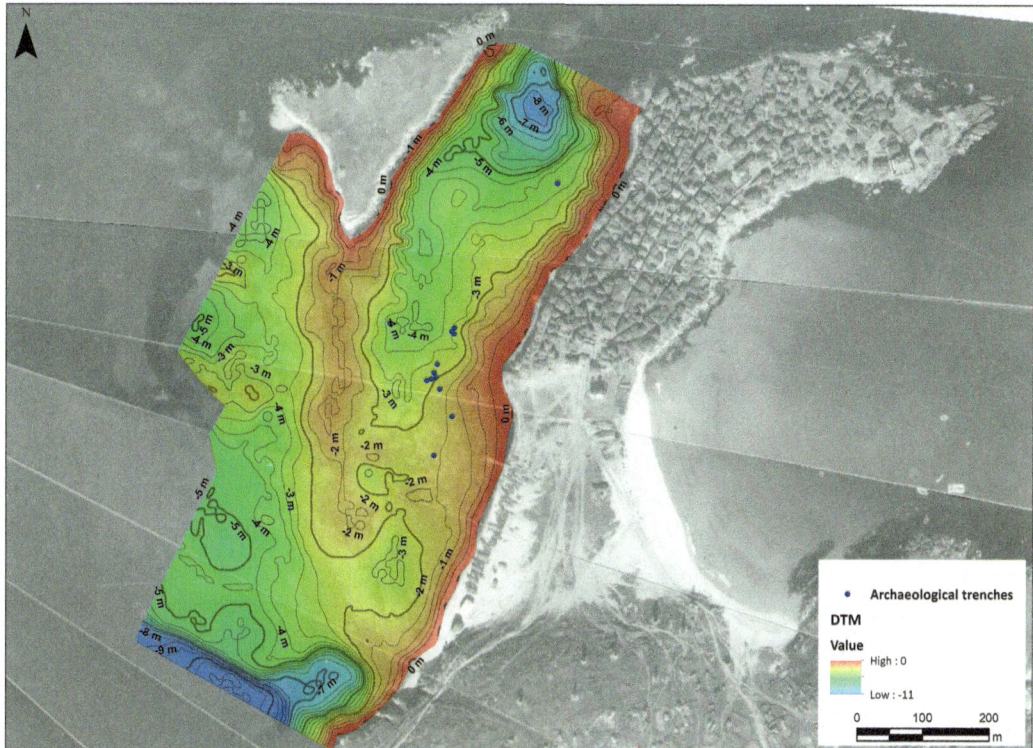

Fig. 9.4 Sozopol harbour. Location of the underwater archaeological trenches (**blue dots**). The extent of the prehistoric settlement, which has been severely disturbed by modern construction activities, can be estimated based on the original terrain (map basis: archive topographic data, presenting the situation prior to the first dredging activities. Aerial photography from 1918. Scientific Archive BAS, Shkorpil archive, no. 165/856; DEM, extracted from bathymetric survey in 1927, State Archive Varna, f. 1K, Български народен морски сговор)

centre of the port, the remains of an ENE settlement were found preserved at a depth of c. 5.0 m bpsl. As well as wooden piles, pottery was recovered, along with artefacts made from flint, stone, bone, and antler.

While the pottery assemblage from Sozopol is clearly made in an ENE style, the technology, shape, and decoration of the pots feature a number of novel, post-ENE elements. This hints at a tradition extending beyond the KGK VI culture complex and into the first half of the Transitional Period (late KGK VI to the beginning of the Cernavodă I culture complex (c. 3800–3700 BC)). Thus, the Sozopol site captures the very latest known moment of the ENE, not only along the Black Sea coast, but also within the supra-regional context (Dimitrov et al. 2020).

A recent reconstruction of the palaeo-terrain has revealed that the ENE settlement was originally built on a flat terrace on the southern shore of a small bay or freshwater or brackish lake, located between the Sozopol peninsula and St. Kirik island, in what is now the northern part of the harbour (Fig. 9.5) (Димитров et al. 2020). The settlement's occupation was terminated by a flood event, caused either by a water level rise or by tectonic subsidence (Draganov 1995). This flood coincided with the Transitional Period. After 3200 BC, the water level dropped, and the bay would have become habitable again.

South of the sector with the ENE remains, an area with a preserved cultural layer from the EBA was discovered during the archaeological campaigns of the 1980s and 1990s (Leshtakov

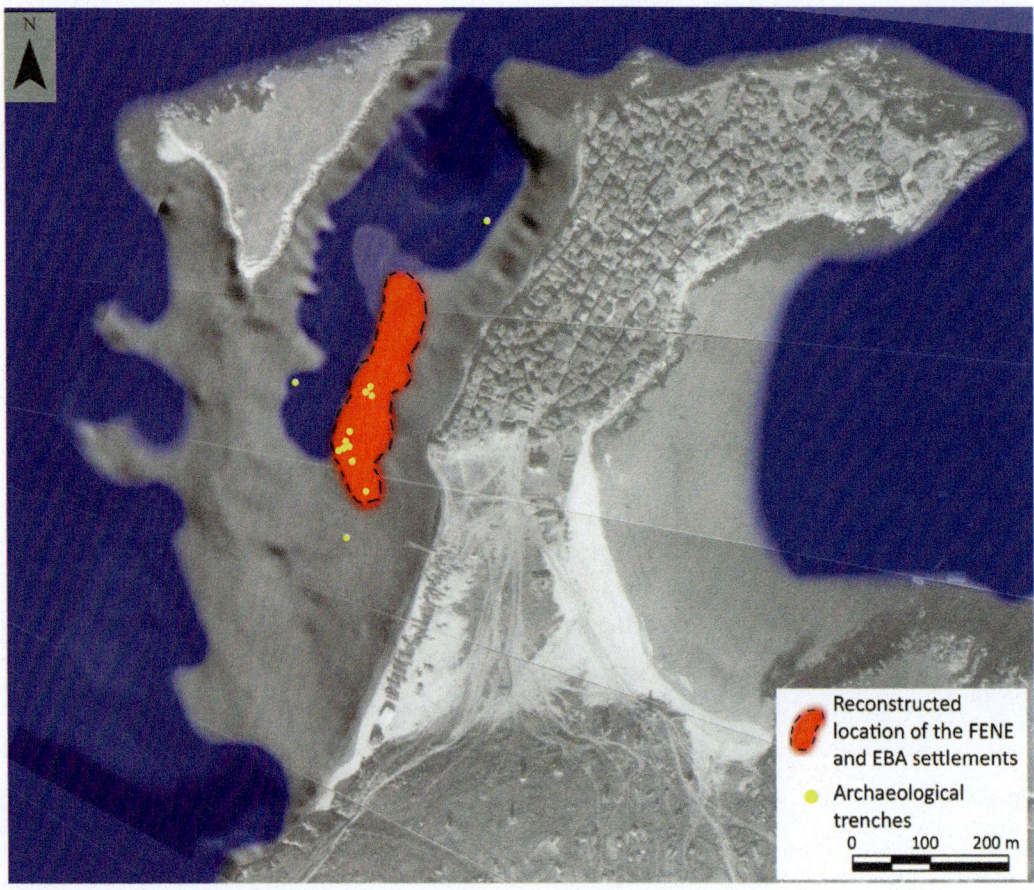

Fig. 9.5 Sozopol harbour. Palaeo-landscape with the estimated extent of late ENE and EBA settlements, based on excavations, test trenches, core drilling, and surface observations (K. Dimitrov)

1994; Draganov 1995). In 2021, these EBA traces were further investigated (Fig. 9.6). Within an area of 5 × 5 m, both vertically inserted wooden posts and horizontally oriented wooden elements were recorded, clearly testifying to pile-dwelling architecture. The pottery from this context is attributed to EBA phases II and III. So far, the EBA settlement of Sozopol represents the most recent known wetland settlement on the southwestern Black Sea coast.

While only a minimum of the pottery finds has been published (for an overview, see Vassileva 2018, 2019; Dimitrov et al. 2020), more effort has been put into the evaluation and publication of the faunal remains (Boev 1995; Spassov and Iliev 1997). At Sozopol, hunting for wild animals seems to have been more important than livestock farming. Marine resources, including dolphins, played a prominent role. Among the remains of domesticated species, an interesting proportion of sheep and goats is noticeable.

Extensive research has been conducted on palaeo-ecological aspects of the settlement. Sediment coring in the harbour was carried out by Filipova-Marinova et al. (2011), followed by targeted analyses related to vegetation and climate. The palynological study indicates the presence of mixed oak forests, which show signs of opening during both occupation phases in response to land appropriation and wood exploitation. The pollen record of both phases points to the cultivation of wheat and barley (Filipova-Marinova et al. 2011).

Fig. 9.6 Sozopol harbour. Remains of the EBA settlement (CUA excavation 2021): pottery and vertical and horizontal timbers (K. Dimitrov)

Three radiocarbon dates were obtained from pieces of charred and uncharred wood from the archaeological layers excavated in the 1980s and 1990s. Ranging between the 42nd- and the 32nd century BC, they fit well with both the late ENE and the EBA find material (Filipova-Marinova et al. 2011) (Table 9.2). Kuniholm et al. (2007) performed dendrochronological analyses on a collection of samples obtained in 1988 from oak-wood piles in an area of Sozopol harbour designated as 'square D', part of a sector that contained only ENE material. From the samples, a 224-year tree-ring chronology was constructed, with an end date in the final stages of the 5th millennium BC, in line with the late ENE occupation. Separate construction phases could not be identified. So far, these are the oldest absolute dates from a wetland settlement along the Bulgarian Black Sea coast.

9.2.2.2 Alepu Lagoon

The Alepu lagoon lies south of Sozopol between the promontories of Sveta Agalina and Sveti Toma (Fig. 9.3). In 2012, Flaux et al. (2016) were able to trace early human presence at the edge of the palaeo-lagoon by means of sediment coring. The archaeological evidence consists of a seemingly anthropogenic layer c. 5.5 m bpsl, i.e. c. 1.5 m underneath the sea bed, containing rich organic materials—charcoal and wood fragments, seeds, fish and shell remains, small bone fragments—as well as lithic fragments and undetermined potsherds. However, evidence of actual dwellings, in the sense of structural features such as wooden posts, was not identified. A small number of oak 'piles' were found vertically fixed into the palaeo-lagoon bottom. Due to their small diameter of 4.5–6 cm, they can hardly be associated with house substructures (especially in the absence of additional, clearer remains of buildings) but are rather to be regarded as stakes for enclosures or fences, fish traps or similar. It will be necessary, in future, to check the lagoon for more substantial evidence of an actual prehistoric settlement (Fig. 9.7 (1)).

Based on radiocarbon dating of the sapwood of one oak stake, the earliest anthropogenic traces from the Alepu lagoon are dated to between 3371 and 3102 cal BC, while further radiocarbon dates from charcoals and organic macro-remains

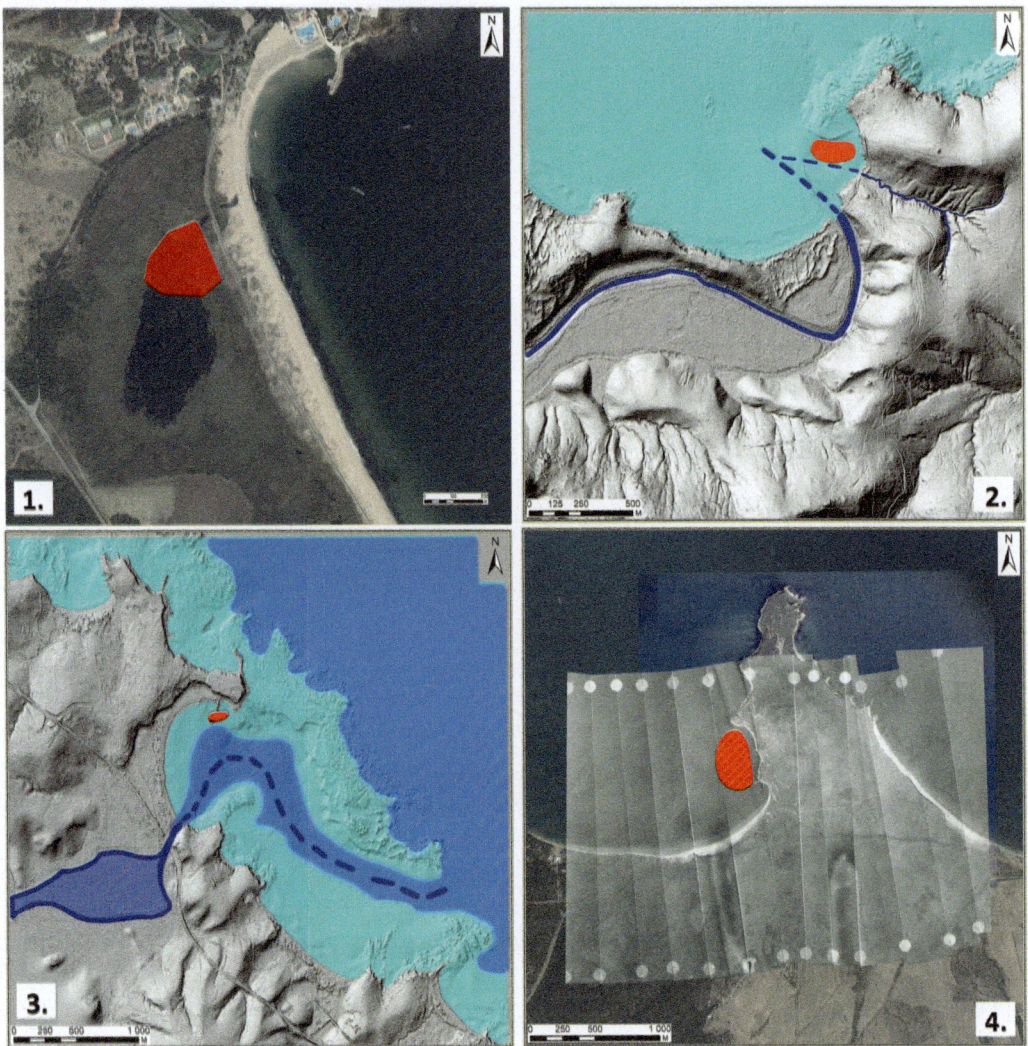

Fig. 9.7 Topography of the archaeological wetland sites south of the Bay of Burgas. The red areas correspond to the estimated extent of settlements. The blue lines refer to the palaeo-rivers. (**1**) **Alepu lagoon** (map basis: satellite image); (**2**) **Ropotamo** (map basis: Lidar DEM); (**3**)

Urdoviza (**Kiten**) (map basis: palaeo-landscape reconstruction, according to Lidar DEM); (**4**) **Atiya** (map basis: aerial photography from 1918. Scientific Archive BAS, Shkorpil archive, no. 165/856) (K. Dimitrov)

indicate human activity between 3341 and 3029 cal BC, the beginning of the EBA (Table 9.2) (Flaux et al. 2016).

9.2.2.3 Ropotamo

The Ropotamo river estuary is situated 5.5 km south of the Alepu lagoon (Fig. 9.3). In prehistoric times, the bay at the mouth of the estuary was protected by Cape St. Dimitar, which then extended further out to sea than it does today.

Archaeological remains from the first millennium BC were recorded at Cape St. Dimitar as early as the end of the 19th century. In 1974, an underwater archaeological site with several anthropogenic layers was detected in the area at the mouth of the estuary. I. Karayotov conducted a series of underwater archaeological investigations in the 1980s, funded by the Maritime Society of Burgas (Карайотов et al. 1989; Georgiev et al. 1991). In 1989, he discovered remains of an EBA settlement (Карайотов 1990; Karayotov 1990, 1992, 2002). At a depth of about 2 m beneath the sea bottom, a layer of mussels and oysters was encountered, from which the tops of vertically inserted wooden posts protruded. Underneath this stratum, a layer containing pottery and other typical EBA materials, including fireplace remains, was reported.

In 2017, the underwater archaeological excavations at the Ropotamo estuary were followed up as part of the international Black Sea Maritime Archaeology Project (Black Sea MAP). Beginning in 2018, annual archaeological excavations were undertaken by the CUA, funded by the Bulgarian Ministry of Culture. In these more recent campaigns, four archaeological trenches with a total surface area of about 80 m^2 were excavated. A sequence of stratified deposits more than 3 m thick was revealed (Вагалински et al. 2018; Димитров et al. 2019, 2020, 2021b). Underneath modern, Ottoman, Byzantine, archaic Greek, and natural deposits, EBA layers were discovered at a depth of between 4.15 and 5.25 m bpsl (Fig. 9.8, Layers Va–c). The finds included EBA pottery, burnt clay, and 75 vertically inserted wooden piles (Fig. 9.9). The latter clearly indicate that the Ropotamo EBA settlement was a pile-dwelling village, featuring buildings constructed on platforms above ground level. While the first and main construction event was characterised by piles driven vertically into the initial ground level VI (Fig. 9.8, left pile), smaller piles were also found in stratigraphically higher positions (Fig. 9.8, right pile). The latter may have originated from later repair work in the course of the settlement occupation, a working hypothesis which requires more detailed investigation.

The preliminary evaluation of the remains from Ropotamo has shown a prevalence of wild animal resources over livestock (Карастоянова 2021), as well as evidence of metallurgical activities. The finds can generally be associated with the Ezero A-Cernavodă III culture complex (Драганов 1990; Leshtakov 1994; Лещаков 2006).

Nine radiocarbon dates from three wooden piles are available for Ropotamo (Grant 2018; for the most recent dates of each of the three piles,

see Table 9.2). While dendrochronological analyses are still ongoing, preliminary Bayesian modelling of the radiocarbon dates allows to chronologically estimate the tree felling, and thus the construction of the settlement, to the end of the 31st century BC, i.e. to EBA stages I and II (Grant 2018; Вагалински et al. 2018). This dating is consistent with the pottery types found at the site.

According to geomorphological observations, the EBA settlement was built on a semi-dry terrace of a tributary of the Ropotamo palaeo-river (Fig. 9.8, Layer VI) that was prone to periodic, small-scale flooding (Fig. 9.7 (2)), leading to the use of pile-dwelling architecture. The covering of the Bronze Age features and finds by alternating sediment layers of marine shells and sand testifies to a water level rise with a lasting impact. The state of preservation of the archaeological material indicates rapid flooding and sedimentation (Dimitrov et al. 2020).

In 2020, several Late Neolithic pottery sherds were found underneath the EBA settlement remains (Fig. 9.8, Layer VII). Based on the sediment properties and the remarkably good state of preservation of the pottery, they probably originated from a terrestrial settlement built near the Ropotamo river about two millennia before the EBA occupation, at a time when the sea level was lower, and were buried in stable terrestrial soil before the settlement was inundated.

9.2.2.4 Urdoviza (Kiten)

The Urdoviza site is located in Kiten Bay, about 20 km south of Sozopol (Fig. 9.3). The archaeological remains lie 8–10 m bpsl. Given the palaeo-terrain and the water level at the time of its occupation, the prehistoric settlement must have been located on the Karaagach river terrace, at a distance of about 500–600 m from the former coastline (Fig. 9.7 (3)) (Draganov 1995; Angelova and Draganov 2003). While scattered stray finds from the Bronze Age had been reported from the 1970s on, the prehistoric settlement at Kiten Bay was only identified in 1986 during the excavation of a merchant ship from the Ottoman period. The site was archaeologically explored between 1986 and 1989. The total

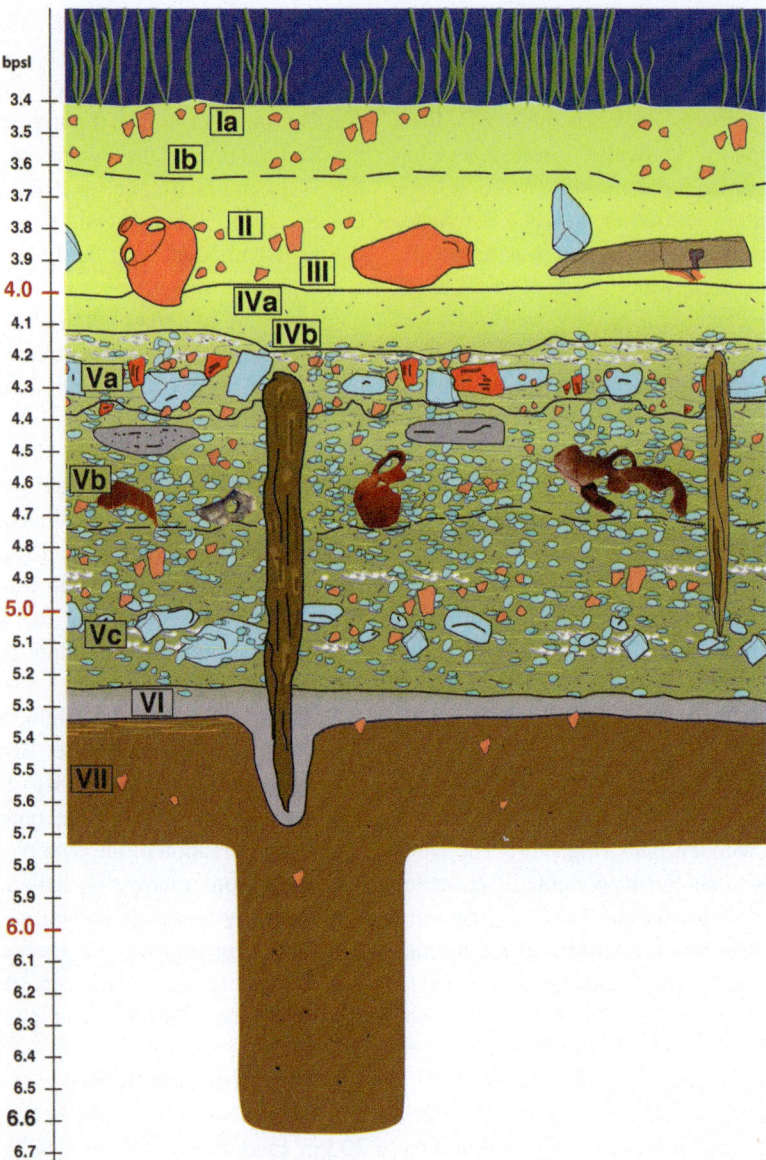

Fig. 9.8 Ropotamo. Schematised stratigraphy (trench 4, CUA excavations 2019–2020). **Layers Ia–b**: 3.4–3.6 m bpsl; 0.2 m thick; mixed sea bottom sediments; accumulated surface finds; Ottoman and mediaeval finds. **Layers II–III**: 3.6–4.0 m bpsl; 0.4 m thick; sea bottom sediments; Roman finds, isolated pre-Roman finds: Hellenistic, Late Archaic, Late Iron Age. **Layers IVa–b**: 4.0–4.15 m bpsl; 0.15 m thick; sea bottom sediments, oysters; no archaeological finds: cultural hiatus. **Layer Va**: 4.15–4.25 m bpsl, 0.1–0.15 m thick; stones, oysters, burnt clay plaster fragments, EBA pottery, tops of vertically inserted wooden posts. **Layer Vb**: 4.25–4.90 m bpsl, 0.65 m thick; marine sediments; EBA pottery, stone querns, wooden posts. **Layer Vc**: 4.90–5.25 m bpsl, 0.3–0.35 m thick; marine sediments, big round stones; EBA pottery, wooden posts. **Layer VI**: 5.25–5.35 m bpsl, 0.1 m thick; grey clay, river mud, free from any marine materials; level into which most larger piles were driven vertically. **Layer VII**: 5.35 m bpsl; excavated to a depth of 6.6 m bpsl without reaching the bottom of the layer (the shape of the lower boundary in the figure corresponds to the excavation boundary, not the bottom of the layer); a few Late Neolithic potsherds, small charcoals (K. Dimitrov)

Fig. 9.9 Ropotamo. EBA wooden piles protruding from the sea bed (CUA excavation 2020) (K. Dimitrov)

Fig. 9.10 a–b Urdoviza (Kiten). Selection of EBA pottery (CUA excavations 1986–1990) (K. Dimitrov)

area investigated covered more than 900 m^2 (Porozhanov 1991; Draganov 1995; Angelova and Draganov 2003, 2003).

The typology of the recovered pottery corresponds to the Western Black Sea coast EBA I and II, i.e. an early stage of the Cernavodă III-Ezerovo culture (2900–2700 BC) (Fig. 9.10). Parallels are found in the archaeological material of Ezerovo II, Ezero VIII–V and Yunatsite XIII–IX (Leshtakov 1994; Angelova and Draganov 2003).

Clay hearths and daub fragments from house floors and walls were recorded (Draganov 1995). In addition to the typical settlement finds, at least 300 wooden posts from buildings, mainly oak, were found protruding vertically from the subsoil. There are indications of timbers that were originally fastened horizontally to the standing posts by means of lacing (Angelova and Draganov 2003). Based on the wooden building remains, it must be assumed that the vertical posts served to raise the house floors from the ground in order to prevent dampness from seeping in from the marshy environment, on the one hand, and water ingress as a result of short-term, periodic flood events on the other.

Wild species dominate the spectrum of fauna prevalent during the time of occupation (Ribarov 1991), with marine resources, comprising tuna and dolphins, standing out in particular. Domesticated animals were also present, as evidenced by remains of cattle and horses. Except for the conspicuous quantity of horse bones as well as the absence of sheep and goats, both the spectrum and the proportions of the animal species found at Urdoviza are comparable with those of Sozopol harbour. Palynological analyses of sediments indicate forest clearance during the settlement period. Unlike at Sozopol, there is no pollen evidence of cereal cultivation in the close environs.

A series of radiocarbon dates from wooden building elements from Urdoviza ranges between c. 2900 and 2300 cal BC (Table 9.2) (Görsdorf and Bojadžiev 1996). Within this range, the older dates are consistent with the typochronology and the dendrochronology of the site. Kuniholm et al. (1998, 2007) analysed 83 of the timbers which were sampled in the 1980s by means of dendrochronology and were able to establish a 285-year tree-ring chronology. Based on this, four to five successive construction phases over a period of 64 years can be distinguished at Urdoviza, dating from between the twenty-ninth and the 28th century cal BC. This absolute time span is consistent with the typochronological classification of the site. The lack of information about where, within the excavation perimeter, the sampled wooden elements were found prevents a more detailed dendroarchaeological assessment regarding individual buildings or the spatial development of the settlement.

The low degree of fragmentation of the Bronze Age pottery was probably due to the last village being rapidly buried in sediments after its abandonment. Following a water transgression, it was apparently covered within a short time by shelly, sandy silt (Angelova and Draganov 2003).

9.2.3 Further Potential Wetland Settlements

Among the other prehistoric submerged sites frequently mentioned in the literature, only Atia, northwest of Sozopol, can be considered to have been an actual wetland settlement, possibly even a pile-dwelling village (Figs. 9.3 and 9.7 (4)).

The site was discovered in 1968, when the navy port of Atia was constructed (Лазаров 1975, 2009). There are many indications that the site has since been completely destroyed. During the dredging of the area, wooden piles, pottery, and other archaeological finds were brought to the surface. The mostly unpublished finds refer to the final ENE and the end of the EBA, and from a chronological point of view appear to be close to those from Sozopol (Vassileva 2018, 2019; Dimitrov et al. 2020).

The 'underwater sites' of Cape Shabla, north of Varna (Peev 2004, 2008; Peev and Slavchev 2018), are likely to have been terrestrial burial sites that were inundated after their abandonment.

9.3 Discussion

While the Varna Lakes sites need to be further explored in order to collect reliable data, enough is already known about the prehistoric wetland settlements of the southwestern Bulgarian Black Sea coast to place them in a broader context. In the following, we discuss aspects of the sites' chronological evolution and their relation to neighbouring bodies of water, as well as their role within the regional settlement topography and economic networks.

9.3.1 Site Chronology

During the wetland occupation phases, settlements were established in sheltered coastal areas with humid environments where the floors of buildings were elevated from ground level by means of wooden substructures ('pile-dwellings'). The settlement phases correlate with bioarchaeological indicators of forest clearance and cereal cultivation. While there is evidence of livestock farming, foraging and the hunting and gathering of marine resources also seem to have been quite important. Between the ENE and the EBA, most settlement activity was interrupted for a period of c. 500 years. This supra-regional hiatus at the very end of the ENE, known as the Transitional Period, witnessed a general abandonment of settlements, accompanied by woodland recovery, indicative of low human impact.

At the present time, several questions are still open regarding the detailed chronology and duration of occupation of the settlements, which can only be answered by additional excavations and targeted analyses. Based on the current state of knowledge, the southwestern Black Sea coast experienced two main phases of prehistoric wetland settlement between c. 4450 and 2525 BC:

- The earliest known wetland settlement, at Sozopol harbour, appeared in the final ENE and seems to have lasted into the earliest stage of the Transitional Period.
- During the EBA, after a hiatus of several hundred years and simultaneously with the post-hiatus resettlement of inland areas, the coastal wetlands were reoccupied. The earliest EBA radiocarbon date comes from the Alepu lagoon, although the nature of the anthropogenic activity at this site is not clear at this stage. The settlements at Ropotamo and Urdoviza (Kiten) appeared as early as the first phase of the EBA and continued to be occupied in EBA phase II. Finally, a later settlement at Sozopol harbour appeared in the second stage of the EBA and continued to be inhabited throughout the third stage and right up to the end of the EBA.

At the moment, post-EBA wetland settlements are not known on the southwestern Black Sea coast.

9.3.2 Settlement Locations and Their Relation to Bodies of Water

The presented wetland settlements were not established on the open seacoast but were set back, on the flat shores of inland waters close to the sea. The proximity of river outlets or lakes would have offered freshwater and/or brackish-water resources (Angelova and Draganov 2003).

For the time being, multiple questions concerning water fluctuations and water levels remain unresolved (Peev et al. 2020). The relics from most of the presented sites consistently indicate that the water level during the ENE and EBA must have been between 5 and 8 m bpsl, or even lower. This archaeological conclusion, however, does not coincide with geological water level reconstructions, which suggest higher water levels in those periods—meaning, in turn, that the settlements would have been considerably inundated at the time of their occupation. This discrepancy can be explained either by undetermined water level fluctuations or by local tectonic subsidence. Further open questions concern the detail of water level fluctuations between the ENE and the EBA, when a major flooding followed by a regression is thought to have taken place (Kislov 2018). Based on the observed taphonomic processes in the presented settlements, it appears that a water level rise must, indeed, have occurred at the time of their abandonment or shortly afterwards. The rapid transgression led to a relatively quick covering of the archaeological traces with sediments, resulting in the good preservation recorded at several sites. There is no doubt that in order to understand the biographies of the wetland settlements and the lives of their inhabitants, on the one hand, and the formation of the archaeological sites on the other, it is crucial to distinguish between supraregional water level changes, resulting in greater and more sustained effects, seasonal water level fluctuations, and micro-events with regional or local impacts.

It must be noted that, in addition to these ecological forces, sociocultural factors should also be included in the discussion about the choice of settlement locations.

9.3.3 Topography and Economy

It is worth mentioning the simultaneous occurrence of two different settlement types on the western Black Sea coast, defined by their locations. During the Late Neolithic, terrestrial settlements were located both on rocky promontories and other places close to the sea (e.g. Budjaka and Akladi Cheiri (Лещаков and Класнаков 2008; Лещаков et al. 2009, 2015; Leshtakov et al. 2020a)), and at a distance from the coast on low ridges (e.g. Garmitsa (Leshtakov et al. 2020b)). As discussed in this review, during both the late and final ENE and the EBA, wetland settlements were established, located in humid environments on the terraces of river estuaries (e.g. Alepu lagoon, Ropotamo, and Urdoviza) and near sea gulfs and bodies of fresh or brackish water (e.g. Sozopol harbour and the Varna Lakes sites). This diversity of dwelling milieus may be explained by increased development and organisation of territory and resources, hand in hand with the subsistence diversification and economic specialisation which is attested at a supra-regional level from the end of the Neolithic onwards.

Within the process of land appropriation, occupation, and use, the precise status of the wetland settlements is unknown. An immediately obvious and plausible explanation for their choice of location is the availability of marine, fluvial, and lagoonal resources. Moreover, the locations of the settlements may also have been related to distribution and exchange networks. The wetland villages of the southern Bulgarian Black Sea coast are located near to the copper ore sources of Medni Rid, known to have been exploited from the late ENE onwards (Pernicka et al. 1997; Dimitrov 2002; Leshtakov 2010; Dimitrov et al. 2020). While copper tool finds in wetland settlements reflect metal consumption, the presence of crucibles, casting moulds, and bellows at the settlements of Sozopol harbour, Urdoviza, and Ropotamo is evidence of on-site copper processing and manufacturing (Dimitrov et al. 2020). Although the settlements' inhabitants may not necessarily have taken on the role of controlling raw material distribution or have been major consumers themselves, given the fact that the western coast of the Black Sea offers one of the most convenient corridors for the transport of goods from the northern Strandzha foothills at Sozopol to the north (Ivanova 2012), they may have been active players in local copper extraction and distribution within a wider exchange

network. With the emergence of a new exchange network at start of the EBA, this complex, interactive dynamic would eventually have collapsed, or rather, been absorbed into a new framework.

9.3.4 Conclusion

From their structural appearance, the Bulgarian wetland settlements are part of the pan-European 'pile-dwelling' phenomenon. While this settlement category is generally co-defined by water level fluctuations, the cases presented here were significantly impacted by the dynamic and sometimes drastic water level changes of the Black Sea. Given the highly favourable environment of the southwestern Black Sea coast during the late ENE and the EBA, both with regard to the availability of resources, for supplying local needs and for supra-regional distribution, and the availability of network opportunities, the simultaneous occurrence of diverse settlement types in the region is most interesting. In addition to their distinctive features, the wetland settlements were obviously an important component of a vibrant prehistoric landscape.

References

Angelova H, Draganov V (2003) Underwater archaeological excavations of submerged late Eneolithic and Early Bronze Age settlements in Kiten and Sozopol (South Bulgarian Black Sea coast). In: Angelova H (ed) Thracia Pontica, VI.2. In Honorem Mihaili Lazarov. Thracia Pontica Series I. Proceeding of the international symposium. Centre for Underwater Archaeology, Sofia, pp 9–22

Benjamin J, Rovere A, Fontana A, Furlani S, Vacchi M, Inglis R, Galili E, Antonioli F, Sivan D, Miko S, Mourtzas N, Felja I, Meredith-Williams M, Goodman-Tchernov B, Kolaiti E, Anzidei M, Gehrels R (2017) Late Quaternary sea-level changes and early human societies in the central and eastern Mediterranean Basin: an interdisciplinary review. Quatern Int 449:29–57

Boev Z (1995) Eneolithic and Early Bronze Age birds from the sunken settlement at Sozopol Bay (Bulgarian Black Sea Coast). Historia Naturalis Bulgarica 5: 51–60

Bojadziev J (1998) Radiocarbon Dates from Southeastern Europe and the Cultural Processes during the fourth millenium B. C. In: Stefanovich H, Todorova H, Hauptmann H (eds) In the steps of James Harvey Gaul.Volume I. James Harvey Gaul. In memoriam. James Harvey Gaul Foundation, Sofia, pp 349–370

Boyadziev Y (1995) Chronology of prehistoric cultures in Bulgaria. In: Alexandrov S, Bailey DW, Panayatov I (eds) Prehistoric Bulgaria. Prehistory Press, Madison, Wisconsin, pp 149–191

Bronk Ramsey C (2009) Bayesian analysis of radiocarbon dates. Radiocarbon 51(1):337–360. https://doi.org/10.1017/S0033822200033865

Dimitrov K (2002) Die Metallfunde aus den Gräberfeldern von Durankulak. In: Todorova H (ed) Durankulak, Band II, Teil 1, Die Prähistorischen Gräberfelder. Publishing House Anubis, Sofia, pp 127–158

Dimitrov K, Draganov V, Prahov N (2020) Submerged prehistoric settlements along the South Bulgarian Black Sea coast. In: Krauss R, Pernicka E, Kunze R, Dimitrov K, Leshtakov P (eds) Prehistoric mining and metallurgy at the Southeast Bulgarian Black Sea Coast. Tübingen University Press, Tübingen, pp 185–245. https://doi.org/10.15496/publikation-51192

Draganov V (1995) Submerged coastal settlements from the Final Eneolithic and the Early Bronze Age in the Sea around Sozopol and Urdoviza Bay near Kiten. In: Alexandrov S, Bailey DW, Panayatov I (eds) Prehistoric Bulgaria. Prehistory Press, Madison, Wisconsin, pp 225–241

Draganov V (1998) The present state of Eneolithic research in northeastern Bulgaria and Thrace. In: Stefanovich H, Todorova H, Hauptmann H (eds) In the steps of James Harvey Gaul. Volume I. James Harvey Gaul. In memoriam. James Harvey Gaul Foundation, Sofia, pp 203–221

Filipova-Marinova M (2007) Archaeological and paleontological evidence of climate dynamics, sea-level change, and coastline migration in the Bulgarian sector of the Circum-Pontic region. In: Yanko-Hombach VV, Gilbert AS, Panin N, Dolukhanov PM (eds) The Black Sea flood question: changes in coastline, climate, and human settlement. Springer, Dordrecht, pp 453–481

Filipova-Marinova M, Giosan L, Angelova H, Preisinger A, Pavlov D, Vergiev S (2011) Palaeoecology of submerged prehistoric settlements in Sozopol Harbour, Bulgaria. In: Benjamin J, Bonsall C, Pickard C, Fischer A (eds) Submerged prehistory. Oxbow Books, Oxford, pp 230–244

Flaux C, Rouchet P, Popova T, Sternberg M, Guibal F, Talon B, Baralis A, Panayotova K, Morhange C, Riapov AV (2016) An Early Bronze Age pile-dwelling settlement discovered in Alepu Lagoon (Municipality of Sozopol, Department of Burgas), Bulgaria. Méditerranée 126:57–70. https://doi.org/10.4000/mediterranee.8203

Georgiev M, Petkov A, Nenov N, Georgiev D, Angelova C (1991) Prospecting of underwater archaeological sites using geophysic methods. In: Lazarov M, Tatcheva M, Angelova C, Georgiev M (eds) Thracia Pontica IV. Quatrième symposium international, Sozopol 6–12 octobre 1988. Галактика – Варна Абагар – Велико Търново, Sofia, pp 451–470

Görsdorf J, Bojadžiev J (1996) Zur absoluten Chronologie der bulgarischen Urgeschichte. Berliner 14C-Datierungen von bulgarischen archäologischen Fundplätzen. Eurasia Antiqua 2:105–173

Grant M (2018) Black Sea MAP dating. Ropotamo. Unpublished report. COARS, University of Southampton

Ivanov I (1993) A la question de la localisation et des études des sites submergés dans les lacs de Varna. Pontica 26:19–26

Ivanova M (2012) Perilous waters: early maritime trade along the western coast of the Black Sea (fifth mill. BC). Oxf J Archaeol 31:339–365

Karayotov I (1990) The antique and the mediaeval port at the mouth of the river Ropotamo. In: Najdenova V, Petrov P, Velkov V (eds) Studies settlement life in Ancient Thrace, vol 5. Terra Antiqua Balcanica, pp 64–66

Karayotov I (1992) Explorations archéologiques sous-marines dans la baie devant l'embouchure du Ropotamo (1985–1986). In: Gyuzelev V (ed) Bulgaria Pontica Medii Aevi. Actes du IIIe colloque international, Nessebar, 27–31 mai 1985. Presses universitaires 'Sveti Kliment Ohridski', Sofia, pp 277–279

Karayotov I (2002) Nouveaux monuments des villes antiques du littoral Ouest de la Mer Noire. In: Гичева Р, Рабаджиев К (eds) Πιτύη (Pitye). Studia in honoren Prof. Ivani Marazov/Изследвания в чест на проф. Иван Маразов. Анубис, Sofia, pp 558–567

Kislov A (2018) On the interpretation of century-millennium-scale variations of the Black Sea level during the first quarter of the Holocene. Quatern Int 465:99–104

Kislov A, Toropov P (2011) Modeling extreme Black Sea and Caspian Sea levels of the past 21,000 years with general circulation models. In: Buynevich I, Yanko-Hombach V, Gilbert SG, Martin RE (eds) Geology and geoarchaeology of the Black Sea Region: beyond the flood hypothesis, vol 473. Special Paper of the Geological Society of America, pp 27–32

Krasteva M, Hristov HS (2014) Unterwasserarchäologie in Bulgarien. Ihre Geschichte und die Erforschung und Erhaltung des unter Wasser liegenden Kulturerbes. Skyllis 14:159–165

Kuniholm P, Kromer B, Tarter S, Griggs C (1998) An early Bronze Age settlement at Sozopol, near Burgas, Bulgaria. In: Stefanovich H, Todorova H, Hauptmann H (eds) In the steps of James Harvey Gaul. Volume I. James Harvey Gaul. In memoriam. James Harvey Gaul Foundation, Sofia, pp 399–409

Kuniholm PI, Newton MW, Kromer B (2007) Dendrochronology of submerged Bulgarian sites. In: Yanko-Hombach V, Gilbert AS, Panin N, Dolukhanov PM (eds) The Black Sea flood question:

changes in coastline, climate, and human settlement. Springer Netherlands, Dordrecht, pp 483–488

Lericolais G, Guichard F, Morigi C, Popescu I, Bulois C, Gillet H, Ryan WBF (2011) Assessment of Black Sea water-level fluctuations since the Last Glacial Maximum. In: Buynevich I, Yanko-Hombach V, Gilbert AS, Martin RE (eds) Geology and geoarchaeology of the Black Sea Region: beyond the flood hypothesis, vol 473. Special Paper of the Geological Society of America, pp 1–18

Leshtakov K (1994) The detachment of the Early Bronze Age ceramics along the South Bulgarian Black Sea coast. 1, Urdoviza EBA pottery. In: Lazarov M, Angelova C (eds) Thracia Pontica V. Actes du symposium international 7–12 octobre 1991. Centre d'archéologie subaquatique, Varna, pp 23–38

Leshtakov P (2010) Two prehistoric sites on the southern Bulgarian Black Sea coast. Am J Archaeol 114 (Special Issue 'Archaeology in Bulgaria, 2007–2009'):734–736

Leshtakov P, Klasnakov M, Samichkova G, McSweeney K (2020a) The prehistoric site Akladi Cheiri, Chernomorets. In: Krauss R, Pernicka E, Kunze R, Dimitrov K, Leshtakov P (eds) Prehistoric mining and metallurgy at the Southeast Bulgarian Black Sea Coast. Tübingen University Press, Tübingen, pp 89–120. https://doi.org/10.15496/publikation-51192

Leshtakov P, Dimitrov K, Abele J, Kunze R, Krauss R (2020b) Catalogue of investigated archaeological sites and mining places. In: Krauss R, Pernicka E, Kunze R, Dimitrov K, Leshtakov P (eds) Prehistoric mining and metallurgy at the Southeast Bulgarian Black Sea Coast. Tübingen University Press, Tübingen, pp 339–405. https://doi.org/10.15496/publikation-51192

Nicholas WA, Chivas AR, Murray-Wallace CV, Fink D (2011) Prompt transgression and gradual salinisation of the Black Sea during the early Holocene constrained by amino acid racemization and radiocarbon dating. Quatern Sci Rev 30:3769–3790

Orcel C, Orcel A (1991) Analyses dendrochronologiques de bois provenant du site lacustre d'Arsenala a Varna (Bulgarie). In: Lazarov M, Tatcheva M, Angelova C, Georgiev M (eds) Thracia Pontica IV. Quatrième Symposium International 6–12 Octobre 1988, Sozopol. Галактика – Варна Абагар – Велико Търново, Sofia, pp 145–160

Parvanov S (2019) Povelyanovo. A site of the Varna culture. Bul e-J Archaeol 7:59–81

Peev P (2004) Submerged prehistoric settlements along the western Black Sea Coast: the problem of situation. In: Dobrzanska H, Erzbet J, Kalichki T (eds) The geoarchaeology of river valleys. Archaeolingua series minor, vol 18. Archaeolingua Alapítvány, Budapest, pp 161–169

Peev P (2008) Underwater sites in the area of cape Shabla (North-East Bulgaria). In: Kostov R, Gaydarska B, Gurova M (eds) Geoarchaeology and archaeomineralogy. Proceedings of the international conference, 29–30 October 2008 Sofia. Publishing House 'St. Ivan Rilski', Sofia, pp 303–304

Peev P, Slavchev V (2018) Bulgaria. Burials and wooden settlement structures. In: Fischer A, Pedersen L (eds) Oceans of archaeology. Jutland Archaeological Society, vol 101. Aarhus University Press, Aarhus, pp 94–99

Peev P, Farr RH, Slavchev V, Grant MJ, Adams J, Bailey G (2020) Bulgaria: sea-level change and submerged settlements on the Black Sea. In: Bailey G, Galanidou N, Peeters H, Jöns H, Mennenga M (eds) The archaeology of Europe's drowned landscapes. Costal Research Library, vol 35. Springer Open, Cham, pp 393–412. https://doi.org/10.1007/978-3-030-37367-2

Pernicka E, Begemann F, Schmitt-Strecker S, Todorova H, Kuleff I (1997) Prehistoric copper in Bulgaria. Eurasia Antiqua 3:41–180

Porozhanov K (1991) Le site submergé d'Ourdoviza. In: Lazarov M, Tatcheva M, Angelova C, Georgiev M (eds) Thracia Pontica IV. Quatrième Symposium International 6–12 Octobre 1988, Sozopol. Галактика – Варна Абагар – Велико Търново, Sofia, pp 109–112

Reimer PJ, Austin WEN, Bard E, Bayliss A, Blackwell PG, Bronk Ramsey C, Butzin M, Cheng H, Edwards RL, Friedrich M, Grootes PM, Guilderson TP, Hajdas I, Heaton TJ, Hogg AG, Hughen KA, Kromer B, Manning SW, Muscheler R, Palmer JG, Pearson C, van der Plicht J, Reimer RW, Richards DA, Scott EM, Southon JR, Turney CSM, Wacker L, Adolphi F, Büntgen U, Capano M, Fahrni SM, Fogtmann-Schulz A, Friedrich R, Köhler P, Kudsk S, Miyake F, Olsen J, Reinig F, Sakamoto M, Sookdeo A, Talamo S (2020) The IntCal20 northern hemisphere Radiocarbon Age calibration curve (0–55 cal kBP). Radiocarbon 62 (4):725–757. https://doi.org/10.1017/RDC.2020.41

Ribarov G (1991) The osteological material from the sunken settlement at Ourdoviza. In: Lazarov M, Tatcheva M, Angelova C, Georgiev M (eds) Thracia Pontica IV. Quatrième Symposium International 6–12 Octobre 1988, Sozopol. Галактика – Варна Абагар – Велико Търново, Sofia, pp 113–118

Spassov N, Iliev N (1997) Animal remains from the submerged Late Eneolithic-Early Bronze Age Settlement in Sozopol (South Bulgarian Black Sea Coast). In: Lazarov M, Angelova C (eds) Thracia Pontica VI.1. Thracia Pontica Series I: La Thrace et les societes maritimes anciennes. Proceeding of the International Symposium, 18–24 September 1994 Sozopol. Centre of Underwater Archaeology, Sozopol, pp 287–314

Stanimirov S (2003) Underwater archaeological sites from ancient and middle ages along the Bulgarian Black Sea coast – classification. Archaeologia Bulgarica 1:1–34

Tsirtsoni Z (ed) (2016) The human face of radiocarbon: reassessing chronology in Prehistoric Greece and Bulgaria, 5000–3000 cal BC. MOM Éditions, Lyon

Todorova H (1995) The Neolithic, Eneolithic and Transitional period in Bulgarian prehistory. In:

Alexandrov S, Bailey DW, Panayatov I (eds) Prehistoric Bulgaria. Prehistory Press, Madison, Wisconsin, pp 79–97

Todorova H (2003) Prehistory of Bulgaria. In: Grammenos D (eds) Recent research in the prehistory of the Balkans. Publications of the Archaeological Institute of Northern Greece, vol 3. Archaeological Institute of Northern Greece, Thessaloniki, pp 257–328

Todorova H, Toncheva G (1975) Die äneolithische Pfahlbausiedlung bei Ezerovo im Varnasee. Germania 53:30–46

Toncheva G (1981) Un habitat lacustre de l'Age du Bronze Ancien dans les environs de la ville de Varna (Ezerovo II). Dacia 27:41–62

Vasileva H (2018) From the bottom of the sea: the Early Broze Age ceramics from Sozopol and Urdoviza. Pontica 51:135–149

Vasileva H (2019) A comparison between the decoration of the Early Broze Age ceramics from Northern and the Southern part of the West Pontic Area, Oltenia. Studii Şi Comunicări. Archeologie-Istorie 26:50–69

Yanko-Hombach V, Gilbert AS, Panin N, Dolukhanov PM (eds) (2007a) The Black Sea flood question: changes in coastline, climate and human settlement. Springer, Berlin

Yanko-Hombach V, Gilbert AS, Dolukhanov P (2007b) Controversy over the great flood hypotheses in the Black Sea in light of geological, paleontological, and archaeological evidence. Quatern Int 167:91–113

Zmeykova I (1991) Certains aspect des problèmes concernant la términologie, la chronologie et la caractéristique culturelle des palafites du bronze ancien très de Varna. In: Lazarov M, Tatcheva M, Angelova C, Georgiev M (eds) Thracia Pontica IV. Quatrième Symposium International 6–12 Octobre 1988, Sozopol. Галактика – Варна Абагар – Велико Търново, Sofia, pp 137–144

Ангелова Х, Драганов В, Димитров К (1994) Потънали селища от финала на халколита и ранната бронзова епоха в созополското пристанище. Археологически открития и разкопки през 1992–1993 г. Велико Търново, рр 17–20

Ангелова Х, Димитров К, Драганов В (1996) Подводни археологически проучвания в акваторията на остров Св. 'Кирик' и пристанището на Созопол. Археологически открития и разкопки през 1995 г. Сандански, рр 113–114

Драганов В (1990) Култура Черна вода III на територията на България и по западното черноморскокрайбрежие. Добруджа 7:156–177

Божков Л (1936) Бележки за Варненското езеро. Известия на българското географско дружество IV:140–167

Божилова Е, Иванов И (1985) Екологични условия в района на Варненското езеро през енеолита и бронзовата епоха според палинологични, палеоботанични и археологически данни. ИНМВ XXC(XXXVI):43–48

Вагалински Л, Джонатан А, Димитров К, Бъчваров К, Пачеко-Руиз Р, Драганов В, Гърбов Д, Рьомби Й,

Педроти Ф, Прахов Н, Георгиева З, Георгиев П (2018) Подводни археологически разкопки в акваторията пред устието на р. Ропотамо. Археологически открития и разкопки през 2017 г. София, рр 720–723

Вайсов И (1992) Състояние на проучванията на т. нар. 'Преходен период в България'. Археология 34 (2):45–49

Димитров К, Драганов В, Прахов Н, Джонатан А, Рьомби Й, Георгиев П, Гърбов Д, Пачеко-Руиз Р, Педроти Ф, Георгиева З (2019) Подводни археологически разкопки в акваторията пред устието на р. Ропотамо. Археологически открития и разкопки през 2018 г. София, рр 743–746

Димитров К, Джонатан А, Рьомби Й, Бъчваров К, Георгиев П, Драганов В (2020) Подводни археологически разкопки (Ранна бронзова епоха, Античност, османски период) в залива пред устието на р. Ропотамо, обл. Бургас. Археологически открития и разкопки през 2019 г., I. София, рр 369–376

Димитров К, Георгиев ПЙ, Прахов Н, Дамянов М, Велковски К (2021a) Спасително подводно археологическо проучване в акваторията на пристанище Созопол. Археологически открития и разкопки през 2020. София, рр 113–118

Димитров К, Георгиев ПЙ, Прахов Н, Гюрова М, Карастоянова Н (2021b) Подводни археологически разкопки в залива пред устието на р. Ропотамо. Археологически открития и разкопки през 2020. София, рр 353–358

Димитров К, Дамянов М, Георгиев ПЙ, Прахов Н, Велковски К (in press) Подводни археологически разкопки в пристанище Созопол, сектор 'Изток'. Археологически открития и разкопки през 2021

Иванов И (1973) Праисторическо наколно селище Страшимирово-2. ИНМВ IX(XXIV):285–288

Иванов И (1980) Подводно изследване на наколно селище при ТЕЦ 'Варна'. Археологически открития и разкопки през 1979 г. София, рр 35–37

Иванов И (1985) Подводни проучвания на наколно селище 'Арсенала'. Археологически открития и разкопки през 1984 г. София, рр 289–291

Иванов И (1987) Подводни археологически проучвания на селище 'Арсенала'. Археологически открития и разкопки през 1986 г. София, рр 281–283

Карайотов И (1990) Подводни археологически проучвания в залива пред устието на р. Ропотамо. Археологически открития и разкопки през 1989, XXXV Национална археологическа конференция по археология. Кюстендил, рр 178–179

Карайотов И, Петков А, Ангелова Х (1989) Подводни археологически проучвания в залива пред устието на река Ропотамо. Археологически открития и разкопки през 1988 година. р 175

Карастоянова Н (2021) Анализ на животински останки от Ропотамо. Национален природонаучен музей, Българска академия на науките. Unpublished report

Лазаров М (1975) Потъналата флотилия. Георги Бакалов, Варна

Лазаров М (2009) Древното корабоплаване по Западното Черноморие. Славена, Варна, pp 196

Лещаков К (2006) Тракия през Бронзовата епоха. ГСУ
ИФ, спец. Арх. 3:141–217

Лещаков П, Класнаков М (2008) Аварийни
археологически разкопки на обект УПИ 8038 в м.
Буджака, гр. Созопол, Бургаска област.
Археологически открития и разкопки през 2007.
София, pp 54–57

Лещаков П, Класнаков М, Недев Д (2009) Спасителни
археологически разкопки на праисторически обект
в местността Аклади чеири, село Черноморец.
Археологически открития и разкопки през 2008
година. София, pp 74–77

Лещаков П, Самичкова Г, Недев Д, Илиева Й,
Славкова Ц (2015) Спасителни археологически
разкопки на късноеолитни и антични структури
в УПИ I–1017, кв. 87, 'Реконструкция на градски
стадион, подобект Подземен паркинг север', гр.
Созопол. Археологически открития и разкопки
през 2014 г. София, pp 76–78

Маргос А (1960) Праисторическите наколни селища
във Варненското езеро. Природа и знание, XIII
(5):15–18

Маргос А (1961а) Към въпроса за датиране на
наколните селища край Варненското езеро. ИВАД
XII:1–6

Маргос А (1961b) Открити следи от нови наколни
селища във Варненското езеро. ИВАД XII:128–131

Маргос А, Тончева Г (1962) Праисторическото
наколно селище при с. Езерово, Варненско. ИВАД
XIII:1–16

Маргос А (1965) Нови находки от наколното селище
при с. Страшимирово, Варненско. Археология VII
(7):57–65

Маргос А (1969) Праисторически наколни селища във
Варненското езеро. In: Савов С (ed) Разказ за
живота на хората от наксолните жилища във
Варненския залив. Народна просвета, София,
pp 85–139

Маргос А (1973) Праисторическо наколно селище
Страшимирово-1. ИНМВ 1973:267–284

Миков В (1950) Следи от наколни селища при
Страшимирово. ИАИ XVII:215–218

Порожанов К (2003) Потъналото селище от Ранната
Бронзова епоха при Урдовиза (разкопки 1986–1988
г.). Изследвания в чест на ст.н.с. I ст. д.и.н.
Хенриета Тодорова. Festschrift für Prof. Dr. Habil.
Henrieta Todorova. Сб. Добруджа, 21, pp 309–322

Прахов Н, Димитров К, Славчев В (2019) Комплексно
археологическо проучване в акваторията и по
бреговете на Варненските езера. Археологически
открития и разкопки през 2018. София, pp 735–737

Тодорова-Симеонова Х (1971) Късноенеолитният
некропол край град Девня – Варненско. ИНМВ
VII(XXII):3–40

Тончева Г (1968) Новооткрити наколни селища край
Варна. Музей и Паметници На Културата VIII
(2):64–65

Тончева Г (1972) Хронология на наколните селища
край Варна. In: Първи конгрес на БИД 1970(I).
София, pp 309–315

Тончева Г, Маргос А (1959) Праисторическото наколо
селище при село Езерово, Варненско. Археология
1959:96–99

Шкорпил Х, Шкорпил К (1921) Наколни постройки в
езерото. ИВАД VI I:79

An Excursus to East Asia: Prehistoric Wetland Settlements of Zhejiang Province, China

10

Haowei Wo and Guoping Sun

Abstract

According to the archaeological discoveries, Zhejiang Province yields the most typical and well-preserved prehistoric wetland settlements in China. In this region, the development of prehistoric cultures is closely related to the natural environment and climate changes. The earliest settlements of the Shangshan culture appeared at about 10,000–8500 BP, probably accompanied by rice cultivation. Whereas rice cultivation was subsequently established, the following Kuahuqiao and Hemudu cultures show strong maritime characteristics. In the Liangzhu period, the huge Liangzhu City and the peripheral water management system were built, which must be understood as symbols of complex society and early state. At the end of the Liangzhu culture, floods submerged most of the land, leaving behind thick sediment deposits. Only after 2000 years, in the Warring States Period (475–221 BC), this land became habitable again. With the establishment of a prehistoric chronology in Zhejiang

province, a lot of multidisciplinary research has been carried out, and natural science and advanced technologies are increasingly applied. The current archaeological research in Zhejiang Province includes the origins of rice cultivation and the emergence of civilisation, geological formation processes and climate change, the spread of maritime cultures and further topics.

Keywords

China · Zhejiang · Prehistoric settlement · Wetland archaeology · Sea-level change · Human adaptation

10.1 Introduction

Zhejiang Province is located along the southeastern coast of China and on the south wing of the Yangtze River Delta. With a surface of 105,500 km^2, Zhejiang is one of the smallest provinces in China. The topography of Zhejiang Province shows a step-shaped slope from southwest to northeast, with alluvial plains in the northeast, hills and coastal plains in the east and mountains and basins in the centre and the southwest. Based on the archaeological discoveries, Zhejiang Province yields the highest number of the prehistoric wetland settlements in China, which are characterised by waterlogged archaeological remains.

H. Wo (✉)
Department of Archaeology, Hangzhou City University, Hangzhou, Zhejiang Province, China
e-mail: wohw@hzcu.edu.cn

G. Sun
Zhejiang Provincial Institute of Relics and Archaeology, Hangzhou, Zhejiang Province, China

© The Author(s) 2025
A. Ballmer et al. (eds.), *Prehistoric Wetland Sites of Southern Europe*, Natural Science in Archaeology, https://doi.org/10.1007/978-3-031-52780-7_10

In 1973, the Hemudu site was discovered and excavated. Hemudu is the first waterlogged prehistoric wetland site excavated in China. It is located in the Ningshao Plain along the eastern coast of Zhejiang Province. Pile-dwellings, well-preserved organic relics and rich remains of animals and plants were found, including cultivated rice. After the excavation of Hemudu, dozens of similar wetland sites were discovered and excavated, such as for instance Zishan, Fujiashan, Tashan, Kuahuqiao or Liangzhu. Most of those sites are located in the flat lands of the Hangjiahu Plain and the Ningshao Plain in the northeast of Zhejiang Province (Fig. 10.1).

Today, the Hangjiahu Plain is the largest alluvial plain in Zhejiang Province. It is a part of the Yangtze River Delta and features the densest river network in China. The terrain of the Hangjiahu Plain (7620 km^2) is very low and flat, with an average elevation of about 3 m a.s.l. The Ningshao Plain (4824 km^2) is a coastal plain on the south bank of the Qiantang River and the Hangzhou Bay. It is located in the northeast of Zhejiang Province, with an average altitude lower than 10 m a.s.l. These two plains are the main distribution areas of the known prehistoric wetland settlements, which were significantly impacted by Holocene sea-level changes.

In the Holocene, the level of the East China Sea fluctuated drastically. During the Last Glacial Maximum, the sea level was more than 100 m bpsl, meaning that a broad area of coastal shelf was exposed. The following postglacial climate was much warmer, which led the sea level to rise rapidly. In early Holocene, most current plains could have been coastal lagoons. Although the relevant studies have yielded slightly different results, the overall trends of the sea-level change are consistent (Fig. 10.2). Hence, the altitude of many Neolithic sites is lower than today's sea level. For example, the original occupation surface of the Jingtoushan site is situated about 5–10 m bpsl (Sun et al. 2021).

The dramatic sea-level change had an important impact on the survival of the region's population and the distribution and development of their settlements, especially relevant for the settlements in the Hangjiahu and Ningshao Plains. Whereas the coastal environment offered abundant marine resources and convenient water transportation ways, the region also suffered from recurrent inundations. Accordingly, most excavated sites in the region are covered by layers of marine or lacustrine sediments. Because of these thick sediments, the archaeological remains are usually well-preserved, meanwhile

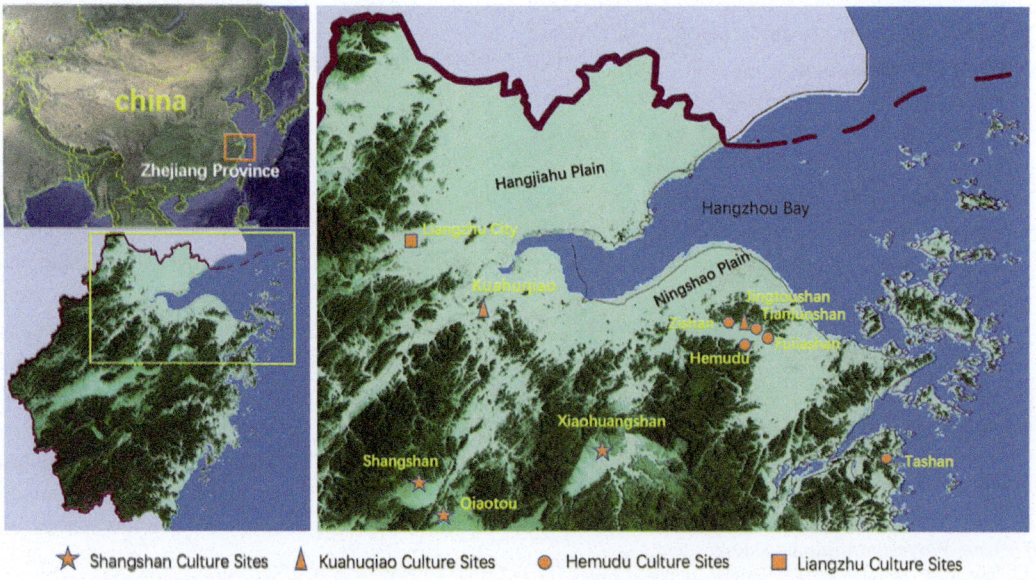

★ Shangshan Culture Sites   ▲ Kuahuqiao Culture Sites   ● Hemudu Culture Sites   ■ Liangzhu Culture Sites

Fig. 10.1 Distribution of the wetland settlements in Zhejiang as mentioned in the text (Authors)

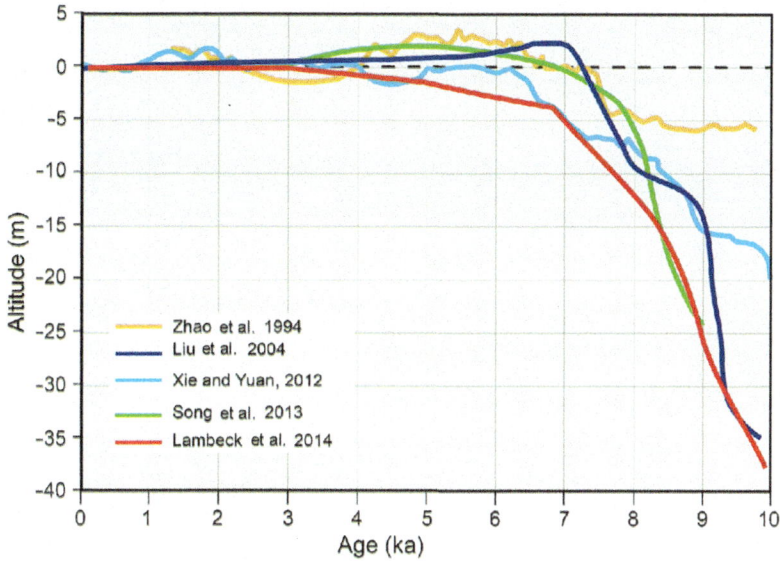

Fig. 10.2 Holocene sea-level changes of the East China Sea (Zheng et al. 2018)

they make it very difficult to discover and access the sites. For instance, the newly excavated Jingtoushan site was covered by 6 m thick marine sediments, and it has only been discovered when some pottery sherds were found in drilling cores taken in the context of construction work (Sun et al. 2021). According to research on the diatoms and paleosalinity, after the Jingtoushan site had initially been covered by marine sediments, the region was mainly a freshwater environment, yet still frequently affected by seawater (Li et al. 2010; Wang et al. 2010).

After decades of archaeological excavations and research, the chronology of the prehistoric periods in Zhejiang Province could be established. From the earliest Shangshan period (10,000–8500 BP) to the Liangzhu period, the chronology indicates a continuous development of prehistoric cultures in Zhejiang Province. These cultures and periods are presented in the following.

10.2 Shangshan Culture (10,000–8500 BP)

The Shangshan culture has its name from the Shangshan site, which is located in today's town of Huangzhai, Pujiang County, along the upper

stretch of the Puyang River. The site was excavated in four campaigns between 2001 and 2008 (Zhejiang Provincial Institute of Relics and Archaeology and Pujiang Museum 2016). A large amount of pottery sherds and stone tools were brought to light. Most of the pottery from the Shangshan site was coated with red colour, and some pottery with painted decoration was found, too. Both polished and chipped stone tools were present. The organic remains were hardly preserved.

Remains of rice, such as husks, rachis and phytoliths, might indicate the cultivation of rice at the time (Zheng and Jiang 2007). In China, evidence of rice is dated as early as 10,000 BP. So far, the most important discoveries in this regard have been made at the sites of Xianrendong and Diaotonghuan in Jiangxi Province (School of Archaeology and Museology of Peking University and Jiangxi Provincial Institute of Relics and Archaeology 2014), Yuchanyan in Hunan Province (Yuan 2013) as well as Shangshan. Whereas the first three sites mentioned are located in caves, Shangshan is an open-air settlement at the border of a valley. The rows of post holes documented at Shangshan may be prove of the earliest form of pile architecture (Fig. 10.3).

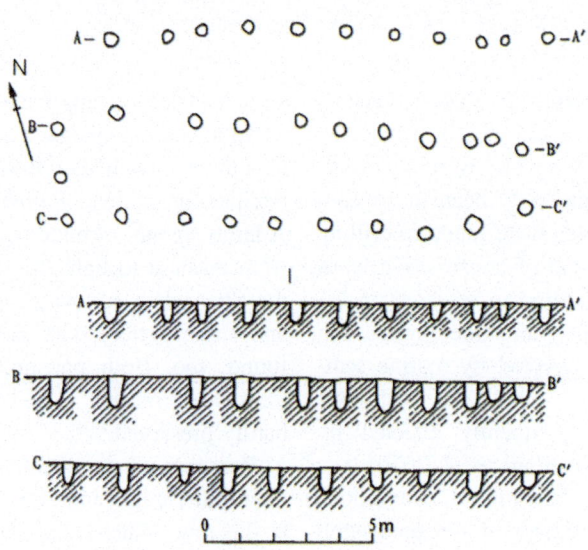

Fig. 10.3 Traces of house substruction F1 at the Shangshan site (Jiang 2007)

Nearly 20 sites associated with the Shangshan culture have been excavated until today, mainly distributed in the Jinqu Basin in the centre of Zhejiang Province. They all represent the earliest Neolithic stage in Zhejiang Province, with radiocarbon dates between 10,000 and 8500 BP. Based on the pottery typology, the Shangshan culture can be divided into three stages, i.e. an early, middle and late periods. The sites of the Shangshan culture originally covered an area of tens of thousands of square metres and featured specific settlement layouts. Whereas the houses and storage pits were consistently located in the south of the villages, the waste pits were situated in the north. In the middle and later stages of the Shangshan period, some settlements began to build ring-shaped trenches around the villages, for instance evidenced at the site of Qiaotou (Jiang 2017), where it measures 3 m in depth and 10 m in width and surrounds an area of about 3000 m^2.

Judging from the archaeological discoveries, the settlements of the Shangshan culture typically are not wetland sites. However, the archaeologists who excavated the relevant sites have also pointed out the presence of thick sediment deposits around some sites which in some cases were burying parts of the sites (Jiang 2007) and hence indicate past flooding events. In the Hangjiahu and Ningshao Plains, no Shangshan culture sites have been discovered yet. There, the earliest sites are attributed to the Kuahuqiao culture. Many settlement remains of the Shangshan culture have also been covered by structures of the subsequent Kuahuqiao Culture.

10.3 Kuahuqiao Culture (8000–7000 BP)

The Kuahuqiao site, located in the Xiaoshan District, Hangzhou City, was discovered by chance in 1990 in the course of soil extraction by a brick factory. At that time, it was the earliest site discovered in the coastal plain area of Zhejiang Province. The site represents the archaeological culture which follows the Shangshan culture.

Low hills are bordering the site in the north and south. The cultural layers were found about 0.5 m bpsl, under about 4.6 m thick grey marine sediments, which helped the site's waterlogged preservation. A large number of pottery sherds, stone tools and organic remains have been excavated at the site, including the earliest dugout in China and bows (Zhejiang Provincial Institute of Relics and Archaeology and Xiaoshan Museum 2004). Several hundreds of carbonised rice remains were recovered, which seem to represent an intermediate state between local wild rice and domesticated rice (Zheng and Zheng 2004).

Judging from the marine sediments that covered the site, the settlement was abandoned due to sea-level fluctuations. Some barnacles, marine creatures that normally live on reefs, were recorded on rocks and pottery in the abandoned settlement. They were mostly found in the southern part of the site, indicating the main direction of the inundations (Zhejiang Provincial Institute of Relics and Archaeology and Xiaoshan Museum 2004).

Since the Kuahuqiao site was the earliest site discovered in the east plains of Zhejiang Province at that time, it was generally believed that the later Hemudu culture was a subsequent spin-off of the Kuahuqiao culture towards the southeast coast. However, the Jingtoushan site, excavated in 2019, has proven that the coastal areas had already been settled during the Kuahuqiao culture period.

In 2013, pottery sherds were discovered in a construction site at Jingtoushan, coming from layers located about 6 m bpsl, which entailed the discovery of the prehistoric settlement of Jingtoushan. According to the archaeological survey, the total extension of the site is about 20,000 m². Today, it is covered by 6–10 m thick marine sediments. Because of these special conditions, the site was excavated in an unusual way. The marine sediments were mostly removed by machines, and the archaeological excavation was performed within large sheet pile walls from steel (50 m × 15 m, enclosed area 750 m²) which was installed around the area to excavate (Fig. 10.4). The latest cultural layers of the Jingtoushan settlement were found in about 6 m depth, the earliest layers in about 10 m depth. The archaeological deposits feature 10 successive layers which in total are more than 2 m thick and refer to a time span of about 500 years.

Many storage pits and organic remains were recovered at Jingtoushan. Some pits were filled with of acorns, razor clam shells, etc. Hundreds of thousands of pottery sherds were collected, including painted ones. The stone artefacts mainly consist of axe blades. Different kinds of bone implements were discovered, and some of them are possibly related to the processing of seafood. A large amount of finely worked and well-preserved woodwares was found, including spears, paddles, axe hafts and even a bowl (Fig. 10.5). The faunal remains are abundant. More than 20 kinds of shellfish and a variety of fish and crabs could be determined. Terrestrial wild species included deer, ox and boar. Among plant remains were acorns, pinecones and peach

Fig. 10.4 Plan of sheet pile walls at the Jingtoushan site (Photo by G. P. Sun)

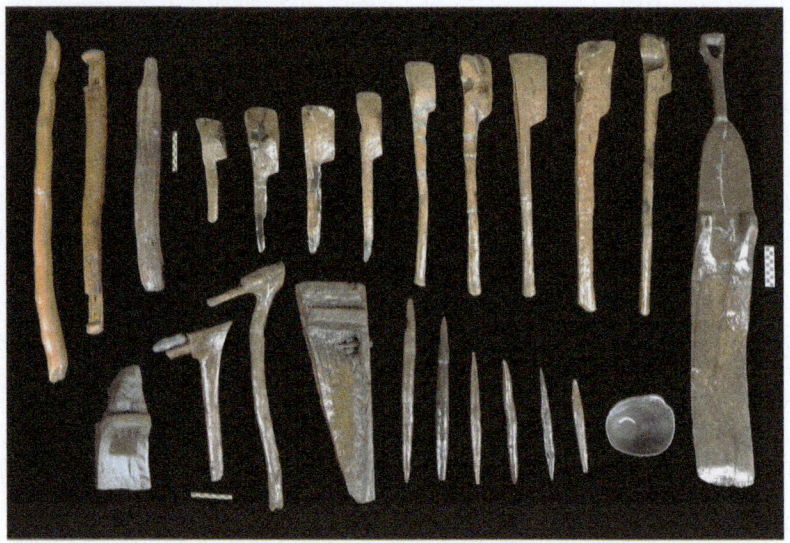

Fig. 10.5 Selection of wooden artefacts excavated at Jingtoushan (Photo by G. P. Sun)

cores. Furthermore, some carbonised rice remains as well as rice rachis were discovered, furthermore some pottery turned out to be tempered with rice husks.

The Jingtoushan archaeological site has provided important information on Holocene sea-level variations and climate change of the Chinese southeast coast. In addition, it features the first shell midden in Zhejiang province and is one of the earliest prehistoric sites at the Chinese southeast coast, dated from 8300 to 7800 BP by radiocarbon dating. After its abandonment, the site was covered by the thick marine sediments. Although the material culture connects the Jingtoushan site to the Kuahuqiao culture, compared to Kuahuqiao sites of the same period, the lifestyle and subsistence at Jingtoushan were obviously maritime in nature.

10.4 Hemudu Culture (7000–5300 BP)

The postglacial eustatic movement of the sea level relatively stabilised around 7000 BP. In the gradual process of regression, the marine sediments turned into swamps or marshland. Although the sea-level movements and the expansion and shrinkage of freshwater bodies still affected the landscape, the discovery of several Hemudu culture sites indicates that humans occupied the coastal areas as soon as the land was exposed. During the Hemudu period, the region was subject to a tropical and southern subtropical monsoon climate, with higher temperatures and more precipitation than in modern times. A fast increase in Hemudu settlements can be noted, the settlers occupying the swampy lowland for about two thousand years until it was no longer habitable.

The eponymous settlement of Hemudu was excavated from 1973 to 1977 (Fig. 10.6). The excavations had a significant impact to the history of archaeology in southern China, since they changed the traditional viewpoint that the Chinese civilisation had originated in the Yellow River valley, and rather revealed the splendid prehistoric cultures on the Chinese southeast coast. The archaeologists found a large amount of carbonised rice and rice husks, which already showed certain characteristics of the domesticated species, and thus made Hemudu a key site in the study of the origins of rice cultivation. The Hemudu site is furthermore the first pile-dwelling site discovered in China. The Hemudu people appear to have been very skilled wood workers, and advanced carpeting techniques were widely applied in the construction of the pile-dwelling houses (Zhejiang Provincial Institute of Relics and Archaeology 2003).

The Tianluoshan site was excavated from 2004 to 2014. Tianluoshan is located in Yuyao City, about 7 km northwest of Hemudu. The site is situated on a slight elevation about 5 m a.s.l. The settlement used to cover a total area of more than 30,000 m², from which an area of about 800 m² is excavated. After the excavation of the Hemudu site, Tianluoshan was another important discovery. At the time of its discovery, it was well preserved in waterlogged conditions. New excavation methods were applied: for example, all the soils from the cultural layers were sieved.

Fig. 10.6 Excavation of the Hemudu site (Zhejiang Provincial Institute of Relics and Archaeology 2003)

Fig. 10.7 Top view of a part of the Tianluoshan site (Photo by G. P. Sun)

The huge amount of all kinds of remains such as animal bones, fish bones, plant seeds or carbonised rice has provided relatively valuable information for diverse studies.

The excavated area corresponds to the settlement's west part (Sun and Huang 2007). In the west, the settlement area was bordered by a river. Whereas a wooden fence separated the settlement from the riverbank, a single-log bridge was installed across the river connecting the settlement with the 'outside world'. Some paddles were found in the river near the bridge, indicating the use of boats (Fig. 10.7). In the residential area of the village, two long houses for everyday purposes, as well as a large ritual building near the village centre, were recognised. The square-shaped pillars from the ritual building, some of them over 50 cm wide, were carefully worked with flat and smooth surfaces.

At Tianluoshan, four different stages of the pile-dwelling building techniques can be recognised, which not only reflect the technical advancement but are also evidence of the settlers' adaptation to the changing environment (Fig. 10.8):

The first stage (about 7000 BP) shows a typical building technique of the early Hemudu culture, manifesting in the archaeological record through simple pile rows. Rows of stakes were directly inserted into the ground to form substructures of pile-dwellings. The diameter of each stake was about 10 cm, and the distance between the stakes was about 10–20 cm. The houses built on these stakes were of a simple architecture but came with a low weight capacity.

In Stage 2 (about 6500 BP), rows of thick pillars were installed in deep pits to serve as building substructures. Most of these wooden pillars have a square cross-section and carry traces of stone axe and adze blades as well as other tools on their surface. The pillars were 30–40 cm wide; the longest pillar reached 3 m. The pillars were mostly arranged in rectangular layouts. In the north of the Tianluoshan excavation area, remains of a long building more than 20 m long and about 10 m wide, and with a surface of nearly 300 m^2, were recovered.

The construction techniques of Stage 3 (about 6000 BP) are based on those from Stage 2. Now, wooden plinths are set underneath the pillars to prevent the poles from further sinking into the ground and hence to increase the weight capacity of the houses. The number of these plinths per pole/pit ranges between 1 and 6 pieces, and some pillars were still standing on these plinths at the moment of their discovery. The diameters of the pits measure 60–100 cm, their depths 50–80 cm. Both rectangular and circular house layouts are

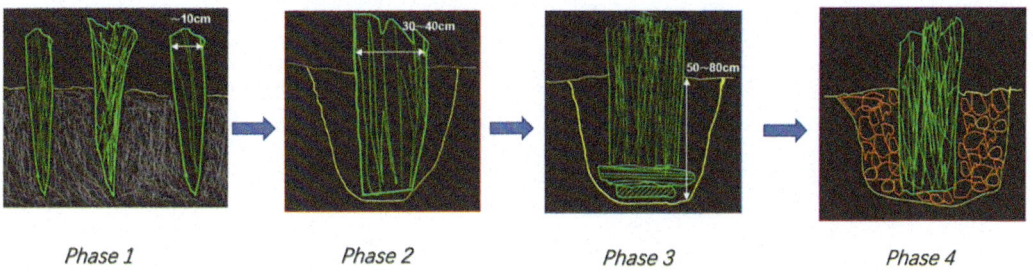

Phase 1   Phase 2   Phase 3   Phase 4

Fig. 10.8 Four stages of pile-dwelling building techniques applied at the Tianluoshan site (Authors)

known. A single-log ladder can be associated to this period, indicating how the houses were accessed.

In Stage 4 (about 5500 BP), the house substructures were built by digging shallow pits, setting stones or wooden plinths into them, and then setting up pillars and filling the gaps with burned soils. The sizes and depths of the foundation pits are different. The house layouts were mainly rectangular. Because of the preservation conditions, no wooden pillars are left.

In the environs of the Tianluoshan settlement, large rice fields must be assumed from three different periods between 7000 and 4500 BP, i.e. the early Hemudu period, the late Hemudu period and the Liangzhu period. According to the archaeological survey, the ancient rice fields around Tianluoshan covered a total area of about 900,000 m^2.

Between the different cultural layers natural sediments were recorded, reflecting the environmental events that affected this area on a regular basis. A large amount of rice husks and rachis as well as weed seeds were found in areas interpreted as former rice fields. The excavation of rice fields of the Liangzhu period revealed a road network between the fields. Some gaps in the roads might be related to irrigation systems.

10.5 Liangzhu Culture (5300–4300 BP)

The Liangzhu culture sites were mainly distributed in the Hangjiahu Plain, the central site of the Liangzhu culture being Liangzhu City

located in the C-shaped basin of the Hangjiahu Plain (see Fig. 10.1).

The presence of Liangzhu City was confirmed on the occasion of the discovery of the city's fortification wall in 2007, which initially was misinterpreted as a dam (Liu 2008). The perimeter of the city wall is about 6000 m, and the width of the rampart ranges between 40 and 60 m. The walls were built of stones and clay. Considering the wall's impressive width as well as the archaeological remains along the foot of the wall, the fortification may also have functioned as a floodwall and a dwelling zone at the same time.

Liangzhu City originally covered an area of about 3,000,000 m^2 and presented a systematically structured layout featuring a palace, noble cemeteries, residential buildings and workshops. Some rivers flowed through the city, and the city walls feature 8 water gates. In the middle of the city is the Mojiaoshan terrace, corresponding to the palace foundations. The terrace covers an area of 300,000 m^2, i.e. one-tenth of the total city area. On the terrace, traces of 35 large houses, arranged in rows, were found. The largest house covers a surface of about 900 m^2 and the smallest building one of about 200 m^2. These buildings were surrounded by a flat square of 70,000 m^2. In the west part of the city are the noble cemeteries of Fanshan and Jiangjiashan, the individuals being buried with hundreds of jade artefacts (Zhejiang Provincial Institute of Relics and Archaeology 2005).

At the time of the city's occupation, a wide freshwater area was to be found northwest of the city. From this region, enormous embankments

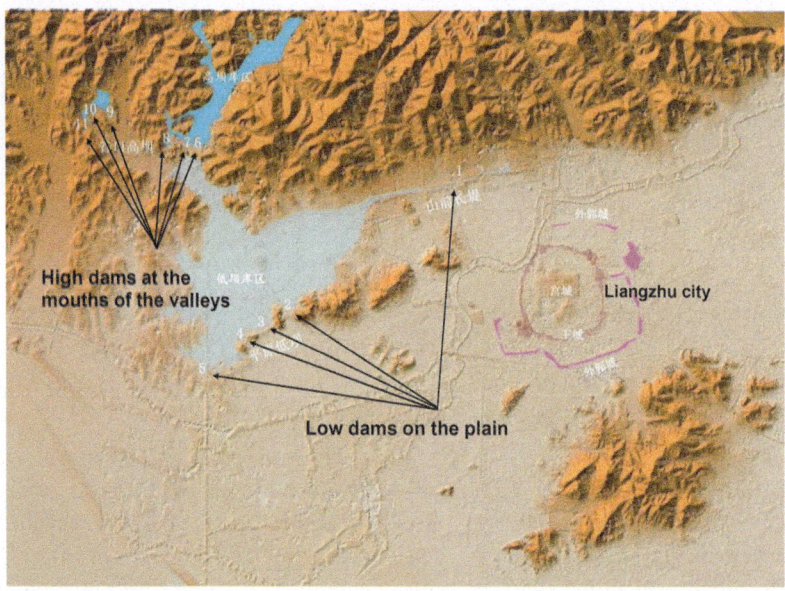

Fig. 10.9 Water reservoir system of Liangzhu City (Zhejiang Provincial Institute of Relics and Archaeology 2019)

made by the Liangzhu people are known (Liu and Wang 2014). Building higher dams at the valley exits and lower dams in the plain, they created a huge reservoir system with a maximal storage capacity of over 40 million m^3 of water (Fig. 10.9). The dams show a similar architecture and were built from grass-wrapped silts, a typical technique of the Liangzhu culture. The construction of Liangzhu City as well as of the dam system demonstrates the Liangzhu culture's complex societal conditions. Of all the known sites of the Liangzhu culture, Liangzhu City constitutes the biggest centre. Though several 'secondary centres' are known from other regions, like in Jiangsu Province and Shanghai, the jade artefacts found in those secondary centres show a high degree of consistency with those from Liangzhu City, indicating the existence of a strong social network across the entire distribution area.

The prosperity of the Liangzhu culture was essentially based on rice-farming economy. In 2017, a large quantity of carbonised rice with an estimated weight of 195 tonnes was found at the Chizhongsi site in the south of the Mojiaoshan terrace (Zhejiang Provincial Institute of Relics and Archaeology 2019).

The highly hierarchised social structure, the advanced technologies, the monopolisation of resources and the social and economic networks indicate that the Liangzhu culture is to be understood as one of the earliest civilisations in China.

At the end of the Liangzhu culture, flooding events caused by the rising sea level submerged most of the land, covering the Hangjiahu Plain with over 1 m thick sediments and causing the Liangzhu culture to disappear. Thereafter, the land became uninhabitable for about 2000 years until the Warring States Period (475–221 BC).

10.6 Conclusion

Zhejiang Province has a prehistory of over ten thousand years. The development of prehistoric cultures was closely related to change in climate and natural environment. In the course of frequent sea level and climate changes, the settlements experienced processes of emergence, development, extinction and relocation. People attempted their best to improve their building and living conditions, making full use of natural resources, domesticating and cultivating rice and constituting complex civilisation.

Although the concept of 'wetland archaeology' is still not popular in China, currently not even as a subdiscipline of Chinese archaeology, the excavations in Zhejiang Province have provided excellent materials for the archaeological research and the reconstruction of these specific settlements and thus demonstrate the characteristics of wetland archaeology and the potential of related research. Meanwhile, the application of natural science and advanced technologies is undoubtedly very important to obtain as much information as possible from the archaeological materials.

References

Jiang LP (2007) Excavation report of the Shangshan site in Pujiang County, Zhejiang Province. Archaeology (Kaogu) 9:7–18 (in Chinese)

Jiang LP (2017) Qiaotou site of Shangshan Culture in Yiwu City. Yearbook Archaeol China (Zhongguo Kaoguxue Nianjian) 2017:250 (in Chinese)

Li ML, Mo DW, Mao LJ, Sun GP, Zhou KS (2010) Paleosalinity in the Tianluoshan site and the correlation between the Hemudu culture and its environmental background. J Geog Sci 20:441–454

Liu B (2008) 2006–2007 Excavation on the Liangzhu City site in Yuhang District, Hangzhou City. Archaeology (Kaogu) 7:4–10 (in Chinese)

Liu B, Wang NY (2014) Findings of archaeological survey of Liangzhu city during 2006–2013. Southeast Culture (Dongnan Wenhua) 2:31–38 (in Chinese)

School of Archaeology and Museology of Peking University, Jiangxi Provincial Institute of Relics and Archaeology (2014) Xianrendong and Diaotonghuan. Cultural Relics Press, Beijing (in Chinese)

Sun GP, Huang WJ (2007) Excavation report of the Tianluoshan site in Yuyao, Zhejiang. Cultural Relics (Wenwu) 11:4–24 (in Chinese)

Sun GP, Mei SW, Lu XJ, Wang YL, Zheng YF, Huang WJ (2021) The Jingtoushan Neolithic site in Yuyao, Zhejiang. Archaeology (Kaogu) 7:3–26 (in Chinese)

Wang SY, Mo DW, Sun GP, Shi CX, Li ML, Zheng YF, Mao LJ (2010) Environmental background analysis of human activities in Tianluoshan site in Yuyao, Zhejiang—phytolith, diatom and other fossil evidence. Quat Sci (Disiji Yanjiu) 02:326–334 (in Chinese)

Yuan JR (2013) The Paleolithic cultures in Hunan Province and the Yuchanyan site. Yuelu Press, Changsha (in Chinese)

Zhejiang Provincial Institute of Relics and Archaeology (2003) Hemudu Site: the archaeological excavation of the Neolithic sites. Cultural Relics Press, Beijing (in Chinese)

Zhejiang Provincial Institute of Relics and Archaeology (2005) Fanshan site. Cultural Relics Press, Beijing (in Chinese)

Zhejiang Provincial Institute of Relics and Archaeology (2019) A comprehensive study of Liangzhu ancient city. Cultural Relics Press, Beijing (in Chinese)

Zhejiang Provincial Institute of Relics and Archaeology, Pujiang Museum (2016) Shangshan Site of Pujiang. Cultural Relics Press, Beijing (in Chinese)

Zhejiang Provincial Institute of Relics and Archaeology, Xiaoshan Museum (2004) Kuahuqiao Site. Cultural Relics Press, Beijing (in Chinese)

Zheng H, Zhou Y, Yang Q, Hu Z, Ling G, Zhang J, Gu C, Wang Y, Cao Y, Huang X, Cheng Y, Zhang X, Wu W (2018) Spatial and temporal distribution of Neolithic sites in coastal China: sea level changes, geomorphic evolution and human adaption. Sci China Earth Sci 61:123–133

Zheng YF, Jiang LP (2007) Remains of ancient rice unearthed from Shangshan site and their significance. Archaeology (Kaogu) 9:19–25 (in Chinese)

Zheng YF, Zheng JM (2004) Study of the remains of ancient rice from the Kuahuqiao site in Zhejiang Province. Chin J Rice Sci (Zhongguo Shuidao Kexue) 18:119–124 (in Chinese)

Part II
Dendrochronology

Dendrochronology and Bayesian Radiocarbon Modelling at the Early Neolithic Site of La Draga (Banyoles, NE Spain)

11

Oriol López-Bultó, Vasiliki Andreaki, Patrick Gassmann, Joan Anton Barceló, Ferran Antolín, Antoni Palomo, Xavier Terradas, and Raquel Piqué

Abstract

La Draga (Banyoles, Spain) is one of the most relevant Early Neolithic site in southern Europe and the Mediterranean region, among other reasons, for its excellent preservation of organic archaeological materials in waterlogged conditions. The site corresponds to a lake dwelling of the early farmers in the region. The goal of this paper is to present the current results of the dendrochronological analysis of the piles and horizontal woods and the ^{14}C dates. In order to understand the relation of the wooden piles with the occupation phases documented, a Bayesian model has been built including the stratigraphy, dendrochronological and ^{14}C data; moreover, a wiggle-matching of the dendrochronology and ^{14}C dates of the piles has also been developed. The dendrochronological results show three phases of tree felling, a single main construction event and establish a minimum duration of the wooden constructions at la Draga of 27 years. Radiocarbon dating combined with dendrochronology confirms these results and dates the construction event around 5310 cal BC.

Keywords

Mediterranean · Neolithic · Dendrochronology · Radiocarbon dating · Bayesian modelling

O. López-Bultó (✉)
Museu d'Arqueologia de Catalunya, Barcelona, Spain
e-mail: joseporiollopez@gencat.cat

V. Andreaki · J. A. Barceló · A. Palomo · R. Piqué
Department of Prehistory, Universitat Autònoma de Barcelona, Barcelona, Spain
e-mail: vasiliki.andreaki@uab.cat

J. A. Barceló
e-mail: juanantonio.barcelo@uab.cat

A. Palomo
e-mail: antoni.palomo@uab.cat

R. Piqué
e-mail: raquel.pique@uab.cat

P. Gassmann
Neuchâtel, Switzerland

F. Antolín
Department of Natural Sciences, German Archaeological Institute, Berlin, Germany
e-mail: ferran.antolin@dainst.de

Department of Environmental Sciences, Integrative Prehistory and Archaeological Science (IPAS), University of Basel, Basel, Switzerland

X. Terradas
Archaeology of Social Dynamics, Spanish National Research Council (CISC)—Mila y Fontanals Institute on Humanities Research (IMF), Barcelona, Spain
e-mail: terradas@imf.csic.es

A. Ballmer et al. (eds.), *Prehistoric Wetland Sites of Southern Europe*, Natural Science in Archaeology, https://doi.org/10.1007/978-3-031-52780-7_11

11.1 Introduction and Objectives

Dendrochronology has been a key methodology in pile-dwelling archaeology to provide new insights on the types of constructions, their organisation and maintenance, and, certainly, their dating and the eventual definition of occupation phases. This technique has been widely developed and applied in central European prehistoric archaeology, as well as in other areas with a dense presence of waterlogged sites such as Scandinavia or the British Isles (Bartholin 1984, 1988; Crumlin-Pedersen 2000; Hillam et al. 1990; Morgan 1988). Nevertheless, there is still a lack of prehistoric dendrochronology tradition in the Mediterranean region and, specifically, in the Iberian Peninsula, where so far, the density of waterlogged sites is much lower than in the above-mentioned areas and therefore the possibilities of obtaining long dating curves are extremely limited. Although, some progress is being done for historical periods in the Iberian Peninsula (Domínguez-Delmás et al. 2015, 2017, 2018, 2020; Domínguez-Delmás and García-González

2015; Ravotto et al. 2016), at the present time the site of La Draga is one of the few prehistoric cases that has provided dendrochronological data (Gassmann 2000).

La Draga (Banyoles, Spain), is one of the most relevant Early Neolithic sites in southern Europe and in the Mediterranean region. It is a pile-dwelling located at the western shore of Lake Banyoles in the North-eastern Iberian Peninsula (Fig. 11.1). It is on a plain, at 170 m a.s.l. and between 35 and 40 km from the Mediterranean Sea. Nowadays the site is partially under the waters of the lake, but the most extensive area is on the mainland. As a partially submerged site, it is characterised by the good preservation of organic remains, which is quite unusual in the Iberian Peninsula. Its chronology (c. 5300–4900 BC) suggests that it was built by some of the first farming groups that settled in the northeast of the Iberian Peninsula (Andreaki et al. 2020; Bogdanovic et al. 2015; Bosch et al. 2000, 2011; Palomo et al. 2014).

Archaeological excavations began in the 90 s and are still on-going. From 1990 to 2021 approximately 1000 m² have been excavated,

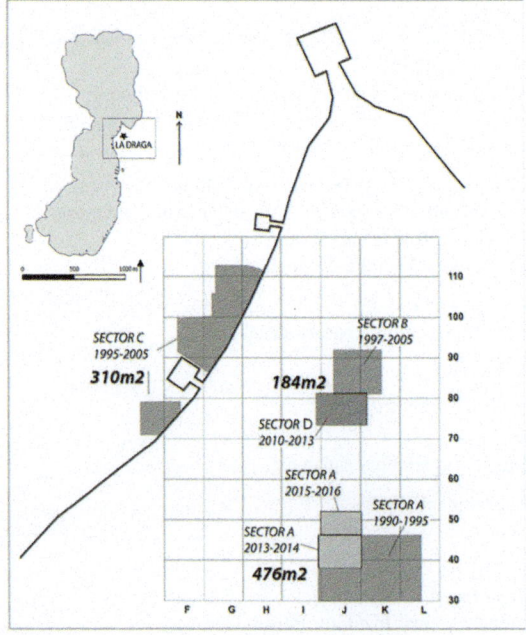

Fig. 11.1 Location of la Draga and sectors excavated (Authors)

representing around 6% of the total area of the site, which is estimated at approximately 15,000 m². The excavated area has been divided into three different sectors—A, B/D, C—that have been defined according to the relationship between the archaeological layers and the water table (Fig. 11.1). Sector A is the driest sector of the excavation and is located on the highest part of the site. In sector A the archaeological layers are well above the water table, therefore, no organic remains have been conserved, except for the tips of the wooden piles driven into the carbonated sands, and thus below the archaeological layers. Even though sector B/D is an inland sector, the lowest archaeological levels, which correspond to the earliest occupation of the site, have remained below the water table since the Neolithic period, favouring the conservation of organic matter. Sector C is strictly an underwater sector located at the current lakeshore edge. This sector, permanently covered with water, has also allowed very good conservation of organic matter.

So far, it has been possible to identify two archaeological horizons that correspond to different occupation phases, with some particularities depending on the sector. The recent excavation of sector B/D has allowed documenting a clear difference between these two archaeological horizons, clearly separated by the presence of a paved surface of travertine slabs that overlap an older layer of timber logs. The layers above this paved surface correspond to the last Neolithic occupations (5216–4981 cal BC). Waterlogged deposits are only found below this horizon, which means that the organic material has not been conserved at this layer and only appears carbonised. Several structures have been identified over this pavement, but no evidence of timber constructions has been linked to it. The travertine slabs which correspond to this late phase cover the entire surface excavated in sector B/D. The layers that appear below this pavement correspond to the earliest occupation (5372–5067 cal BC). This oldest phase is characterised by the exceptional conservation of organic matter which has allowed recovering an extraordinary sample of wood remains, among them hundreds of piles and beams used for the construction of the dwellings as well as wooden tools related to daily life practices.

In sector A several pit structures, with irregular forms and reduced or extensive dimensions have been identified. These features are filled out with large quantities of diverse archaeological remains such as charred seeds, faunal bones, fragments of ceramics, quartz, flint, bone tools, pieces of ornaments, and grinding tools. Because of the material they contain, the pits have been interpreted as landfills for food waste and manufactured objects that were once considered useless (Palomo et al. 2014). Hearths or combustion features are numerous in this sector. About forty have been found with different morphologies. Finally, pavements of possible roofed structures have also been documented. Deep into the carbonated sands, under the archaeological layers piles ends have been recovered too although it remains unclear their relationship with the stone structures. In sector C, the wooden collapse of the oldest occupation is preserved although the travertine pavement loses its continuity and is consequently less well documented. Wooden piles are also present in the underwater sector C.

Regarding artisanal productions, the site has provided more than 200 plant-based artefacts associated with this earliest phase (Bosch et al. 2006), which constitute a valuable record of some technical productions barely documented in the Early Neolithic sites of the western Mediterranean. Among these, abundant evidence of wooden agricultural tools such as digging sticks (López-Bultó et al. 2020) and sickle handles (Palomo et al. 2011), woodworking tools (Palomo et al. 2013), and bows (Piqué et al. 2015) exists. Moreover, the site has provided direct evidence of the use of vegetable fibres to produce ropes and basketry and indirect evidence of textile production with vegetal fibres (de Diego et al. 2017; Herrero-Otal et al. 2021; Piqué et al. 2021a, b; Romero-Brugués et al. 2021).

The remains of artefacts made with bone materials, shells, and minerals of different natures, provide a complete picture of the diversity of production processes (Bosch et al. 2000, 2006, 2011; Terradas et al. 2017).

Regarding subsistence, agriculture and livestock practices are also well documented. The community of La Draga cultivated several species of cereals and, probably, legumes, in addition to the opium poppy (Antolín 2016; Antolín et al. 2018). Among the cereals, the naked wheat stands out and, to a lesser extent, barley. Hulled cereals have also been found, such as einkorn and emmer wheat. Domestic animals are represented by pig, cattle, goat, sheep, and dog, representing in total more than 97% of fauna remains recovered during archaeological excavations (Antolín et al. 2014, 2018; Saña 2011). Fishing, hunting, and gathering practices are also documented and would constitute an important complement to the food supplied by domestic animals and plants at certain times of the year, in addition to providing raw materials.

Regarding the wooden remains, so far it has been possible to carry out a first approach to the construction of the dwellings based on the morphology, dimensions, and GIS analysis (López-Bultó 2015). Oak (*Quercus* sp. Deciduous) was the main raw material used for piles and horizontal woods. Most of the piles are trunks with bark, with diameters between 20 and 200 mm, among the piles some of them preserved more than one metre long driven in the carbonated sands, whilst between the horizontal wood elements of more than 5 m have been recovered. The excellent preservation conditions in La Draga allowed a thorough and wide study of the Early Neolithic dwellings. However, despite the good preservation of wooden remains, it has not been possible to carry out, so far, a dendrochronological dating of the wooden structures. The first dendrochronological analysis of samples recovered in the field seasons 1991–1998 was carried out on a sample of 233 wooden piles (Gassmann 2000). This first study allowed identifying some characteristics of the trees and forests exploited, mainly young individuals (only 11 of them had more than 65 years). Only few of them could be cross-dated and provided a dendrochronological assembly. The reduced number of growth rings of the individuals analysed, together with the lack of dendrochronological reference sequences for the northeast of the Iberian Peninsula did not allow going further. The sampling of wooden piles and horizontal wood continued systematically which has allowed elaborating a floating dendrochronological curve of the site (Piqué et al. 2020).

The main objective of this work is to present and discuss the current results of the dendrochronological analysis to provide new insights on settlement development and wooden constructions typology, as well as palaeoecology and woodland management through the dendrotypology. One of the main questions that pose the site of La Draga is the relation of the wooden piles with the occupation phases observed during excavation. With this objective, we have carried out a Bayesian model including the stratigraphy as well as dendrochronological and ^{14}C data. The dendrochronological and chronological results presented in this work are a milestone in the research of La Draga.

11.2 Materials

In the different sectors of La Draga, hundreds of structural wooden elements and samples have been recovered and sampled, considering piles and horizontal wood. Since the beginning of the excavations in 1990, 1271 piles and 494 horizontal woods have been recovered and sampled, amounting to 1765 structural timbers that can potentially be dendrochronologically measured. The sector with the higher number of structural elements is sector B/D, where 494 horizontal structural elements and 771 piles have been recovered. In the archaeological levels of sector A, no horizontal woods have been preserved, as stated before. Nevertheless, the tips of 458 wooden piles driven in the carbonated sands have been recovered and sampled in this sector. In sector C, the total number of piles recovered and sampled is 42.

A sample of 12 wooden elements was selected for radiocarbon dating (Table 11.1). Sample selection included mainly wooden piles, but also one horizontal wood. Among piles, a selection was made, based on dendrochronological analyses, and samples of trees cut down in different years were used.

Table 11.1 Samples used for radiocarbon dating, 11 vertical piles (PT) and 1 horizontal wood (TT)

Sample number	Sector/s	Species	Age	Felling season	Dendrochronological dating (internal floating chronology)
PT-1311	A	QU	22 tree-rings	Winter	216–237
PT-0986	B/D	QU	36 tree-rings	Winter	202–237
PT-0089	A	QU	56 tree-rings	Winter	182–237
PT-0738	B/D	QU	85 tree-rings	Winter	153–237
PT-0605	B/D	QU	57 tree-rings	Winter	181–237
TT-0468	B/D	QU	168 tree-rings	Winter	38–237
PT-1450	A	QU	19 tree-rings	Winter	223–241
PT-1441	A	QU	20 tree-rings	Winter	228–247
PT-0153	A	QU	26 tree-rings	Winter	223–248
PT-0191	A	QU	28 tree-rings	Winter	229–256
PT-0584	B/D	QU	43 tree-rings	Spring	217–259
PT-0582	B/D	QU	28 tree-rings	Winter	238–265

QU Quercus sp. (Authors)

11.3 Methods

The dendrochronological study of these materials is still on-going; so far, more than eleven hundred samples have been measured. The dendrochronological approach has focused first on the characterisation of the trees exploited taking into account the taxa, age and diameters of the wood selected.

Secondly, the analysis has focused on the determination of the year and season of cut of the oaks identified in order to build a floating dendrochronological sequence for the site of la Draga and an internal chronology of the dwellings according to the wooden structural elements. Finally, the growing pattern of the oaks has been analysed to identify the forest stands represented in the sample.

In order to relate the internal chronology of the dwellings obtained from the dendrochronology with the absolute chronology of the site, we have radiocarbon dated a sample of wooden piles and horizontal woods. The calendar chronology of the local floating tree-ring sequence was fixed using available radiocarbon estimates of the last growth ring in 12 logs. A wiggle-matching Bayesian model using the radiocarbon dates and the dendrochronological gap between

installation piles and the ones assigned to later repairs has been estimated using OxCal 4.4 and ChronoModel 2.0. Concerning the radiocarbon-dated samples a chronological model was built by defining an initial event (*terminus post quem*) in terms of the statistical combination of all radiocarbon-dated wooden samples. The *posterior* probability was calculated based on phases with a hypothetical duration of 2 years, and with known gaps between them depending on the difference in tree-rings.

A Bayesian model with all the radiocarbons dates of the site, including the wooden samples, has been calculated. INTCAL2020 calibration curve was used for the calibration of the dates (Reimer et al. 2020), and the posterior probability of the model has been calculated using both OxCal 4.4 and ChronoModel 2.0 (Lanos and Dufresne 2019).

11.4 Results and Discussion

11.4.1 The Timber Acquisition and Use

The majority of the structural timber analysed has been determined as oak (*Quercus* sp. Deciduous) (Bosch et al. 2006; López-Bultó and Piqué

2018), these taxa represent more of the 95% of the identified wood. Other taxa identified are hazel (*Corylus avellana*), laurel (*Laurus nobilis*), dogwood (*Cornus* sp.), *Acer* sp., and Maloideae but their presence is minimal.

One of the main characteristics of the archaeological wooden timbers from La Draga is their relatively small average diameter (Fig. 11.2). In respect to the number of tree rings of the samples, it is noteworthy the low average number of rings. The number of samples with less than 30 tree rings supposes more than 70% of the total, but this is especially significant regarding the piles, among which almost 80% count less than 30 tree rings (López-Bultó and Piqué 2018; Piqué et al. 2021a, b) (Fig. 11.3). In addition, among the samples with a higher number of tree rings, there are a significant number of samples with actual narrow rings and almost no latewood, which difficult their measuring (Gassmann 2000).

Most logs used for the piles correspond to young trees (less than 30 years) with diameters range between 50 and 100 mm, which indicates a clear selection of sizes for this purpose. Among the horizontal wood, due to their transformed cross-section, the original diameter has been calculated from the tree-ring curvature. This estimation shows a higher variability in diameter than the vertical piles; although most diameters are smaller than 100 mm, there are more trees in the highest ranks. The trend observed in the tree-rings is similar: a higher number of old trees are present among the horizontal wood.

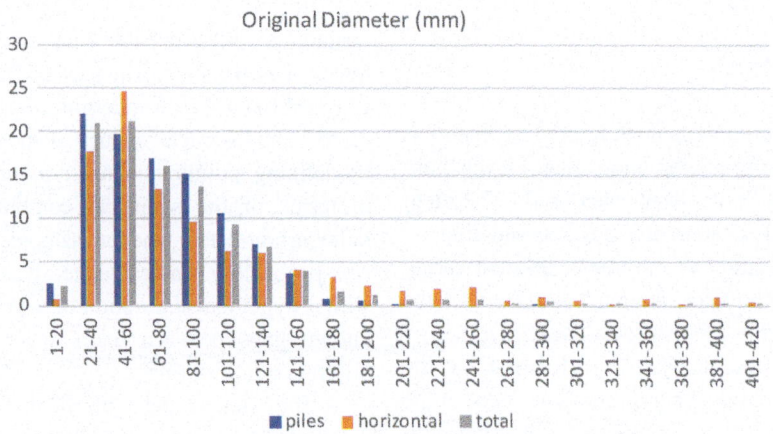

Fig. 11.2 Trunk diameters between the piles and horizontal wood, in percentages (Authors)

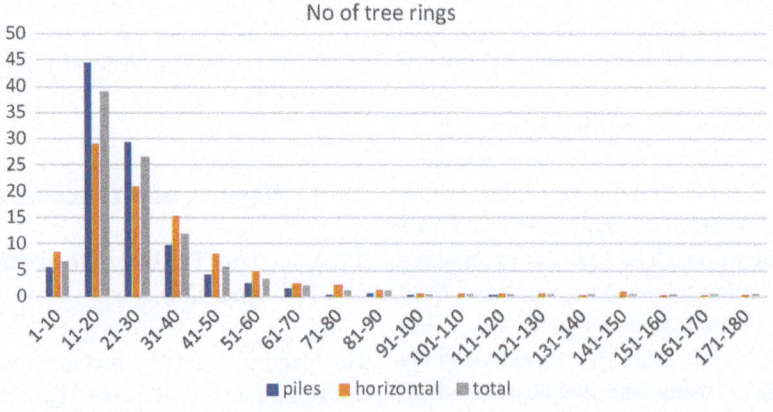

Fig. 11.3 Number of tree rings between the piles and horizontal wood, in percentages (Authors)

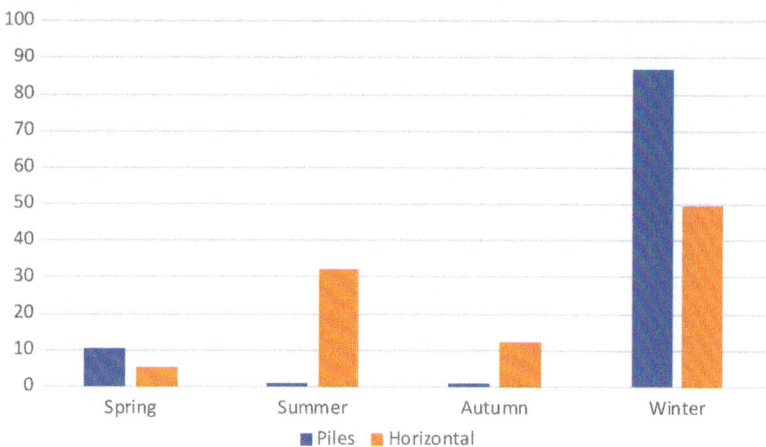

Fig. 11.4 Logging seasons documented among piles and horizontal woods, in percentages (Authors)

The acquisition seasonality is also relevant for architectonic timbers. Although wood was acquired at any time of the year, a clear predilection can be seen for the procurement of timber during winter (López-Bultó and Piqué 2018) (Fig. 11.4). There are some differences in the procurement time between piles and horizontal wood; among the latter, a higher percentage of trees were cut down in summer and autumn.

Although is unclear the shape and size of the dwellings or structures, several elements allow to propose that dwellings were built over wooden of platforms, slightly elevated. Some of the piles preserved a fork in their upper part; these forks were over the carbonated sands and were more or less aligned (Barceló et al. 2019; López-Bultó 2015). Moreover, long horizontal woods, some of near 5 m. long, were associated to these forks, which suggest they could be structural parts of the platforms. Despite the walls are not preserved, it has been possible to observe that some of the piles were inclined, which suggest that they were used as wall and roof.

11.4.2 Construction and Repair: Dendrochronological Results

The total number of wooden samples that have been measured is more than eleven hundred (on-going analysis) of which it has been possible to cross-date 136 samples, all of them well-dated and with high synchronisation (Fig. 11.5). These 136 dated samples come from the three different sectors of the pile-dwelling and correspond to piles and horizontal timber as well. The creation of the mean curves has been possible due to the horizontal timber, which has a higher average number of tree rings per sample. For the first time, it has been possible to create a floating dendrochronological curve for la Draga that covers 265 years. This floating curve has allowed building an internal chronology of the site's dwellings in respect of the moment of construction and further repairs.

According to the year and season of cut of the piles, the dendrochronology has established three different phases or moments on the felling of the trees: Phase I is a phase previous to the main construction phase and it is composed by approximately 8% of the dated samples, Phase II or construction phase with approximately 66% of the dated samples, and Phase III or reparation phase including approximately 26% of the dated samples.

Phases I and II correspond to the foundational moment of the settlement. Phase I of logging includes piles from different sectors and a land plank that were made from oaks felled between winter 233/234 and winter 236/237. The oldest ones may come from reuse, stored wood, or dead standing trees. The youngest ones (year 236) may have the same origins as the oldest ones or, more plausibly, have been part of a preparatory falling for the main site that was constructed the following year. Phase II of logging is marked by the construction moment. It includes piles as well as horizontal timbers that were felled on winter 236/237 and probably also during the year 238. Analysis of the global diagram shows that the three sectors (A, B/D, C) are affected by 237/238 logging. Meaning that the entire extension of the site was built at the same time. The diagram block shows that generally, the piles of sector B/D are younger than the piles of sector A, except for two piles from sector B (332 and 1137), which were also much larger (above 15 cm in diameter).

The dated piles are present in all the sectors of the site (Fig. 11.5). This would imply that dwellings and structures of the first village of la Draga were built at the same time, during a very important construction project, building wooden constructions all around the different sectors. This strict contemporaneity of most of the wooden elements used for construction is quite unusual compared with other lakeside settlements. Throughout the alpine region, concerning the Early Neolithic excavations and even in later chronologies, sites are generally smaller, and the construction of built structures hardly occurred simultaneously, settlements tend to expand in temporal phases relatively slowly expanding to neighbouring areas (Arnold 1990). Two examples from the Neolithic period are Hornstaad-Hörnle IA, built from 3910 BC onwards (see Billamboz 2006), and Sutz-Lattrigen/Riedstation, built between 3393 and 3389 BC (Hafner 2005).

The organisation of several simultaneous logging episodes has already been suggested based on species and morphologies selected, as well as on the falling season (López-Bultó 2015; López-Bultó and Piqué 2018), but with the identification of a great foundation phase, this hypothesis can be strongly supported. The simultaneous installation of piles and planks on the entire surface of the village reinforces the hypothesis, previously suggested based on archaeological evidence (Barceló et al. 2019;

Campana 2018; López-Bultó 2015) of construction of platforms where quadrangular structures have been raised. As a hypothesis, the use of an elevated platform above the carbonated sands and the surface of the water would not have prevented the installation of poles belonging to the wooden structures.

Phase III of felling trees stands for at least 27 years, from year 238 to year 265 of the local chronology. This phase also extends around the three sectors of the site, and it is interpreted as reparations or reinforcements of the wooden constructions. Following the dynamics of the construction phases, bear witness to a minimum occupation of the village of 28 years.

Taking into consideration that the site is still been excavated nowadays and the dendrochronological analysis is still on-going, there are a few points worthy to discuss.

The piles bear witness to a minimum occupation of the village of 27–28 years (from 237–238 to 265 of the floating chronology curve). As it can be observed in Fig. 11.5, the later dendrochronological phases described tend to use younger oak trees. Therefore, it is likely that some of the non-dated youngest piles could extend the occupation period, as the inhabitants of la Draga tend to make use of already exploited and cleared oak complexes with young and fast-growing oak trees.

The dendrotypological analysis allowed the identification of the exploitation of different oak complexes or woodlands. We use the concept of 'oak complexes' because, besides the logging dates that allow for the identification of logging phases over time, the regular spread of oak ages makes it difficult to isolate forest stands in a sufficiently relevant way. Based on the age of the samples and their period of exploitation, it has been possible to identify at least two oak forests exploited. The first oak forest is exclusively represented among the piles. It would correspond to a young and clear oak forest where trees have a fast-growing rate, with little but wide tree rings. This oak complex is more widely exploited during the reparation (Phase III) than during the construction (Phase II). The second oak complex exploited is composed mainly of horizontal

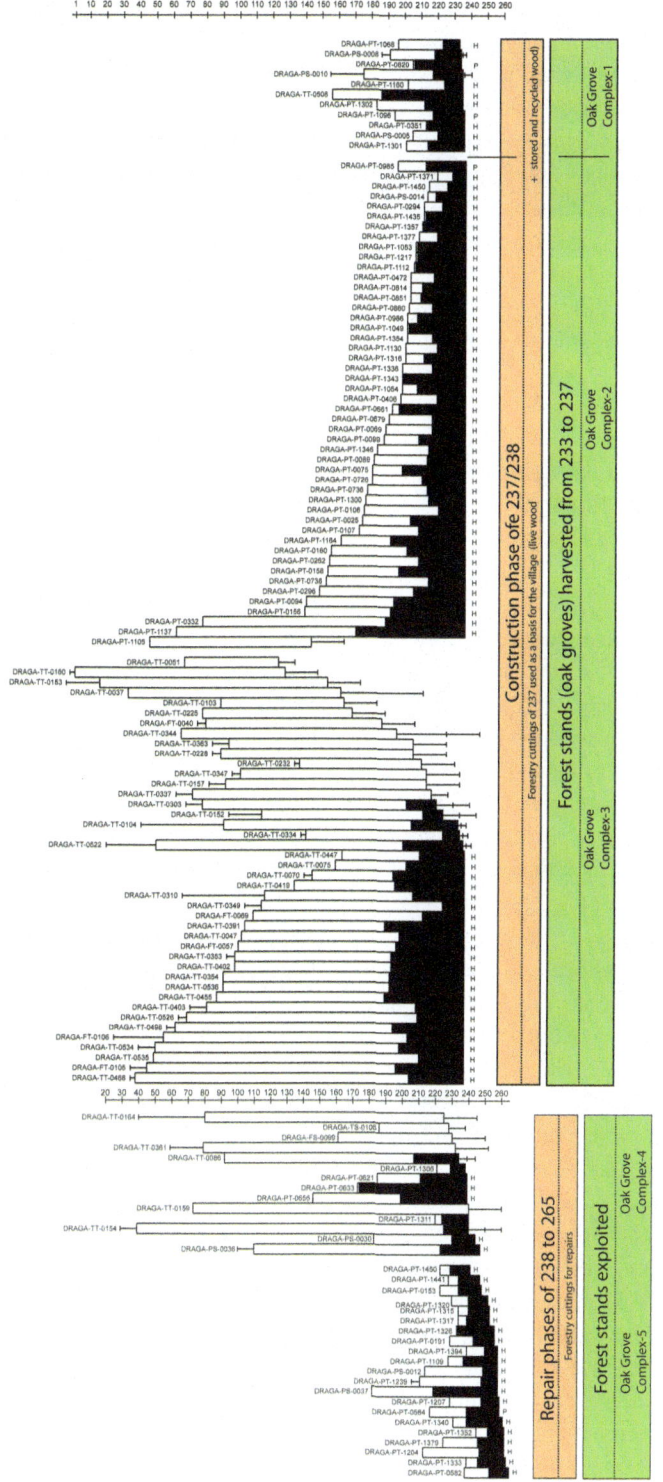

Fig. 11.5 Dendrochronological diagram of La Draga (Authors)

timber and a few piles. It would correspond to an old and dense oak forest. Given the extremely slow growth of these old oaks, it is possible to argue that these old oak trees have hardly benefited from thinning cuts during the last fifty years for the youngest and the last two centuries for the oldest. The samples from this old oak complex are characterised by many narrow tree rings, with almost no latewood formation. This oak complex is exploited mainly during the construction (Phase II), but also during the reparations (Phase III).

11.4.3 Wiggle-Matching of Piles and Chronological Model of La Draga Occupation

The dendrochronological tree-ring estimation and radiocarbon dates for the last growth ring (*cambium*) were obtained for 13 samples. The foundational tree felling is represented by seven samples from different timber logs, all of them were cut down during the year 237/238 of the local floating sequence. Six additional dates come from samples of logs that were cut down later, in the years 241, 247, 248, 256, 259, and 265 of the local sequence (Table 11.1). In one case (TT-0468), the radiocarbon date obtained seems too recent considering that it was cut down the same year as the foundational tree according to the dendrochronology. It is the thickest trunk with the longest tree ring sequence. Two samples from the same log were processed at Uppsala laboratory (UA62943 and UA65467), and both provided an estimated radiocarbon date that seems nearly 300 years younger when compared to the estimated dates of the other sampled piles. Unlike the rest of the logs, used as vertical piles, it is the only dated horizontal board, and therefore, its reaction to waterlogged conditions would have been different, allowing possible contamination by microorganisms—it is a water-saturated wooden sample—which could have altered the original radiocarbon content, and probably 'rejuvenated' the dating. More data are necessary to support this hypothesis. It is also the

one nearest to the actual lake shoreline, and we have not yet any other sample in the same spatial proximity to evaluate the post-depositional effects in this area of the site. Both dated samples from the same tree-ring have been discarded from the chronological models (Andreaki et al. 2020, 2022).

Estimation of calendar year for the tree-ring 237/238 has been calculated using the statistical combination of all dates for the same year coming from different trees. Until now, radiocarbon dating has been applied only to the outer rings of the samples in the present analysis. Until more rings of the same sample are dated, the intervals calculated below serve only as an initial hypothesis (Andreaki et al. 2022). Most radiocarbon dates associated with the dendrochronological year of tree felling (Beta481571, UA62942, UA62940, UA62941, UBAR314, UBAR1308) pass the Ward and Wilson (1978) test.[1] Sample Beta505910 appears to be a clear outlier and has not been considered. That means those samples are strictly contemporaneous, and the position in the calendar scale of the depositional event can be calculated in terms of the statistical combination (average) of the uncalibrated ^{14}C ages of the samples contained in the spatial unit. Using IntCal20 calibration curve, the statistical combination of the 6 radiocarbon dates that passed the test gives an estimate of 6311 ± 17 BP, and a calibrated confidence interval between 5313 cal BC and 5222 cal BC (68.3% interval) for the year 237/238 of the local floating tree-ring sequence (Andreaki et al. 2022).

The wiggle-matching Bayesian model using OxCal 4.4 has a very high agreement after the elimination of outliers (Andreaki et al. 2022). The results suggest an estimate for tree felling (year 237/238 of the floating sequence) at the time range 5313–5222 (mean 5293 cal BC of the 68.3% confidence interval), and the time range

---

[1] Ward and Wilson (1978) test: probability test known also as x2 test available in the OxCal program for calibration. It offers the possibility to compare the contemporaneity of dates between them. If the test is passed, it means that the dates are considered contemporary.

5291–5212 cal BC (mean 5272 cal BC of the 68.3% confidence interval) for the last measured tree-ring (year 265 of the floating sequence) (Andreaki et al. 2022).

We have reproduced the same model using ChronoModel 2.0, using also IntCal20 as a calibration curve (Fig. 11.6). We have defined 6 phases, in which the first one integrates the six isotopic events related with the ring 237/8 of the local tree-ring sequence, and the other five phases represent the tree rings 247, 248, 256, 259, and 265, for which we have a single radiocarbon

date. The *posterior* probability is calculated based on phases with two years of hypothetical duration and with known gaps between the phases depending on the difference in tree-rings. This is the same assumption used in OxCal 4.4 ChronoModel. The results suggest a calendar date for year 237/238 of the tree-ring floating sequence between 5340 and 5288 cal BC (68.3%); mean of 5315 cal BC. The calendar estimate for the most recent tree-ring (year 265 of the floating sequence) is 5184–4995 cal BC; mean of 5072 cal BC. Given that we have a

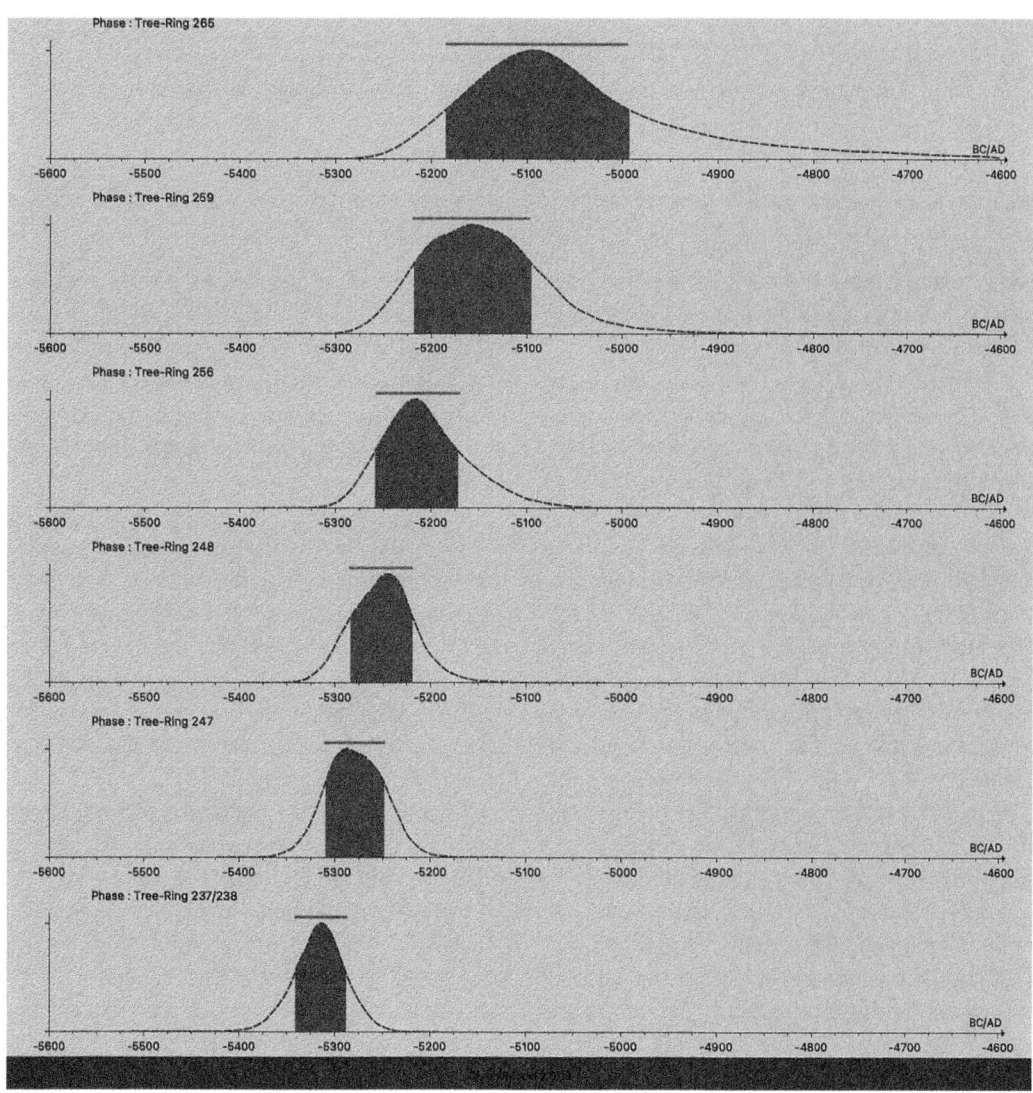

Fig. 11.6 Wiggle-matching of dendrochronological ordered piles. Calculated using ChronoModel 2.0.18. IntCal20 (Authors)

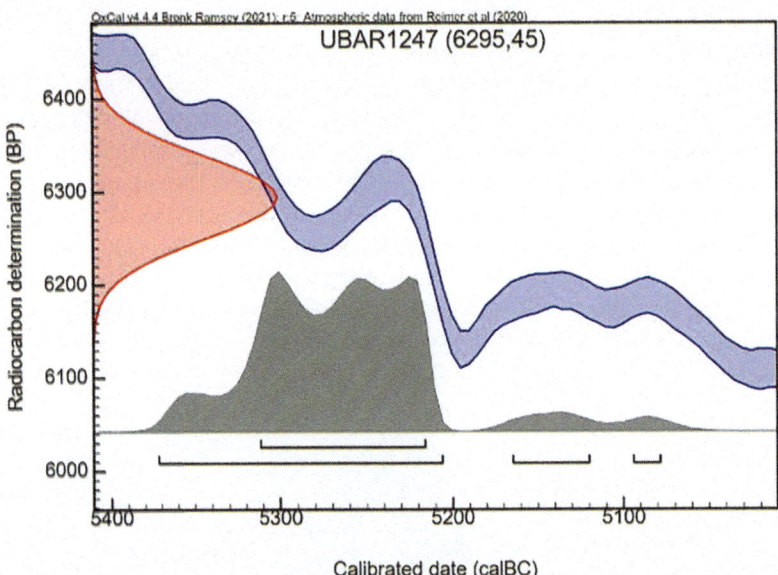

Fig. 11.7 Calibration of single date UBAR1247 (Tree-ring 265 of the local dendrochronological sequence) (Authors)

single dated sample for most of the tree-rings, specific intervals for start and end boundaries have not been calculated.

Combining the results of both calculations, we have hypothetically fixed the depositional event of wooden platforms construction as a Gaussian distributed interval with mean date of −5315 cal BC and a hypothetical standard deviation of 5 years considering the possible errors in such estimate. It is considered as the *terminus post quem (TPQ)* of Neolithic occupations at the site. The intrinsic deformation of calibration curve (IntCal20) after 7100 BP (Reimer et al. 2020) adds also statistical noise to the estimation of the latest moments of first occupation. See, for instance, the case of radiocarbon date UBAR1247, dating tree-ring 265 (Fig. 11.7). This has also been considered in other publications (Manen et al. 2019; Oms et al. 2016).

Prior probability is clearly divided into two well-differentiated subintervals, before and after 5200 cal BC, where calibration curve IntCal20 changes abruptly its direction. The subinterval 5370–5207 cal BC concentrates the maximum confidence (90.2%), and therefore, a much shorter temporal duration is to be expected for the first occupation, between 5315 and 5250 cal BC. Wiggle-matching allows to fix the most probable estimate around 5280 cal BC, and we have used this figure as a *terminus ante quem (TAQ)* for dating first occupation, as well as for estimating the beginning of the short-duration abandonment of the settlement that followed first occupation.

A more detailed chronological model has been calculated using ChronoModel 2.0 based on the stratigraphic constraints and the functional, spatial, and chronological ordering among 33 depositional events, and the conclusions from wiggle-matching model, fixing tree-ring 237/238 at 5315 cal BC and tree-ring 265 at 5280 cal BC. Dates signalled as outliers by OxCal 4.4 have been retained here (Fig. 11.8).

Results (Fig. 11.9) suggest a phase temporal range for the first occupation within the interval of 5315–5180 cal BC (HPD Region 95%), and an estimated duration coinciding with tree-ring information, less than 30 years. The second occupation would have started around 5100 cal BC (mean of the HPD 95%: 5163–4984 cal BC) and ending around 4948 (mean of the HPD 95%: 5028–4524 cal BC). Up to three different occupations can be distinguished in this second phase, being the first (Phase Travertine I) the most

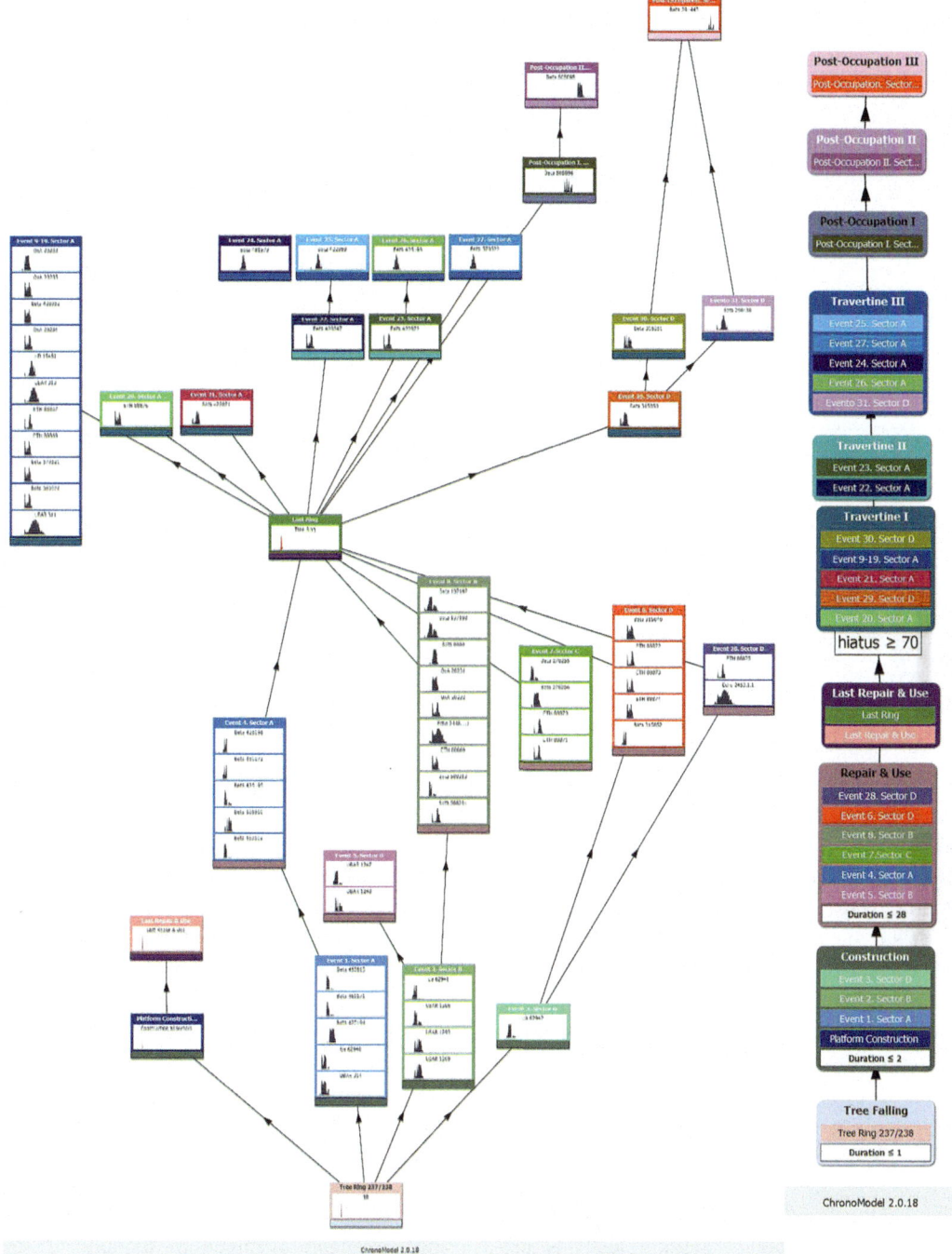

Fig. 11.8 Diagram of the 33 depositional events and their stratigraphic relationships on the left and their organisation into phases on the right of the figure. ChronoModel 2.0. IntCal20 Calibration curve (Andreaki et al. 2022)

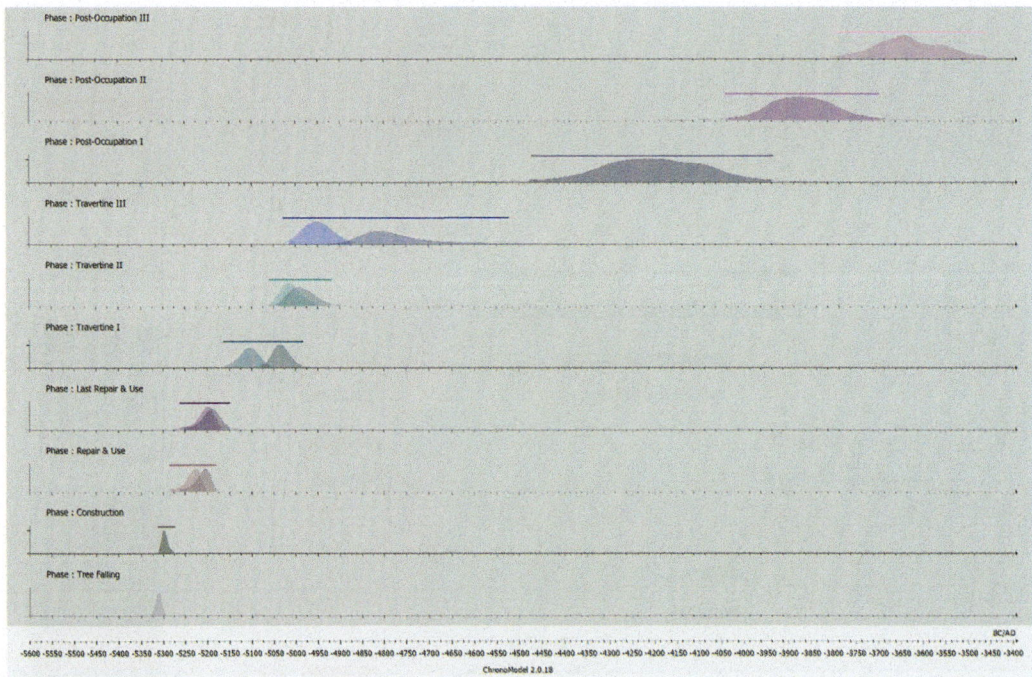

Fig. 11.9 Results of a chronological model based on 9 phases organising 33 depositional events. ChronoModel 2.0. Intcal20 Calibration curve (Authors)

clearly defined chronologically, with around 70 years of duration, and the other occupations (Travertine II and Phase Travertine III) less well defined because only a few dates could be assigned to each one based on stratigraphic ordering and relationships.

Using the above stratigraphic ordering and radiocarbon estimates for the duration of depositional events, we have defined a general temporal sequence (Fig. 11.10), based on Allen algebra estimated spatial relationships.

11.5 Conclusions

The dendrochronological results, as well as the Bayesian radiocarbon modelling analysis presented in this paper, have supposed a milestone on La Draga excavation site, as it implied a greater understanding of its chronological phases, wooden constructions and even palaeoecology during the Early Neolithic.

For the first time, it has been possible to create a floating dendrochronological curve for the Early Neolithic in southwestern Europe. Based on this mean curve, it has been possible to date 136 oak samples from the earliest phase of occupation at La Draga. The results obtained so far state that the site was mainly constructed in a great construction phase which lasted 1–2 years and can be identified in all the sectors of the excavation. It has also been possible to identify a short number of oaks felled 4–5 years before the great foundation, which could represent the use of stored wood, reuse or even preparations of the foundation. The felling of oaks the years after the great foundation would represent repairs or reinforcements of the wooden constructions. The results establish a minimum duration of the wooden constructions at la Draga of 27 years.

Radiocarbon dating combined with dendrochronology allowed to establish that the La Draga site would have been erected around 5315 cal BC, and to confirm that it took place in

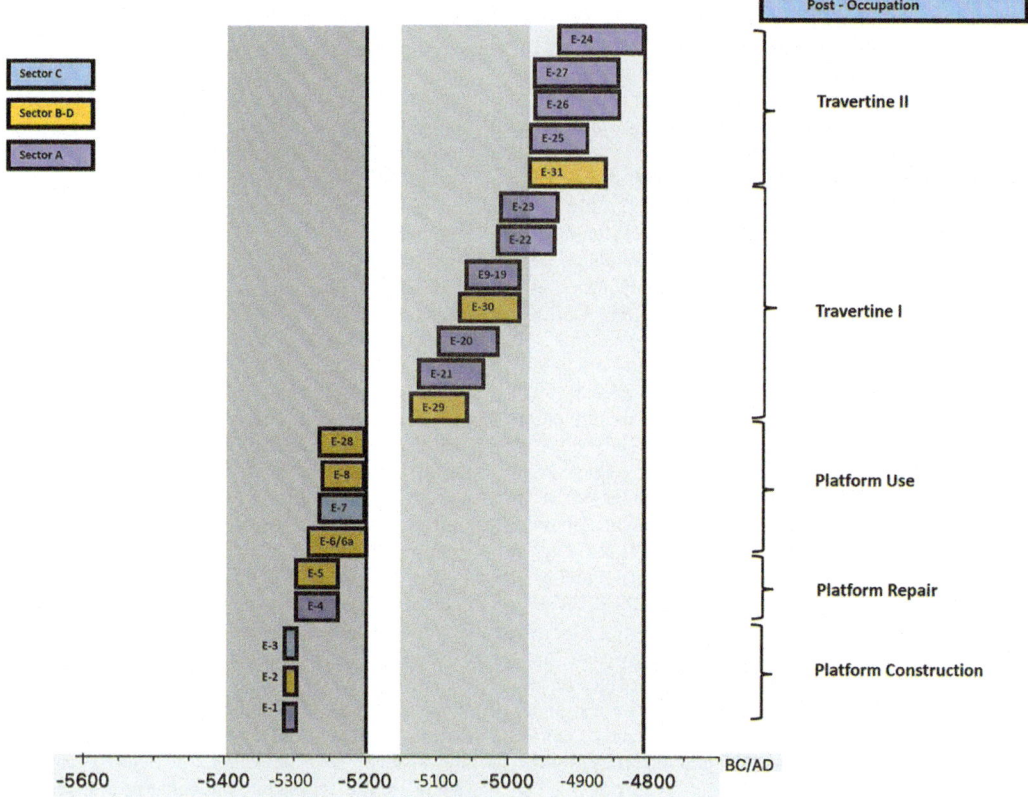

Fig. 11.10 Allen Diagram for site occupational temporal sequence with depositional events E-1 to E-31 and their respective phases (Authors)

a single construction event all over the site, after some years of exploration and management of forests around the Lake Banyoles. This first occupation would have been ended 30 years after, according to preserved tree-rings, but given the difficulties of measuring the thinnest logs; this date might be extended some years with future work. Radiocarbon-dated samples estimate the end of the first occupation some years before 5200 cal BC. After a short gap, in which the site may have been abandoned, a new occupation began around 5170 cal BC until the first centuries of 5th millennium. This most recent phase involved the reorganisation of the habitat, preparing pavements with travertine slabs on which the daily activities in the settlement were developed.

References

Andreaki V, Barcelo JA, Antolín F, Bogdanovic I, Gassmann P, López-Bultó O, Morera N, Palomo A, Piqué R, Revelles J, Terradas X (2020) Un modelo bayesiano para la adiocarbo del yacimiento neolítico de La Draga (Banyoles. Girona). Un caso de estudio con ChronoModel 2.0. In: Barcelo JA, Morell B (eds) Métodos cronométricos en arqueología, historia y paleontología. Dextra Editorial, Madrid, pp 403–418

Andreaki V, Barcelo JA, Antolín F, Gassmann P, Hajdas I, López-Bultó O, Martínez-Grau H, Morera N, Palomo A, Piqué R, Revelles J, Rosillo R, Terradas X (2022) Absolute Chronology at the waterlogged site of La Draga (Lake Banyoles, NE Iberia). Bayesian chronological models integrating tree-ring measurement, Radiocarbon dates and micro-stratigraphical data. Radiocarbon 64(5):907–948. https://doi.org/10.1017/RDC.2022.56

Antolín F (2016) Local, intensive and diverse? Early farmers and plant economy in the North-East of the

Iberian Peninsula (5500–2300 cal BC). Barkhuis Publishing, Groningen

Antolín F, Buxó R, Jacomet S, Navarrete V, Saña M (2014) An integrated perspective on farming in the early Neolithic lakeshore site of La Draga (Banyoles, Spain). Environ Archaeol 19:241–255. https://doi.org/10.1179/1749631414Y.0000000027

Antolín F, Berihuete M, Blanco À, Buxó R, Garcia L, Marlasca R, Navarrete V, Saña M, Verdún E (2018) El rebost domèstic i el rebost salvatge. In: Palomo A, Piqué R, Terradas X (eds) La Revolució Neolítica. La Draga, El Poblat Dels Prodigis. Ajuntament de Banyoles, Diputació de Girona, MAC, UAB, CSIC-IMF, Banyoles, pp 45–50

Arnold B (1990) Cortaillod-Est et les villages du lac de Neuchâtel au Bronze final. Structure de l'habitat et proto-urbanisme, Archéologie neuchâteloise. Editions du Ruau, Saint-Blaise

Barceló JA, Calvano M, Campana I, Piqué R, Palomo A, López-Bultó O (2019) Rebuilding the past: 3D reconstruction and BIM analysis of a Neolithic House at La Draga (Girona, Spain). In: Kremers H (ed) Digital cultural heritage. Springer International Publishing, Cham, pp 157–168. https://doi.org/10.1007/978-3-030-15200-0_11

Bartholin TS (1984) Dendrochronology in Sweden. In: Mörner N-A, Karlén W (eds) Climatic changes on a yearly to millennial basis: geological, historical and instrumental records. Springer Netherlands, Dordrecht, pp 261–262. https://doi.org/10.1007/978-94-015-7692-5_28

Bartholin TS (1988) Årstal i træ: dendrokronologi i Lund. Ugeskr Jordbrug 133:282–287

Billamboz A (2006) Dendroarchäologische Untersuchungen in den neolithischen Ufersiedlungen von Hornstaad-Hörnle. In: Dieckmann B, Harwalth A, Hoffstadt J (eds) Siedlungsarchäologie im Alpenvorland, vol 9. Theiss, Stuttgart, pp 297–414

Bosch À, Chinchilla J, Tarrús J (eds) (2000) El poblat lacustre neolític de La Draga. Excavacions de 1990–1998. Monografies del CASC, vol 2. Museu d'Arqueologia de Catalunya, Girona

Bosch À, Chinchilla J, Tarrús J (eds) (2006) Els objectes de fusta del poblat neolític de la Draga. Excavacions 1995–2005. Monografies del CASC, vol 6. Museu d'Arqueologia de Catalunya, Girona

Bosch À, Chinchilla J, Tarrús J (eds) (2011) El poblat lacustre del neolític antic de La Draga. Excavacions 2000–2005. Monografies del CASC, vol 9. Museu d'Arqueologia de Catalunya, Girona

Bogdanovic I, Bosch À, Buxó R, Chinchilla J, Palomo A, Piqué R, Saña M, Tarrús J, Terradas X (2015) La Draga en el contexto de las evidencias de ocupación del lago de Banyoles. In: Gonçalves VS, Diniz M, Sousa AC (eds) 5.º Congresso do Neolítico Peninsular. Actas. Estudos e memórias, vol 8. Centro de Arqueologia de Lisboa, Lisboa, pp 228–235

Campana I (2018) Prehistoric house and 3D reconstruction: towards a BIM archaeology. TDX (Tesis

Doctorals en Xarxa). Ph.D. thesis. Universitat Autònoma de Barcelona

Crumlin-Pedersen O (2000) To be or not to be a cog: the Bremen Cog in perspective. Int J Naut Archaeol 29:230–246. https://doi.org/10.1111/j.1095-9270.2000.tb01454.x

de Diego M, Piqué R, Saña M, Clemente I, Mozota M, Palomo A, Terradas X (2017) Fibre production and emerging textile technology in the early Neolithic Settlement of La Draga (Banyoles, Northeast Iberia; 5300 to 4900 cal BC). In: Bravermanová M, Březinová H, Malcolm-Davies J (eds) Archaeological textiles. Links between past and present NESAT XIII. Technical University of Liberec, Faculty of Textile Engineering, Institute of Archaeology of the CAS, Prague, Liberec, Praha, pp 293–302

Domínguez-Delmás M, García-González I (2015) Investigación dendrocronológica de maderas del pecio Delta II (Cádiz, España). Zenodo. https://doi.org/10.5281/zenodo.4480243

Domínguez-Delmás M, Alejano-Monge R, Van Daalen S, Rodríguez-Trobajo E, García-González I, Susperregi J, Wazny T, Jansma E (2015) Tree-rings, forest history and cultural heritage: current state and future prospects of dendroarchaeology in the Iberian Peninsula. J Archaeol Sci 57:180–196. https://doi.org/10.1016/j.jas.2015.02.011

Domínguez-Delmás M, Trapaga-Monchet K, Nayling N, García-González I (2017) Natural hazards and building history: roof structures of Segovia cathedral (Spain) reveal its history through tree-ring research. Dendrochronologia 46:1–13. https://doi.org/10.1016/j.dendro.2017.09.002

Domínguez-Delmás M, van Daalen S, Alejano-Monge R, Wazny T (2018) Timber resources, transport and woodworking techniques in post-medieval Andalusia (Spain): Insights from dendroarchaeological research on historic roof structures. J Archaeol Sci 95:64–75. https://doi.org/10.1016/j.jas.2018.05.002

Domínguez-Delmás M, Rich S, Traoré M, Hajj F, Poszwa A, Akhmetzyanov L, García-González I, Groenendijk P (2020) Tree-ring chronologies, stable strontium isotopes and biochemical compounds: towards reference datasets to provenance Iberian shipwreck timbers. J Archaeol Sci Rep 34:102640. https://doi.org/10.1016/j.jasrep.2020.102640

Gassmann P (2000) Premiers résultats dendrochronologiques concernant l'exploitation du chêne sur le site littoral de La Draga (Banyoles): In: Bosch À, Chinchilla J, Tarrús J (eds) El Poblat Lacustre Neolític de La Draga, Excavacions de 1990 a 1998. Monografies del CASC, vol 2. Museu d'Arqueologia de Catalunya, Girona

Hafner A (2005) Sutz-Lattrigen, Hauptstation. Erosionsschutzmassnahmen 2000–04: neolithische Ufersiedlungen. In: Suter PJ, Ramstein M (eds) Archäologie im Kanton Bern. Fundberichte und Aufsätze. Archeologie dans le canton de Berne. Chronique archeologique et textes. Archäologie im Kanton Bern, vol 6A. Archäologischer Dienst des Kantons Bern, Bern, pp 49–52.

https://boris.unibe.ch/143187/1/ADB_00_2005_Jahrbuch_AKBE_6A.pdf

Herrero-Otal M, Romero-Brugués S, Piqué Huerta R (2021) Plants used in basketry production during the early Neolithic in the north-eastern Iberian Peninsula. Veget Hist Archaeobot 30:729–742. https://doi.org/10.1007/s00334-021-00826-1

Hillam J, Groves CM, Brown DM, Baillie MGL, Coles JM, Coles BJ (1990) Dendrochronology of the English Neolithic. Antiquity 64:210–220. https://doi.org/10.1017/S0003598X00077826

Lanos P, Dufresne P (2019) ChronoModel version 2.0: software for chronological modelling of archaeological data using Bayesian statistics

López-Bultó O (2015) Processos d'obtenció, transformació i ús de la fusta en l'assentament neolític antic de la Draga (5320–4800 cal BC). Universitat Autònoma de Barcelona, Bellaterra

López-Bultó O, Piqué R (2018) Wood Procurement at the early Neolithic site of La Draga (Banyoles, Barcelona). J Wetland Archaeol 18:56–76. https://doi.org/10.1080/14732971.2018.1466415

López-Bultó O, Piqué R, Antolín F, Barceló JA, Palomo A, Clemente I (2020) Digging sticks and agriculture development at the ancient Neolithic site of la Draga (Banyoles, Spain). J Archaeol Sci Rep 30:102193. https://doi.org/10.1016/j.jasrep.2020.102193

Manen C, Perrin T, Guilaine J, Bouby L, Bréhard S, Briois F, Durand F, Marinval P, Vigne J-D (2019) The Neolithic transition in the Western Mediterranean: a complex and non-linear diffusion process—the radiocarbon record revisited. Radiocarbon 61:531–571. https://doi.org/10.1017/RDC.2018.98

Morgan RA (1988) Tree-ring studies of wood used in Neolithic and Bronze age trackways from the Somerset levels, parts i and ii. In: BAR British series, vol 184. BAR Publishing, Oxford. https://doi.org/10.30861/9780860545262

Oms FX, Martín A, Esteve X, Mestres J, Morell B, Subirà ME, Gibaja JF (2016) The Neolithic in Northeast Iberia: chronocultural phases and 14C. Radiocarbon 58:291–309. https://doi.org/10.1017/RDC.2015.14

Palomo A, Gibaja JF, Piqué R, Bosch A, Chinchilla J, Tarrús J (2011) Harvesting cereals and other plants in Neolithic Iberia: the assemblage from the lake settlement at La Draga. Antiquity 85:759–771

Palomo A, Piqué R, Terradas X, López-Bultó O, Clemente I, Gibaja JF (2013) Woodworking technology in the Early Neolithic site of La Draga (Banyoles, Spain). In: Anderson PC, Cheval C, Durand A (eds) Regards croisés sur les outils liés au travail des végétaux. An interdisciplinary focus on plant-working tools. XXXIII Rencontres Internationales d'archéologie et d'histoire d'Antibes. APDCA, Antibes, pp 383–396

Palomo A, Piqué R, Terradas X, Bosch À, Buxó R, Chinchilla J, Saña M, Tarrús J (2014) Prehistoric occupation of Banyoles Lakeshore: results of recent excavations at La Draga Site, Girona, Spain. J Wetland Archaeol 14:58–73. https://doi.org/10.1179/1473297114Z.00000000010

Piqué R, Palomo A, Terradas X, Tarrús J, Buxó R, Bosch À, Chinchilla J, Bodganovic I, López-Bultó O, Sana M (2015) Characterizing prehistoric archery: technical and functional analyses of the Neolithic bows from La Draga (NE Iberian Peninsula). J Archaeol Sci 55:166–173

Piqué R, Gassmann P, López-Bultó O, Andreaki V, Barceló-Álvarez JA, Palomo A, Tarrús J, Terradas-Batlle X (2020) Dendrocronología, C14 y arquitectura del yacimiento neolítico antiguo de La Draga (Banyoles). In: VII Congreso Internacional sobre el Neolítico en la Península Ibérica. Universidad de Sevilla, Sevilla

Piqué R, Alcolea M, Antolín F, Berihuete-Azorín M, Berrocal A, Rodríguez-Antón D, Herrero-Otal M, López-Bultó O, Obea L, Revelles J (2021a) Mid-holocene palaeoenvironment, plant resources and human interaction in Northeast Iberia: an archaeobotanical approach. Appl Sci 11:5056. https://doi.org/10.3390/app11115056

Piqué R, Berihuete-Azorín M, Franch A, Gassmann P, Girbal J, Herrero-Otal M, López-Bultó O, Palomo A, Rageot M, Revelles J, Romero-Brugués S, Terradas X (2021b) Woody and non-woody forest raw material at the early Neolithic site of La Draga (Banyoles, Spain). In: Berihuete-Azorín M, Seijo MM, López-Bultó O, Piqué R (eds) The missing woodland resources. Archaeobotanical studies of the use of plant raw materials. Advances in archaeobotany, vol 6. Barkhuis Publishing, Eelde, pp 41–58

Ravotto A, Mestres i Torres JS, Pugès i Dorca M, Segura de Yebra J (2016) Estudi dendroarqueològic del pou de Foneria. Quarhis: Quaderns d'Arqueologia i Història de la Ciutat de Barcelona 12:171–179

Reimer PJ, Austin WEN, Bard E, Bayliss A, Blackwell PG, Ramsey CB, Butzin M, Cheng H, Edwards RL, Friedrich M, Grootes PM, Guilderson TP, Hajdas I, Heaton TJ, Hogg AG, Hughen KA, Kromer B, Manning SW, Muscheler R, Palmer JG, Pearson C, van der Plicht J, Reimer RW, Richards DA, Scott EM, Southon JR, Turney CSM, Wacker L, Adolphi F, Büntgen U, Capano M, Fahrni SM, Fogtmann-Schulz A, Friedrich R, Köhler P, Kudsk S, Miyake F, Olsen J, Reinig F, Sakamoto M, Sookdeo A, Talamo S (2020) The IntCal20 Northern hemisphere radiocarbon age calibration curve (0–55 cal kBP). Radiocarbon 62:725–757. https://doi.org/10.1017/RDC.2020.41

Romero-Brugués S, Piqué Huerta R, Herrero-Otal M (2021) The basketry at the early Neolithic site of La Draga (Banyoles, Spain). J Archaeol Sci Rep 35:102692. https://doi.org/10.1016/j.jasrep.2020.102692

Saña M (2011) La gestió dels recursos animals. In: Bosch À, Chinchilla J, Tarrús J (eds) El poblat lacustre del neolític antic de La Draga. Excavacions 2000–2005. Monografies del CASC, vol 9. Museu d'Arqueologia de Catalunya, Girona, pp 177–212

Terradas X, Piqué R, Palomo A, Antolín F, López-Bultó O, Revelles J, Buxó R (2017) Farming practices in the early neolithic according to agricultural tools: evidence from La Draga Site (Northeastern Iberia). In: García-Puchol O, Salazar-García DC (eds) Times of neolithic transition along the Western Mediterranean. Springer International Publishing, Cham, pp 199–220. https://doi.org/10.1007/978-3-319-52939-4_8

Ward GK, Wilson SR (1978) Procedures for comparing and combining radiocarbon age determinations: a critique. Archaeometry 20:19–31. https://doi.org/10.1111/j.1475-4754.1978.tb00208.x

Dendrochronology of Italian Pile-Dwellings: The Challenge of Filling the Gaps Between 5000 and 1000 BC

12

Nicoletta Martinelli

Abstract

The paper deals with dendrochronological analysis on pile-dwelling sites in northern Italy. After a brief history of research and a general overview of the chronological framework, the author describes the elaboration of the main regional chronologies based on cross-dated tree-ring series from wood samples obtained from pile-dwelling settlements. The site series have been dendrochronologically cross-dated and the chronologies dated by radiocarbon wiggle-matching. They belong to the Bronze Age: the timespan of GARDA 1 is 2204–1829 ± 10 cal BC and the timespan of GARDA 3 is 1897–1678 ± 14 cal BC. They allow high-precision dating of building activities in the Bronze Age pile-dwelling villages in the Lake Garda region during a period from the 21st to 17th centuries BC, from the 1st phase of the Early Bronze Age (EBA1) until the transition between the last phase of the Early Bronze Age (EBA2) and the beginning of the Middle Bronze Age. Later dendrochronological single-site sequences span from the 16th century until the end of the 14th century BC and highlight asynchronous trends in settling in wetland environments on both sides of the Alps.

Keywords

Italy · Pile-dwellings · Bronze Age · Dendrochronology · Oak chronologies · Wiggle-matching

12.1 Dendrochronology of Italian Pile-Dwellings: A Brief History

The first application of dendrochronology on prehistoric wetland sites in Italy goes back to the 1970s: Elio Corona, the pioneer of Italian dendrochronology, together with D'Alessandro and Follieri examined posts from the Neolithic site of Fimon-Molino Casarotto in the Berici Hills region; samples from alder, oak, ash, beech, and maple piles were analysed (Corona et al. 1974). More extensive studies on pile-dwelling settlements began to be conducted only in the 1980s, thanks to the foundation of the first Italian dendrochronological laboratory at the Museum of Natural History in Verona, where samples from the Bronze Age villages of Bande di Cavriana, Peschiera-Setteponti and Lucone di Polpenazze were investigated. The Istituto Italiano di Dendrochronologia, founded in Verona in 1983, undertook a study of pile-dwelling villages in the

N. Martinelli (✉)
Laboratorio Dendrodata, Verona, Italy
e-mail: nicoletta.martinelli@dendrodata.it

Lake Garda area, both on the lake shore and in the surrounding morainic zone, which led in about ten years to the creation of the oak Bronze Age regional chronology GARDA 1 (Martinelli 1996).

The first published version of the curve included the series from the sites of Porto di Cisano (1986 excavation), Lazise-La Quercia (1986–1990 underwater investigations) Lavagnone (1974 and 1993 excavations), Cavaion Veronese (1980–1984 research), Barche di Solferino (1937–1938 excavations), together with those from the above-mentioned sites of Bande di Cavriana (1981 and 1983 excavations) and Lucone D (1986 excavation) (for references see Table 12.1). As no teleconnection was found with the millennia-long oak standard curves, the absolute dating of the GARDA 1 regional chronology was achieved by means of the wiggle-matching method, which combines radiocarbon and tree-ring data. This enabled the identification of many different felling phases in the 7 sites investigated with an error of ± 10y, from 2049–2042 BC to 1844–1835 BC[1] and established for the first time the existence of Bronze Age wooden structures built before 2000 BC in Alpine pile-dwellings.

Since then, over almost three decades of dendrochronological studies, the number of pile-dwelling villages investigated has increased substantially, especially in northern Italy, including sites located in almost every region, from Friuli in the east to Piedmont in the west. Investigations have also been carried out in central Italy, in the volcanic lakes of Albano, Bracciano (Rome), Mezzano (Viterbo) and in the marshy area of Celano Paludi (L'Aquila).

In northern Italy during the 1990s, the underwater mapping and sampling campaigns promoted by STAS (*Servizio Tecnico per l'Archeologia Subacquea*—Technical Underwater Archaeology Service of the Ministry for Cultural Heritage), together with the local Archaeology Superintendencies, gave a new impulse to dendrochronological analysis and revitalised research on the absolute dating of

prehistoric pile-dwellings. More recently, thanks to the increase in interdisciplinary research[2]—and the inscription of the transnational serial site 'Prehistoric Pile Dwellings around the Alps' in the UNESCO World Heritage List in 2011—many new dendrochronological samples suitable for further elaboration have been collected. Lastly, the Ministry of Cultural Heritage funded a further dendrochronological campaign as part of work for an upgrade of the UNESCO site's Italian Management Plan.[3]

Despite this abundance of studies, there has unfortunately been a lack of specific research projects for the creation of master curves for dating purposes. Two main regional oak chronologies have currently been elaborated, GARDA 1 and GARDA 3, which span a period of about five centuries from the Early Bronze Age to the beginning of the Middle Bronze Age, although a group of local site chronologies date to the Late Bronze Age (see Sect. 12.2).

12.2 Dendrochronology of Italian Pile-Dwellings: State of the Art

The number of prehistoric sites subjected to more or less extensive dendrochronological surveys is substantial (more than 50), but the total number of samples analysed remains rather small compared to the situation on the northern side of the Alps. For instance, at present, Lavagnone and Lucone di Polpenazze (Lucone D) are the only two Bronze Age sites where the number of timbers studied reaches several hundred, with in-progress dendrochronological analysis programmed to accompany the excavations that are currently in course.

---

[1] Intervals obtained from sapwood estimates (Fasani and Martinelli 1996).

[2] The research, carried out at the laboratory of the company Dendrodata s.a.s. until 2015, and later at the Laboratorio Dendrodata in Verona, was mainly funded by the Lombardy Regional Authority, the Lombardy Superintendencies (Varese, Brescia, Mantova, Cremona), the Superintendency of the Autonomous Province of Trento, the Friuli-Venezia Giulia Superintendency, and also by a number of local authorities and the Universities of Bern (Switzerland) and Bradford (UK).

[3] Under Law 77/2006 E.F. 2013.

Table 12.1 List of the dates of the Bronze Age pile-dwelling sites in northern Italy from dendrochronological series elaborated by the author: for each site the first and last attested felling date is reported

Site	Date of last ring cal BC	Felling date cal BC	WM error ± y	Last ring	Source of absolute dating	Bibliography
Corno di Sotto BS	2128	> 2115	10	Hearthwood	Site sequence of Lucone D	Baioni et al. (2015)
	2092	> 2085	10	Hearthwood	Site sequence of Lucone D	
Ronchi del Garda VR	2079	> 2067	10	Hearthwood	GARDA 1-I chronology	Fozzati et al. (2006) and Martinelli (2007b)
	2069	2059–2039	10	Sapwood	GARDA 1-I chronology	
Belvedere VR (1st village)	2061	2060–2040	10	Sapwood	GARDA 1-I chronology	Martinelli (2007a, b) and Fozzati et al. (2006)
	2006	2006	10	*Waldkante*	GARDA 1-I chronology	Fozzati et al. (2015) and Capulli et al. (2014)
Barche di Solferino MN	2049	2048–2037	10	Sapwood	GARDA 1-I chronology	Martinelli (1996, 2007b)
	1837	1836–1835	10	Sapwood	GARDA 1-I chronology	
Lucone di Polpenazze BS	2034	2034–2033	10	*Waldkante*	GARDA 1-I chronology	Martinelli (1985–1988, 2007a) and Baioni et al. (2015, 2021), Chap. 2 in this book
	1967	1967	10	*Waldkante*	GARDA 1-I chronology	
Bande di Cavriana MN	2012	2012	10	*Waldkante*	GARDA 1-I chronology	Martinelli (2007a, b) and Baioni et al. (2018)
	1961	1960–1953	10	Sapwood	GARDA 1-I chronology	
Oppeano VR (site 4C)	2001	2000–1999	10	*Waldkante* uncertain	GARDA 1-I chronology	Gonzato et al. (2021)
	1990	1989–1988	10	*Waldkante* uncertain	GARDA 1-I chronology	
Ca' Nova di Cavaion VR	2000	1999–1979	10	Sapwood	GARDA 1-I chronology	Martinelli (1996, 2007b) and Fasani and Martinelli (1996)
	1967	1966–1954	10	Sapwood	GARDA 1-I chronology	
Lavagnone BS	2037	2030–2010	10	Sapwood	GARDA 1-I chronology	Fasani and Martinelli (1996) and Martinelli (2007b)
	2019	2015–1995	10	Sapwood	GARDA 1-I chronology	
Dossetto di Nogara VR	1973	1963–1943	10	Hearthwood	GARDA 1-I chronology	Martinelli (2007b) and Martinelli (unpublished data)
	1933	1932–1912	10	Sapwood	GARDA 1-I chronology	
San Francesco BS	1967	> 1955	10	Hearthwood	GARDA 1-I chronology	Poggiani Keller et al. (2005) and Martinelli and Pignatelli (2018)
	1848	1847–1845	10	Sapwood	GARDA 1-II chronology	
Frassino VR (lakeshore site)	1954	1948–1928	10	Sapwood	GARDA 1-I chronology	Gonzato et al. (2014)
	1890	1888–1868	10	Sapwood	GARDA 1-I chronology	

(continued)

Table 12.1 (continued)

Site	Date of last ring cal BC	Felling date cal BC	WM error ± y	Last ring	Source of absolute dating	Bibliography
Porto di Cisano VR	1913	1913	10	*Waldkante*	GARDA 1-II chronology	Martinelli and Tinazzi (1990) and Martinelli (2007b)
	1879	1878–1868	10	Sapwood	GARDA 1-II chronology	
La Quercia/Lazise VR (1st chronology)	1939	1938–1935	10	Sapwood	GARDA 1-II chronology	Martinelli and Tinazzi (1992), Martinelli (2007b) and Fozzati et al. (2015)
	1844	1843–1835	10	Sapwood	GARDA 1-II chronology	
San Sivino-Gabbiano BS	1887	1875–1855	10	Sapwood	GARDA 1-II chronology	Martinelli (2020)
	1850	1849–1848	10	Sapwood	GARDA 1-II chronology	
Canàr di S. Pietro Polesine RO	1931	> 1919	10	Hearthwood	Independent wiggle-match	Martinelli et al. (1998)
	1871	> 1859	10	Hearthwood	Independent wiggle-match	
Lugana Vecchia BS	1859	> 1847	10	Hearthwood	GARDA 1-II chronology	Martinelli and Pignatelli (2018)
Peschiera-Setteponti VR	1837	> 1825	9	Hearthwood	GARDA 3 chronology	Martinelli (unpublished data)
Frassino I VR (underwater site)	1776	1776	9	*Waldkante*	GARDA 3 chronology	Fozzati et al. (2012, 2015), Martinelli (2022) and unpublished data
	1703	1703	9	*Waldkante*	GARDA 3 chronology	
Belvedere VR (2nd village)	1751	1751	9	*Waldkante*	GARDA 3 chronology	Fozzati et al. (2006), Capulli et al. (2014), Martinelli (2022) and unpublished data
	1677	1655–1635	9	Sapwood	GARDA 3 chronology	
Bosca di Pacengo VR	1735	1727–1720	9	Sapwood	GARDA 3 chronology	Fozzati et al. (2012, 2015), Martinelli (2022) and unpublished data
	1678	1677–1658	9	Sapwood	GARDA 3 chronology	
Pezzalunga MN	1715	1707–1687	9	Sapwood	GARDA 3 chronology	Martinelli (2007b) and Martinelli (unpublished data)
Bodio centrale VA	1693	1693–1692	22	*Waldkante*	Independent wiggle-match	Martinelli (2014a, b)
	1674	1674–1672	22	*Waldkante* uncertain	Independent wiggle-match	
Gaggio-Keller VA	1674	1666–1646	29	Sapwood	Site sequence of Ponti-o-Cazzago	Cermesoni et al. (in preparation) and Martinelli (unpublished data)
Ponti o Cazzago VA	1724	1724	29	*Waldkante*	Independent wiggle-match	Cermesoni et al. (in preparation) and Martinelli (unpublished data)
	1657	1657	29	*Waldkante*	Independent wiggle-match	
Il Sabbione VA	1674	1674	32	*Waldkante*	Independent wiggle-match	Martinelli (2017)
	1605	1605	32	*Waldkante*	Independent wiggle-match	

(continued)

Table 12.1 (continued)

Site	Date of last ring cal BC	Felling date cal BC	WM error ± y	Last ring	Source of absolute dating	Bibliography
Castellaro del Vho CR	1583	1577–1557	29	Sapwood	Independent wiggle-match	Martinelli (2001) and Martinelli (unpublished data)
Tombola di Cerea VR	1433	1433–1432	15	*Waldkante* uncertain	Independent wiggle-match	Martinelli (2005) and Salzani et al. (2018)
	1411	1402–1392	15	Sapwood	Independent wiggle-match	
Viverone Vi1-Emissario (BI-TO)	1425	1424	41	*Waldkante* uncertain	Independent wiggle-match	Rubat Borel et al. (2022)
	1402	1401	41	*Waldkante* uncertain	Independent wiggle-match	
Oppeano VR (site 4E)	1384	1384	14	*Waldkante*	Site sequence of Tombola	Gonzato et al. (2021)
Iseo—ex Resinex BS	1343	> 1331	17	Hearthwood	Independent wiggle-match	Poggiani Keller et al. (2005)

WM wiggle-match; method for sapwood estimates in Fozzati et al. (2015, footnote 8)

The samples analysed generally come from submerged sites in small Alpine lakes (e.g. Lake Monate), larger ones (e.g. Lake Garda), or volcanic lakes (e.g. Lake Albano), from wetland sites along the shores of former lakes in morainic regions, or shallow areas near rivers, and also from sites in ancient lagoons, like San Gaetano di Caorle (Venice) (Bianchin Citton and Martinelli 2005). The chronological range is ample too: sites are known from the early Neolithic (La Marmotta on Lake Bracciano) (Fugazzola Delpino et al. 1993) to the transition between the Late Bronze Age (Final BA) and Iron Age (Stagno di Livorno) (Zanini and Martinelli 2005).

However, this paper focuses on northern Italian pile-dwellings, which belong to the 'Alpine pile-dwelling phenomenon' (Fig. 12.1) and form part of the UNESCO transnational site 'Prehistoric Pile Dwellings around the Alps', of which 19 of the constituent sites are in Italy. Its principal goal is not to describe the results obtained for each site, but to illustrate the results as a whole, focusing on several achievements and also on the outstanding problems.

Due to the widespread occurrence of deciduous oak forests in the Po Valley and the nearby hilly regions in the Holocene, oak was the most important wood during prehistoric and proto-historic times in northern Italy. Given its particular suitability for building purposes, especially in wet environments, it was the most widely used timber in pile-dwellings: more than 90% of the vertical posts in Bronze Age stilt-houses were made of deciduous oak, namely *Quercus* sp. Sect. *ROBUR*, as described by Cambini (1967). Only in the high-Alpine villages in Trento province, Ledro and Fiavè, were coniferous woods used. Even though oak is a very suitable wood for dendrochronology, in Italy, we have to cope with the absence of millennia-long Italian oak chronologies that would allow the series to be dated absolutely with annual precision. The lack of old living oak forests and the rarity of samples from recent times, as a consequence of deforestation and extensive landscape clearance in proto-historic and historical times, are the main reasons why this important achievement has not yet been attained. Currently,

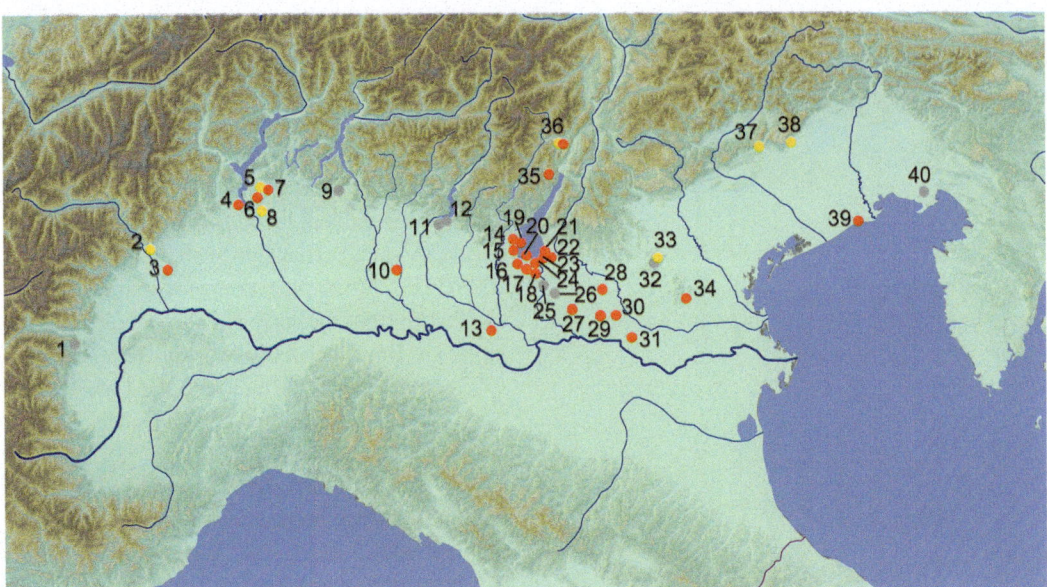

Fig. 12.1 Distribution map of the main Italian pile-dwellings. **Yellow**: Neolithic sites; **red**: Bronze Age sites; **grey**: sites not mentioned in the text. For reasons of scale, some points indicate more than one settlement. **1** Avigliana (Torino), Torbiera di Trana; **2** Montalto Dora (Torino), Lago Pistono; **3** Viverone (Biella)/Azeglio (Torino), Lago di Viverone (sites Vi1, Vi2, Vi3); **4** Arona (Novara), Lagone di Mercurago; **5** Biandronno (Varese), Isolino Virginia; **6** Cadrezzate (Varese), pile-dwellings in Lake Monate; **7** Bodio Lomnago (Varese), Bodio centrale, Desor Maresco, Keller-Gaggio; Bardello (Varese), Palude Ranchet, Bardello Stoppani; Cazzago Brabbia (Varese), Ponti o Cazzago; **8** Besnate (Varese), Lagozza; **9** Bosisio Parini (Lecco/Como), Cascina del Pascolo; **10** Sergnano (Cremona); **11** Corte Franca (Brescia), Valle delle Paiole and other sites; **12** Iseo (Brescia), Torbiere di Iseo (various sites); **13** Piadena Drizzona (Cremona), Lagazzi del Vho; **14** Polpenazze del Garda (Brescia), Lucone; **15** Moniga del Garda (Brescia), Porto; Padenghe del Garda (Brescia), West Garda and La Cà; Desenzano del Garda (Brescia), Corno di Sotto; **16** Desenzano del Garda (Brescia), Lavagnone; **17** Cavriana (Mantova), Bande-Corte Carpani; Solferino (Mantova), Barche; **18** Monzambano (Mantova), Castellaro Lagusello; **19** Manerba del Garda (Brescia), San Sivino-Gabbiano; **20** Sirmione (Brescia), San Francesco, Porto Galeazzi, Lugana Vecchia, La Maraschina; **21** Cisano (Verona), Porto; **22** Cavaion Veronese (Verona), Cà Nova; **23** Lazise (Verona), La Quercia and pile-dwellings of Pacengo (Porto, Bor, Bosca); **24** Peschiera del Garda (Verona), Frassino, Belvedere and historical pile-dwelling of Peschiera (Imboccatura del Mincio, Bacino Marina; Palafitta del Mincio or Setteponti); **25** Volta Mantovana (Mantova), Isolone del Mincio; **26** Roverbella (Mantova), Prestinari and minor pile-dwellings; **27** Isola della Scala (Verona), pile-dwelling sites; **28** Oppeano (Verona), Feniletto, 4C, 4D, 4E sites; **29** Nogara (Verona), Dossetto; **30** Cerea (Verona), Tombola; **31** San Pietro Polesine (Rovigo), Canàr; **32** Arcugnano (Vicenza), Fondo Tomellero, and other sites; **33** Arcugnano (Vicenza), Molino Casarotto, Le Fratte; **34** Arquà Petrarca (Padova), Laghetto della Costa; **35** Ledro (Trento), Molina di Ledro; **36** Fiavé (Trento), Torbiera Carera; **37** Colmaggiore di Tarzo (Treviso), Revine Lago; **38** Polcenigo/Caneva (Pordenone), Palù di Livenza; **39** Caorle (Venezia), San Gaetano; **40** Terzo di Aquileia (Udine), Canale Anfora. Graphics by Marco Baioni; courtesy of M. Baioni, C. Mangani, R. Micheli

the only supra-regional oak chronology elaborated, based on trees living in 4 stands in northern Italy, reaches back only to the year 1815 AD (Martinelli unpublished data).

Lengthy collaboration with colleagues Katarina Čufar (Slovenia) and André Billamboz (Germany) has made it clear that the absence of very long local regional oak tree-ring

chronologies also plays an important role with regard to the possibility of determining absolute dates to the year, through cross-dating against existing oak master curves: the nearest one from southern Germany reaches back to the 9th millennium BC. Research on living oak trees growing in Slovenia has shown that successful trans-Alpine teleconnections from south to north can be performed only on well-replicated chronologies more than 400 years long (Martinelli et al. 2018). This was shown to be true for prehistoric times too, when the Eneolithic 442-year-long oak chronology created from series of 6 pile-dwelling villages in the Ljubljana marsh was dated to between 3771 and 3330 BC against the German/Swiss reference chronology (Čufar et al. 2015).

Unfortunately, the timing of pile-dwelling settlement in Slovenia and northern Italy is different, and cross-dating between chronologies from these two regions south of the Alps cannot be performed. There is only one exception: a teleconnection established between the Slovenian HOC-QUSP1 chronology, and the chronology established for Building 1 at the Neolithic pile-dwelling of Palù di Livenza in the first half of the 4th millennium BC.

At present, Palù di Livenza, where samples from both vertical and horizontal timbers have been taken (see Chap. 3 in this book), is the Neolithic site that has been dendrochronologically investigated in most detail. On the other hand, from the ancient village of Isolino di Varese, on the isle near the eastern shore of Lake Varese, only a few samples have been investigated from structures built in the 5th and 4th millennia cal BC (Banchieri et al. 2004–2009). Neolithic samples also come from Padenghe sul Garda (Brescia) on the western shore of Lake Garda, dated to the early centuries of the 4th millennium cal BC (Poggiani Keller et al. 2005). Slightly later are the samples analysed from Fimon-Le Fratte (Vicenza), a site in the Berici Hills region, which might constitute the only series available for the Late-Neolithic/Copper Age transition in the latter centuries of the first half of the 4th millennium BC (Martinelli and Pignatelli 2016).

During the Copper Age (from about 3500 to 2200 BC), Italian lake shores seem to have been almost abandoned. Only two sites are known for preserved wooden elements of the period, both in Lake Varese: Isolino di Varese, on the south-eastern part of the island, samples from which were analysed at the Dendroarchaeologisches Labour in Hemmenhofen (Banchieri et al. 2015),

and the submerged lake dwelling of Bodio centrale. At Bodio, the larger settlement belongs to the Early/Middle Bronze Age transition and dates mainly to the 17th century BC, but in the eastern part of the settlement area, some elm and ash posts were sampled, and radiocarbon dated to the 3rd millennium BC.

As the above list indicates, dendrochronologically investigated Late-Neolithic and Copper Age structures are few; moreover, their timbers are made from very young trees of various species and oak components make up less than 50%, which is not very suitable for dendrochronology. The situation at the beginning of the Bronze Age was quite different, when the load-bearing vertical posts of stilt-houses were made almost exclusively from oak trees. Moreover, the number of pile-dwelling sites increases notably, with main concentrations in two areas: Lake Garda and its morainic hills, and Lake Varese, together with the small lakes nearby. In these settlements, the dominant building type is the stilt-house; these are often, especially in the Early Bronze Age, made from old oak trees. Therefore, since the 1990s studies have been focused on this kind of village for the purpose of creating well-replicated multi-centennial prehistoric chronologies. Most of them, however, have yielded only a few samples (in some cases just 1 or 2!) and research campaigns involving the analysis of more than 100 samples have been conducted in only 10 pile-dwelling settlements.[4]

Although this paper focuses on chronological issues, we must recall the important contribution of dendrochronology to the identification of pile-dwelling buildings and the interpretation of the development of village structures. Some examples are given in the paper by Baioni et al. (see Chap. 3 in this book), which also show the difficulties of applying the method in very small areas of investigation, and on multi-phase settlements. Here, we would like only to highlight the strong similarities with north Alpine sites that

---

[4] These are (from east to west) Palù di Livenza, Lazise-La Quercia, Cisano-Porto, Bosca di Pacengo, Belvedere di Peschiera, Bande di Cavriana, Lavagnone, Lucone D, Il Sabbione, Viverone Vi1-Emissario.

arose from the study of the 'Il Sabbione' village in Lake Monate with regard to building dimensions and supporting structures; the construction of successive palisades, that were continually lengthened and approached ever closer to the shore, resembles those of almost contemporaneous Swiss villages.

Dendrotypology plays a fundamental role in the identification of the wooden elements felled at the same time, which is essential for building reconstruction, and gives information about forest exploitation. Comparison between the data from north and south of the Alps not only highlighted asynchronous trends in settling in wetland environments on both sides but allowed investigation of the dendrotypological framework defined for the regions of Upper Swabia, Lake Constance, and northern Italy (Billamboz and Martinelli 2015). Based mainly on the data from northern Italy, the impact on woodland cover between 2100 and 1900 BC seems to present discrepancies between the different settlements, reflecting micro-regional human behaviour. On the contrary, the period 1900–1800 BC in both regions (Germany and Italy) shows a phase of afforestation, maybe due to a decrease in settlement activities, attested also by the almost complete absence of dendro-dates between 1850 and 1750 BC in northern Italy. Later, until 1500 BC, a period of wider settlement development is attested, with a stronger impact on forests than before: dendrotypology documents a phase of landscape opening and woodland thinning in both regions. In northern Italy, the impoverishment of the environment is suggested by the exploitation of various unusual tree species (among them Turkey oak) towards the end of the villages' habitation.

12.3 Regional Oak Chronologies from Pile-Dwellings

The regional oak chronologies currently available for absolute dating were elaborated from samples coming from Bronze Age settlements, where the presence of large groups of suitable oak timbers allowed the creation of well-

replicated site chronologies from almost contemporaneous sites. The cross-dating of these site chronologies led to the elaboration of regional chronologies which offer accurate dates for sequencing the history of wetland occupation, often with one-year precision on a relative scale. Because of the lack of teleconnections with European oak standard curves (see Sect. 12.2), they were absolutely dated by means of the radiocarbon wiggle-matching technique.

At present, the respective mean tree-ring series from various settlements in the Lake Garda region and the nearby Po Plain to the east (Media Pianura Veronese) have been cross-dated, enabling the construction of two main oak regional chronologies, GARDA 1 (22nd–19th centuries cal BC, 13 settlements) and GARDA 3 (19th–17th centuries cal BC, 4 settlements) (Martinelli 2020). These oak regional chronologies were first elaborated between 1994 and 2005 (Martinelli 1994, 2007a), but have been recently upgraded—both in terms of length and replication—thanks to still-ongoing tree-ring investigation at the Dendrodata Laboratory in Verona (Martinelli 2020).[5] Since also for the new versions of the series it was not possible to find any teleconnections with master curves north of the Alps, their absolute dates still rely on radiocarbon. Recently, both the wiggle-matches were tested with the new calibration curve Intcal20 (Reimer et al. 2020) and a recent version of the program Oxcal (4.4.4—*D-Sequence* model) (Bronk Ramsey 2009); the results for GARDA 1 are only slightly different from those obtained previously, so it was decided not to alter the published absolute dates associated with their relative scale until proper dendro-dates are achieved.

In more detail, the oak series GARDA 1 is split into two sequences for dating purposes, GARDA 1—1st part (2171–1961 ± 10 cal BC) and GARDA 1—2nd part (2061–1837 ± 10 cal BC), because of their weak overlap between the

---

[5] This recent elaboration and upgrading of the oak chronologies was performed as independent personal research, not supported by public or private funding.

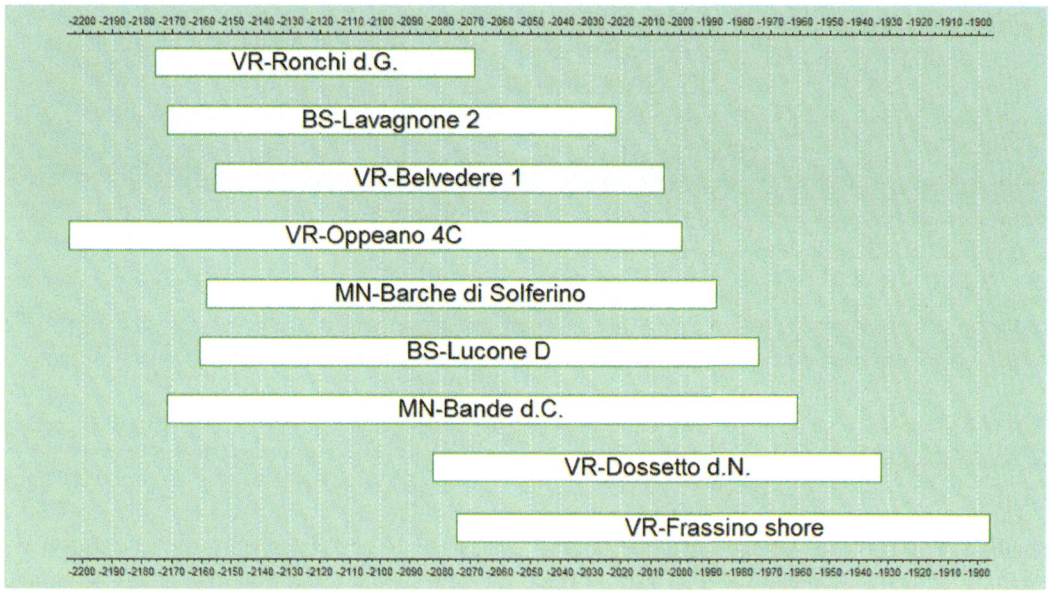

Fig. 12.2 Time spans of the 9 tree-ring site chronologies from pile-dwelling villages in the regional chronology GARDA 1—1st part (Author)

years 2000 and 1960 cal BC (Fig. 12.2). At present, the new two series are as follows:

The GARDA 1—1st part spans the period 2204–1896 BC and includes 163 component series from 9 pile-dwelling sites in the provinces of Brescia, Mantua, and Verona; many of the samples come from the sites of Belvedere di Peschiera del Garda (1st village), Bande di Cavriana, and Lucone D. The main contribution to the recent upgrading of the series comes from the settlements of Oppeano-site 4C (Gonzato et al. 2021) and a new site investigated in 2014 on the western shore of the small Lake Frassino, near Peschiera del Garda, facing the well-known submerged site Frassino 1 (Gonzato et al. 2014); a large group of new sample series comes from the recent excavation at the site Lucone D (Baioni et al. 2021).

Felling dates were identified both due to the presence of the waney-edge on some samples and by sapwood estimates: from the oldest (2060–2040 BC) in Belvedere—1st village, to the youngest (1888–1868 BC) in the newly discovered Frassino village (Table 12.1). These absolute dates were, and still are, a turning

point for Early Bronze Age absolute chronology in central Europe, as the settlements of Bande di Cavriana, Barche di Solferino, Belvedere di Peschiera, Lavagnone, and Lucone D, established in the 21st century BC, can be considered the oldest Bronze Age pile-dwellings in the Alpine region (Billamboz and Martinelli 2015).

Other series from sites on the southern shore of Lake Garda, Corno di Sotto and San Francesco (Baioni et al. 2015; Martinelli and Pignatelli 2018), or in the morainic region (Ca' Nova di Cavaion) are dated using the GARDA 1 —1st part chronology. Moreover, the inclusion of series from the sites of Oppeano-site 4C and Dossetto di Nogara, settlements near ancient rivers in the Media Pianura Veronese, demonstrates that the regional chronology is suitable for the absolute dating of oak samples both from the region around Lake Garda and the eastern plain south of Verona.

The GARDA 1—2nd part spans the period 1993–1829 cal BC and includes 59 component series from 4 underwater pile-dwelling sites situated along the shore of Lake Garda, both in the

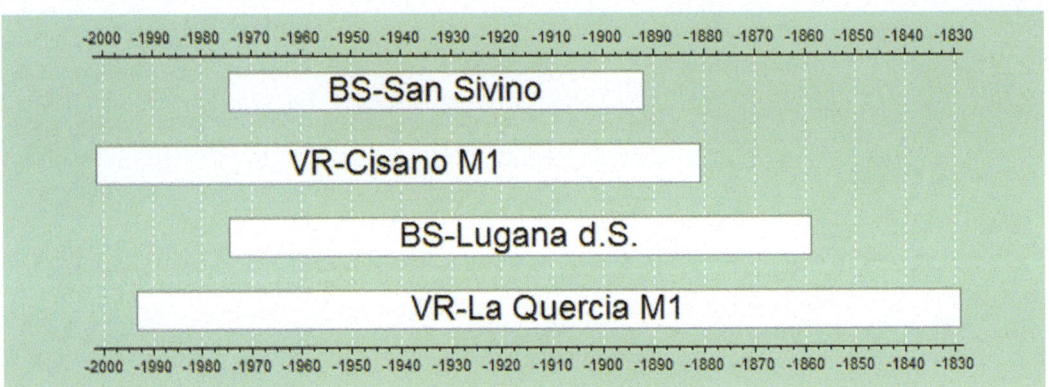

Fig. 12.3 Time spans of the 4 tree-ring site chronologies from pile-dwelling villages in the regional chronology GARDA 1—2nd part (Author)

south-eastern part (in Veneto) and the south-western part (in Lombardy) (Fig. 12.3). The main contribution to the upgrading of the series comes from a submerged site investigated in 2018–2019 near the town of Manerba: San Sivino-Gabbiano. The oldest felling date (1938–1935 BC) and the youngest one (1848–1828 BC) were identified in the long-lasting site of Lazise–La Quercia and might be related to the oldest village, attributable to Phase 2 of the Early Bronze Age (Table 12.1) (Martinelli 2020).

A similar problem, due to the poor replication of the series, could explain the absence of overlapping also between the GARDA 1—2nd part and the GARDA 3 chronology, which is dated from 1897 to 1678 ± 9 cal BC, based on wiggle-matching results elaborated on the IntCal20 calibration curve (Martinelli 2022). The GARDA 3 series spans the period 1897–1678 BC and includes 46 component series from 4 pile-dwelling sites, in the south-eastern part of the Lake Garda region, in the province of Verona (Fig. 12.4). The chronology includes the series from the well-known submerged site Frassino 1, in Lake Frassino, near Peschiera del Garda. The oldest felling date (1776 BC) was identified at Frassino 1, and the youngest one (1655–1635 BC) in Belvedere del Garda—2nd village. The results given by this chronology might be of great importance for the chronological assessment of the transition from the Early Bronze Age to Middle Bronze Age, but unfortunately a secure link between the dendro-dates and the archaeological finds is lacking at the moment. Some series from two other sites in the southern part of the Lake Garda area (Pezzalunga and Peschiera-Setteponti) are dated using the GARDA 3 chronology.[6]

At Frassino 1, posts of *Quercus* sp. Sect. *CERRIS* and *Quercus* sp. Sect. *ROBUR* (*sensu Cambini*) have been discovered; only the Sect. *ROBUR* series have been integrated into the chronology, although both the oak woods cross-date. The presence of Turkey oak is exclusive to the last of the village's felling episodes and could indicate that a different timber supply was used at the time—or the depletion of forests, which were sources of better-quality woods.

The region of Lake Varese, with its nearby small lakes, contains the second main concentration of pile-dwelling settlements. At the end of last century, dendrochronological analyses were part of a project of underwater investigation and interdisciplinary research on the submerged site 'Il Sabbione' in Lake Monate; the construction of a well-replicated site chronology dated by wiggle-matching allows the identification of felling episodes between 1674 and 1605 ± 32 cal BC (Martinelli 2017). At the beginning of this century, in 2006–2012, the Bodio Centrale site

---

[6] Recently, Leone Fasani kindly has made available the sample series from the site of Castellaro Lagusello. Their elaboration is still in progress, but some series seem to suggest the possibility of a further upgrading of the GARDA 3 chronology.

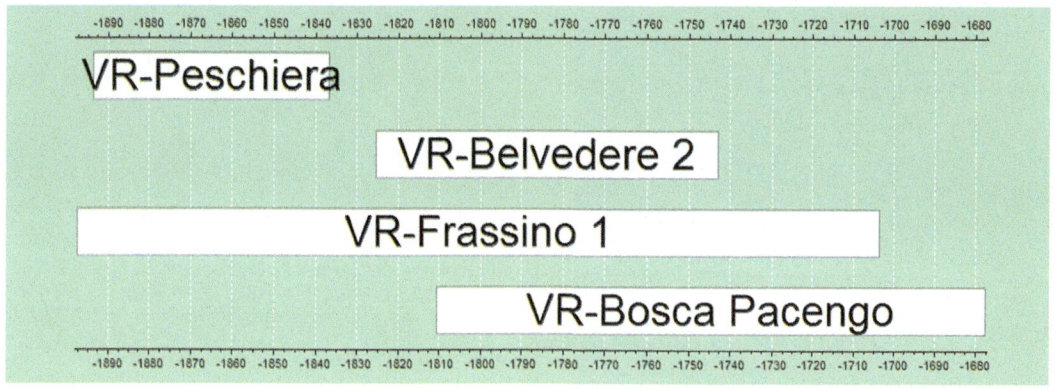

Fig. 12.4 Timespans of the 4 tree-ring site chronologies from pile-dwelling villages in the regional chronology GARDA 3 (Author)

was excavated in Lake Varese; due to the characteristics of the woods employed, which resemble those from Late-Neolithic structures, only a very short site chronology, with felling dates between 1693 and 1672 ± 22 cal BC, was created (Martinelli 2014a, b).

More recently (in 2019 and 2020), two new submerged sites were investigated in the lake: Gaggio-Keller and Ponti o Cazzago, both discovered during the 19th century AD; only a few timbers were analysed, but the two site-series cross-date each other and are likely to constitute the basis of a new Bronze Age chronology (Cermesoni et al. in preparation). The absolute dates obtained by wiggle-matching for the series from Ponti-o-Cazzago suggest that the site was founded a little earlier than the 'Il Sabbione' village (Table 12.1).

Only single-site chronologies have been elaborated for the period between 1650 and 1300 BC. Their absolute dating relies on independent wiggle-matches and involves an error of ± 15y to ± 41y. Two long series of great archaeological importance (from Viverone Vi1-Emissario and Tombola di Cerea) have been created. Although they date to the same period, they do not cross-date—probably because of the great distance between them—and thus do not furnish a regional chronology.

12.4 Conclusion

For the period when pile-dwelling villages are attested in northern Italy, between 5000 and 1000 BC, dendrochronological series are not evenly and adequately distributed across the centuries. Investigated Neolithic sites are rare, and less in number in Italy compared to north of the Alps. Therefore, the goal of creating a master curve for this period will not be achieved in the near future —also because of the characteristics of the wood types, which are not that suitable for dendrochronological dating (see Sect. 12.2). The absence of tree-ring series for the period 3200–2300 is likely to reflect the absence of pile-dwellings in the period, thus confirming the marked difference in lake settlement timing between north and south of the Alps.

As a consequence, the challenge of 'filling the gaps' seems to be achievable only for the 2nd millennium BC. A glance at the timespans of the various Bronze Age site chronologies (Fig. 12.5) might give the false impression of a continuous succession of series, but unfortunately this is not the case. Only the site chronologies that belong to the regional chronologies cross-date each other, but not the others, even when they partly belong to the same period.

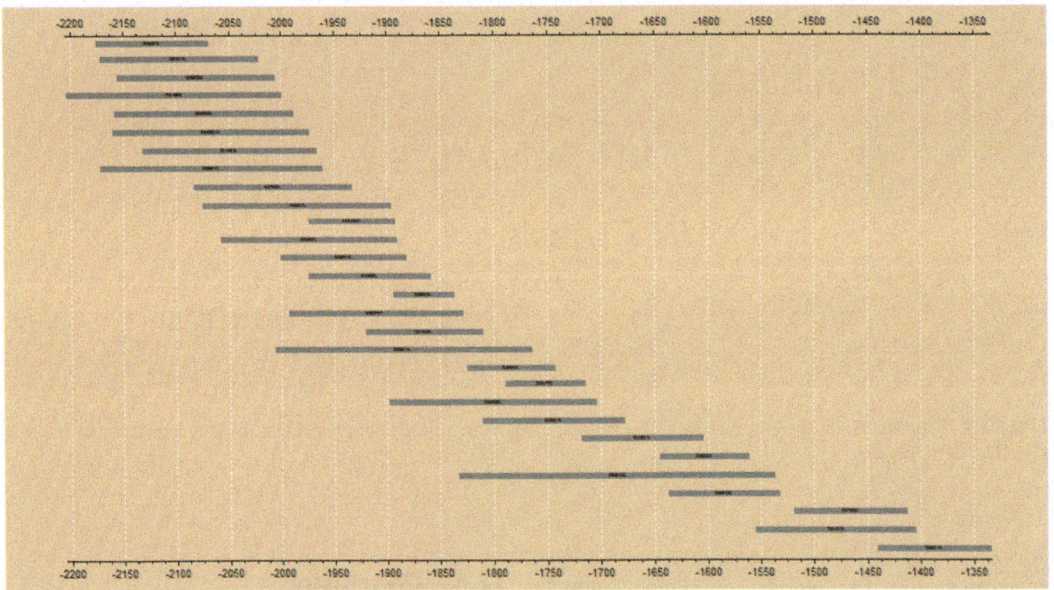

Fig. 12.5 Timespans of all the site chronologies from Bronze Age pile-dwelling settlements in northern Italy (Author)

As already stated, the reasons for this circumstance are the low number of samples taken in certain sites and the poor replication of some site chronologies. But other factors could affect the result: different environmental provenance of the samples, and different behaviour of the oak trees, since timber sources might be both woods on morainic hills and lowland forests. The lack of trans-regional cross-dating could be ascribed to ancient woodland management too: prehistoric people obtained timber from different stand structures and particularly from the understorey. Cross-dating is difficult when analysing very narrow tree rings from both suppressed trees grown in almost natural forests and very young trees from coppice-managed forests.

Given the difficulties in proceeding with this research because of the problems involved in creating master curves anchored to the present in northern Italy, teleconnection with the European oak references remains the only possible way of obtaining absolute dates for the floating Italian regional chronologies. In order to meet this challenge, regional chronologies should be enlarged and reinforced with new series; this is essential, because advances in dendrochronological research in northern Italy are of great importance for the study of the development of the pile-dwelling phenomenon as a whole in the Alps and surrounding areas, due to the differing periods of settlement in wetland environments.

Pile-dwelling development in northern Italy coincides with two major hiatuses in wetland occupation in the northern Alpine region, from the 22nd to 19th centuries BC, and from the 15th to 12th centuries BC. The latter interval, probably strongly linked to climate changes and lake-level fluctuations, seems to have had significantly less influence on northern Italian Bronze Age communities.[7] This finding opens new possible scenarios concerning lake settlers' resilience, thus suggesting interesting new research topics—and possible inspirations for tackling current societal and climate challenges.

Acknowledgements Although the creation of the oak chronologies was performed as personal, independent research, not supported by public or private funds, I'm sincerely grateful to all the institutions and archaeologists who have entrusted me with investigations of pile-dwelling sites: the local Archaeology Superintendencies,

---

[7] For a deeper insight into relationships between the Bronze Age population and the environment in northern Italy, see Magny et al. (2009) and Leonardi et al. (2015).

museums, and universities—including foreign ones; especially the Lombardy Regional Authority, the municipalities of Desenzano, Gavardo, Manerba, Bodio Lomnago, and Viverone, the superintendencies of Lombardy (provinces of Varese, Brescia, Mantova, Cremona), Veneto (provinces of Verona, and Vicenza), Friuli-Venezia Giulia, and the Autonomous Province of Trento, and the Ministry of Cultural Heritage which funded a dendrochronological campaign under Law 77/2006 E.F. 2013 (see also footnote 2).

References

Baioni M, Longhi C, Mangani C, Martinelli N, Nicosia C, Ruggiero MG, Salzani P (2015) La palafitta del Corno di Sotto (Desenzano del Garda, Brescia) nell'ambito dello sviluppo dei primi insediamenti palafitticoli del lago di Garda. In: Leonardi G, Tiné V (eds) Preistoria e Protostoria del Veneto. XLVIII Riunione Scientifica dell'Istituto Italiano di Preistoria e Protostoria. Studi di Preistoria e Protostoria, vol 2. Istituto Italiano di Preistoria e Protostoria, Firenze, pp 177–186

Baioni M, Leonardi G, Fozzati L, Martinelli N (2018) Le palafitte: definizione e caratteristiche di un fenomeno complesso attraverso alcuni casi di studio. In: Baioni M, Mangani C, Ruggiero MG (eds) Le Palafitte: Ricerca, conservazione, valorizzazione. Atti del Convegno Internazionale. Palafitte/Palafittes/Pfahlbauten/Pile dwellings, vol 0. SAP Società Archeologica, Quingentole (Mantova), pp 27–42

Baioni M, Bona F, Mangani C, Martinelli N, Nicosia C, Perego R, Quirino T, Saletta E (2021) Daily life in a north Italian Early Bronze Age pile dwelling: Lucone di Polpenazze del Garda (Italy—Brescia). In: Jallot L Peinetti A (eds) Use of space and domestic areas: functional organisation and social strategies. Proceedings of the XVIII UISPP world congress. Archaeopress, Paris, pp 53–66

Banchieri D, Martinelli N, Pignatelli O (2004–2009) Nuove indagini sui resti lignei dell'Isolino Virginia. Sibrium XXV:179–184

Banchieri DG, Bini A, Mainberger M (2015) Isolino Virginia, a waterlogged Tell in a south pre-alpine Lake: preservation and erosion problems. In: Brem H, Ramseyer D, Rouliére-Lambert M-J, Schifferdecker F, Schlichtherle H (eds) Archéologie & Érosion 3, Actes de la troisiéme rencontre internationale. Mêta Jura, Lons-le-Saunier, pp 183–190

Bianchin Citton E, Martinelli N (2005) Cronologia relativa e assoluta di alcuni contesti veneti dell'età del Bronzo recente, finale e degli inizi dell'età del Ferro. Nota preliminare. In: Bertoloni G, Delpino F (eds) Oriente e Occidente: metodi e discipline a confronto. Riflessioni sulla cronologia dell'età del Ferro italiana. Atti dell'Incontro di studi, Roma, 30–31 ottobre 2003. Mediterranea, vol I. Istituti Editoriali e Poligrafici Internazionali, Pisa, pp 239–253

Billamboz A, Martinelli N (2015) Dendrochronology and Bronze Age pile-dwellings on both sides of the Alps: from chronology to dendrotypology, highlighting settlement developments and structural woodland changes. In: Menotti F (ed) The end of the lake-dwellings in the Circum-Alpine region. Oxbow Books, Oxford, Philadelphia, pp 68–84

Bronk Ramsey C (2009) Bayesian analysis of radiocarbon dates. Radiocarbon 51(1):337–360

Cambini A (1967) Micrografia comparata dei legni del genere Quercus. In: Contributi scientifico-pratici per una migliore conoscenza ed utilizzazione del legno X, pp 7–49

Capulli M, Fozzati L, Martinelli N, Pellegrini A (2014) La palafitta sommersa di Peschiera—Belvedere sul lago di Garda (VR). Le ricerche archeologiche subacquee e l'utilizzo della tecnologia GIS come supporto per le analisi spaziali e la ricostruzione planimetrica delle strutture palafitticole. In: Leone D, Turchiano M, Volpe G (eds) Atti del III Convegno di Archeologia Subacquea. Edipuglia, Bari, pp 103–110

Cermesoni B, Locatelli D, Luglietti S, Martinelli N, Rottoli M, Nelle O, Billamboz A (in preparation) Nuovi dati per lo studio delle modificazioni ambientali e della composizione forestale tra Neolitico ed età del Bronzo nel comprensorio del lago di Varese. In: Le scienze della Preistoria e Protostoria: Paleoecologia, Archeobiologia, applicazioni digitali e archeometria. LVI Riunione Scientifica dell'Istituto Italiano di Preistoria e Protostoria

Corona E, D'Alessandro R, Follieri M (1974) I pali lignei dell'abitato neolitico di Fimòn-Molino Casarotto. Ann Bot XXXIII:237–258

Čufar K, Tegel W, Merela M, Kromer B, Velušček A (2015) Eneolithic pile dwellings south of the Alps precisely dated with tree-ring chronologies from the north. Dendrochronologia 35:91–98

Fasani L, Martinelli N (1996) Cronologia assoluta e relativa dell'antica età del bronzo nell'Italia settentrionale (dati dendrocronologici e radiometrici). In: Cocchi Genick D (ed) L'antica età del Bronzo in Italia, Atti del Congresso. Octavo editore, Firenze, pp 19–32

Fozzati L, Bressan F, Martinelli N, Valzolgher E (2006) Underwater archaeology and prehistoric settlement dynamic in a great alpine lake: the case study of Lake Garda. In: Hafner A, Niffeler U, Ruoff U (eds) Die neue Sicht. Unterwasserarchäologie und Geschichtsbild. Akten des 2. Internationalen Kongresses für Unterwasserarchäologie. Antiqua, vol 40. Archäologie Schweiz, Basel, pp 78–91

Fozzati L, Martinelli N, Valzolgher E (2012) Re-emerged maps: investigating the topography of the pile-dwellings offshore at Pacengo (Lake Garda, Northern Italy). In: Henderson JC (ed) IKUWA 3. Beyond boundaries. Proceedings of the 3rd international congress on underwater archaeology. Kolloquien zur Vor- und Frühgeschichte, vol 17. Habelt, Bonn, pp 345–352

Fozzati L, Leonardi G, Martinelli N (2015) Wetlands. Palafitte e siti umidi nell'età del Bronzo del Veneto: territori e cronologia assoluta. In: Leonardi G, Tiné V (eds) Preistoria e Protostoria del Veneto. XLVIII Riunione Scientifica dell'Istituto Italiano di Preistoria e Protostoria. Studi di Preistoria e Protostoria, vol 2. Istituto Italiano di Preistoria e Protostoria, Firenze, pp 241–250

Fugazzola Delpino MA, D'Eugenio G, Pessina A (eds) (1993) "La Marmotta", Anguillara Sabazia (RM). Scavi 1989. Un'abitato perilacustre d'età neolitica. Bull Paletnol Ital N.S. II 84:181–342

Gonzato F, Baioni M, Balista C, Mangani C, Martinelli N, Nicosia C, Pignatelli O, Voltolini D (2014) Peschiera del Garda, laghetto del Frassino. Indagini 2014. Notizie Archeol Veneto 3:130–141

Gonzato F, Mangani C, Martinelli N, Nicosia C (2021) Different ways to handle the domestic space by comparison: the case of Bronze Age villages in Vallese di Oppeano (Verona—ITA). In: Jallot L, Peinetti A (eds) Use of space and domestic areas: functional organisation and social strategies. Proceedings of the XVIII UISPP world congress. Archaeopress, Paris, pp 67–76

Leonardi G, Cupitò M, Baioni M, Longhi C, Martinelli N (2015) Northern Italy around 2200 BC—from Copper Age to Early Bronze Age: continuity and/or discontinuity? In: Meller H, Arz WH, Jung R, Risch R (eds) 2200 BC—Ein Klimasturz als Ursache für den Zerfall der alten Welt? 7. Mitteldeutscher Archäologentag. Tagungen des Landesmuseums für Vorgeschichte Halle, vol 12(I). Landesmuseum für Vorgeschichte, Halle, pp 283–304

Magny M, Galop D, Bellintani P, Desmet M, Didier J, Haas JN, Martinelli N, Pedrotti A, Scandolari R, Stock A, Vannière B (2009) Late-Holocene climatic variability south of the Alps as recorded by lake-level fluctuations at Lake Ledro (Trentino, Italy). The Holocene 19(4):575–589

Martinelli N (1985–1988) Le strutture lignee dell'abitato di Lucone di Polpenazze (BS). Indagine dendrocronologica e tecnomorfologia. Ann Museo Gavardo 16:45–60

Martinelli N (1994) Aspetti culturali e cronologici dell'antica e media età del Bronzo nell'area benacense alla luce di nuove ricerche dendrocronologiche. Postgraduate specialization thesis, Scuola di Specializzazione in Archeologia dell'Università degli Studi di Pisa

Martinelli N (1996) Datazioni dendrocronologiche per l'età del Bronzo dell'area alpina. In: Randsborg K (ed) Absolute chronology. Archaeological Europe 2500–500 BC. Proceedings of the conference. Acta Archaeologica, vol 67. Acta Archaeologica Supplementa, vol I. Munskgaard, Copenhagen, pp 315–326

Martinelli N (2001) Le indagini dendrocronologiche e le datazioni radiometriche. In: Frontini P (ed) Castellaro del Vhó. Campagne di scavo 1996–1999. Scavi delle Civiche Raccolte Archeologiche di Milano. Raccolte Archeologiche e Numismatiche, Milano, pp 215–223

Martinelli N (2005) Dendrocronologia e Archeologia: situazione e prospettive della ricerca in Italia. In: Attema P, Nijboer A, Zifferero A (eds) Communities and settlements from the Neolithic to the Early Medieval period. Proceedings of the 6th conference of Italian archaeology. Papers in Italian archaeology, VI. BAR international series, vol 1452(I). Archaeopress, Oxford, pp 437–448

Martinelli N (2007a) Gli insediamenti palafitticoli dell'antica età del Bronzo nell'area benacense: studio stratigrafico e strutturale su scala cronologica ad alta precisione. Ph.D. thesis, Dipartimento di Archeologia dell'Università degli Studi di Padova

Martinelli N (2007b) Dendrocronologia delle palafitte dell'area gardesana: situazione delle ricerche e prospettive. In: Morandini F, Volonté M (eds) Contributi di archeologia in memoria di Mario Mirabella Roberti. Atti del XVI Convegno Archeologico Benacense. Annali Benacensi, vol XIII–XIV, pp 103–120

Martinelli N (2014a) Le indagini dendroarcheologiche. In: Grassi B, Mangani C (eds) Storie sommerse. Ricerche alla palafitta di Bodio Centrale a 150 anni dalla scoperta. Soprintendenza per i Beni Archeologici della Lombardia, Cremona, pp 83–91

Martinelli N (2014b) La datazione radiocarbonica delle strutture lignee. In: Grassi B, Mangani C (eds) Storie sommerse. Ricerche alla palafitta di Bodio Centrale a 150 anni dalla scoperta. Soprintendenza per i Beni Archeologici della Lombardia, Cremona, pp 93–96

Martinelli N (2017) Capitolo VII. Gli insediamenti palafitticoli del lago di Monate. Il contributo della dendrocronologia allo studio dell'antica e media età del Bronzo. In: Harari M (ed) Il territorio di Varese in età preistorica e protostorica. La Storia di Varese. International Research Center for Local Histories and Cultural Diversities dell'Università degli Studi dell'Insubria, Busto Arsizio (Varese). Nomos Edizioni, Busto Arsizio, pp 172–195

Martinelli N (2020) Multicentennial regional oak chronologies for northern Italy: an updating. In: 2020 IMEKO TC-4 international conference on metrology for archaeology and cultural heritage. Proceedings. IMEKO, Trento, pp 575–578

Martinelli N (2022) Dendrocronologia dell'età del Bronzo in Italia settentrionale: stato dell'arte e aggiornamenti. In: de Marinis RC, Rapi M (eds) Preistoria e Protostoria in Lombardia e Canton Ticino. Atti LII Riunione Scientifica Istituto Italiano di Preistoria e Protostoria. Rivista di Scienze Preistoriche LXXII, pp. 431–439

Martinelli N, Pignatelli O (2016) Le strutture lignee nell'insediamento di Le Fratte: indagini dendroarcheologiche, xilotomiche e radiocarboniche. In: Bianchin Citton E (ed) Nuove ricerche nelle Valli di Fimon. L'insediamento del tardo Neolitico de Le Fratte di Arcugnano. Editrice Veneta SAS. Provincia di Vicenza, Vicenza, pp 213–241

Martinelli N, Pignatelli O (2018) Dagli anelli degli alberi un aiuto all'archeologia: la dendrocronologia. In: Roffia E (ed) Sirmione in età antica. Il territorio del comune dalla Preistoria al Medioevo. Edizioni ET, Milano pp 65–66

Martinelli N, Tinazzi O (1990) Le strutture lignee dell'abitato di Cisano: analisi dendrocronologica e tecnomorfologica (Indagine preliminare). In: Salzani L (ed) Nuovi Scavi nella palafitta di Cisano. Comune di Bardolino, Vago di Lavagno (Verona), pp 69–81

Martinelli N, Tinazzi O (1992) Lazise—La Quercia: indagine dendrocronologica sui campioni prelevati nel corso delle campagne di scavo dal 1986 al 1990. In: C'era una volta Lazise. Exposition Catalogue. Neri Pozza Editore, Vicenza, pp 102–105

Martinelli N, Pappafava M, Tinazzi O (1998) Datazione dendrocronologica dei resti strutturali. In: Balista C, Bellintani P (eds) Canàr di San Pietro Polesine. Ricerche archeo-ambientali sul sito palafitticolo. Padusa Quaderni, vol 2. Centro Polesano di Studi Storici, Archeologici ed Etnografici, Stanghella (Padova), pp 105–113

Martinelli N, Čufar K, Billamboz A (2018) Dendroarchaeology between teleconnection and regional patterns. In: Baioni M, Mangani C, Ruggiero MG (eds) Le Palafitte: Ricerca, conservazione, valorizzazione. Atti del Convegno Internazionale. Palafitte/Palafittes/Pfahlbauten/Pile dwellings, vol 0. SAP Società Archeologica, Quingentole (Mantova), pp 67–77

Poggiani Keller R, Binaghi Leva MA, Menotti E, Roffia M, Pacchieni T, Baioni M, Martinelli N, Ruggiero MG, Bocchio G (2005) Siti di ambiente umido della Lombardia: rilettura di vecchi dati e nuove ricerche. In: Della Casa P, Trachsel M (eds) WES'04 —wetland economies and societies. Proceedings of the international conference. Collectio Archeologica, vol X. Chronos, Zurich, pp 233–250

Reimer PJ, Austin WEN, Bard E, Bayliss A, Blackwell PG, Bronk Ramsey C, Butzin M, Cheng H, Lawrence Edwards R, Friedrich M, Grootes PM, Guilderson TP, Hajdas I, Heaton TJ, Hogg AG, Hughen KA, Kromer B, Manning SW, Muscheler R, Palmer JG, Pearson C, van der Plicht J, Reimer RW, Richards DA, Marian Scott E, Southon JR, Turney CSM, Wacker L, Adolphi F, Büntgen U, Capano M, Fahrni SM, Fogtmann-Schulz A, Friedrich R, Köhler P, Kudsk S, Miyake F, Olsen J, Reinig F, Sakamoto M, Sookdeo A, Talamo S (2020) The IntCal20 northern hemisphere radiocarbon age calibration curve (0–55 cal kBP). Radiocarbon 62:725–757

Rubat Borel F, Martinelli N, Köninger J, Menotti F (2022) Un contributo per la cronologia assoluta del Bronzo Medio: l'abitato perilacustre di Viverone Vi1-Emissario e l'Italia nordoccidentale. In: de Marinis RC, Rapi M (eds) Preistoria e Protostoria in Lombardia e Canton Ticino. LII Riunione Scientifica dell'Istituto Italiano di Preistoria e Protostoria. Rivista di Scienze Preistoriche LXXII, pp. 441–458

Salzani L, Balista C, Butta P, Martinelli N, Torri P, Bosi G, Mazzanti M, Mercuri AM, Accorsi CA, Bertolini M, Thun Hohenstein U (2018) La palafitta di Tombola di Cerea (Verona) (Scavo 1999). IpoTesi Preist 10:51–142. https://ipotesidipreistoria.unibo.it/article/view/9094

Zanini A, Martinelli N (2005) New data on the absolute chronology of late Bronze Age in Central Italy. In: Le Secrétariat du Congrès (eds) The Bronze Age in Europe and the Mediterranean. General sessions and posters. Section 11. Acts of the XIVth UISPP congress. BAR international series, vol 1337. Archaeopress, Oxford, pp 147–155

Pile-Dwellings at Ljubljansko Barje, Slovenia: 25 Years of Dendrochronology

Anton Velušček and Katarina Čufar

Abstract

Interdisciplinary research on the pile-dwellings in the Ljubljansko barje, Slovenia, has been carried out, with brief interruptions, since their discovery in 1875. Since 1995 these efforts have been coordinated by the Institute of Archaeology of the ZRC SAZU. Systematic excavations and interdisciplinary research were carried out on prehistoric pile-dwelling sites, and dendrochronology was introduced as a basic method for determining the time frame of their existence. To this end, wood was collected from 16 sites for wood identification, dendrochronology and radiocarbon dating. Between 1995 and 2017, nearly 8800 samples of waterlogged wood, mainly from the piles the dwellings were built on, were collected and examined. Approximately 20% of the samples were oak (*Quercus* sp.) and ash (*Fraxinus* sp.), with more than 45 tree rings selected for dendrochronological

study. Oak and ash tree-ring chronologies were established for most of the sites. Site chronologies that overlapped were merged into longer chronologies. Dating was carried out using ^{14}C dating supported by a wiggle-matching procedure, and for the 4th millennium BC settlements with the help of teleconnection with German-Swiss reference chronology from sites approximately 500 km away north of the Alps. For the oldest settlement Resnikov prekop, which was already inhabited around 4600 BC, we could not establish a chronology due to the insufficient number of wood samples. The most important tree-ring chronologies of oak are: BAR-3330 (time span 3771–3330 BC) dated by dendrochronology, as well as SG-VO (3285–3109 ± 14 cal BC) and ZA-QUSP1 (2659–2417 ± 18 cal BC) both dated by radiocarbon wiggle-matching). BAR-3330 helped us date eight sites, SG-VO two sites, and ZA-QUSP1 three sites indicating the end of the Copper Age on the Ljubljansko barje. Slovenian oak chronologies from different periods have the potential to be teleconnected with those from other regions.

A. Velušček (✉)
Research Centre of the Slovenian Academy of Sciences and Arts, Institute of Archaeology, Ljubljana, Slovenia
e-mail: anton.veluscek@zrc-sazu.si

K. Čufar
University of Ljubljana, Biotechnical Faculty, Department of Wood Science and Technology, Ljubljana, Slovenia
e-mail: katarina.cufar@bf.uni-lj.si

Keywords

Ljubljansko barje · Slovenia · Pile-dwellings · Dendrochronology · Dating · Neolithic · Eneolithic

A. Ballmer et al. (eds.), *Prehistoric Wetland Sites of Southern Europe*, Natural Science in Archaeology, https://doi.org/10.1007/978-3-031-52780-7_13

13.1 Introduction

Ljubljansko barje is the only area with preserved archaeological remains of prehistoric pile-dwellings discovered in Slovenia (Fig. 13.1). It is located southwest of the capital Ljubljana on the south-eastern edge of the Alps. The Ljubljansko barje is a shallow wetland of tectonic origin covering an area of 163 square kilometres. It can be flooded at times due to snowmelt or heavy rainfall. From the Late Pleistocene to the early Late Holocene the area was covered by a shallow lake.

The first pile-dwelling sites were discovered in 1875 by Karl Deschmann near the village of Studenec, now Ig. During field research, Deschmann discovered ornamental pottery and many traces of prehistoric metallurgical activity. After the discovery and pioneering excavations in

1875–1877, several research campaigns were carried out, e.g. 1907–1908 (W. Schmid), 1931–1945 (R. Ložar), 1953–1989 (J. Korošec, T. Bregant) and several campaigns after 1992. In 2011, the two different groups of pile-dwellings from the surroundings of the village of Ig were recognised by UNESCO with the inscription of a serial nomination 'Prehistoric Pile-Dwellings around the Alps' on the World Heritage List. We present the results of the activities coordinated by the Institute of Archaeology, Scientific Research Centre of the Slovenian Academy of Science and Art (IA ZRC SAZU), which began in 1995 and are still ongoing. This campaign introduced new sampling methods developed by teams working on pile-dwellings around the Alps. Particular attention was paid to the collection of archaeological wood and the introduction of dendrochronology. The interdisciplinary

Fig. 13.1 Maps showing the location of Ljubljansko barje on the southeastern edge of the Alps, and Ljubljansko barje basin with the locations of 16 pile-dwellings from the 5th, 4th and 3rd millennia BC investigated between 1995 and 2017 (Drawing by Tamara Korošec, ZRC SAZU Institute of Archaeology)

approach also included the development of palynology, archaeobotany and archaeozoology, and collaboration with numerous partners from other fields such as archaeometry and textile engineering.

The introduction of dendrochronology was crucial and is now an important foundation for archaeological research in wet environments. The dendrochronology laboratory was established in the early 1990s at the Department of Wood Science and Technology, Biotechnical Faculty, at the University of Ljubljana. Since then, the laboratory has been involved, among other things, in the construction of tree-ring chronologies (e.g. Čufar et al. 2008a) as well as basic research in dendroecology (e.g. Čufar et al. 2014), dendroclimatology (e.g. Čufar et al. 2008b) and preservation of waterlogged wood (e.g. Čufar et al. 2008c), all of which contributed to the application of dendrochronology in archaeology.

Dendrochronology was needed to revise the chronology of the Ljubljansko barje, where over 40 pile-dwelling sites were recorded (Velušček 2004). In order to obtain a sufficient amount of waterlogged wood, archaeological excavations were carried out on previously excavated areas, riverbeds, drainage ditches and by small-scale excavations or trenching at various sites.

13.2 Excavations and Wood Analyses

Between 1995 and 2017, we performed almost 30 field studies on 16 sites of prehistoric pile-dwellings (Table 13.1, Fig. 13.1). The archaeological investigations aimed for systematic acquisition and examination of waterlogged wood, mainly from the posts on which the dwellings were built.

Table 13.1 Pile-dwellingsettlements at the Ljubljansko barje, year of excavation, number of collected wood samples, percentage of oak (*Quercus*) and ash (*Fraxinus*) samples, end dates of the tree-ring chronologies based on dendrochronology (D), radiocarbon (^{14}C), or radiocarbonwiggle-matching (W) or with a combination of methods (Authors)

Site	Code	Year of excavation	No. of wood samples	*Quercus* %	*Fraxinus* %	End date BC	Dating method
Resnikov prekop	RP	2002	34	0	24	∼ 4600	^{14}C
Črnelnik	CEN	2014	39	72	26	∼ 3694	D
Trebež	TR	2017	83	25	47	3649	D
Strojanova voda	STV, SV	2012	351	16	62	3578	D
Hočevarica	HOC, HO	1995, 1998	361	37	51	3570	D
Maharski prekop	MP	2005	234	35	30	3487	D
Črešnja pri Bistri	CR	2003	124	49	20	3407	D
Spodnje Mostišče	SM1 + 2	1996, 1997	690	59	22	3351	D
Stare Gmajne	SG	2002, 2004, 2006, 2007	932	36	44	3330 3109 ± 14	D W
Veliki Otavnik	VO	2006	30	57	20	3108 ± 14	W
Blatna Brezovica	BB	2003	170	51	32	∼ 3071	
Parte-Iščica	PI	1997, 1998	1265	2	70	∼ 2610	^{14}C
Parte	PAR	1996	242	33	62	2458 ± 18	W
Založnica	ZAL, ZA	1995, 1999, 2001, 2009	1465	30	55	2417 ± 18	W
Dušanovo	DU, CG	2010, 2013, 2017	305	2	72	∼ 2490	
Špica	SPC	2010, 2011	2452	21	60	∼ 2450	

At each of the sites, we collected samples of all woody elements and conducted wood identification. Of the 8777 wood samples (Table 13.1), more than 99% came from the poles on which the dwellings had been built. The most common species were ash (*Fraxinus excelsior*), oak (*Quercus robur* and *Quercus petraea*), followed by alder (*Alnus glutinosa*), maple (*Acer* sp.), willow (*Salix* sp.), poplar (*Populus* sp.), hazel (*Corylus avellana*), hornbeam (*Carpinus betulus*), beech (*Fagus sylvatica*), silver fir (*Abies alba*) and elm (*Ulmus* sp.) found in smaller quantities (Čufar et al. 2013; Out et al. 2023).

Overall, ash accounted for more than half and oak for more than a quarter of all wood samples. The proportions of oak and ash, as well as other wood species, varied considerably from site to site. At Resnikov prekop, there was no oak and 24% ash, while alder was the leading species at 53%. At the 4th millennium BC sites, the leading wood species was either oak or ash, while at the 3rd millennium BC sites, ash was the leading species (Table 13.1). The proportions of alder, willow, poplar, maple, hazel, hornbeam, beech, silver fir and elm were generally lower (Čufar et al. 2013).

Wood samples were approximately 10 cm thick slices or cuts from the piles. For each sample, we identified the wood species, measured the diameters and counted the tree-rings. Samples of oak, ash and beech that contained more than 45 rings were selected for tree-ring width measurements. For each sample and site, the tree-ring series were cross-dated and assembled into floating chronologies. For each of the chronologies, we collected representative samples for radiocarbon dating (e.g. Čufar et al. 2010). We also attempted to cross-date each new chronology with previously established chronologies of the area and attempted to teleconnect them with existing chronologies of the remote sites.

13.3 Dendrochronology and Dating

Among the site chronologies produced, those of oak were the most interesting, as oak is the most common and important wood in European archaeology for which a network of long reference chronologies exists (Haneca et al. 2009). The chronologies of ash were also considered.

As the excavations progressed, the number of chronologies increased. Most of them were first dated by ^{14}C. Some chronologies of different settlements overlapped, and we were able to combine them into longer chronologies. This increased the possibilities of applying wiggle-matching and improving dating (Čufar et al. 2010).

Oak (*Quercus* sp.) chronologies from the 4th millennium BC could be joined to form a 442-year chronology BAR-3330 (Fig. 13.2). This was based on 106 cross-dated tree-ring series of wood from six pile-dwelling sites. BAR-3330 was cross-dated with a combined German-Swiss chronology from sites approximately 500 km to the north of the Alps and its time span of 3771–3330 BC was established (Čufar et al. 2015).

Other chronologies with end dates younger than 3330 BC were dated using radiocarbon wiggle-matching only, as their dendrochronological dating has not yet been successful. Such 4th millennium BC chronologies are those from Stare gmajne young and Veliki Otavnik—their combined chronology SG-VO spans the period 3285–3108 ± 14 cal BC. At the sites where we have produced oak and ash chronologies, some ash chronologies could be dated by heteroconnection with the oak chronologies.

In the pile-dwellings of the 3rd millennium BC in Ljubljansko barje, ash chronologies predominated (Table 13.1, Fig. 13.2). Their dendrochronological dating by teleconnection with oak chronologies from the north of the Alps is not very probable, as the ash chronologies are mainly short, and they reflect local conditions and disturbances (cf. Capano et al. 2020). The only longer and well reproduced oak chronology is the one from Založnica, ZAL-QUSP1 which spans the period 2659–2417 ± 18 cal BC. It allowed us to cross-date the ash chronologies of Založnica, Dušanovo and Parte and the oak chronology of Parte. The sites of Založnica with Parte and Dušanovo mark the end of the Copper Age at Ljubljansko barje (Velušček and Čufar

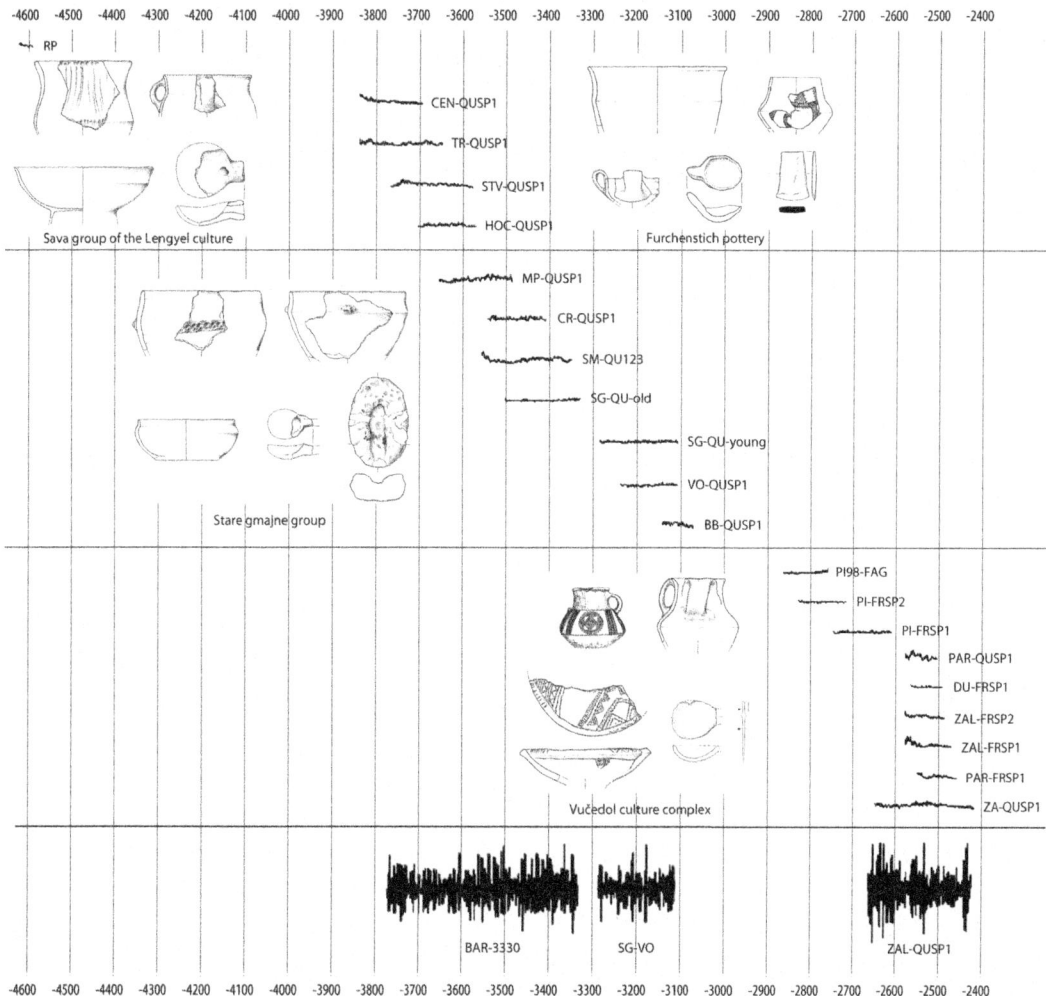

Fig. 13.2 Time spans of main oak (QUSP), ash (FRSP) and one beech (PI98-FAG) chronologies of individual pile-dwellings dated with three oak chronologies: BAR-3330 (time span 3772–3330 BC), dendrochronologically dated, as well as SG-VO (3285–3108 ± 14 cal BC), and ZAL-QUSP1 (2659–2417 ± 18 cal BC), both dated by radiocarbon wiggle-matching. The drawings are representing typical cultural groups (not in scale) after Trampuž-Orel and Heath 2008, Pl. 1.2,12; Velušček 2009, Pl. 3.19,15; Velušček and Čufar 2014, Figs. 13.1 and 13.2 (Authors)

2003; Velušček et al. 2011). The chronologies presented allowed us to date about 10–20% of the wood samples from most settlements.

The end dates of the site chronologies denote the last (dated) felling activity on the sites (Table 13.1, Fig. 13.2). Based on the felling dates of individual trees, we were also able to determine multiple construction phases at most sites. Building activities or repairs could be inferred from a large number of trees felled in the same year or within a narrow period of a few years. The years of building activities and the duration of the settlements have been discussed in several publications (e.g. Čufar et al. 2010; Velušček and Čufar 2003). Some pile-dwelling settlements were occupied for about 60 years or even longer, while the others lasted only 20 years or less (Velušček 2005). Apart from the Špica site, the investigated pile-dwelling settlements are located in the southern part of the

Ljubljansko barje and form three groups at a distance of about 10 km from each other (Fig. 13.1). In almost all the periods studied, we were able to detect contemporaneous building activities at different sites. The pairs of simultaneous dwellings are, for example Strojanova voda-Hočevarica, Spodnje mostišče-Stare gmajne old and Parte-Založnica. In some cases, the same sites were settled more than once, e.g. Hočevarica and Stare gmajne (Fig. 13.2).

13.4 The Chronology and Cultural Characteristics of the Ljubljansko Barje Pile-Dwellings

The pile-dwellings from the Ljubljansko barje studied by dendrochronology were built in a time frame between c. 4600 and 2400 BC. During this long period, the occupation of the Ljubljansko barje basin was not continuous, and several hiatuses have been detected. The longest among them lasted almost a thousand years—between c. 4600 and 3700 BC. The reasons why the area was probably not permanently settled cannot be fully explained, but could have been caused by cultural, economic, or climatic factors (Velušček 2004). The explanation that the hiatuses that have been detected are merely a consequence of the state of research seems to be the least likely according to the current knowledge.

The dendrochronological findings complement each other with the findings on material culture, where several cultural horizons or settlement periods could be delineated based on the pottery (Fig. 13.2), such as:

1. the Sava group of the Lengyel culture (Resnikov prekop),
2. the culture of Furchenstich pottery (Trebež, Črnelnik, Hočevarica, etc.),
3. the Stare gmajne cultural group (Maharski prekop, Črešnja pri Bistri, Spodnje mostišče, Stare gmajne, Veliki Otavnik, Blatna Brezovica, etc.),
4. the Vučedol culture complex of the early and middle 3rd millennium BC (e.g. Parte-Iščica, Parte, Založnica, Dušanovo, Špica),

5. the Litzen pottery horizon (excavation sites of K. Deschmann and R. Ložar),
6. the horizon of pile-dwellings from the middle and the beginning of the Late Bronze Age (e.g. Blato and Šivčev prekop).

In the settlements where the simultaneity was proved by dendrochronology, this is also confirmed by the almost identical finds. A typical example is pottery finds from the pile-dwelling settlements of Parte and Založnica (Fig. 12.2).

13.5 The Impact of Wood Research on Slovenian Wetland Archaeology

Field excavations and examinations helped us to date the time of existence and building activity on the pile-dwellings and obtain information about earlier forests and the environment, as well as about human life and the influence of ancient populations on the environment (Out et al. 2020, 2023; Tolar et al. 2011). The wood species used to build the pile-dwellings still grow in the Ljubljansko barje (ash, oak, alder, willow, poplar) or on the nearby hills and mountains bordering the Ljubljansko barje to the south (beech, silver fir, oak, maple, elm, hazel, hornbeam). We suspect that the choice of wood species used for poles generally depended on the location of the pile-dwelling, whether it was closer to a lake or floodplain or closer to the mountains (Tolar et al. 2011). The choice was probably also influenced by the availability of wood, which was collected in the vicinity of the settlement if possible. Although many sources report that the pile-dwellers are said to have managed forests due to their high demand for wood (e.g. Bleicher and Staub 2023), we have not yet been able to prove this for the Ljubljansko barje, as shown by two studies of Out et al. (2020, 2023).

The few wooden artefacts discovered, such as the about 5600-year-old bow made of yew (*Taxus baccata*), several logboats made of oak and the 5150-year-old wooden wheel made of ash with an oak axle (the world's oldest known wheel and axle) show that the pile-dwellers knew how to select, process and use the wood properly (Velušček et al. 2009). They were able to select, process and use the wood from the immediate surroundings of the settlements or from more distant locations in the best possible way.

Wood research was only a small but essential part of the multidisciplinary investigations of the Ljubljansko barje wetland. The multidisciplinary approach helped us to reconstruct a holistic picture of human life, environment and connections with other sites. During the almost three decades of research, we examined sites from different periods and with different cultural backgrounds. Pollen analyses demonstrated that the earliest pile-dwellers were arable farmers, not just hunters and farmers occupied with domestic animals (Andrič et al. 2008; Tolar et al. 2011; Toškan et al 2020; Velušček et al. 2011).

Archaeometric studies have placed the beginning of local metallurgy in the first half of the 4th millennium BC, and the use of arsenical copper in the 4th millennium BC and antimony copper in the early 3rd millennium BC (Trampuž-Orel and Heath 2008). The study of dog coprolite has also provided insights into the role of humans' first animal companion in the community (Tolar et al. 2021).

For the representation or understanding of prehistoric everyday life in the Ljubljansko barje, the reconstruction of a real house ground plan in the pile-dwelling settlement Parte-Iščica (Velušček et al. 2000, Fig. 6) was crucial. This was done by analysing the wood from vertical piles in the riverbed, where a complete lack of stratigraphic data was registered. Our survey has shown at least three rectangular houses of about 25 m^2 each, built side by side in a row. On the Ljubljansko barje the SW–NE direction of the houses was predominant. According to the dendrochronological data, it was possible to correct an earlier reconstruction of the layout of the settlement, proposed in the 1970s which assumed the existence of one or two huge wooden platforms made up of various house structures in different orientations and shapes. Our work, however, showed that the village consisted of several detached rectangular houses, rather than one big platform. The reinterpretation of old data based on the dendrochronological investigations allowed us to gain some insights into the prehistoric village zones with distinct economic activities. The distribution of finds, artefacts and especially archaeozoological features associated with the reinterpreted house plans revealed some important differences between them. Based on this, a specialisation within the small prehistoric community was proposed (Toškan et al. 2020).

13.6 The Potential of Dendrochronological Teleconnection

In Slovenia, an area south of the Alps, the dendrochronological dating of the first pile-dwelling tree-ring chronologies could only be done by teleconnection, i.e. using remote reference chronologies from the area north of the Alps. Attempts to do so, however, have long failed (Čufar et al. 2010), as chronologies from the Ljubljansko barje pile-dwellings were short and poorly replicated. However, after the shorter chronologies could be combined into a 442-year-long chronology BAR-3330, the teleconnection with the German-Swiss reference was successful and a time span of 3771–3330 BC was confirmed with high and significant statistical parameters (Čufar et al. 2015). Teleconnection basically suggests that tree-ring variation in geographically distant areas is determined by common (climatic) factors. It depends on quality (i.e. length and replication), and it appears that BAR-3330, with a length of 442 years and a replication of 106 (i.e. containing tree-ring series from 106 trees), reached the quality level for successful teleconnection. Prior dating by radiocarbon wiggle-matching played an important role here, and the previously estimated end date of BAR-3330 (3332 ± 10 cal BC) was shifted by only two years after dendrochronological dating (Čufar et al. 2010, 2015).

The possibility of teleconnection of chronologies from Slovenia with those from the north of the Alps, e.g. the Hohenheim oak

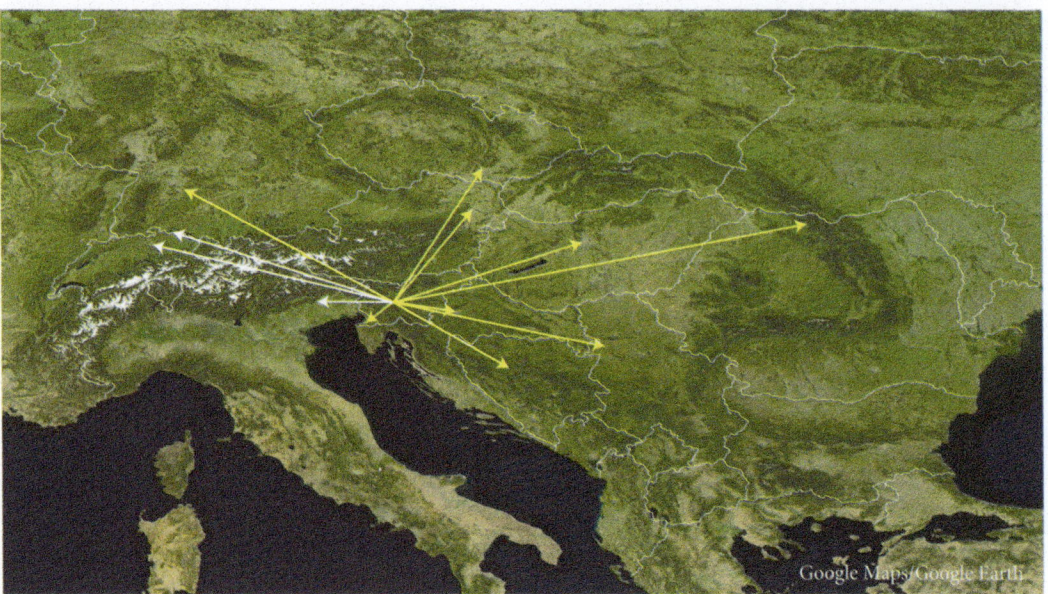

Fig. 13.3 Teleconnection of Slovenian oak chronologies: the arrows connect Slovenia with the locations of chronologies that could be cross-dated. Yellow arrows: recent trees and historic structures; white arrows: waterlogged wood from prehistoric sites (Authors)

chronology (Friedrich et al. 2004), was also demonstrated by the modern regional oak chronology of Slovenia (length 548 years, time span 1549–2003, replication183) (Čufar et al. 2008a). Moreover, this chronology could be teleconnected with chronologies from regions up to 700 km away from the east, north and southeast of Slovenia, including Austria, Czech Republic, Hungary, Croatia, Serbia and Romania (Čufar et al. 2008a, 2014; Kolar et al. 2012; Nechita et al. 2018) (Fig. 13.3).

In addition, Pearson et al. (2014) and Wazny et al. (2014), who have worked with oak chronologies from southeast Europe and Turkey and subfossil oaks from the Balkans, report that tree-ring series from the Balkans could make an important contribution to improve the dendrochronological network and bridge the temporal and spatial gaps for dating prehistoric sites.

This gives hope that chronologies from Slovenia could help to fill the gaps to increase the likelihood of dating wood from the southeast, including North Macedonia, Albania and Greece (Maczkowski et al. 2021; Maczkowski et al. this issue).

Comparisons of recent chronologies from Slovenia with the few existing from Italy showed that teleconnection of oaks from both countries should be possible. However, the only successful cross-dating of prehistoric chronologies from Slovenia and Italy to date includes the chronologies from Hočevarica and the site of Palù di Livenza in Pordenone, NE Italy, about 150 km away (Čufar and Martinelli 2004). Other prehistoric chronologies from Slovenia and Italy could not be cross-dated because they do not cover the same time period (Martinelli et al. 2011).

13.7 Concluding Remarks

The introduction of dendrochronology into the study of the pile-dwellings of Ljubljansko barje was a significant achievement for Slovenian prehistoric archaeology. The most important and decisive result is the absolute chronology for the area, which can be also applied to the entire southeast Alpine region and even beyond. It revealed a non-continuous occupation and long settlement gaps between the different cultural phases. For example, between the oldest pile-dwelling settlement of Resnikov prekop and the following one there was almost 1000 years of interruption in settlement, while the other gaps are of shorter duration. Simultaneous settlements have also been recorded in different parts of the Ljubljansko barje. So far mainly short-lived settlements have been confirmed, which may pose a challenge to interpretations of the presumably long-lasting occupations of dryland settlements elsewhere, where chronologies of sites are based solely on stratigraphy, typological analysis of finds and mostly sparse ^{14}C dating.

It is now possible to compare absolute dates for the 4th millennium BC in calendar years with all kinds of archaeological finds, which was unimaginable before the introduction of dendrochronology. However, for the 2nd millennium BC, when the last pile-dwellers presumably lived on Ljubljansko barje, we still lack wood research. The main reason for this is that the sites of this period have not yet been sufficiently explored, if at all, and it seems that they were also less numerous. Therefore, for the time being the chronology for this part of the history of Ljubljansko barje is mainly based on typological analyses and chronologies from neighbouring areas.

In addition, dendrochronology has offered the possibility to observe developments within centuries or even decades and years. During the research, we discovered that within the 36th century BC, after the abandonment of the Hočevarica settlement, the most recent site with the so-called *Furchenstich* (stab-and-drag) pottery, a new pile-dwelling settlement appeared on the other side of the Ljubljansko barje at a distance of about 10 km, now with a different type of pottery. A cultural change is suggested with the probable introduction of a completely new group of settlers in the area. We also confirmed that the dwellings consisted of smaller rectangular houses, which is contrary to earlier interpretations describing huge settlement platforms.

Dendrochronology also helped to obtain reliable dates for specific finds. The famous wooden wheel from the Stare gmajne site was dated using C14, and additionally by dating the settlement obtained through dendrochronology, typological analysis and stratigraphy.

The use of dendrochronology in the study of pile-dwellings in Slovenia provides several new cues to think about various topics, such as prehistoric woodland management, which has been studied by analysing the age and diameter of roundwood from all pile-dwelling sites of Ljubljansko barje (Out et al. 2020, 2023). Among other things, these works point to new possible future directions in the study of Ljubljansko barje as an archive of natural and anthropogenic history and human–environment interactions.

Acknowledgements This research was funded by the Slovenian Research and Innovation Agency ARIS, programmes P4-0015 and P6-0064. We thank the staff and students of the Department of Wood Science and Technology, Biotechnical Faculty, University of Ljubljana, and the Institute of Archaeology, ZRC SAZU for their immense help in the field, in the laboratory and in all phases of the research. We also thank Tamara Korošec for her help with illustrations and Paul Steed for language editing.

References

Andrič M, Kroflič B, Toman MJ, Ogrinc N, Dolenec T, Dobnikar M, Čermelj B (2008) Late quaternary vegetation and hydrological change at Ljubljansko barje (Slovenia). Palaeogeogr Palaeoclimatol Palaeoecol 150–165(1–2):150–165

Bleicher N, Staub P (2023) A question of method and place? A critical reappraisal of the methods of dendroarchaeology, anthracology, archaeobotany and roundwood analysis on the question when systematic woodland management began in Europe. Quat Int. https://doi.org/10.1016/j.quaint.2022.05.006

Capano M, Martinelli N, Baioni M, Tuna T, Bernabei M, Bard E (2020) Is the dating of short tree-ring series still a challenge? New evidence from the pile dwelling of Lucone di Polpenazze (northern Italy). J Archaeol Sci 121:105190

Čufar K, de Luis M, Eckstein D, Kajfež-Bogataj L (2008b) Reconstructing dry and wet summers in SE Slovenia from oak tree-ring series. Int J Biometeorol 52:607–615

Čufar K, Gričar J, Zupančič M, Koch G, Schmitt U (2008c) Wood anatomy, cell-wall structure and topochemistry of waterlogged archaeological wood aged 5200 and 4500 years. IAWA J 29(1):55–68

Čufar K, Kromer B, Tolar T, Velušček A (2010) Dating of 4th millennium BC pile-dwellings on Ljubljansko barje, Slovenia. J Archaeol Sci 37:2031–2039

Čufar K, Velušček A, Kromer B (2013) Two decades of dendrochronology in the pile dwellings of the Ljubljansko barje, Slovenia. In: Bleicher N, Schlichtherle H, Gassmann P, Martinelli N (eds) Dendro-Chronologie-Typologie-Ökologie, Festschrift für André Billamboz zum 65. Beier und Beran Verlag, Langenweiflbach, Geburtstag, pp 35–40

Čufar K, Grabner M, Morgos A, Martinez del Castillo E, Merela M, de Luis M (2014) Common climatic signals affecting oak tree-ring growth in SE Central Europe. Trees 28:1267–1277. https://doi.org/10.1007/s00468-013-0972-z

Čufar K, Tegel W, Merela M, Kromer B, Velušček A (2015) Eneolithic pile dwellings south of the Alps precisely dated with tree-ring chronologies from the north. Dendrochronologia 35:91–98

Čufar K, Martinelli N (2004) Teleconnection of chronologies from Hočevarica and Palù di Livenza, Italy. In: Velušček A (ed) Hočevarica, an Eneolithic Pile Dwelling in the Ljubljansko Barje. Opera Instituti Archaeologici Sloveniae, vol 8. Inštitut za arheologijo ZRC SAZU and Založba ZRC, Ljubljana, pp 286–289. https://doi.org/10.3986/9789612545055

Čufar K, de Luis M, Zupančič M, Eckstein D (2008a) A 548-year long tree-ringchronology of oak (Quercus sp.) for SE Slovenia and its significance as dating tool and climate archive. Tree-Ring Res 64(1):3–15

Friedrich M, Remmele S, Kromer B, Hofmann J, Spurk M, Kaiser KF, Orcel C, Küppers M (2004) The 12,460-year Hohenheim oak and pine tree-ring chronology from Central Europe—a unique annual record for radiocarbon calibration and paleoenvironment reconstructions. Radiocarbon 46(3):1111–1122

Haneca K, Čufar K, Beeckman H (2009) Oaks, tree-rings and wooden cultural heritage: a review of the main characteristics and applications of oak dendrochronology in Europe. J Archaeol Sci 36(1):1–11

Kolar T, Kyncl T, Rybnicek M (2012) Oak chronology development in the Czech Republic and its teleconnection on a European scale. Dendrochronologia 30 (3):243–248

Maczkowski A, Bolliger M, Ballmer A, Gori M, Lera P, Oberweiler C, Szidat S, Touchais G, Hafner A (2021) The early bronze age dendrochronology of Sovjan (Albania): a first tree-ring sequence of the 24th – 22nd c BC for the Southwestern Balkans. Dendrochronologia 66:125811

Martinelli N, Čufar K, Billamboz A (2011) Dendroarchaeology between teleconnection and regional patterns. In: Baioni M, Mangani C, Ruggiero MG (eds) Le palafitte: Ricerca, conservazione, valorizzazione. Atti del Convegno Internazionale. Palafitte/Palafittes/Pfahlbauten/Pile dwellings, vol 0. SAP Società Archeologica, Quingentole (Mantova), pp 67–77. https://repozitorij.uni-lj.si/IzpisGradiva.php?id=127093&lang=slv. Accessed 18 May 2021

Nechita C, Eggertsson O, Badea ON, Popa I (2018) A 781-year oak tree-ring chronology for the middle ages archaeological dating in Maramureş (Eastern Europe). Dendrochronologia 52:105–112

Out WA, Baittinger C, Čufar K, López-Bultó O, Hänninen K, Vermeeren C (2020) Identification of woodland management by analysis of roundwood age and diameter: neolithic case studies. For Ecol for Manag 467:118136

Out WA, Hänninen K, Merela M, Velušček A, Vermeeren C, Čufar K (2023) Evidence of woodland management at the Eneolithic pile dwellings (3700–2400 BCE) in the Ljubljansko barje, Slovenia? Plants 12(2):291. https://doi.org/10.3390/plants12020291

Pearson CL, Wazny T, Kuniholm PI, Botić K, Durman A, Seufer K (2014) Potential for a new multimillennial tree-ring chronology from subfossil Balkanriver oaks. Radiocarbon 56(4); Tree-Ring Res 70(3):51–59

Tolar T, Jacomet S, Velušček A, Čufar K (2011) Plant economy on a Late Neolithic lake dwelling site in Slovenia at the time of the Alpine Iceman. Veg Hist Archaeobotany 20:207–222

Tolar T, Galik A, Le Bailly M, Dufour B, Caf N, Toškan B, Bužan E, Zver L, Janžekovič F, Velušček A (2021) Multi-proxy analysis of waterlogged preserved late Neolithic canine excrements. Veg Hist Archaeobotany 30(3):107–118

Toškan B, Achino KF, Velušček A (2020) Faunal remains mirroring social and functional differentiation? The copper age pile-dwelling site of Maharski prekop (Ljubljansko barje, Slovenia). Quatern Int 539:62–77

Trampuž-Orel N, Heath D (2008) Copper finds from the Ljubljansko barje (Ljubljana Moor) A contribution to the study of prehistoric metallurgy. Arheološki Vestnik 59:17–29

Velušček A (2004) Past and present lake-dwelling studies in Slovenia: Ljubljansko barje (the Ljubljana Marsh). In: Menotti F (ed) Living on the lake in prehistoric Europe: 150 years of lake-dwelling research. Routledge, London, New York, pp 69–82

Velušček A, Čufar K (2003) Založnica near Kamnik pod Krimom on the Ljubljansko barje (Ljubljana Moor): a settlement of the Somogyvár-Vinkovci Culture. Arheološki Vestnik 54:123–158

Velušček A, Čufar K, Levanič T (2000) Parte-Iščica, archaeological and dendrochronological investigations. Arheološki Vestnik 51:83–107

Velušček A, Toškan B, Čufar K (2011) The decline of pile-dwellings at Ljubljansko barje. Arheološki Vestnik 62:51–82

Velušček A, Čufar K (2014) Pile-dwellings at Ljubljansko barje. In: Tecco-Hvala S (ed) Studia Praehistorica in Honorem Janez Dular. Opera Instituti Archaeologici Sloveniae, vol 30. Inštitut za arheologijo ZRC SAZU and Založba ZRC, Ljubljana, pp 39–64

Velušček A, Čufar K, Zupančič M (2009) Prehistoric wooden wheel with an axle from the pile-dwelling Stare gmajne at the Ljubljansko barje. In: Velušček A (ed) Stare gmajne pile-dwelling settlement and its era. Opera Instituti Archaeologici Sloveniae, vol 16. Inštitut za arheologijo ZRC SAZU and Založba ZRC, Ljubljana, pp 197–222

Velušček A (2005) The Kras plateau in southwestern Slovenia and the Ljubljansko barje in the Neo-

Eneolithic period: a comparative study. In: Mihevc A (ed) Kras: water and life in a rocky landscape. Založba ZRC, Ljubljana, pp 199–219

Velušček A (2009) Stare gmajne pile-dwelling settlement near Verd. In: Velušček A (ed) Stare gmajne pile-dwelling settlement and its era. Opera Instituti Archaeologici Sloveniae, vol 16. Inštitut za arheologijo ZRC SAZU and Založba ZRC, Ljubljana, pp 49–121

Wazny T, Lorentzen B, Köse N, Akkemik Ü, Boltryk Y, Güner T, Kyncl J, Kyncl T, Nechita C, Sagaydak S, Kamenova Vasileva J (2014). Bridging the gaps in tree-ring records: creating a high-resolution dendrochronological network for southeastern Europe. Radiocarbon 56(4); Tree-Ring Research 70(3):39–50

Wetland Dendrochronology: An Overview of Prehistoric Chronologies from the Southwestern Balkans

14

Andrej Maczkowski, Matthias Bolliger, and John Francuz

Abstract

In the past few decades, a number of prehistoric wetland archaeological sites have been detected in the south-central part of the Balkan Peninsula. However, only a few of them have been excavated. In this study, we discuss the characteristics of the wooden remains and selected tree-ring width chronologies from the archaeological sites of Sovjan, Ploča Mičov Grad and Dispilio. They represent the first prehistoric centennial and multi-centennial tree-ring chronologies from the region, covering various periods of the Neolithic, Chalcolithic and the Bronze Age. The dominant wood species utilised on the sites are members of the genus *Quercus*, but significant numbers of *Juniperus* and *Pinus* were also recovered, in addition to some other deciduous species. Through radiocarbon dating and wiggle-matching, we were able to anchor these floating tree-ring chronologies on the calendar scale with high temporal resolution.

Keywords

Dendrochronology · Neolithic · Chalcolithic · Bronze age · Pile-dwellings · Balkan prehistory · Wiggle-matching · Dendroarchaeology · Absolute dating · EXPLO

14.1 Introduction

The fundamental framework within which archaeological research operates is the chronological ordering of objects and events. Initially building upon mythological-religious timescales (Renfrew 1973), through the development of seriation and relative archaeological chronologies (Montelius 1903), up to the twentieth century when radiocarbon-dating was discovered (Libby 1952)—the chronology of human prehistory has become ever more precise. From its advent in the early 20th century (Douglass 1929), dendrochronology has had a foundational role in the construction of precise relative and absolute calendric scale chronologies for cultures and periods for which detailed written sources are lacking (Menotti 2004; Schweingruber 1988; Haneca et al. 2009).

The long history of research on prehistoric wetland sites in Central Europe (Hafner et al. 2020), particularly in Switzerland and Southern

A. Maczkowski (✉) · M. Bolliger
Institute of Archaeological Sciences and Oeschger Center for Climate Change Research (OCCR), University of Bern, Bern, Switzerland
e-mail: andrej.maczkowski@unibe.ch

M. Bolliger
e-mail: matthias.bolliger@be.ch

Dendrochronological Laboratory, Archaeological Service of the Canton of Bern, Bern, Switzerland

J. Francuz
Institute of Archaeological Sciences, University of Bern, Bern, Switzerland
e-mail: john.francuz@faculty.unibe.ch

A. Ballmer et al. (eds.), *Prehistoric Wetland Sites of Southern Europe*, Natural Science in Archaeology, https://doi.org/10.1007/978-3-031-52780-7_14

Germany (Huber 1963; Ruoff 1979; Billamboz 2014), has contributed to the construction of millennia-long regional tree-ring chronologies (Becker et al. 1985; Friedrich et al. 2004). This enabled an almost historical approach to prehistoric finds, features and settlement development. In the Balkans, the great potential for constructing millennia-long tree-ring chronologies has been recognised and explored since the 1970s (Kuniholm and Striker 1983). However, robust dendrochronological references for archaeological dating have been completed only relatively recently (e.g. Čufar et al. 2010; Westphal et al. 2010; Pearson et al. 2014; Roibu et al. 2021). One of the main reasons for this is the lack of suitable archaeological wooden material, which can be preserved for millennia only in very specific conditions.

The region in the southwestern Balkan Peninsula, centred around several Quaternary tectonic lakes (Fig. 14.1), represents the most

suitable ground for multidisciplinary investigations of wetland sites. The lakes provide an excellent environment for long-term wood preservation. This paper summarises the dendrochronological analyses of wooden material sampled from constructions remains on wetland sites on Lake Ohrid (the site of Ploča Mičov Grad), Lake Kastoria (Dispilio) and on the now-extinct Lake Maliq (Sovjan and Maliq).

14.1.1 Previous Excavations and Dating

The wood samples from the three different sites discussed in this paper were sampled in various campaigns between 2018 and 2019. The archaeological site of Sovjan was discovered in 1988 and parts of it have been systematically excavated between 1993 and 2006. The long stratigraphic sequence of the site spans from the

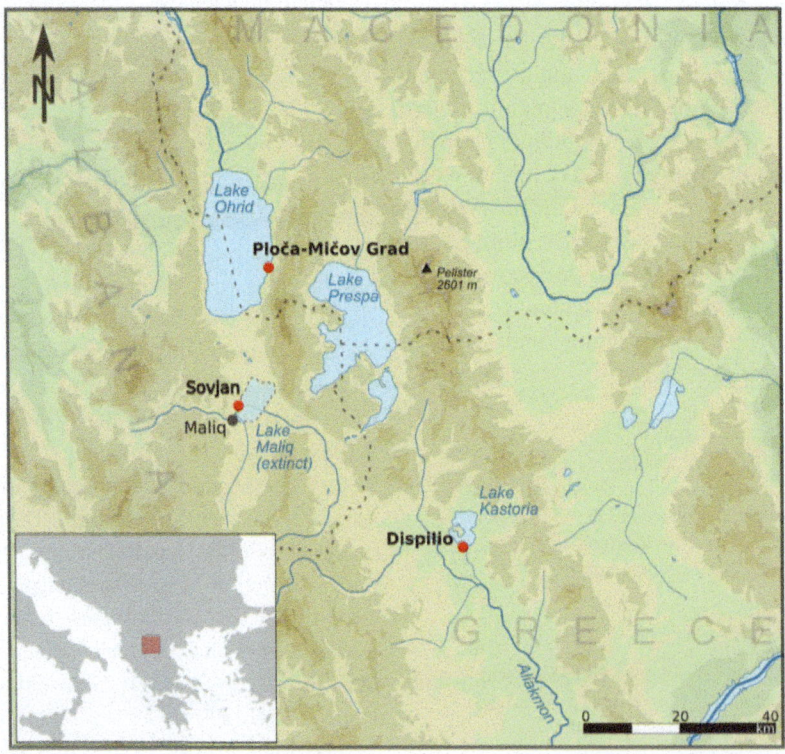

Fig. 14.1 Geographical position of the sites discussed in the text (Reconstruction of the former Lake Maliq according to Fouache et al. (2010) and Reich et al. (2021); background map based on Wikimedia file 'Macedonia topography (texts).svg')

Neolithic up to the Iron Age. Radiocarbon dating of organic material from different layers provided a coarse absolute timeframe for the chronology of the site. The oldest samples yielded calibrated dates in the range of 7300–7000 cal BC (Oberweiler et al. 2020). However, the finds and contexts of this oldest layer are somewhat ambiguous (Oberweiler et al. 2020; Lera and Touchais 2003). On the other hand, the youngest part of the sequence is chronologically delimited by a relative typological dating to the 9th–8th century BC. Located in the middle of the stratigraphic sequence are the most exceptionally preserved architectural wooden remains from Sovjan, belonging to the Bronze Age layer 8. For this layer, one radiocarbon date was available, placing it in the range of 2303–2040 cal BC (Gori and Krapf 2015). The main features in this layer consist of the so-called *Maison du Canal* (House on the canal), a partially excavated trackway, and the corner of another building. Although the wooden remains were previously sampled for dendrochronological analysis, the results were never published, and those samples are not available anymore (Oberweiler et al. 2020). Thus, parts of the site were re-excavated in 2018 and 34 samples of wooden remains were recovered for dendrochronological analysis. The results of this analysis were published in detail elsewhere (Maczkowski et al. 2021), and only a brief overview will be presented here.

Some 30 km north of Sovjan is the underwater site of Ploča Mičov Grad (henceforth Ploča), on the eastern shore of Lake Ohrid (Fig. 14.1). First underwater investigations and documentation of the pile field took place between 1997 and 2005 (Kuzman 2009) during which more than 6000 wooden piles were documented at a depth of 3–5 m from the contemporary lake level. Based on the pottery finds recovered from the lake bottom the use of the site was relatively dated to the Late Bronze Age and Iron Age. However, no systematic sampling of the wooden piles was performed.[1] In 2018, a first systematic pile sampling took place, which continued in 2019 within the

EXPLO project (Hafner et al. 2021). The material discussed in the present paper was extracted during these two latter campaigns.

The southern shore of Lake Kastoria, about 60 km to the south-east of Lake Ohrid (Fig. 14.1), lies the wetland archaeological site of Dispilio. The first excavations took place in 1940, but the modern systematic investigation began in 1992 and, with sporadic hiatuses, continues until today. The prehistoric layers of Dispilio were identified in three large trenches covering an area of more than 2000 m^2 (Facorellis et al. 2014). The relative typological dating and the radiocarbon dating places the oldest anthropogenic deposits of the site in the Middle and Late Neolithic, however, indications of the late Early Neolithic phase were also identified (Facorellis et al. 2014). The site was inhabited continuously until probably the end of the 3rd millennium BC (Early to Middle Bronze Age), with a notable hiatus in the second half of the 4th millennium BC (Karkanas et al. 2011). However, no systematic sampling of the wooden remains was performed before 2019.[2] In the course of the EXPLO Project, the remains of the piles which were uncovered in previous excavation in the Western trench were sampled.

14.1.2 Environmental Setting

The topography of the southwestern Balkans is characterised by mountain ranges flanked by flat valleys. The mountain ranges usually extend in a NW to SE direction, rising frequently above 2000 m a.s.l. Despite the relatively short distance from both the Aegean and the Adriatic Sea, the environmental conditions are not dominated by the Mediterranean influences. The relief, the orographic barriers, and the altitude, together influence the local climate which is characterised as continental to sub-Mediterranean.

The three archaeological sites presented in this paper are situated on similar altitudes, with Sovjan being the highest, at an altitude of c.

---

[1]A handful of samples from Ploča were analysed at the University of Arizona in 2013, unpublished.

[2]Around ten wooden samples from Dispilio were measured by P. Kuniholm in the late 1990s, also unpublished.

820 m, Ploča at 695 m and Dispilio at 630 m a.s.
l. July and August are the hottest months with
least precipitation, while the wettest months are
November and December. The annual precipitation averages between 600 and 750 mm in the
lowlands, with significant increase in precipitation at higher altitudes.

The tree genera at lower altitudes are generally represented by *Ulmus*, *Salix*, *Platanus*,
Fraxinus, *Carpinus*, *Acer*, *Quercus*. At mid-
altitudes different species of oak (*Q. cerris*, *Q.
frainetto*, *Q. petraea*, *Q. pubescens*, *Q. trojana*,
Q. ilex), together with *Carpinus*, *Ostrya
carpinifolia*, and different juniper species
(*Juniperus excelsa*, *J. foetidissima*, *J. oxycedrus*)
form mixed stands, while the high altitudes are
covered by beech and coniferous forests and
alpine shrubs and meadows (*Fagus*, *Abies*,
Pinus, *Picea*, *J. communis*, etc.). The forests in
the region have been heavily impacted by millennia of human influence, culminating with the
demographic expansion during the 19th and first
decades of the 20th century. Proper forest management and protection were established only
well into the 20th century (Matevski et al. 2011;
Saratsi 2003). However, despite the long-term
human influence on the forests in the region, a
common environmental setting ensures a comparable regional climatic signal preserved in the
tree-rings.

14.2   Materials and Methods

In the course of various excavations and sampling campaigns, more than 1600 samples of
subfossil wood (Hafner et al. 2021; Oberweiler
et al. 2020; Reich et al. 2021; see also Chaps. 6,
7 and 8 in this book) were collected from the
archaeological sites of Ploča, Dispilio and Sovjan. To date, 864 individual wood samples have
been analysed. The prehistoric material came in
various forms of preservation, depending on its
provenance, species and depositional environment. On all sites, during an initial macroscopic
field examination, samples were grouped
according to species (oaks, other deciduous and
conifers). The wooden samples from all sites

were kept constantly in wet conditions, while for
transport and storage they were put in airtight
plastic bags or vacuum-sealed.

After the initial field-lab analyses at the sites
of Ploča and Dispilio in 2018 and 2019, further
measuring and wood anatomical analyses were
performed at the Dendrochronological Laboratory of the University of Bern. The sample surfaces were prepared for measuring using razor
blades and, except for some deciduous species,
the radii were treated with chalk to increase the
contrasts between different wood anatomical
features. Dendrochronological measurement was
carried out under a stereo microscope (Leica
M50) on a mechanical *Isel LES 4* measuring
platform equipped with a *Mitutoyo Digimatic*
calliper with an accuracy of 0.01 mm. The
measuring of the samples and the statistical
cross-dating were performed in the *Dendroplus*
software (Ulrich Ruoff, version 28 Nov. 2013,
unpublished). The cross-dating of the samples
from Ploča was further tested and verified with
PAST5 (Version 5.0.610, http://www.sciem.
com/). The visual cross-dating was checked in
a vector graphics software. For the statistical
assessment of the relative cross-dating positions,
the standard dendrochronological parameters
were followed: the *t-value* (Hollstein 1980
[henceforth t_{HO}]) and the coefficient of parallel
variation *Gleichläufigkeit* (Eckstein and Bauch
1969 [henceforth GLK]). If not otherwise noted,
all the t_{HO} values mentioned in this text refer to
the non-detrended tree-ring sequences.

During the fieldwork, samples at Ploča and
Dispilio were also selected according to dendrochronological potential: presence of last
growth ring (henceforth waney edge), regular ring
growth and higher number of tree-rings (> 40).
This selection was maintained in the initial cross-
dating process for the establishment of robust
chronologies. The dendrotypological features
such as growth trend, cambial age, mean ring
width and presence/absence of sapwood (Billamboz 2008) of the resulting tree-ring sequences
were also considered. The archaeological plan
was not consulted during the cross-dating process
to ensure the quality and independence of the
cross-dating.

14.3   Results

14.3.1   Sovjan–Korçë

14.3.1.1   Wood Species and Characteristics

The wood anatomical analysis of all the measured wooden samples ($n = 34$) from Sovjan indicated the presence of 4 distinguishable taxa (Fig. 14.2). *Quercus* sp. represents the majority of the samples ($n = 30$), while the rest consists of *Ulmus* sp ($n = 2$), *Fraxinus* sp. ($n = 1$) and *Abies* ($n = 1$). Out of 12 different oak species present in today's flora in the region (Dida 2003), recent historical dendrochronological studies (Westphal et al., 2010) have identified at least seven of these species as suitable construction material. All oak samples from Sovjan belong to a deciduous type of oak. Therefore, the most probable taxa represented in the archaeological material are *Quercus cerris, Q. trojana, Q, petraea, Q. pubescens*, with *Q. frainetto* being the most widespread species. These oaks are classified either as belonging to the section *Cerris* (red oaks) or *Quercus* (white oaks).

Based solely on the wood anatomical characteristics, it is possible to taxonomically distinguish different oaks only to a section level (Akkemik and Yaman 2012; Merela and Čufar 2013). However, previous studies in the region have combined and used historical tree-rings sequences from different oak sections (e.g. Hughes et al. 2001; Griggs et al. 2007), and therefore in this study, we did not separate the oak tree-ring sequences according to sections. We broadly define the oak samples as *Quercus* spp.

An important observed feature in the oak samples from Sovjan is the high occurrence of tyloses in most of the sapwood rings, except the last 1–3 rings. The sapwood is the physiologically active part of the tree, forming its outer part, and in those species where it is distinguishable (as in oaks) it has a lighter colour than the physiologically inactive, inner part–the heartwood. Identifying the remains of sapwood on oaks opens the possibility for an estimate of the missing rings. However, the depositional conditions in wetland sites in some cases impair the possibility of colour distinction on wooden

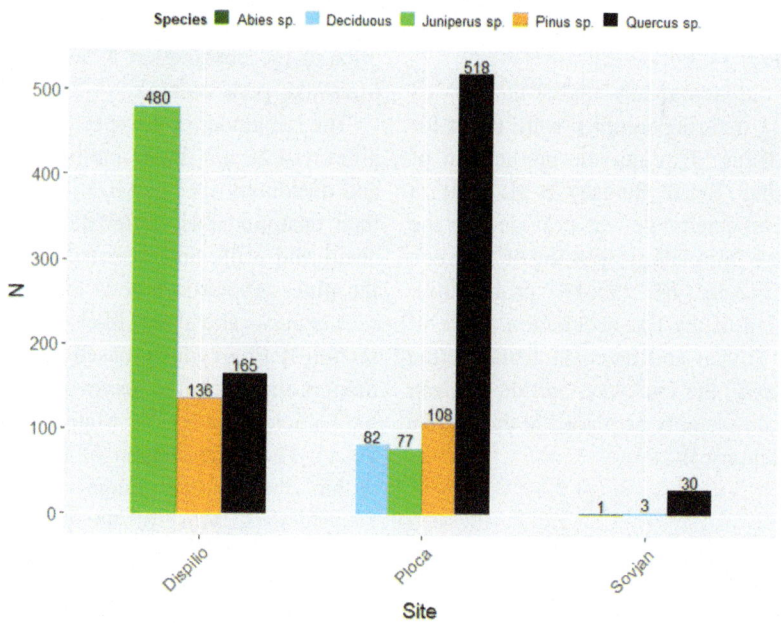

Fig. 14.2 Bar chart with the species composition of the wood samples from the three sites. Numbers on top of each bar represent the absolute specimen counts for each species. Species colours correspond to the colours in Fig. 14.5 (Authors)

material. Considering the above and taking into account the low number of pith-to-bark samples from Sovjan, sapwood estimation (e.g. Bleicher et al. 2020) was not attempted.

14.3.1.2 Chronology and Radiocarbon Dating

The samples were first averaged to sub-chronologies (mean curves) according to the cross-dating parameters (GLK and t_{HO}) and their dendrotypological characteristics. Six mean curves were constructed in this way for Sovjan, consisting of 2 to 9 samples. Four of these mean curves (17 samples in total) were averaged together into the main site chronology SOV18 spanning 269 years. After the cross-dating process, the inspection of the archaeological plan supported the construction of the initial sub-chronologies (Maczkowski et al. 2021). Most of the samples belong to the last phase of occupation of the Early Bronze Age level 8 of Sovjan, while the mean curves SOV7 and SOV12 represent earlier phases of construction or repairs (Fig. 14.3). The means SOV-3 and -4 are not included in the main chronology; however, their most probable cross-dating position, as suggested by both dendrochronology and wiggle-matching, is shown on Fig. 14.3.

In order to chronologically anchor the SOV18 chronology, 11 tree-ring samples were taken for radiocarbon dating. Through the application of wiggle-matching (Bronk Ramsey et al. 2001) a very narrow end-date range was provided for the last ring of the chronology, placing it between *2158 and 2142 cal BC* (95.4% probability). Thus, the dating of the last occupation phase of level 8 from Sovjan and its main features (the *Maison du Canal*, the trackway, and the *Maison du Pêcheur*) can securely be placed in the middle of the 22nd century BC.

14.3.2 Ploča Mičov Grad–Ohrid

14.3.2.1 Wood Species and Characteristics

From the total number of 800 wooden elements sampled in Ploča, 65% ($n = 517$) are oaks (Fig. 14.2). All of these samples belong to ring-porous species of oak. Today, in the surroundings of the site there are at least five different species of oak (*Quercus trojana, Q. frainetto, Q. cerris, Q. pubescens* and *Q. petraea*) (Matevski et al. 2011). Of these, *Q. pubescens* and *Q. frainetto* grow from the lakeside up until 900 m. a.s.l. In the mid-altitudes *Q. frainetto* forms mixed stands with *Q. cerris, Fraxinus ornus* and *Ostrya carpinifolia. Q. petraea,* together with *F. ornus,* reach the upper range limit of thermophilous trees, below the beech forests, which grow up until the treeline at around 1900 m. *Q. trojana* has azonal distribution on Galičica Mountain, forming stands on the warmest and driest slopes. Since all the oak species on Galičica belong to the sections *Cerris* (red oaks) or *Quercus* (white oaks) and most of them grow even today within a reasonable distance from the site, the species from Ploča are also broadly referred to as *Quercus* spp. (cf. *supra*). Around 10% of the sample is represented by non-oak ring- to diffuse-porous species, of which *Ostrya carpinifolia, Carpinus* sp., *Fraxinus* sp. and *Acer* sp. have been identified to date. Although protected as a National Park since 1958, it must be noted that Mt. Galičica was severely deforested prior to the designation as a protected area (V. Matevski, pers. comm).

The remainder of the species is represented by pines (14%, $n = 108$), junipers (10%, $n = 77$) and deciduous species (10%, $n = 82$). Based on their anatomical characteristics (dentate ray tracheid end-walls and fenestriform cross-pitting), the pine samples are determined as *Pinus* subsect. *Pinus*. The more likely species from this section is *Pinus nigra*, based on its ecology and modern-day vicinity; however, it is not present on Galičica today (V. Matevski, pers. comm. 2021). The wider region of Lake Ohrid is also within the today's range of *P. mugo* and *P. sylvestris*. Most of the pine samples from Ploča exhibit regular, concentric secondary growth, and no significant scarring has been observed (e.g. fire scars).

After the pines, the other prominent conifers are the juniper woods. Considering the great similarity of the anatomical characteristics of

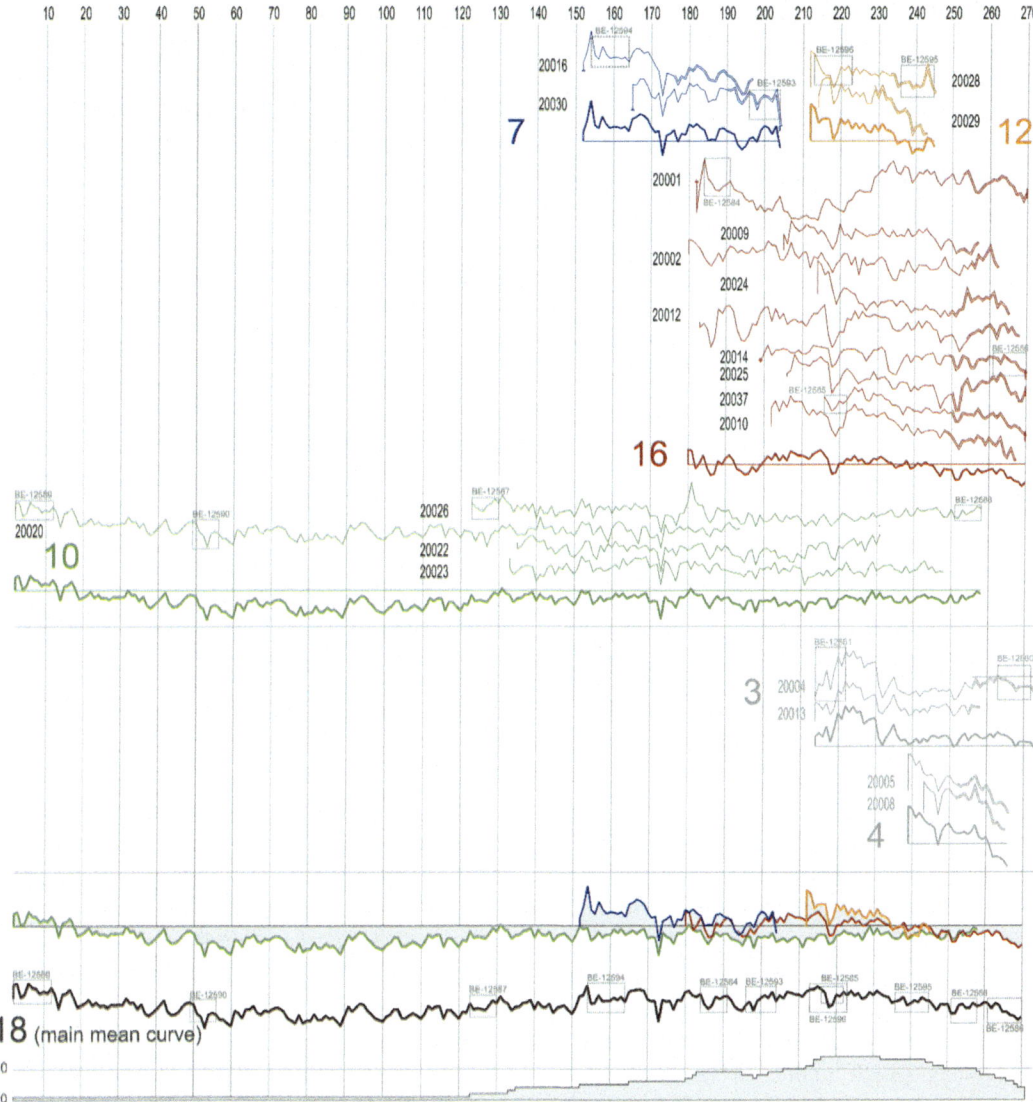

Fig. 14.3 All the cross-dated raw tree-ring sequences from Sovjan. Different curve colours correspond to each sub-chronology (mean curve) grouped as described in the text. Thinner curves with smaller-font 5-digit numbers represent individual samples, thick curves with 1–2-digit numbers correspond to the averaged mean curve of the samples group above it. Small dotted-line rectangles along the curves represent the positions of radiocarbon samples and corresponding labels. Groups 3 and 4 were not included in the main chronology. Straight vertical lines represent relative years decades (Authors)

different juniper species (Akkemik and Yaman 2012; Crivellaro and Schweingrube 2013), it is impossible to distinguish them based on those parameters. Today, on the slopes of Galičica, three tree-like species of junipers can be found (*J. excelsa*, *J. foetidissima* and *J. oxycedrus*) in addition to the shrub-like common juniper (*J. communis*). Based on the diameter of the samples' stems and the modern ecology of these junipers, the choice of species could be narrowed down to *J. excelsa* and/or *J. foetidissima* (therefore *Juniperus* spp.).

14.3.2.2 Chronology and Radiocarbon Dating

A total of 665 samples have been processed dendrochronologically from Ploča. All reliable *Quercus* sp., *Pinus* sp. and *Juniperus* sp. have been measured. Additionally, *Fraxinus* sp. and *Ostrya carpinifolia* samples have also been measured. Tree-ring sequences of 261 samples could be cross-dated and averaged into several working chronologies at the current state. Three of the main oak site chronologies are briefly presented here as examples. Details of the other chronologies will be provided in site-dedicated publications.

The main the oak chronologies (MKO100, MKO102 and MKO 103) span more than 200 years with replication of 92, 12 and 78 samples, respectively. These major chronologies are the backbone for further dating work on prehistoric oaks. The felling dates of the well-replicated chronology MKO100 concentrate in its second half, together with samples that have some preserved sapwood, which suggests that most felling dates would fall in the last 70 years of the chronology. It is important to note, however, that the first 68 and the last 35 years of the chronology are represented by only one sample. Further excavations and analyses of the already measured tree-ring sequences should address the insufficient replication in parts of this chronology.

On the other hand, the first half of the MKO103 chronology is very densely replicated containing a high number of samples with preserved waney edge, making the chronology very suitable for further dendroarchaeological analysis. There is a notable drop in replication around the relative year—1380, separating the felling dates grouping in the older part of the chronology from the felling dates grouping in the younger half by more than 70 years (Fig. 14.4).

In addition to the oak chronologies, four conifer TRW chronologies were constructed (Fig. 14.5). In total 49 pine samples (*Pinus* spp.) were cross-dated into three main chronologies: MKO104, MKO105 and MKO107, with a replication of 7, 35 and 6 samples. The measuring and correlation process demanded a special focus on missing or wedging rings. The high occurrence of intra-annual density fluctuations (IADFs) in the *Juniperus* genus (e.g. Esper 2000; Sass-Klaassen et al. 2008) also posed a problem in our study. In spite of this, of the 68 measured ones, 40 juniper tree-ring sequences were cross-dated and averaged into one main juniper chronology (MKO101) spanning 273 years.

Since the measuring efforts were in the first place concentrated on the dendrochronologically more proven species like oaks and pines, only a limited number of other deciduous samples have been measured. Nevertheless, it was possible to establish an *Ostrya carpinifolia* (European hop-hornbeam) chronology (MKO106), consisting of 7 samples and a length of 71 years (Fig. 14.5).

A series of samples for radiocarbon measurement were taken from the cross-dated woods to establish a temporal anchor for the chronologies. The wiggle-matching of the resultant radiocarbon dates points to a calibrated age range end-date of *4362–4344 cal BC* (95.4 % probability) for MKO100 and *1277–1252 cal BC* (95.4 %) for MKO103 (Table 14.1; Fig. 14.4). Thus, the periods covered by the most replicated chronologies are from mid-48th to mid-44th century cal BC for MKO100 and from the 16th to mid-13th century cal BC for MKO103. The last ring of the third longest oak chronology MKO102, with a replication of 13 tree-ring sequences, is calibrated to the *1819–1749 cal BC* (95.4 %). Additionally, the three shorter oak chronologies consisting of a couple of samples each, do not cross-date reliably to the existing robust chronologies, although wiggle-matched dates indicate some or complete temporal overlap (cf. Figure 14.5).

At an earlier stage of the correlation work, first results of wiggle-matching indicated that chronologies of different species are contemporaneous. The growing number of correlated samples made it possible to cross-date these chronologies. For the 5th millennium BC, the main chronologies of oak, pine and juniper could be correlated dendrochronologically. While from the 2nd millennium BC (Late Bronze Age) oak, pine and hop-hornbeam chronologies were correlated (Table 14.2). For the statistical tests, in some cases, the chronologies were truncated if the replication in the overlapping parts was low

Fig. 14.4 Bar charts of the chronologies MKO100 and MKO103 from Ploca Mičov Grad. Each bar represents one wood sample. The black part of the bars corresponds to heartwood, orange to sapwood. Small black rectangles present at the end of some bars indicate the presence of the last growth rings. Green rectangles on the black lines at the bottom of the two graphs represent the positions of tree-rings sampled for radiocarbon dating. In insets in the upper right corner of each graph represent the highest posterior density probability distribution for the calibrated date of the last growth ring, according to the wiggle-matching model implemented in OxCal v.4.4.3 (Bronk Ramsey et al., 2001; atmospheric data from Reimer et al., 2020). The time scale is in relative years BC based on the output of the OxCal model. *Note* the year scale is not continuous, 2900 relative calendric years separate the two plots (Authors)

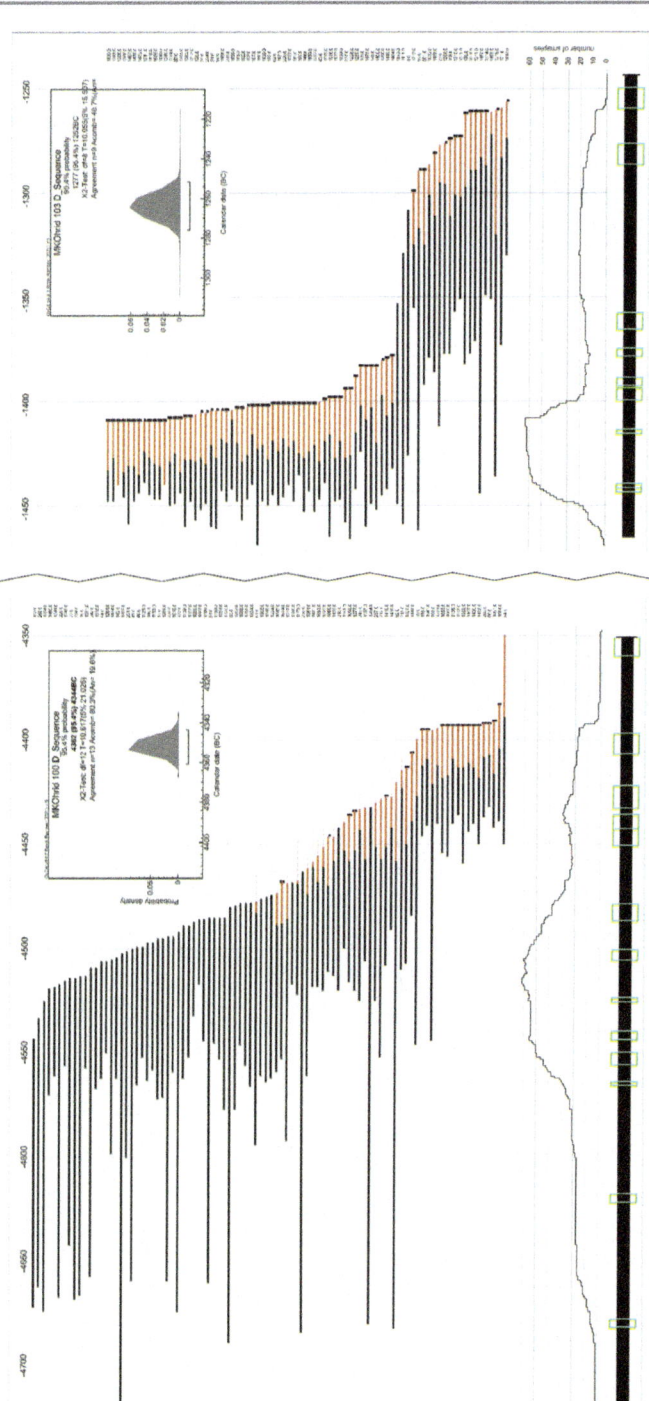

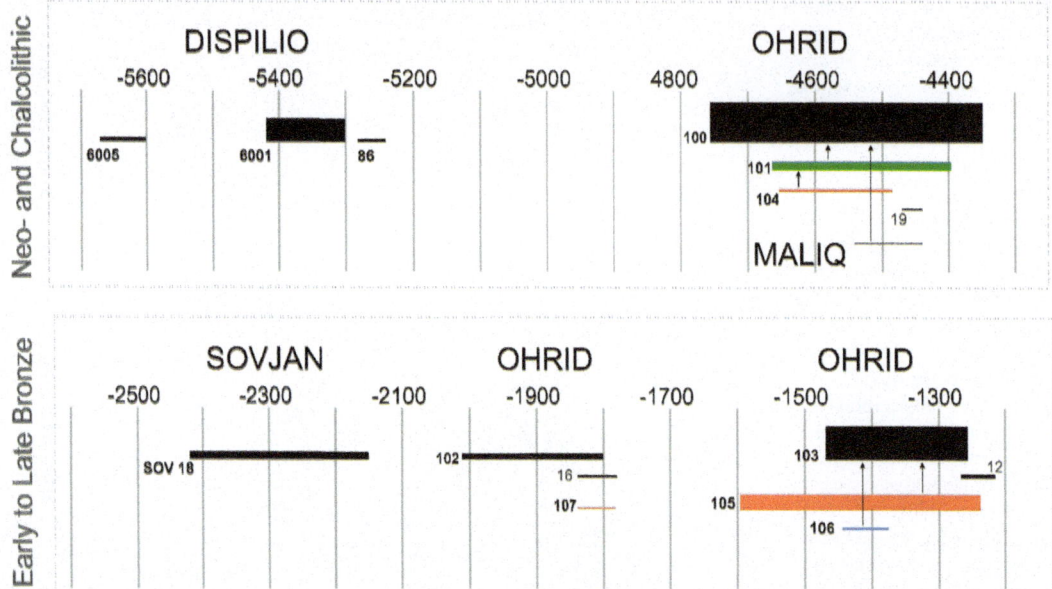

Fig. 14.5 Conceptual graphical representation of the temporal span of the tree-ring width chronologies, positioned according to radiocarbon dating. Coloured horizontal bars represent the chronologies; bar thickness corresponds to the relative replication of a chronology; length corresponds to the chronological extent of each chronology or sub-chronology; different colours represent different species (see Fig. 14.2). Numbers beside each bar are the codes of the chronologies, as discussed in the text. Black vertical arrows indicate dendrochronological cross-dating between the chronologies. Time scale in relative years BC (Authors)

Table 14.1 Radiocarbon wiggle-matching (Bronk Ramsey et al. 2001) of the EXPLO tree-ring chronologies from the archaeological sites of Sovjan, Ploča Mičov Grad and Dispilio

Site/Chronology	Last ring modelled end-date 68.3% probability	Last ring modelled end-date 95.4% probability	No. of samples	No. of 14C dates
Sovjan				
SOV18	2154–2146 BC cal	2158–2142 BC cal	17	11
Ploca Micov Grad				
MKO100	4357–4348 BC cal	4362–4344 BC cal	92	13
MKO102	1817–1754 BC cal	1821–1750 BC cal	13	6
MKO103	1271–1258 BC cal	1277–1252 BC cal	78	9
Dispilio				
DISP6005	5606–5599 BC cal	5611–5595 BC cal	8	6
DISP6001	5309–5300 BC cal	5316–5296 BC cal	57	11
DISP6003	5273–5246 BC cal	5293–5205 BC cal	2	3
DISP86	5212–5201 BC cal	5278–5199 BC cal	6	3

'No. of samples' refers to the replication of each chronology, i.e. number of cross-dated individual wood samples (Authors)

Table 14.2 Dendrochronological cross-dating matrix between tree-ring chronologies of different species from Ploča Mičov Grad

Chronology (MKO)	Reference chronology	t_{HO}	GLK (%)	Overlap	Notes
101 (*Juniperus* sp.)	100 (*Quercus* sp.)	8.5	71	272	
104 (*Pinus* sp.)	101 (*Juniperus* sp.)	7.7	74	143	Last 50 rings ignored (low replication)
104 (*Pinus* sp.)	100 (*Quercus* sp.)	5.4	61	163	Last 50 rings ignored (low replication)
105 (*Pinus* sp.)	103 (*Quercus* sp.)	7.4	70	141	Years 200–360. First rings of reference 103 ignored
106 (*Ostrya carpinifolia*)	103 (*Quercus* sp.)	6.9	80	70	

'tHO' represents t-values after Hollstein (1980); 'GLK' is the Gleichläufigkeit or coefficient of parallel variation (Eckstein and Bauch 1969); 'Overlap' refers to the overlapping number of years between chronologies. Values presented here refer to overlaps of chronologies where they have sufficient replication (> 5), i.e. chronologies were truncated for these comparisons in areas where replication was low
'No. of samples' refers to the replication of each chronology, i.e. number of cross-dated individual wood samples (Authors)

or affected by irregular growth patterns. All these dendrochronological heteroconnections were confirmed by radiocarbon dating and wiggle-matching besides the visual inspection. The work on the wooden material from Ploča is still ongoing and more detailed overview of the results will be presented in the near future.

Additionally, in collaboration with the French School at Athens (cf. Oberweiler et al. 2020), two oak and one pine sample from the well-known prehistoric site of Maliq, Albania (Prendi 2018), were obtained. The two oak samples from Maliq were averaged together to form a mean, which in turn could be cross-dated to the 5th millennium BC oak chronology from Ploča—MKO100 (Fig. 14.5), thus confirming the possibility of dendrochronological teleconnections in the region.

14.3.3 Dispilio

14.3.3.1 Wood Species and Characteristics

A total of 781 wood samples were taken from the excavated wood posts from the site of Dispilio. The species composition is dominated by juniper (*Juniperus* L., 61%, $n = 480$, cf. Figure 14.2). Several juniper species are present in the wider region today (*J. communis,*

J. excelsa, J. foetidissima, J. oxycedrus). Considering the average diameter of round juniper samples (≥ 12 cm) and the more shrub-like nature of *J. communis*, the latter species could be tentatively excluded from the list of possible species represented in the samples. The differentiation of *Juniperus* species based solely on wood anatomical characteristics is not possible (Schweingruber 1990; Akkemik and Yaman 2012). However, the observation of distinct macroscopic features, such as sap-heartwood contrast and transition, amount of IADFs occurrence (Campello et al. 2007; Akkemik and Yaman 2012) and tendency to secondary growth in protruding lobes ('strip-bark growth', e.g. Esper 2000) hint to a possible presence of at least two species. But even such marked macroscopic differences of the stem wood do not allow the exclusion of the phenotypical plasticity of the genus as a possible reason for these differences.

The second most abundant genus represented in Dispilio woods assemblage is the pines. Based on the observed anatomical characteristics (dentate ray tracheid end-walls in radial sections and large fenestriform pits), the pine woods are defined as *Pinus* subsect. *Pinus* (cf. *P. nigra, P. sylvestris, P. mugo*).

Oak samples (*Quercus* spp., deciduous) account for 21% ($n = 165$) of the sampled woods.

In the surroundings of Dispilio, the present-day environmental conditions and oak species are very similar to the other two sites described above (cf. *supra*, 14.3.2.1), therefore the presence of species from both taxonomical oak sections, *Quercus* and *Cerris*, may be expected.

14.3.3.2 Chronologies and Radiocarbon Dating

Measurement of the tree-rings sequences of the oak samples was again the priority at this site. Therefore, a total of 165 oak wood samples were measured, whereas the measuring of the junipers and pines is still in progress. With the secure cross-dating of 73 samples (A-dated, according to Francuz (1980)) four mean curves were constructed, with pending cross-dating of another 14 B-dated samples. Three of these mean curves (Table 14.1; Fig. 14.5) have a relatively low replication, while the chronology DISP6001 has a replication of 57 samples. Sapwood is preserved on many samples, but only a few have the waney edge retained. Working traces on the surface of the wood were not observed on most samples and their outer surface was frequently visibly eroded. Therefore, the lack of waney edge on the oaks can be attributed to poor preservation conditions. Given the hiatus of a couple of years between the excavation of the pile field trench and the dendrochronological sampling campaign, some of the wood degradation can be attributed to open air drying and erosion.

From each Dispilio (sub-) chronology, several tree-ring samples were taken for radiocarbon dating. The wiggle-matching of the radiocarbon dates point to the second half of the 6th millennium cal BC for most of the cross-dated samples. However, one of the shorter mean curves DISP6005 has calibrated end-date range in the period of *5612–5594 cal BC* (95.4 % probability). The calibrated end-date range of the main Dispilio chronology DISP6001 dates to an earlier period, in the interval of *5322–5293 cal BC* (95.4 %). Only five samples from this chronology have a preserved waney edge, making it difficult to assess the different felling phases or settlement duration. The last measured rings of the mean curves DISP6003 and DISP86 point to

the ranges *5291–5209 cal BC* (95.4 %) and *5278–5200 cal BC* (95.4 %), respectively. The two mean curves do cross-date visually, but because of the low ring number and different growth patterns, the cross-dating remains tentative.

Furthermore, two horizontal wood samples found during the excavation were also sampled for radiocarbon dating since their dendrochronological cross-dating failed (not shown in Fig. 14.5). Considering the calibrated radiocarbon results *(5727–5661 cal BC* (95.4 % probability) and *5471–5404 cal BC* (95.4 %)), it is important to note that both samples do not have any preserved sapwood. In the case of the latter one, around 100 more rings were not measurable and were not radiocarbon dated. Even if these horizontal pieces could belong to any of the above-mentioned phases, it is impossible to determine their actual relation to the other vertical woods or archaeological layers.

14.4 Discussion

The dendrochronological results presented in this study provide the first high-resolution prehistoric dating for the southwestern Balkans. The data presented here is not limited to the sites covered in this paper. For instance, the tree-ring chronologies can be used for independent dating of wooden objects from other sites or for annual calibration of other environmental proxies, among other applications.

In the past few years, the high potential for dendrochronological teleconnections across the Balkans has been demonstrated by several investigations (e.g. Čufar et al. 2015; Roibu et al. 2021). We were able to confirm this also for the prehistoric periods in the southwestern Balkans, by cross-dating the Chalcolithic period TRW sequences from the sites of Maliq (Korçë Plain) and Ploča (Lake Ohrid). This validates the robustness and quality of our chronologies. Therefore, a dendrochronological connection is to be expected with historical or absolutely dated chronologies from the pile-dwelling sites in Central Europe or the Alpine region as sample replication increases, new sites are investigated, and more subfossil trunks are recovered.

The noncontemporaneity of the chronologies from the three sites discussed in this study impedes the cross-dating between them. Nevertheless, these chronologies are the first radiocarbon-anchored multi-centennial prehistoric tree-ring chronologies from the region, thus enabling a within-site annually resolved dating of different features. Moreover, they are a firm starting point for the diachronic study of woodland ecology, use and management in prehistory.

Through the combination of dendrochronological correlations and wiggle-matching radiocarbon dates, we were able to date level 8 of Sovjan with a very high chronological resolution to the middle of the 22nd century cal BC (*2158–2142 cal BC* (95.4 % probability)). The most important features of level 8 are represented by a very well-preserved 15-m-long apsidal house (*Maison du Canal*), part of a trackway, and a corner of another dwelling (*Maison du Pêcheur*) (Gori 2015). Within the *Maison du Canal*, at least three construction/repair events could be identified (Fig. 14.3). The construction of the two dwellings and the trackway in the same felling event is a rare evidence of a possible communal effort in settlement construction in the Early Bronze Age. The 269-year-long chronology SOV18 did not show any reliable cross-dating positions with the only other possibly contemporaneous oak chronology from the wider region, Orašje 3 in Bosnia (cf. Pearson et al. 2014). Such an outcome is not unexpected at an early stage of the prehistoric dendrochronological research in the Balkans and may be influenced by low replication (almost a half of SOV18 is represented by one sample), the large distance between the sites (>500 km), but also the uncertain provenance of the subfossil river oaks from Orašje.

For Ploča Mičov Grad (Ohrid), it was possible to build robust mean curves for different species. Heteroconnection between the oak, juniper, pine and hop-hornbeam chronologies, in combination with wiggle-matching of radiocarbon dates, led to a very well-replicated and unique dataset from a pile-dwelling site in the southwestern Balkans. The results show that the pile field, with a density of almost 7 piles per m^2 (Hafner et al. 2021), is the result of several settlement phases between the middle of the 5th and the end of the 2nd millennium BC. From the currently cross-dated samples with a preserved waney edge from the middle of the 5th millennium, a felling dates span of around 90 years can be defined. Further work on sapwood reconstruction and evaluation of the archaeological material should clarify the chronological span of the settlement in the 5th millennium and provide more detailed data for regional comparisons with other important sites (e.g. Bulatović and Vander Linden 2017; Boyadzhiev et al. 2021). In the second half of the 2nd millennium BC, the current state of data suggests two settlement phases with a gap of around 70 years.

Our results from the site of Ploča Micov Grad conform well with published Holocene palynological data by Wagner et al. (2009) and the tree-species composition of the sampled material is in line with it. A notable exception is the absence of *Abies* sp. wood samples, despite the presence of *Abies* sp. macro-remains on-site in Ploča (Naumov et al. 2019). According to Wagner et al. (2009), *Abies* sp. was abundant in the Chalcolithic period around the lake. More detailed insights from pollen into the arboreal composition of the surroundings of Lake Ohrid is anticipated in the near future (Wagner et al. 2014; Chap. 15 in this book).

In Dispilio, one main chronology could be constructed, with additional shorter (sub-) chronologies. According to the archaeological evidence, the occupation of the site of Dispilio started somewhere in the Middle Neolithic (c. 5800–5300 BC) (Facorellis et al. 2014; Karkanas et al. 2011). Our results also cover most of this time range, with one felling phase around 57–56th century cal BC, and another group of felling events around 54th–53rd century cal BC. The end-date of our main chronology DISP6001 (*5322–5293 cal BC* (95.4 %)) falls within the transitional period (Kotsakis 2007; Facorellis et al. 2014; Maniatis and Pappa 2020) from Middle to Late Neolithic. One of the most prominent Neolithic finds from the Balkans, the so-called Dispilio tablet featuring script-like characters incised on its surface, has previously been radiocarbon-dated to 5324–5079 cal BC (Facorellis et al. 2014). This date suggests the approximately same period as the end-date range of DISP6001, and it is plausible that it belongs to the same occupation event.

At this stage, it is not possible to discern with an annual resolution the felling dates within the Dispilio chronologies. This is owed to the low incidence of oak samples with waney edge in the Dispilio assemblage. The possible reasons for this outcome are the great calendar age of the samples, the depositional environment and the fluctuating water table of Lake Kastoria. Additionally, some of the preservation issues are due to the initial post-excavation strategy which did not account for protecting the wooden remains in situ. However, the measuring and analyses of the juniper and pine samples from the site are still ongoing. Junipers are known for their very hard and resistant wood and owing to this many of the samples have an intact last growth ring. As the heteroconnections from Ploča show, the cross-dating of different species is possible and reliable. Such cross-dating among different species will hopefully elucidate annually resolved phases of Dispilio.

The chronologies of Sovjan, Ploča and Dispilio provide an important contribution to the construction of a regional chronological framework for dating archaeological wood samples. With increasing replication and additional investigations of other sites in the region, a connection with historical tree-ring chronologies or with absolutely dated chronologies from the pile-dwelling sites in the Alpine region can be expected. Furthermore, the parallel evaluation with other lines of research will yield insights into the intra-site, relative chronological processes with a yearly precision.

Acknowledgements The dendrochronological analysis was performed at the Institute of Archaeological Sciences of the University of Bern. This project has received funding from the European Research Council (ERC) under the European Union's Horizon 2020 research and innovation programme (grant agreement No. 810586). The Franco-Albanian Korçë Basin Archaeological Expedition is a joint project of the Archaeological Institute of Tirana, the University of Paris I, CNRS UMR 7041 laboratory and the French School at Athens. It was supported by the French Ministry of Foreign Affairs. The excavations on Lake Ohrid were conducted in collaboration with the NI Museum, Ohrid, and the Centre for Prehistoric Research, Skopje. We would like to thank all the students from the Universities of Thessaloniki and Bern that were involved in the fieldwork and wood sample curation, for their great commitment and support. We also thank Charlotte Pearson for kindly providing the Orašje 3 chronology and Ünal Akkemik for his advice on wood anatomy. For the valuable information on the modern arboreal flora on Galičica, we would like to thank Vlado Matevski.

References

Akkemik Ü, Yaman B (2012) Wood anatomy of eastern mediterranean species. Kessel Publishing House, Remagen-Oberwinter

Becker B, Billamboz A, Egger H, Gassman P, Orcel A, Orcel C, Ruoff U (1985) Dendrochronologie in der Ur- und Frühgeschichte: die absolute Datierung von Pfahlbausiedlungen nördlich der Alpen im Jahrringkalender Mitteleuropas. Antiqua, vol 11. Verl. Schweizerische Gesellschaft für Ur- und Frühgeschichte, Basel

Billamboz A (2008) Dealing with heteroconnections and short tree-ring series at different levels of dating in the dendrochronology of the Southwest German pile-dwellings. Dendrochronologia 26:145–155. https://doi.org/10.1016/j.dendro.2008.07.001

Billamboz A (2014) Regional patterns of settlement and woodland developments: dendroarchaeology in the Neolithic pile-dwellings on lake constance (Germany). Holocene 24:1278–1287. https://doi.org/10.1177/0959683614540956

Bleicher N, Walder F, Gut U, Bolliger M (2020) The Zurich method for sapwood estimation. Dendrochronologia 64:125776. https://doi.org/10.1016/j.dendro.2020.125776

Boyadzhiev Y, Boyadzhiev K, Brandtstätter L, Krauß R (2021) Chronological modelling of the Chalcolithic settlement layers at Tell Yunatsite, Southern Bulgaria. Doc Praehistorica XLVIII:252–275. https://doi.org/10.4312/dp.48.5

Bronk Ramsey C, van der Plicht J, Weninger B (2001) 'Wiggle matching' radiocarbon dates. Radiocarbon 43:381–389. https://doi.org/10.1017/S0033822200038248

Bulatović A, Vander Linden M (2017) Absolute dating of copper and early bronze age levels at the eponymous archaeological site Bubanj (Southeastern Serbia). Radiocarbon 59:1047–1065. https://doi.org/10.1017/RDC.2017.28

Campelo F, Nabais C, Freitas H, Gutiérrez E (2007) Climatic significance of tree-ring width and intra-annual density fluctuations in Pinus pinea from a dry mediterranean area in Portugal. Ann for Sci 64:229–238. https://doi.org/10.1051/forest:2006107

Crivellaro A, Schweingruber FH (2013) Atlas of wood, bark and pith anatomy of eastern mediterranean trees and shrubs: with a special focus on cyprus. Springer, Berlin. https://doi.org/10.1007/978-3-642-37235-3

Čufar K, Kromer B, Tolar T, Velušček A (2010) Dating of 4th millennium BC pile-dwellings on Ljubljansko barje, Slovenia. J Archaeol Sci 37:2031–2039. https://doi.org/10.1016/j.jas.2010.03.008

Čufar K, Tegel W, Merela M, Kromer B, Velušček A (2015) Eneolithic pile dwellings south of the Alps precisely dated with tree-ring chronologies from the north. Dendrochronologia 35:91–98. https://doi.org/10.1016/j.dendro.2015.07.005

Dida M (2003) State of forest tree genetic resources in Albania (No. FGR/62E), Forest Genetic Resources Working Papers

Douglass AE (1929) Secret of the Southwest solved by Talkative tree rings. Natl Geogr Mag 737–770. https://ltrr.arizona.edu/sites/ltrr.arizona.edu/files/bibliodocs/Douglass%2C%20AE_Secret%20of%20the%20Southwest%20Solved%20by%20Talkative%20Tree%20Rings_1929.pdf

Eckstein D, Bauch J (1969) Beitrag zur Rationalisierung eines dendrochronologischen Verfahrens und zur Analyse seiner Aussagesicherheit. Forstwissenschaftliches Cent 88:230–250. https://doi.org/10.1007/BF02741777

Esper J (2000) Long-term tree-ring variations in Juniperus at the upper timber-line in the Karakorum (Pakistan). Holocene 10:253–260. https://doi.org/10.1191/095968300670152685

Facorellis Y, Sofronidou M, Hourmouziadis G (2014) Radiocarbon dating of the neolithic lakeside settlement of Dispilio, Kastoria, Northern Greece. Radiocarbon 56:511–528. https://doi.org/10.2458/56.17456

Fouache E, Desruelles S, Magny M, Bordon A, Oberweiler C, Coussot C, Touchais G, Lera P, Lézine AM, Fadin L, Roger R (2010) Palaeogeographical reconstructions of Lake Maliq (Korça Basin, Albania) between 14,000 BP and 2000 BP. J Archaeol Sci 37:525–535. https://doi.org/10.1016/j.jas.2009.10.017

Francuz J (1980) Dendrochronologie. In: Furger AR (ed) Die Siedlungsreste Der Horgener Kultur. Die Neolithischen Ufersiedlungen von Twann. Archäologischer Dienst des Kantons Bern, Bern, pp 197–210

Friedrich M, Remmele S, Kromer B, Hofmann J, Spurk M, Kaiser KF, Orcel C, Küppers M (2004) The 12,460-Year Hohenheim Oak and pine tree-ring chronology from central Europe—A unique annual record for radiocarbon calibration and paleoenvironment reconstructions. Radiocarbon 46:1111–1122. https://doi.org/10.1017/S003382220003304X

Gori M, Krapf T (2015) The bronze and iron age pottery from Sovjan. Iliria 39:91–135. https://doi.org/10.3406/iliri.2015.2500

Gori M (2015) Along the Rivers and Through the Mountains: A revised chrono-cultural framework for the south-western Balkans during the late 3rd and early 2nd millennium BCE. Universitätsforschungen zur Prähistorischen Archäologie, vol 268. Dr. Rudolf Habelt GmbH, Bonn

Griggs C, Degaetano A, Newton M (2007) A regional high-frequency reconstruction of May—June precipitation in the north Aegean from oak tree rings, A. D. 1089–1989. Int J Climatol 27:1075–1089. https://doi.org/10.1002/joc.1459

Hafner A, Hinz M, Mazurkevich AN, Dolbunova EV, Pranckenaite E (2020) Introduction: Neolithic and Bronze Age pile dwellings in Europe. An outstanding archaeological resource with a long research tradition and broad perspectives. In: Hafner A, Dolbunova EV, Pranckenaite E, Mazurkevich AN, Hinz M (eds) Settling waterscapes in Europe: the archaeology of neolithic and bronze age pile-dwellings. Propylaeum, Heidelberg, pp 1–6

Hafner A, Reich J, Ballmer A, Bolliger M, Emmenegger L, Fandré J, Francuz J, Hostettler M, Lotter A, Naumov G, Nedeljkovic D, Stäheli C, Szidat S, Todorova V, Wieser A, Bogaard A (2021) First absolute chronologies of neolithic and bronze age settlements at lake ohrid based on dendrochronology and radiocarbon dating. J. Archaeol. Sci. Reports 38:1–30. https://doi.org/10.1016/j.jasrep.2021.103107

Haneca K, Čufar K, Beeckman H (2009) Oaks, tree-rings and wooden cultural heritage: a review of the main characteristics and applications of oak dendrochronology in Europe. J Archaeol Sci 36:1–11. https://doi.org/10.1016/j.jas.2008.07.005

Hollstein E (1980) Mitteleuropäische Eichenchronologie: trierer dendrochronologische Forschungen zur Archäologie und Kunstgeschichte. von Zabern, Mainz am Rhein

Huber B (1963) Jahrringchronologische Synchronisierung der Jungsteinzeitlichen Siedlungen Thayngen-Weier und Burgäschisee-Süd und -Südwest. Germania 41:1–9

Hughes MK, Kuniholm PI, Eischeid JK, Garfin G, Griggs CB, Latini C (2001) Aegean tree-ring signature years explained. Tree-Ring Res 57:67–73. http://hdl.handle.net/10150/262557

Karkanas P, Pavlopoulos K, Kouli K, Ntinou M, Tsartsidou G, Facorellis Y, Tsourou T (2011) Palaeoenvironments and Site Formation Processes at the Neolithic Lakeside Settlement of Dispilio, Kastoria. Northern Greece 26:83–117. https://doi.org/10.1002/gea.20338

Kotsakis K (2007) Prehistoric macedonia. In: Koliopoulos I (ed) The history of macedonia. Museum of the Macedonian Struggle Foundation, Thessaloniki, pp 1–15

Kuniholm PI, Striker CL (1983) Dendrochronological investigations in the Aegean and neighboring regions, 1977–1982. J f Archaeol 10:411–420. https://doi.org/10.1179/009346983791504165

Kuzman P (2009) Pile house settlements at the Ohrid Lake. In: Kuzman P, Tričkovska J, Pavlov Z (eds) Ohrid—World heritage site. Ministry of Culture of the Republic of Macedonia, Cultural Heritage Protection Office, Skopje, pp 16–29

Lera P, Touchais G (2003) Sovjan (Albanie). Bull Corresp Hellén 127:578–609. https://doi.org/10.3406/bch.2003.7337

Libby WF (1952) Radiocarbon dating. The University of Chicago Committee on Publications in the Physical Sciences. University of Chicago Press, Chicago

Maczkowski A, Bolliger M, Ballmer A, Gori M, Lera P, Oberweiler C, Szidat S, Touchais G, Hafner A (2021) The Early Bronze Age dendrochronology of Sovjan (Albania): A first tree-ring sequence of the 24th–22nd c. BC for the Southwestern Balkans. Dendrochronologia 66:125811. https://doi.org/10.1016/j.dendro.2021.125811

Maniatis Y, Pappa M (2020) Radiocarbon dating of the neolithic settlement at Makriyalos, Pieria, North Greece. Radiocarbon 62:467–483. https://doi.org/10.1017/RDC.2020.12

Matevski V, Čarni A, Avramoski O, Juvan N, Kostadinovski M, Košir P, Marinšek A, Paušič A, Šilc U (2011) Forest vegetation of the Galičica mountain range in Macedonia. Шумската вегетација на планината Галичица во Македонија. Gozdna vegetacija gorovja Galičica v Makedoniji. Biološki institut Jovana Hadžija ZRC SAZU, Makedonska akademija na naukite i umetnostite, Ljubljana. https://doi.org/10.3986/9789610502906

Menotti F (ed) (2004) Living on the lake in prehistoric Europe. 150 years of lake-dwelling research. Routledge, London

Merela M, Čufar K (2013) Gustoća i mehanička svojstva drva bjeljike hrasta u usporedbi s drvom srži. Drv Ind 64:323–334. https://doi.org/10.5552/drind.2013.1325

Montelius O (1903) Die typologische methode. Im Selbstverlage des Verfassers, Stockholm

Naumov G, Hafner A, Taneski B, Ballmer A, Reich J, Hostettler M, Bolliger M, Francuz J, Machkovski A, Bogaard A, Antolin F, Charles M, Tinner W, Morales Del Molino C, Lotter A (2019) Istraživanje na lokalitetot Ploča-Mićov Grad kaj Gradište (Ohridsko Ezero) vo 2019 godina (Research in 2019 at the site of Ploča-Mićov Grad near Gradište (Lake Ohrid)) (in Macedonian). Patrimonium.mk. Periodical for cultural Heritage—Monuments, Restoration, Museums 12 (17):11–46https://doi.org/10.7892/boris.140479

Oberweiler C, Lera P, Kurti R, Touchais G, Aslaksen OC, Blein C, Elezi G, Gori M, Krapf T, Maniatis Y, Wagner S (2020) Mission archéologique franco-albanaise du bassin de Korçë. Bull Archéologique Écoles Françaises à L'étranger. https://doi.org/10.4000/baefe.1660

Pearson CL, Ważny T, Kuniholm PI, Botić K, Durman A, Seufer K (2014) Potential for a new multimillennial tree-ring chronology from subfossil Balkan river Oaks. Radiocarbon 56:S51–S59. https://doi.org/10.2458/azu_rc.56.18342

Prendi F (2018) Vendbanimi Prehistorik i Maliqit. The prehistoric settlment of maliq. Academy of Albanological Studies—Institute of Archaeology. Botimet M&B, Tirana

Reich J, Steiner P, Ballmer A, Emmenegger L, Hostettler M, Stäheli C, Naumov G, Taneski B, Todoroska V, Schindler K, Hafner A (2021) A novel structure from motion-based approach to underwater pile field documentation. J Archaeol Sci Rep 39:103120. https://doi.org/10.1016/j.jasrep.2021.103120

Reimer PJ, Austin WEN, Bard E, Bayliss A, Blackwell PG, Bronk Ramsey C, Butzin M, Cheng H, Edwards RL, Friedrich M, Grootes PM, Guilderson TP, Hajdas I, Heaton TJ, Hogg AG, Hughen KA, Kromer B, Manning SW, Muscheler R, Palmer JG, Pearson C, van der Plicht J, Reimer RW, Richards DA, Scott EM, Southon JR, Turney CSM, Wacker L, Adolphi F, Büntgen U, Capano M, Fahrni SM, Fogtmann-Schulz A, Friedrich R, Köhler P, Kudsk S, Miyake F, Olsen J, Reinig F, Sakamoto M, Sookdeo A, Talamo S (2020) The IntCal20 Northern Hemisphere radiocarbon age calibration curve (0–55 cal kBP). Radiocarbon 62:725–757. https://doi.org/10.1017/rdc.2020.41

Renfrew C (1973) Before civilization: the radiocarbon revolution and prehistoric Europe. Jonathan Cape Ltd., London

Roibu CC, Ważny T, Crivellaro A, Mursa A, Chiriloaei F, Ştirbu MI, Popa I (2021) The Suceava oak chronology: a new 804 years long tree-ring chronology bridging the gap between central and south Europe. Dendrochronologia 68:125856. https://doi.org/10.1016/j.dendro.2021.125856

Ruoff U (1979) Neue dendrochronologische daten aus der Ostschweiz. Z für Schweiz Archäologie Kunstgeschichte 36:94–96

Sass-Klaassen U, Keuschner H, Buerkert A, Helle G (2008) Tree-ring analysis of Juniperus excelsa from the northern Oman mountains. In: TRACE. Tree Rings in Archaeology, Climatology and Ecology. Proceedings of the Dendrosymposium 2007. May 3rd - 6th 2007. Riga, Latvia, pp 99–108

Saratsi E (2003) Landscape history and traditional management practices in the Pindos Mountains, Northwest Greeece, c. 1850–2000. University of Nottingham, Nottingham

Schweingruber FH (1988) tree rings: basics and applications of dendrochronology. Kluwer Academic Publishers, Dordrecht. https://doi.org/10.1007/978-94-009-1273-1

Schweingruber FH (1990) Mikroskopische Holzanatomie. Formenspektren mitteleuropäischer Stamm- und Zweighölzer zur Bestimmung von rezentem und subfossilem Material. Anatomie microscopique du bois. Microscopic wood anatomy, 3rd ed. Eidgenössische Forschungsanstalt für Wald, Schnee und Landschaft, Birmensdorf

Wagner B, Lotter AF, Nowaczyk N, Reed JM, Schwalb A, Sulpizio R, Valsecchi V, Wessels M, Zanchetta G (2009) A 40,000-year record of environmental change from ancient Lake Ohrid (Albania and Macedonia). J Paleolimnol 41:407–430. https://doi.org/10.1007/s10933-008-9234-2

Wagner B, Wilke T, Krastel S, Zanchetta G, Sulpizio R, Reicherter K, Leng MJ, Grazhdani A, Trajanovski S, Francke A, Lindhorst K, Levkov Z, Cvetkoska A, Reed JM, Zhang X, Lacey JH, Wonik T, Baumgarten H, Vogel H (2014) The SCOPSCO drilling project recovers more than 1.2 million years of history from Lake Ohrid. Sci. Drill. 17:19–29. https://doi.org/10.5194/sd-17-19-2014

Westphal T, Tegel W, Heussner KU, Lera P, Rittershofer K-F (2010) Erste dendrochronologische Datierungen historischer Hölzer in Albanien. Archaeol Anz 2:75–95. https://doi.org/10.34780/8ib4-c616

Part III

Palaeoecology and Bioarchaeology

A Palaeoecological Perspective on the Environmental Impact of Prehistoric Societies in Europe with Emphasis on Greece and the Southern Balkans

César Morales-Molino, Lieveke van Vugt,
Ariane Ballmer, Sarah Brechbühl,
Kathrin Ganz, Sylvia Gassner, Erika Gobet,
Albert Hafner, André F. Lotter, Carolina Senn,
Antoine Thévenaz, and Willy Tinner

Abstract

Here, we provide an overview on the environmental impact of Europe's first farmers, focusing on the vegetation shifts that occurred in the southern Balkans during the Neolithisation. First, we draw on recent methodological developments in palaeoecology that contribute to tighten its linkages with archaeology. We start highlighting the importance of highly precise and accurate lake sediment chronologies to enhance comparison with dendrochronologically dated archaeological settlements. Then, we assess modern pollen-vegetation relationships to better interpret the fossil pollen records. The results reveal (i) an overall good match between the main vegetation types of the southern Balkans and their soil pollen assemblages, and (ii) that pollen assemblages from lake surface samples reflect reliably the surrounding vegetation. Afterwards, we summarise our latest results from the region. In Limni Zazari (Greece), continuous pollen records of cereals and ruderal plants allow dating early farming activities around 6250 BC. At Ploča Mičov

C. Morales-Molino (✉) · L. van Vugt ·
S. Brechbühl · K. Ganz · S. Gassner · E. Gobet ·
A. F. Lotter · C. Senn · A. Thévenaz · W. Tinner
Institute of Plant Sciences and Oeschger Centre
for Climate Change Research (OCCR),
University of Bern, Bern, Switzerland
e-mail: cesar.morales@unibe.ch

L. van Vugt
e-mail: lieveke.vanvugt@unibe.ch

S. Brechbühl
e-mail: sarah.brechbuehl@unibe.ch

K. Ganz
e-mail: kathrin.ganz@unibe.ch

E. Gobet
e-mail: erika.gobet@unibe.ch

A. F. Lotter
e-mail: andre.lotter@unibe.ch

C. Senn
e-mail: carolina.senn@unibe.ch

A. Thévenaz
e-mail: antoine.thevenaz@unibe.ch

W. Tinner
e-mail: willy.tinner@unibe.ch

A. Ballmer
Independent Researcher, Bern, Switzerland
e-mail: mail@arianeballmer.com

A. Ballmer · A. Hafner
Institute of Archaeological Sciences and Oeschger
Centre for Climate Change Research (OCCR),
University of Bern, Bern, Switzerland
e-mail: albert.hafner@unibe.ch

© The Author(s) 2025
A. Ballmer et al. (eds.), *Prehistoric Wetland Sites of Southern Europe*,
Natural Science in Archaeology, https://doi.org/10.1007/978-3-031-52780-7_15

Grad (Lake Ohrid, North Macedonia), pollen evidence of cereals and weeds place the onset of the Neolithic at 5500–5100 BC, i.e. significantly earlier than the tree-ring inferred age of the settlement. Current efforts aim at producing new palaeoecological records, refining the available ones, and adding palaeoclimatic reconstructions that allow determining which role climate variability played in the Neolithisation.

Keywords

Pollen analysis · Land use · Forest clearance · Farming · Neolithic · Metal ages · EXPLO

15.1   Introduction

The southern Balkans have traditionally been a hotspot for palaeoenvironmental research, particularly because of the presence of ancient tectonic lakes preserving remarkably long sedimentary records. Thus, several of the most renowned long continuous terrestrial pollen records in Europe, some of them spanning several glacial-interglacial cycles, are from this region, such as Tenaghi Philippon (Fletcher et al. 2013; Milner et al. 2016; Tzedakis et al. 2006; van der Wiel and Wijmstra 1987a, b; Wijmstra 1969; Wijmstra and Smit 1976), Ioannina (Roucoux et al. 2011; Tzedakis 1993; Tzedakis et al. 2002), Lake Prespa (Panagiotopoulos et al. 2014) and Lake Ohrid (Donders et al. 2021; Lézine et al. 2010; Panagiotopoulos et al. 2020; Sadori et al. 2016). A notable number of postglacial and particularly Holocene palaeoecological records have also been published during the past few decades (Bottema 1974; Gassner et al. 2020; Jahns 2005; Kouli and Dermitzakis 2010; Lawson et al. 2004, 2005; Masi et al. 2018; Panagiotopoulos et al. 2013; Willis 1992, 1994). Nevertheless, a substantial proportion of these sequences lack reliable and precise chronologies due to potential issues arising from radiocarbon dating of bulk sediment, particularly on calcareous bedrock settings (Finsinger et al. 2019;

Deevey et al. 1954). Likewise, such studies often provide very good overviews on the regional vegetation dynamics during the Late Glacial and the Holocene, but they rarely address more specific ecological aspects of vegetation change. In particular, although some previous research has investigated the interactions of the different prehistoric cultures with the environment, in particular during the Neolithisation (Glais et al. 2016, 2017; Krauß et al. 2018; Masi et al. 2018), detailed multi-proxy studies can still shed new light on the environmental consequences of this process.

From the Fertile Crescent, Neolithic farming spread rapidly into Europe via Anatolia, arriving in the Balkans first to follow later two pathways into Central Europe (Ammerman and Cavalli-Sforza 1971; Guilaine 2003; Hofmanová et al. 2016). Although some uncertainties still persist regarding the routes followed by the initial spread of Neolithic farming into Europe (e.g. Kotsakis 2014; Guilaine 2018), the Neolithisation in the southern Balkans was undoubtedly crucial for the later expansion of farming into Central and Northern Europe. The adoption of farming in the southern Balkans around 6500 BC brought along major socio-economic changes that in turn modified to a large extent the local ecosystems. Understanding the interactions between the first Neolithic settlers of the southern Balkans and its environment is particularly relevant because it represents the first time that Neolithisation arrived into a submediterranean (i.e. rather cool-temperate and moist) setting, very different from that where farming developed in the Fertile Crescent (Gassner et al. 2020).

In this chapter, our overarching goal is to assess the impact of prehistoric communities inhabiting the southern Balkans and their surrounding environment, with a particular focus on the advent of the Neolithisation. First farming communities were established in our study area, which includes northern Greece and southern North Macedonia, by the mid-7th millennium BC (Chrysostomou et al. 2015), making it particularly well-suited to understand the expansion of Neolithic farming from Anatolia into Central Europe (Tringham 2000). First, we summarise recent advances in high-precision chronologies for lake sediment sequences that allow more direct comparison with the archaeological evidence. Then, we stress the importance of calibration studies addressing the relationship between pollen assemblages and vegetation for more precise reconstructions of vegetation dynamics over long timescales, illustrating this with a recent case study from northern Greece. Later, we present our most recent work on vegetation, fire, and land use dynamics in the southern Balkans, i.e. the multi-proxy palaeoecological records of Limni Zazari (northern Greece) and Ploča Mičov Grad (Lake Ohrid, North Macedonia), emphasising its connections with archaeology. The main approaches used to achieve this goal are pollen, spore, and charcoal analysis. Pollen analysis allows reconstructing vegetation dynamics over long timescales, whereas selected anthropogenic pollen indicators enable tracking changes in land use through time. Changes in the abundances of dung fungal spores (e.g. *Sporormiella* type, *Sordaria* type, *Podospora* type) can be associated with varying densities of herbivores, in particular of livestock. Microscopic charcoal particles are used as a proxy for regional fire activity. Finally, we provide an outlook for developments of the palaeoecology-archaeology interactions in the study region resulting from our ongoing work.

15.2 Strengthening the Palaeoecology-Archaeology Connection: High-Precision Chronology of Lake Sediment Sequences

Palaeoecological off-site records provide valuable information about long-term ecosystem change; in addition, they offer detailed insights on prehistoric and historical land use (Behre 1981; Deza-Araujo et al. 2020, 2021). Palaeoecology evolved mainly from plant macrofossil analysis (19th century AD) and palynology (20th century AD; Lang et al. 2023), and its latter

developments include multi-proxy approaches aimed at capturing ecological and environmental complexity (e.g. Pedrotta et al. 2021). Much attention has been paid to developing multiple lines of evidence and thus avoiding inferring different processes from the same proxy. The latter could result in circular reasoning, which occurs for instance when pollen is used as proxy for both climate and vegetation change. The emergence of new biological (e.g. diatoms, chironomids, aDNA) and geological (e.g. X-ray fluorescence and hyperspectral imaging) proxies archived in sedimentary records has enabled moving forward in this direction. Recent efforts in microscopy, biochemistry and palaeogenetics have also increased taxonomic resolution, for instance to better distinguish crops from wild relatives. During the past c. 20 years, high-resolution contiguous sampling has been applied to generate uninterrupted palaeoecological time series to decipher ecosystem, fire and land use dynamics at 5–20 year intervals (Ammann et al. 2000; Gobet et al. 2003; Lotter 1999; Pedrotta et al. 2021; Rey et al. 2019a; Tinner et al. 1999). Nevertheless, high-precision chronologies allowing thorough comparisons with the archaeological record have been missing till very recently. While in archaeology chronological precision was recognised as a crucial prerequisite already decades ago, modern palaeoecology still suffers conspicuous dating uncertainties (Finsinger et al. 2019). Owing to the long duration of palaeoecological records (Holocene and even beyond), the scarcity of terrestrial plant macrofossils in the sediments, and the costs of radiocarbon dating, chronological uncertainties are usually > 100–200 years (Tinner et al. 2003). Another major issue in palaeoecology is that many researchers still date bulk samples to save time, since searching for well-preserved short-lived terrestrial plant macrofossils (e.g. seeds, fruits, needles, leaves, periderm) is time-consuming. Bulk dating may introduce dating errors of up to several centuries because of the hard-water effect, the in-built age of charcoal of wood, and reworking (Finsinger et al. 2019; Gavin 2001; Oswald et al. 2005).

To overcome such chronological limits, two varved lake sediment sequences from Moossee and Burgäschisee (Swiss Plateau) were subsampled for short-lived and well-preserved terrestrial plant macrofossils (Rey et al. 2019b). The subsamples spanned constant time intervals (annual lamination counting) to reduce chronological variability. Wiggle-matching of the probability distributions of calibrated radiocarbon dates allowed constructing robust, high-precision chronologies for the sediment records. Specifically, age uncertainties of the 95% confidence interval of the modelled ages were 19 yr for Moossee and 54 yr for Burgäschisee (always calibrated ages) over the entire study period of 3000 years. Further, uncertainties of only 13 and 18 yr were achieved for shorter time intervals at Moossee and Burgäschisee, respectively (Rey et al. 2019b). The latter precisions not only improved any previous chronology based on varve counting but even allowed direct comparison with dendrochronological data. Combining such palaeoecological high-precision chronologies with high-resolution analyses that delivered uninterrupted time series of agricultural activities and environmental change at regular intervals of c. 10 years opens novel and exciting avenues to link palaeoecology-based land use reconstructions and archaeological insights (Rey et al. 2019b). For instance, the latest palaeoecological Moossee and Burgäschisee time series showed synchronous episodes of forest clearance during the Neolithic that suggest coeval slash-and-burn activities around the two lakes, which might be correlated to environmental change such as climate variability (Rey et al. 2019a). These forest openings were connected to arable and pastoral farming activities and correspond to phases with increased abundances of archaeological finds at the sites. Moreover, when taking into account the dating uncertainties of previous high-resolution palaeo-ecological studies, they are synchronous with forest opening and subsequent reforestation phases found at other sites on the Swiss Plateau, in southern Germany and the southern Alpine forelands, suggesting large-scale environmental triggers of European Neolithic land use dynamics (Rey et al. 2019a).

15.3 Modern Pollen-Vegetation Relationships for Enhanced Interpretation of Fossil Pollen Records

Taxon-specific differences in pollen production, dispersal, deposition, and preservation make interpretation of fossil pollen assemblages challenging. Consequently, the 'correct' interpretation of palynological records requires investigating the pollen representation of the extant vegetation types. Since comprehensive studies assessing the modern pollen representation of the diverse vegetation types of the southern Balkans were still lacking, we addressed this question across the large environmental gradients present in this area of the north-eastern Mediterranean region (Senn et al. 2022). We first conducted vegetation surveys across the various vegetation types of northern Greece, from coastal evergreen maquis to alpine meadows, and collected moss and topsoil samples to analyse their (modern) pollen assemblages. In a second step, we analysed the pollen composition of the topmost samples of surface cores retrieved from lakes located in the same region at different elevations and relatively diverse environmental settings with the aim of assessing the correspondence between modern terrestrial and lake surface samples. More specifically, we sampled 61 terrestrial sites (vegetation survey + surface sample) assigned to eight major vegetation types as well as 12 lakes (Fig. 15.1; Senn et al. 2022). The relationships between plant and pollen assemblages in terrestrial surface samples as well as the comparison between terrestrial and lake surface sample pollen assemblages were quantified using several numerical techniques such as ordination (detrended correspondence analysis DCA), Procrustes analysis and multivariate classification trees (Senn et al. 2022).

We found an overall good correspondence between plant and pollen assemblages in terrestrial surface samples. However, some cautionary messages emerge from the results that should be born in mind when interpreting pollen records

Fig. 15.1 a Map showing the location of the terrestrial (numbered coloured symbols) and lake surface samples (yellow stars) in Greece. **b** Pictures of the vegetation types to which the terrestrial surface samples were assigned according to the local vegetation. Photograph frames and dots in the map are in the same colour to denote the vegetation type (Modified from Senn et al. 2022)

from the southern Balkans. The most important is that pollen assemblages do not reflect accurately the composition of the local vegetation when this is open or covers small surfaces and is surrounded by extensive forests dominated by trees producing large loads of wind-dispersed pollen (Senn et al. 2022). Additionally, attainable taxonomic resolution of pollen identification can be limiting in certain circumstances in areas of particular plant diversity such as the Mediterranean region (e.g. *Pinus*, *Quercus*). On the other hand, the data show that pollen assemblages of lake surface samples largely agree with those of the terrestrial surface samples collected in their

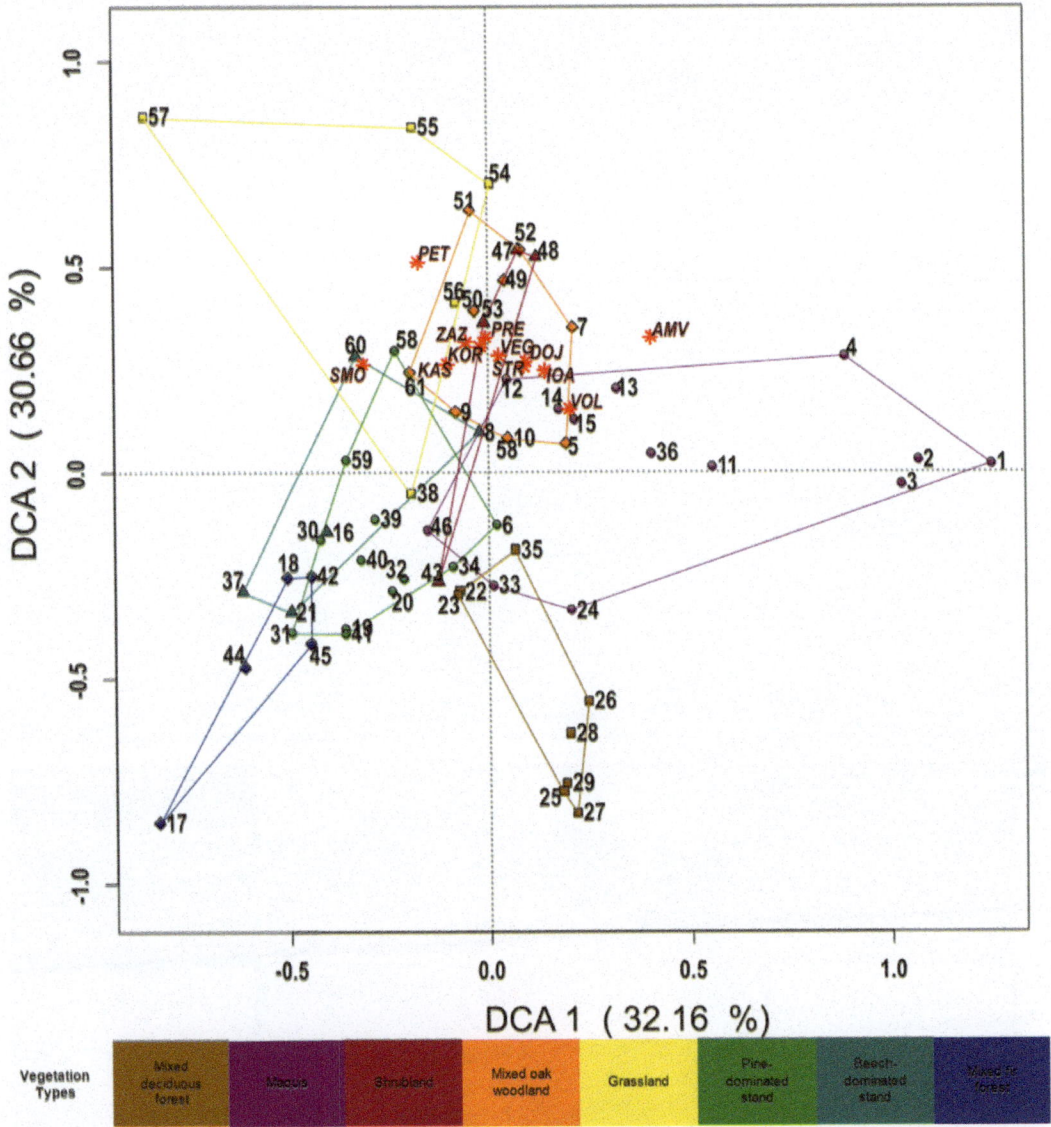

Fig. 15.2 Ordination diagram showing the results of the detrended correspondence analysis (DCA) on the terrestrial surface sample pollen dataset of northern Greece (numbered symbols coloured according to the vegetation type). The twelve lake surface samples from Greek lakes (red asterisks) were added passively to assess their relationship with the terrestrial surface samples.

Abbreviations of the lakes studied: AMV = Limni Amvrakia; DOJ = Limni Doirani; IOA = Limni Ioannina; KAS = Limni Kastoria; KOR = Limni Koroneia; PET = Limni Petres; PRE = Limni Megali Prespa; SMO = Drakolimni Smolika; STR = Limni Strofilia/Prokopou; VEG = Limni Vegoritis; VOL = Limni Volvi; ZAZ = Limni Zazari (Reproduced from Senn et al. 2022)

catchments or of the vegetation types found there, thus reflecting the composition of the surrounding vegetation (Fig. 15.2; Senn et al. 2022). These results allow outlining the main

conclusion of this study: Vegetation composition and structure in the study area can be reliably reconstructed through time using down core pollen records.

15.4 Tracking Ancient Human Impact on the Landscapes of the Southern Balkans

15.4.1 Limni Zazari (Greece)

Limni Zazari is a small (c. 200 ha) and shallow (mean water depth 5.0 m, maximum water depth 7.6 m) lake located in northern Greece (40°37′31″N, 21°32′50″E, 600 m a.s.l.; Fig. 15.3). The climate is warm-temperate mediterranean according to the Köppen-Geiger classification (Csa), albeit with continental features. Mean annual precipitation and temperature are c. 550 mm and 12.5 °C, respectively, and the mean temperature of the coldest (January) and warmest (July) months are c. 3 and 23 °C, respectively. The lake is eutrophic, probably because the rocks dominant in the catchment are calcareous and due to the anthropogenic activities—such as fertilisation of fields. Around the lake, the vegetation is dominated by agricultural fields, grasslands, shrublands (mostly *shibljak*), and overgrazed oak woodlands (*Quercus frainetto*, *Quercus cerris*, *Carpinus orientalis*, *Cornus mas*). *Fagus sylvatica* becomes increasingly frequent and even dominant in the woodlands located at higher elevation in the mountains bordering the catchment to the northwest. The presence of several prehistoric archaeological sites in relative vicinity to the lake (Limnochori, Anarghiri, Agios Panteleimon, Dispilió, Mavropigi; Gassner et al. 2020) makes this site particularly suitable to investigate the impact of prehistoric societies on the surrounding ecosystems.

Close to the deepest part of the lake (5.3 m water depth), a coring team of the Palaeoecology section of the University of Bern took two parallel cores with a UWITEC piston corer operated from a floating platform in April 2016. The cores were correlated visually according to their lithostratigraphy and the resulting composite sedimentary sequence was c. 6 m long (Gassner

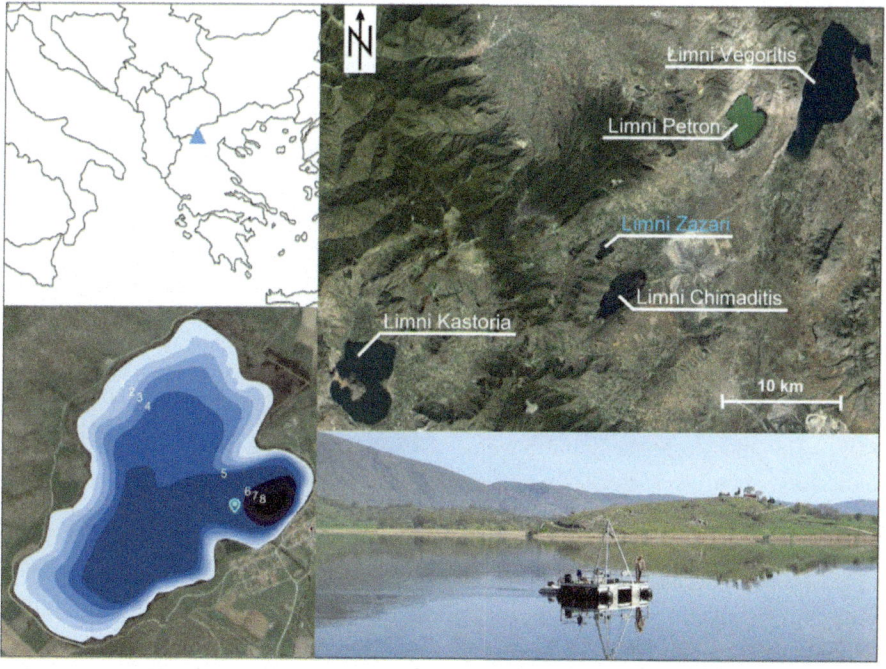

Fig. 15.3 (**Top left**) Location of Limni Zazari in northern Greece. (**Top right**) Satellite picture of the Limni Zazari catchment alongside some of the other major lakes in the region. (**Bottom left**) Bathymetric map of Limni Zazari. Figures indicate water depth. (**Bottom right**) Picture of the UWITEC coring platform on Limni Zazari (photo credit: André F. Lotter) (Reproduced from Gassner et al. 2020)

et al. 2020). Radiocarbon dating of terrestrial plant macrofossils and modelling of the age-depth relationship show that the sediment record of Limni Zazari spans the past c. 20,000 years (Gassner et al. 2020). Particular attention was paid to date as precisely as possible the section around the presumable introduction of farming in the area, that is c. 7000–6000 BC (Gassner et al. 2020). To study the vegetation-fire-land use interactions, we used a multi-proxy palaeoecological approach including pollen, spore, and charcoal analyses.

Although the palynological record of Limni Zazari allows reconstructing vegetation, fire and land use history since almost the Last Glacial Maximum (c. 20,000 years ago; Gassner et al. 2020), here we will focus on the phases corresponding to the Holocene (i.e. the past c. 11,500 years; Fig. 15.4). At the beginning of the Holocene (i.e. c. 9550 BC), mixed deciduous oak forests (*Quercus frainetto* type, *Quercus cerris* type) were the dominant vegetation, benefitting from the warm and relatively dry climate, although meadows were also present (*Sanguisorba minor* type, *Rumex*, *Centaurea nigra* type; Gassner et al. 2020). Forests continued dominating the landscape until c. 6650 BC, when they experienced a significant disruption (particularly *Quercus cerris* type) simultaneous with notable expansions of *Artemisia* and *Centaurea nigra* type and increasing fire activity (Gassner et al. 2020). Woodland decline and the simultaneous expansion of steppic grasslands might have been triggered by a cold and/or dry period which may have extended to the so-called 8.2 ka event, characterised by lower moisture availability in the eastern Mediterranean (Kotthoff et al. 2008). The most reliable anthropogenic pollen indicators such as Cerealia type and *Plantago lanceolata* type are rare or lacking, which prevents the unambiguous attribution of this episode of forest clearance to human activities like farming (Gassner et al. 2020). However, the continuous records of Cerealia type and *Plantago lanceolata* type pollen, alongside the consistent record of *Sporormiella* type dung fungal spores and the spread of Poaceae and *Polygonum aviculare* type (ruderal plant) point

to the establishment of farming activities at c. 6250 BC (Gassner et al. 2020). This is in agreement with the dates of the first Early Neolithic archaeological sites in the Four Lakes District area (6500–5800 BC; Chrysostomou et al. 2015) and the close Mavropigi (c. 20 km away) site (6550 BC; Karamitrou-Mentessidi et al. 2015). Forests dominated by pines and deciduous oaks recovered in the region at c. 6150 BC after this transient deforestation event, reaching their maximum development at 5450–4550 BC (Gassner et al. 2020), during the Late Neolithic (Chrysostomou et al. 2015). Two major maxima in charcoal influx and finds of cultural indicators (Cerealia type, *Plantago lanceolata* type) indicate enhanced fire activity at c. 5050 and 3350 BC related to farming (Gassner et al. 2020). Interestingly, *Fagus sylvatica* only spread following the latter, whereas taxa typical of open vegetation increased at approximately 2250 BC following a conspicuous increase in *Plantago lanceolata* type related to farming (Gassner et al. 2020). *Fagus sylvatica* only spread at c. 850 BC, when climate got cooler and moister and land use intensified (cereal cultivation; Gassner et al. 2020). Agricultural activities were also responsible for the forest clearance episode dated at 1150 BC (Gassner et al. 2020), i.e. the onset of the Greek Iron Age (Chrysostomou et al. 2015). The last three millennia were characterised by increasingly intensified agricultural activities, the spread of *Fagus*, grasslands and cereal fields, and high regional fire activity (maxima at 550 BC and AD 1350; Gassner et al. 2020).

15.4.2 Ohrid Ploča Mičov Grad (North Macedonia)

Lake Ohrid (40°59′38.8″N, 020°47′51.7″E, 693 m a.s.l.; Fig. 15.5) is a large (358 km^2) and deep (maximum water depth 288 m) lake located between Macedonia and Albania. The lake is of tectonic origin and the surrounding bedrock is calcareous (Panagiotopoulos et al. 2014; Popovska and Bonacci 2007). The local climate is humid warm-temperature or subtropical

Limni Zazari (600 m a.s.l., Greece)

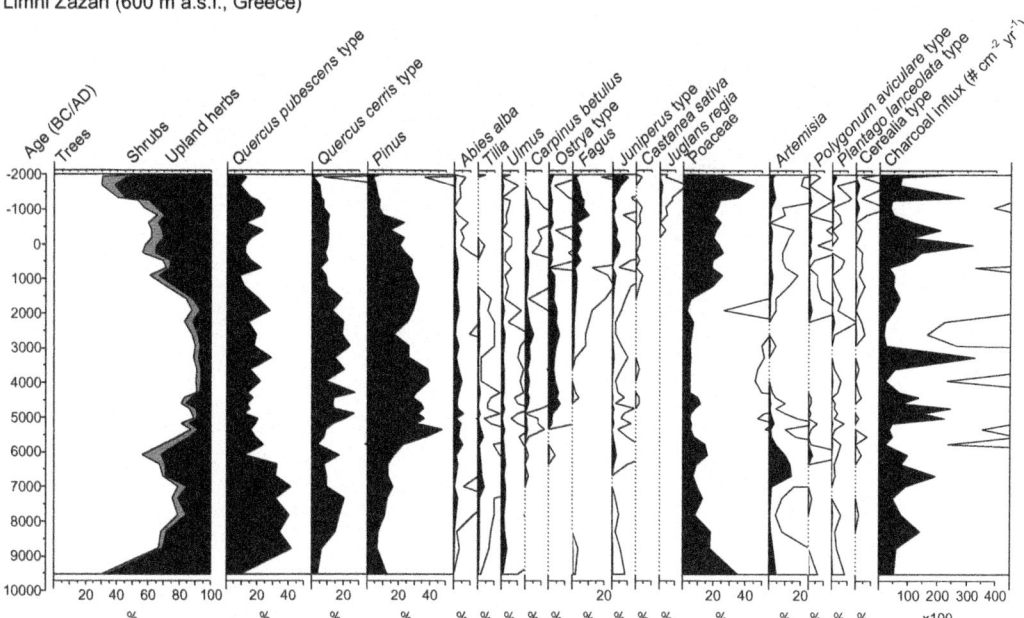

Fig. 15.4 Summary overview pollen percentage and charcoal influx diagram of Limni Zazari (northern Greece). Hollow curves represent 10 × exaggerations (Modified from Gassner et al. 2020)

according to the Köppen-Geiger classification (Cfa), although humid cool-temperate (Cfb) and continental climates (Dfb) are widespread in the lake catchment. On the lakeshore, mean annual temperature is c. 11.5 °C and the mean temperature of the coldest (January) and the warmest (July) are c. 2.5 °C and 22 °C, respectively, whereas the average annual precipitation is c. 550–650 mm. Temperature decreases, and rainfall increases steeply with elevation in the surrounding mountains. Forests in the area are mainly dominated by deciduous oaks (*Quercus frainetto, Quercus cerris*) in the lowlands and by beech (*Fagus sylvatica*), fir (*Abies borisii-regis*) and pine (*Pinus sylvestris, P. heldreichii, P. peuce*) in the upper vegetation belts (Acevski and Mandžukovski 2019). Several thermophilous species (including *Carpinus orientalis, Ostrya carpinifolia, Quercus trojana, Juniperus excelsa*) can be found along the lake shore (Matevski et al. 2011).

In 2019, we took sediment cores from the prehistoric lake settlement site Ploča Mičov Grad; the settlement is dated to the Neolithic and Bronze Age (4600–1200 BC; Hafner et al. 2021;

Chaps. 6 and 14 in this book). Vegetation, fire, and land use history were reconstructed by analysing pollen, spores, stomata and microscopic charcoal particles. The Lake Ohrid sedimentary sequences record the vegetation and environmental history from c. 11,500 to 3050 BC, but here we only discuss the preliminary results from the Holocene section (i.e. c. 9550–3050 BC; Fig. 15.6).

During the Mesolithic (9350–5550 BC) the area around Ploča Mičov Grad was covered by closed diverse temperate forests with deciduous oaks (*Quercus pubescens* type), pines (*Pinus*), firs (*Abies*), alder (*Alnus glutinosa* type), linden (*Tilia*), elm (*Ulmus*) and manna ash (*Fraxinus ornus*). At the onset of the Neolithic (5550–5150 BC), the forest started to open and mainly *Pinus* declined, while open land expanded. The occurrence of pollen of crops and weeds (e.g. Cerealia type, *Polygonum aviculare* type) suggests agricultural activities in the area. Around 5100 BC, forests strongly declined, while open land and agricultural activities increased. After 4850 BC, the forest recovered, and fields and meadows declined for a short period of about 200 years. At

Fig. 15.5 Overview of the study site and its surroundings (Google Maps 2021) **a** Aerial photograph of Lake Ohrid and Lake Prespa. **b** Aerial picture of the Ploča Mičov Grad area with the location of the coring locations (red dots) next to the modern reconstruction of the Neolithic settlement (Authors)

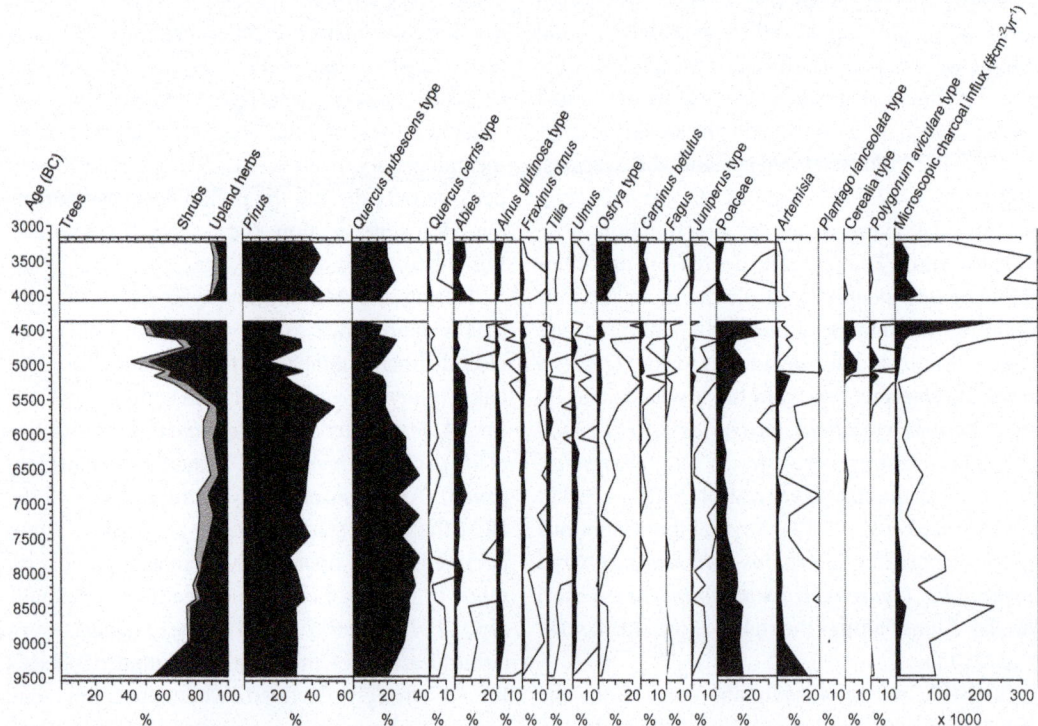

Fig. 15.6 Preliminary pollen percentage and charcoal influx diagram of Ploča Mičov Grad (Lake Ohrid, North Macedonia). Hollow curves represent 10 × exaggerations (Authors)

4650 BC forests declined and arable and pastoral activities recovered. The renewed deforestation and increase of agricultural activities were associated with increasing fire activity. After a short hiatus (4450–4050 BC) our record starts again with the region around Ploča Mičov Grad covered again with dense forest. However, the composition of the forest changed and *Abies, Quercus, Ostrya* and *Pinus* became dominant, mixed with *Fagus.*

Taken together, our preliminary palaeoecological record suggests there were two phases of Neolithic agricultural activities at Ploča Mičov Grad. The earliest evidence of agricultural activities around the site dates to 5550 BC. This date falls within the time range for the onset of agricultural activities in the region, as the first Neolithic societies appeared around 5500 BC in the Lake Ohrid and Prespa region (Krauß et al. 2018). However, these results indicate farming activities nearly 1000 years earlier than the first archaeologically inferred settlement phase at Ploča Mičov Grad (Hafner et al. 2021). After this first phase of agricultural activities, land use declined significantly as indicated by the low amounts of cultural indicators and the recovery of the forests. This phase of land abandonment was short-lived and agricultural activities resumed around 4500 BC at even higher intensity than before. The clear increase in fire activity suggests that people were using fire to open the forests and create arable land (Gassner et al. 2020). Due to low sample resolution and low chronological precision, the exact timing and full extent of the land use activities are not clear yet. Further analysis of the vegetation history at Ploča Mičov Grad is necessary to improve our knowledge of Neolithic land use. Nevertheless, our findings highlight the importance of multi-proxy studies and how palaeoecology can assist archaeological research. Thus, palaeoecology not only sheds light on the environmental context in which humans lived but also allows dating the first significant interactions of human settlers with their environment.

15.5 Conclusions and Outlook

In this chapter, we have outlined recent progresses in palaeoecological research that contribute significantly to strengthen the linkages with archaeology. Further, we have shown how palaeoecological data may complement archaeology providing more accurate estimates for the arrival of the different cultures in a region thanks to the potentially continuous time coverage. However, the potential of well-planned collaborative research involving archaeologists, bioarchaeologists and palaeoecologists has certainly not been fully exploited yet. Ongoing research within the frame of the EU-funded ERC Synergy project EXPLO aims at developing this in several ways, as summarised briefly below. First, we are working on producing proxy-based palaeoclimatic reconstructions of both temperature (branched glycerol dialkyl glycerol tetraethers–brGDGTs-) and precipitation (leaf wax δD) with the aim of disentangling the likely role of climate in the Neolithisation of the southern Balkans. Second, we are analysing several additional lake sediment records located along an east–west gradient to better understand how Neolithisation occurred in the region. Finally, high-resolution analyses of particularly precisely dated sediment records will provide further insights into the interplay between human activities in different prehistoric periods and vegetation change.

Acknowledgements We would like to express our gratitude to Sandra Brügger, Willi Tanner, Peter Ruprecht, Sebastian Eggenberger and Tryfon Giagkoulis for their crucial help with the coring of Limni Zazari and Lake Ohrid. Thanks to the co-authors of the research case studies presented here, that is Christoph Schwörer, Jacqueline van Leeuwen, Hendrik Vogel, Tryfon Giagkoulis, Stamatina Makri, Martin Grosjean, Vivian Felde and Joseph Volery. The project EXPLO (ERC Synergy Grant ID 810586) funded most of the research presented in this chapter. We also thank the Hellenic Ministry of Environment and Energy as well as the Hellenic Survey of Geology and Mineral for the permits to collect surface samples in several Greek National Parks and core the lakes studied.

References

Acevski J, Mandžukovski D (2019) Dendroflora of the Galičica mountain range and island of Golem grad in the Republic of Macedonia. Contrib, Sect Nat, Math Biotechnical Sci, MASA 40:261–271. https://doi.org/10.20903/csnmbs.masa.2019.40.2.152

Ammann B, Birks HJB, Brooks SJ, Eicher U, von Grafenstein U, Hofmann W, Lemdahl G, Schwander J, Tobolski K, Wick L (2000) Quantification of biotic responses to rapid climatic changes around the Younger Dryas—a synthesis. Palaeogeogr Palaeoclimatol Palaeoecol 159:313–347

Ammerman AJ, Cavalli-Sforza LL (1971) Measuring the rate of spread of early farming in Europe. Man 6:674–688

Behre K-E (1981) The interpretation of anthropogenic indicators in pollen diagrams. Pollen Spores 23:225–245

Bottema S (1974) Late quaternary vegetational history of northwest Greece. Ph.D. dissertation. University of Groningen, The Netherlands

Chrysostomou P, Jagoulis T, Maeder A (2015) The culture of four lakes: prehistoric lakeside settlements (6th–2nd mill. BC) in the Amindeon Basin, Western Macedonia, Greece. Archäologie Schweiz 38:24–32

Deevey ES, Gross MS, Hutchinson GE, Kraybill HL (1954) The natural ^{14}C contents of materials from hard-water lakes. Proc Natl Acad Sci 40:285–288

Deza-Araujo M, Morales-Molino C, Tinner W, Henne PD, Heitz C, Pezzatti GB, Hafner A, Conedera M (2020) A critical assessment of human-impact indices based on anthropogenic pollen indicators. Quatern Sci Rev 236:106291

Deza-Araujo M, Morales-Molino C, Conedera M, Pezzatti GB, Pasta S, Tinner W (2021) Influence of taxonomic resolution on the value of anthropogenic pollen indicators. Veg Hist Archaeobotany 31:67–84. https://doi.org/10.1007/s00334-021-00838-x

Donders T, Panagiotopoulos K, Koutsodendris A, Bertini A, Mercuri AM, Masi A, Comborieu-Nebout N, Joannin S, Kouli K, Kousis I, Peyron O, Torri P, Florenzano A, Francke A, Wagner B, Sadori L (2021) 1.36 million years of Mediterranean forest refugium dynamics in response to glacial-interglacial cycle strength. Proc Natl Acad Sci 118:e2026111118

Finsinger W, Schwörer C, Heiri O, Morales-Molino C, Ribolini A, Giesecke T, Haas JN, Kaltenrieder P, Magyari EK, Ravazzi C, Rubiales JM, Tinner W (2019) Fire on ice and frozen trees? Inappropriate radiocarbon dating leads to unrealistic reconstructions. New Phytol 222:657–662

Fletcher WJ, Müller UC, Koutsodendris A, Christanis K, Pros J (2013) A centennial-scale record of vegetation and climate variability from 312 to 240 ka (Marine Isotope Stages 9c-a, 8 and 7e) from Tenghi Philippon, NE Greece. Quatern Sci Rev 78:108–125

Gassner S, Gobet E, Schwörer C, van Leeuwen J, Vogel H, Giagkoulis T, Makri S, Grosjean M,

Panajiotidis S, Hafner A, Tinner W (2020) 20,000 years of interactions between climate, vegetation and land use in Northern Greece. Veg Hist Archaeobotany 29:75–90

Gavin DG (2001) Estimation of inbuilt age in radiocarbon ages of soil charcoal for fire history studies. Radiocarbon 43:27–44

Glais A, López-Sáez JA, Lespez L, Davidson R (2016) Climate and human-environment relationships on the edge of the Tenaghi-Philippon marsh (Northern Greece) during the Neolithization process. Quatern Int 403:237–250

Glais A, Lespez L, Vannière B, López-Sáez JA (2017) Human-shaped landscape history in NE Greece. A palaeoenvironmental perspective. J Archaeol Sci Rep 15:405–422

Gobet E, Tinner W, Hochuli PA, van Leeuwen JFN, Ammann B (2003) Middle to Late Holocene vegetation history of the Upper Engadine (Swiss Alps): the role of man and fire. Veg Hist Archaeobotany 12:143–163

Guilaine J (2003) De la vague à la tombe, la conquête néolitique de la Méditerranée (8000–2000 avant J.-C.). Le Seuil, Paris

Guilaine J (2018) A personal view of the neolithisation of the Western Mediterranean. Quatern Int 470:211–225

Hafner A, Reich J, Ballmer A, Bolliger M, Antolín F, Charles M, Emmenegger L, Fandré J, Francuz J, Gobet E, Hostettler M, Lotter AF, Maczkowski A, Morales-Molino C, Naumov G, Stäheli C, Szidat S, Taneski B, Todoroska V, Bogaard A, Kotsakis K, Tinner W (2021) First absolute chronologies of Neolithic and Bronze Age settlements at Lake Ohrid based on dendrochronology and radiocarbon dating. J Archaeol Sci Rep 38:103107

Hofmanová Z, Kreutzer S, Hellenthal G, Sell C, Diekmann Y, Díez-del-Molino D, van Dorp L, López S, Kousathanas A, Link V, Kirsanow K, Cassidy LM, Martiniano R, Strobel M, Scheu A, Kotsakis K, Halstead P, Triantaphyllou S, Kyparissi-Apostolika N, Urem-Kotsou D, Ziota C, Adaktylou F, Gopalan S, Bobo DM, Winkelbach L, Blöcher J, Urtenländer M, Leuenberger C, Çilingiroğlu Ç, Horejs B, Gerritsen F, Shennan SJ, Bradley DG, Currat M, Veeramah KR, Wegmann D, Thomas MG, Papageorgopoulou C, Burger J (2016) Early farmers from across Europe directly descended from Neolithic Aegeans. Proc Natl Acad Sci 113:6886–6891

Jahns S (2005) The Holocene history of vegetation and settlement at the coastal site of Lake Voulkaria in Acacarnia, western Greece. Veg Hist Archaeobotany 14:55–66

Karamitrou-Mentessidi G, Efstratiou N, Kaczanowska M, Kozlowski JK (2015) Early Neolithic settlement of Mavropigi in western Greek Macedonia. Eurasian Prehistory 12:47–116

Kotsakis K (2014) Domesticating the periphery. Pharos 20:41–73

Kotthoff U, Müller UC, Pross J, Schmiedl G, Lawson IT, van de Schootbrugge B, Schulz H (2008) Lateglacial and Holocene vegetation dynamics in the Aegean region: an integrated view based on pollen data from marine and terrestrial archives. The Holocene 18:1019–1032. https://doi.org/10.1177/095968360809 5573

Kouli K, Dermitzakis MD (2010) Contributions to the European Pollen database 11. Lake Orestiás (Kastoria, northern Greece). Grana 49:154–156

Krauß R, Marinova E, De Brue H, Weninger B (2018) The rapid spread of early farming from the Aegean into the Balkans via the Sub-Mediterranean-Aegean Vegetation Zone. Quatern Int 496:24–41

Lang G, Ammann B, Behre K-E, Tinner W (Eds) (2023) Quaternary vegetation dynamics of Europe. Haupt, Bern, Switzerland

Lawson I, Frogley M, Bryant C, Preece R, Tzedakis PC (2004) The Lateglacial and Holocene environmental history of the Ioannina basin, north-west Greece. Quatern Sci Rev 23:1599–1625

Lawson IT, Al-Omari S, Tzedakis PC, Bryant CL, Christanis K (2005) Lateglacial and Holocene vegetation history at Nisi Fen and the Boras mountains, northern Greece. The Holocene 15:873–887

Lézine A-M, von Grafenstein U, Andersen N, Belmecheri S, Bordon A, Caron B, Cazet J-P, Erlenkeuser H, Fouache E, Grenier C, Huntsman-Mapila P, Hureau-Mazaudier D, Manelli D, Mazaud A, Robert C, Sulpizio R, Tiercelin J-J, Zanchetta G, Zeqollari Z (2010) Lake Ohrid, Albania, provides an exceptional multi-proxy record of environmental changes during the last glacial-interglacial cycle. Palaeogeogr Palaeoclimatol Palaeoecol 287:116–127

Lotter AF (1999) Late-glacial and Holocene vegetation history and dynamics as evidenced by pollen and plant macrofossil analyses in annually laminated sediments from Soppensee (Central Switzerland). Veg Hist Archaeobotany 8:165–184

Masi A, Francke A, Pepe C, Thienemann M, Wagner B, Sadori L (2018) Vegetation history and paleoclimate at Lake Dojran (FYROM/Greece) during the Late Glacial and Holocene. Clim past 14:351–367

Matevski V, Čarni A, Avramovski O, Juvan N, Kostadinovski M, Košir P, Marinšek A, Paušič A, Šilc U (2011) Forest vegetation of the Galičica mountain range in Macedonia. ZRC SAZU, Ljubljana

Milner AM, Roucoux KH, Collier REL, Müller UC, Pross J, Tzedakis PC (2016) Vegetation responses to abrupt climatic changes during the last interglacial complex (Marine Isotope Stage 5) at Tenaghi Philippon, NE Greece. Quatern Sci Rev 154:169–181

Oswald WW, Anderson PM, Brown TA, Brubaker LB, Hu FS, Lozhkin AV, Tinner W, Kaltenrieder P (2005) Effects of sample mass and macrofossil type on radiocarbon dating of arctic and boreal lake sediments. Holocene 15:758–767

Panagiotopoulos K, Aufgebauer A, Schäbitz F, Wagner B (2013) Vegetation and climate history of the Lake Prespa region since the Lateglacial. Quatern Int 293:157–169

Panagiotopoulos K, Böhm A, Leng MJ, Wagner B, Schäbitz F (2014) Climate variability over the last 92 ka in SW Balkans from analysis of sediments from Lake Prespa. Clim past 10:643–660

Panagiotopoulos K, Holtvoeth J, Kouli K, Marinova E, Francke A, Cvetkoska A, Jovanovska E, Lacey JH, Lyons ET, Buckel C, Bertini A, Donders T, Just J, Leicher N, Leng MJ, Melles M, Pancost RD, Sadori L, Tauber P, Vogel H, Wagner B, Wilke T (2020) Insights into the evolution of the young Lake Ohrid ecosystem and vegetation succession from a southern European refugium during the early Pleistocene. Quatern Sci Rev 227:106044

Pedrotta T, Gobet E, Schwörer C, Beffa G, Butz C, Henne PD, Morales-Molino C, Pasta S, van Leeuwen JFN, Vogel H, Zwimpfer E, Anselmetti FS, Grosjean M, Tinner W (2021) 8000 years of climate, vegetation, fire and land-use dynamics in the thermo-mediterranean vegetation belt of northern Sardinia (Italy). Veg Hist Archaeobotany 30:789–813. https://doi.org/10.1007/s00334-021-00832-3

Popovska C, Bonacci O (2007) Basic data on the hydrology of Lakes Ohrid and Prespa. Hydrol Process 21:658–664

Rey F, Gobet E, Schwörer C, Wey O, Hafner A, Tinner W (2019a) Causes and mechanisms of synchronous succession trajectories in primeval Central European mixed fagus sylvatica forests. J Ecol 107:1392–1408

Rey F, Gobet E, Szidat S, Lotter AF, Gilli A, Hafner A, Tinner W (2019b) Radiocarbon wiggle matching on laminated sediments delivers high-precision chronologies. Radiocarbon 61:265–285

Roucoux KH, Tzedakis PC, Lawson IT, Margari V (2011) Vegetation history of the penultimate glacial period (Marine isotope stage 6) at Ioannina, north-west Greece. J Quat Sci 26:616–626

Sadori L, Koutsodendris A, Panagiotopoulos K, Masi A, Bertini A, Comborieu-Nebout N, Francke A, Kouli K, Joannin S, Mercuri AM, Peyron O, Torri P, Wagner B, Zanchetta G, Sinopoli G, Donders TH (2016) Pollen-based paleoenvironmental and paleoclimatic change at Lake Ohrid (south-eastern Europe) during the past 500 ka. Biogeosciences 13:1423–1437

Senn C, Tinner W, Felde VA, Gobet E, van Leeuwen JFN, Morales-Molino C (2022) Modern pollen-vegetation-plant diversity relationships across large environmental gradients in northern Greece. The Holocene 32:159–173

Tinner W, Hubschmid P, Wehrli M, Ammann B, Conedera M (1999) Long-term forest fire ecology and dynamics in southern Switzerland. J Ecol 87:273–289

Tinner W, Lotter AF, Ammann B, Conedera M, Hubschmid P, van Leeuwen JFN, Wehrli M (2003) Climatic change and contemporaneous land-use phases north and south of the Alps 2300 BC to 800 AD. Quatern Sci Rev 22:1447–1460

Tringham R (2000) Southeastern Europe in the transition to agriculture in Europe: bridge, buffer, or mosaic. In: Price TD (ed) Europe's first farmers. Cambridge University Press, Cambridge, pp 19–56

Tzedakis PC (1993) Long-term tree populations in northwest Greece through multiple quaternary climatic cycles. Nature 364:437–440. https://doi.org/10.1038/364437a0

Tzedakis PC, Lawson IT, Frogley MR, Hewitt GM, Preece RC (2002) Buffered tree population changes in a quaternary refugium: evolutionary implications. Science 297:2044–2047

Tzedakis PC, Hooghiemstra H, Pälike H (2006) The last 1.35 million years at Tenaghi Philippon: revised chronostratigraphy and long-term vegetation trends. Quatern Sci Rev 25:3416–3430

van der Wiel AM, Wijmstra TA (1987a) Palynology of the lower part (78–120 m) of the core Tenaghi Philippon II, Middle Pleistocene of Macedonia, Greece. Rev Palaeobot Palynol 52:73–88

van der Wiel AM, Wijmstra TA (1987b) Palynology of 112.8–197.8 m interval of the core Tenaghi Philippon III, Middle Pleistocene of Macedonia. Rev Palaeobot Palynol 52:89–117

Wijmstra TA (1969) Palynology of the first 30 m of a 120 m deep section in northern Greece. Acta Bot Neerl 18:511–527

Wijmstra TA, Smit A (1976) Palynology of the middle part (30–78 m) of the 120 m deep section in northern Greece (Macedonia). Acta Botanica Neerlandica 25:297–312

Willis KJ (1992) The late quaternary vegetational history of northwest Greece III. a comparative study of two contrasting sites. New Phytol 121:139–155

Willis KJ (1994) The vegetational history of the Balkans. Quatern Sci Rev 13:769–788

Isolino Virginia (Lake Varese, Italy): New Archaeobotanical Research at the Earliest Pile-Dwelling of the Circumalpine Area

16

Bigna L. Steiner, Ferran Antolín,
Raül Soteras, Mauro Rottoli,
and Daria G. Banchieri

Abstract

The results of the archaeobotanical analysis and radiocarbon dating programme of the eight cores retrieved from Isolino Virginia (Lake Varese) in 2018 are here presented. We could identify at least two phases of occupation, between 5000 and 4700 and 4250 and 3650 cal BC, with excellent preservation conditions and hence yielding abundant plant macroremains (c. 15,000). The main crops during the 5th millennium cal BC are naked wheat, naked barley, flax, opium poppy and possibly also pea. This crop assemblage connects the site with the Western Mediterranean area instead of the Eastern Italian sites, where glume wheats were the most important crops. Possible changes around 4000 BC are observed. Wild fruit gathering was an important activity during the whole Neolithic occupation of the island.

Keywords

Northern Italy · Neolithic · Waterlogged preservation · Plant macroremains · Prehistoric agriculture · Gathering

Supplementary Information The online version contains supplementary material available at https://doi.org/10.1007/978-3-031-52780-7_16.

B. L. Steiner (✉) · F. Antolín
Integrative Prehistory and Archaeological Science (IPAS), Department of Environmental Sciences, University of Basel, Basel, Switzerland
e-mail: bigna.steiner@unibas.ch

F. Antolín
e-mail: ferran.antolin@dainst.de

F. Antolín · R. Soteras
Department of Natural Sciences, German Archaeological Institute, Berlin, Germany
e-mail: raul.soteras@dainst.de

M. Rottoli
Laboratorio di Archeobiologia, Musei Civici di Como, Como, Italy

D. G. Banchieri
Varese, Italy

16.1 Introduction

Isolino Virginia is a small island (c. 250 m long, c. 100 m of maximum width and c. 15,000 m^2 of surface) located in the Lake Varese (Lombardy, Italy) (Fig. 16.1). Archaeological investigations started in the 19th century and have continued until today (Banchieri 2017). In more than four metres of sediments, human occupations from c. 6th to 3rd millennia cal BC are recorded, covering the Early Neolithic to the Bronze Age. New occupations compressed older deposits, thus submerging them under the water table and allowing waterlogged preservation conditions (Banchieri et al. 2015; Bini and Zuccoli Bini 2016).

Archaeobotanical research at the site has been previously attempted, mainly focusing on the central trench opened by M. Bertolone during the 1950s, but also recently from the organic layers between wooden piles found at the north-eastern shore of the island. This work allowed first insights into the main crops grown at the site during the Early Neolithic period (c. 5000–4600 cal BC) (namely barley (*Hordeum distichon/vulgare*), emmer (*Triticum dicoccum*), einkorn (*Triticum monococcum*), naked wheat (*Triticum aestivum/durum*) and opium poppy (*Papaver somniferum*)), but also the main gathered plants (hazel (*Corylus avellana*), oak (*Quercus* sp.), wild vine (*Vitis vinifera* subsp. *sylvestris*), wild strawberry (*Fragaria vesca*), bramble (*R. fruticosus*), raspberry (*R.*

idaeus), dewberry (*R. caesius*) and crab apple (*Malus sylvestris*)) (Banchieri and Rottoli 2009; Banchieri et al. 2009, 2020). Nevertheless, well-preserved layers from the period between 4500 and 3500 cal BC had not yet been studied.

In the framework of the Swiss National Science Foundation Professorship project Agri-Change (www.agrichange.duw.unibasel.ch), new investigations at key sites with waterlogged preservation in the NW Mediterranean area were planned and Isolino Virginia was included as an essential case study for the project questions on farming and resilience during the Neolithic period. During the fieldwork undertaken in June–July 2018, monolith profile samples were taken, as well as sediment cores of various diameters obtained in different parts of the island. The thinner cores allowed a first assessment of the different Neolithic occupations at the site, along with their associated archaeobotanical data, and

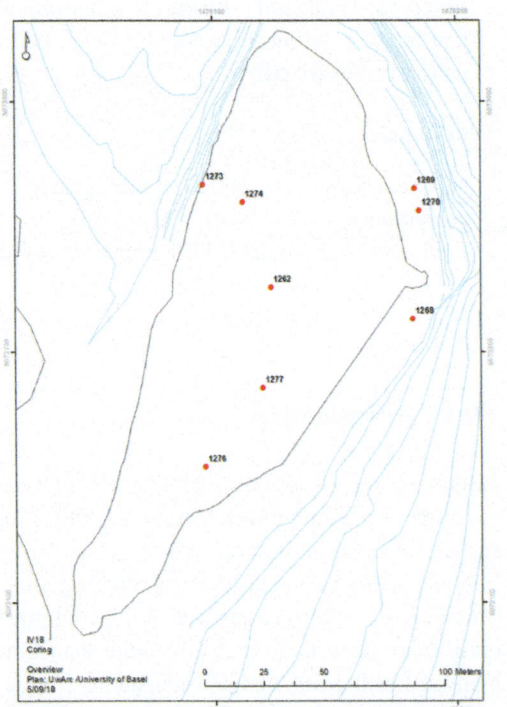
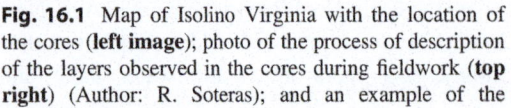

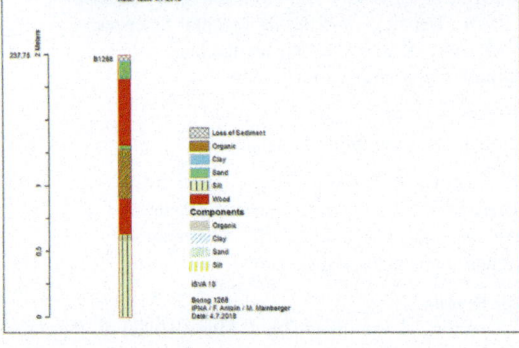

Fig. 16.1 Map of Isolino Virginia with the location of the cores (**left image**); photo of the process of description of the layers observed in the cores during fieldwork (**top right**) (Author: R. Soteras); and an example of the representation of one core following the field description based on the two main sedimentary components (**bottom right**) (Author: M. Mainberger)

the preservation conditions of the different layers. This paper focuses on these thinner cores, while the analyses of the thicker cores are currently being finished within a GroundCheck-funded project at the German Archaeological Institute (https://www.dainst.blog/groundcheck/isolino-virginia/).

The aims of this paper are to discuss:

- The stratigraphic sequence at the site together with the corresponding radiocarbon dates;
- The quality of the preservation of the archaeological deposits in the different cores;
- The archaeobotanical results per core and settlement phase in the broader context of the NW Mediterranean area.

16.2 Materials and Methods

A total of eight sediment cores were obtained at Isolino Virginia (Fig. 16.1). These were extracted with a manual borer (c. 3 cm of width and 1 m of length) aiming to record different areas of the island and shore areas where waterlogged Early Neolithic deposits were known to exist. The maximum depth needed to reach the lake bottom sediments was 3 m. A first sedimentary description (first and second component) was done in the field and cores were presented accordingly (ESM 16.1). Each core received a

number (i.e. B1262), and for the sample labelling an indication of the core depth (1, 2 or 3 meaning 1st, 2nd or 3rd metre of depth) and a correlative sample number for the site was added at the end (i.e. ISVA18_B1262_1_202).

Organic layers observed were sampled in the field and sieved with the wash-over technique (i.e. Steiner et al. 2015) with sieves of 2, 1 and 0.35 mm mesh size at the IPAS, University of Basel, where the plant remains identification was also done with the help of the seed reference collection and literature (Cappers et al. 2006; Pignatti 1982). Preservation parameters and other types of remains were also semiquantified and evaluated following previous work (Antolín et al. 2017a). The results of the seed and fruit identification were introduced into ArboDat (Kreuz and Schäfer 2014), with modified ecological groupings (according to Brombacher and Jacomet 1997; Oberdorfer 2001; www.infoflora.ch).

A total of 46 samples were sieved, between 2 and 10 samples per core were analysed, depending on the number of organic layers observed. The total amount of sediment was 2386 L, and the average sample volume was about 50 ml (Table 16.1).

Radiocarbon dates performed on archaeobotanical remains (seeds and fruits) from these cores have already been published (Antolín et al. 2022) and they will be used here to analyse in greater detail the single cores. OxCal 4.4.2. (Bronk Ramsey 2017) and the atmospheric

Table 16.1 Number of samples analysed per core and volume of sediment investigated, densities of archaeobotanical finds (r/L = remains per litre of sediment) (Authors)

Core	Number of samples	Total volume (ml)	Average volume per sample (ml)	Number of seed and fruit remains (*of which charred)	Average density (r/L) of identified seed/fruit remains
1262	8	420	52	876 (*4)	2086
1268	6	330	55	1236 (*6)	3746
1269	3	210	70	3135 (*15)	14,929
1270	4	350	87	2027 (*18)	5791
1273	2	18	9	100	5556
1274	9	410	45	3687 (*299)	8993
1276	10	480	48	3977 (*174)	8285
1277	4	170	42	2528 (*394)	14,871

curve IntCal20 (Reimer et al. 2020) have been used for the radiocarbon date calibration and analysis.

Since not all analysed samples have been dated, these preliminary results will be presented per core and not per chronological phase.

16.3 Results

16.3.1 Stratigraphy and Chronology of the Cores

A detailed description of the cores, along with the radiocarbon dates available for each of them is presented in ESM 16.1 and 16.2. The cores showed a very different sedimentary nature. Core 1262, at the centre of the island, is different from all other cores because clear layers of loam were present between organic layers and wood finds. It dates to the Early Neolithic (5000–4800 cal BC) and it seems that each organic layer accumulated in a short period of time of less than 50 years. This area is known as Q I, from the excavations of the 1950s by Bertolone, where 7 wooden platforms were documented. In 2007, Banchieri reanalysed these deposits and confirmed their nature as wooden floors and their attribution to the Early Neolithic (Banchieri 2017). Cores 1268, 1269 and 1270 were very similar and since they only yielded dates from the Early Neolithic phase and they only seemed to have one organic layer, they have been grouped together as Early Neolithic deposits from the north-eastern shore of Isolino Virginia. They showed more water influence and sandy deposits. Core 1274 yielded at least two clear phases, one dated to the final phase of the Early Neolithic (c. 4800–4600 cal BC) and another one dated to the late 5th millennium cal BC (c. 4200–4000 cal BC). Sand bands were observed between organic layers. Core 1273 was taken between the north-western shore of the island and the main lakeshore, with the hope of capturing any underlying deposits. Nothing but two reduced slightly organic layers were observed. Finally, Cores 1276 and 1277 did not yield any Early Neolithic dates, but mostly dates from the period between 4300–3700 cal

BC. Organic layers were separated by sand bands.

The Kernel Density Estimation (KDE) modelling of the radiocarbon dates suggests two phases of occupation (5000–4600 cal BC and 4300–3800 cal BC, approximately) with a possible hiatus between them (Fig. 16.2).

16.3.2 General Results

A total of 15,424 identifiable remains were recovered, mostly (94%) in a subfossil or uncharred state (the complete results per sample can be found in ESM 16.3). The number of charred finds is in any case relatively high, considering the total amount of sediment processed is only slightly above 2 L of sediment. This is due to high concentrations in samples B1274_1_224 and B1277_2_297 and 298. The density of finds per core fluctuated between 2000 r/L and 15,000 r/L (Table 16.1). No particular patterns were observed in the general distribution of finds per core, other than the small amount of finds in core 1273 ($N = 100$), which was already expected due to its location in the stream of water between the island and the main lakeshore. Charred material was more abundant in the cores taken on the island in comparison with those on the shoreline.

A very broad diversity of taxa has been identified, with 94 different taxa (either at species, genus, or family level; taxa also appearing in a charred state have been written in bold) (Figs. 16.3, 16.4 and 16.5). Aquatic and shoreline/wetland vegetation is represented by 19 taxa, including *Potamogeton crispus*, *Mentha aquatica/arvensis*, *Chara* sp., *Najas flexilis*, *N. intermedia/marina*, ***Trapa natans***, *Nuphar* sp., *Alnus glutinosa*, *Cladium mariscus*, ***Lycopus europaeus***, *Phragmites australis*, *Schoenoplectus* sp., *Typha* sp., *Bidens* sp., *Cyperus flavescens*, *C. fuscus*, *Polygonum hydropiper*, ***P. lapathifolium*** and *P. minus*. Woodland and woodland edge vegetation was represented by 19 taxa, among which we can mention: ***Abies alba***, *Acer* sp., *Betula* sp., *Fagus* sp., ***Malus sylvestris***, *Physalis alkekengi*, ***Quercus* sp.**, *Rubus caesius*,

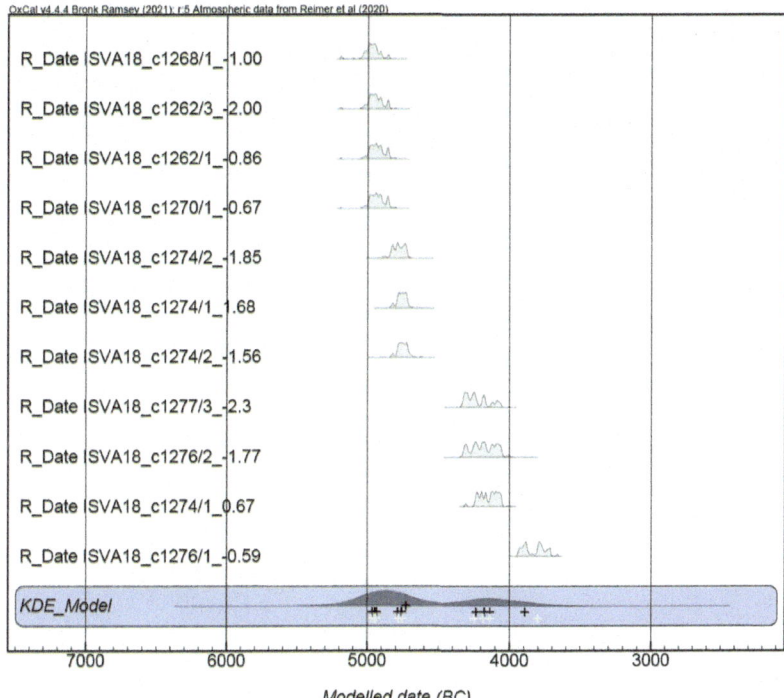

Fig. 16.2 KDE model of the radiocarbon dates obtained from the sediment cores (Authors)

R. fruticosus, **R. idaeus***, Viscum album, Vitis vinifera* subsp. *sylvestris, Cornus sanguinea,* **Corylus avellana***, Clinopodium vulgare,* **Fragaria vesca***, Hypericum perforatum* and *Origanum vulgare*. Weeds, ruderals and meadow plants are represented by 29 taxa. These comprise: **Plantago lanceolata***, Lythrum salicaria, Fallopia convolvulus, Silene cretica, Aethusa cynapium, Arenaria serpyllifolia, Brassica rapa, Capsella* sp., *Chenopodium album, Galeopsis* sp., *Polygonum aviculare,* **P. lapathifolium/persicaria, Solanum nigrum***, Arctium* sp., *Artemisia* sp., *Daucus carota, Eupatorium cannabinum, Hyoscyamus niger, Lapsana communis, Nepeta cataria, Plantago major, Ranunculus repens, Reseda luteola, Silene pratensis, Urtica dioica, Verbascum* sp., *Verbena officinalis, Malva* sp., *Sonchus* sp. Finally, the group of cultivars includes several cereals, and oil plants: **Hordeum distichon, H. vulgare/distichon, Triticum dicoccum, T. monococcum, T. aestivum/durum/ turgidum, T. durum/turgidum, T. timopheevii, Linum usitatissimum***, Papaver somniferum*.

Possible remains of pea pod (Fabaceae cf. *Pisum sativum*) were found as well, similar to those described in previous work (Antolín and Schäfer 2020). Modern contaminations have been observed, particularly *Salix* sp. fruits and *Phytolacca* sp. fruit stones.

Regarding useful plants in particular, cereals are represented both in charred and uncharred form. Chaff remains (c. 1200 uncharred, 730 charred) were more abundant than charred grains ($N = 59$), as expected given the context and the small sample size (Antolín et al. 2017b). Among the chaff remains, both barley (*Hordeum distichon/vulgare*) and naked wheat (*Triticum aestivum/durum/turgidum*) are dominant. Oil plants were mostly represented in an uncharred state, particularly opium poppy (*Papaver somniferum*), with no charred remains found, but also for flax, with only 3% of charred remains found. This is common in sites with good, waterlogged preservation conditions since oil plants do not survive well the charring process (Jacomet et al. 1989; Märkle and Rösch 2008).

Fig. 16.3 Photos of seed and fruit remains from cultivated plants found in the analysed core samples: **a** *Triticum durum/turgidum* type, charred rachis segment; **b** *T. durum/turgidum* type, uncharred rachis segment; **c** *T. monococcum*, charred spikelet fork; **d** *T. dicoccum*, charred spikelet fork; **e** *T. timopheevii*, charred spikelet fork; **f** *Hordeum distichon*, charred ear segment; **g** *H. vulgare* s.l., uncharred rachis segment; **h** *L. usitatissimum*, uncharred capsule segment tip; **i** *L. usitatissimum*, uncharred seed (Photos by R. Soteras)

Fig. 16.4 Photos of seed and fruit remains from gathered plants found in the analysed core samples: **a** *Vitis vinifera* subsp. *sylvestris*, uncharred pip; **b** *Quercus* sp., uncharred hilum; **c** *Maloideae*, uncharred pericarp; **d** *Trapa natans*, uncharred fruit spinules; **e** *T. natans*, uncharred fruit; **f** *T. natans*, charred coronary disc; **g** *Rubus fruticosus* uncharred fruit; **h** *Solanum nigrum* uncharred seed; **i** *Fragaria vesca* uncharred fruit (Photos by R. Soteras)

Fig. 16.5 Photos of seed and fruit remains from other wild plants found in the analysed core samples: **a** *Silene cretica*, uncharred seed; **b** *Lycopus europaeus*, uncharred seed; **c** *Aethusa cynapium*, uncharred fruit; **d** *Phytolacca* sp., uncharred fruit (modern intrusion) (Photos by R. Soteras)

16.3.2.1 Results Per Core (for All Graphs, See ESM 16.4)

Core B1262

According to the available radiocarbon dates, this core dates to the Early Neolithic phase. Taken in the central trench of the island, the core shows a strong trend towards higher density values (c. 5000 r/L) in the lowermost samples, along with a significantly better representation of indicators for good preservation conditions (Fig. 16.6). Nevertheless, this does not seem to affect the proportions between the main represented ecological groups, being weeds and ruderals often dominant, with a significant additional presence of cultivars and plants from woodland areas. The composition of most samples was dominated by potentially useful plants such as *Trapa natans, Malus sylvestris, Chenopodium album, Fragaria vesca* and *Rubus fruticosus*. Regarding cultivated plants, naked wheat and barley were the best represented cereals, along with opium poppy. No flax remains were recovered. Some remains of possible pea pods (Fabaceae) were present in the lowermost sample of the sequence.

Cores B1268, B1269, B1270

This group of cores at the north-east of the island are considered to belong to a roughly contemporary occupation phase dated to the Early Neolithic period (c. 5000–4800 cal BC).

Core B1268 (Fig. 16.7) yielded samples of variable composition. The samples at the top of the sequence presented better results (densities of c. 4000–10,000 r/L), but indicators of bad preservation related to erosive processes were present. The evaluation of ecological groups indicated an important presence of aquatic and wetland plants at the base of the core, and their continued presence in other parts of the core. Among the best represented aquatic and wetland plants at the base of the stratigraphy was *Najas intermedia/marina*, while at the top levels *Polygonum lapathifolium* and *Cyperus fuscus* were dominant. Plants from woodland areas and cultivars were usually the best represented groups. Regarding cultivars, naked wheat and barley are the best represented ones, particularly at the top of the core. Among the remaining wild species, plants with edible uses appeared in higher densities: *Rubus* species, *Fragaria vesca, Corylus avellana, Quercus* sp. and Amaranthaceae.

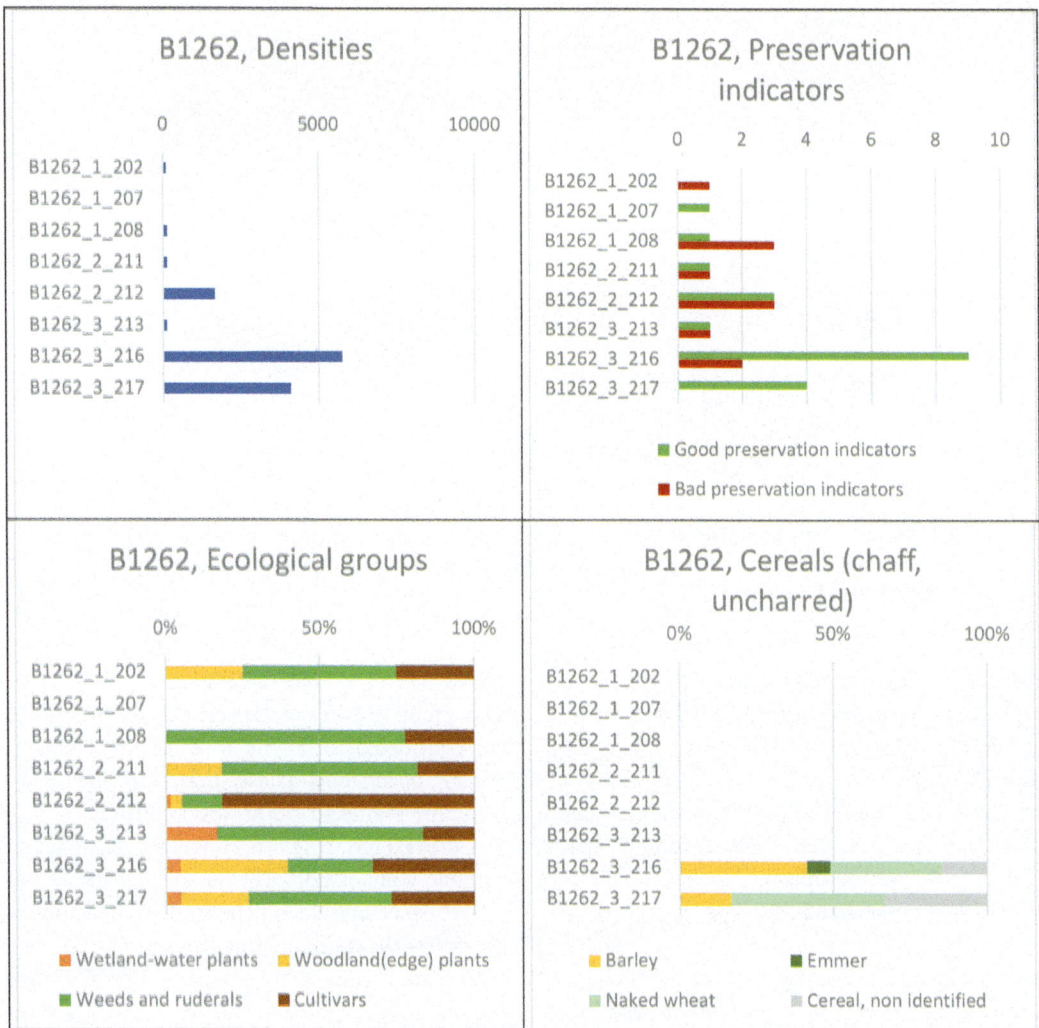

Fig. 16.6 Results of the archaeobotanical analysis of Core B1262 (Authors)

The samples analysed from core B1269 were rich in plant macroremains (12,000–23,000 r/L). Their composition changes, presenting better preservation towards the top, where also more woodland plants were present, while the bottom layers provided large numbers of seeds from the group of weeds and ruderals (mainly from the Amaranthaceae family). Aquatic and wetland plants were rare, mainly represented by *Polygonum lapathifolium*. Among the woodland edge plants, *Fragaria vesca* and the *Rubus* species were quantitatively dominant, along with *Corylus avellana* and *Vitis vinifera* subsp. *sylvestris*.

Core B1270 was quite similar. Only one sample of loamy nature yielded very few plant remains, but the remaining three samples showed densities of c. 5000 to 10,000 r/L. Better preservation seemed to be observed towards the top, where more cultivars were also preserved. Aquatic/wetland plants (mainly *Najas intermedia/marina*) were more numerous in the loamy sample. Weeds and ruderals were well represented in the high-density samples. Among the wild plants, *Abies*, *Rubus* and *Fragaria* were the most common taxa.

Among the cultivars, barley and naked wheat were the best represented cereals in the three

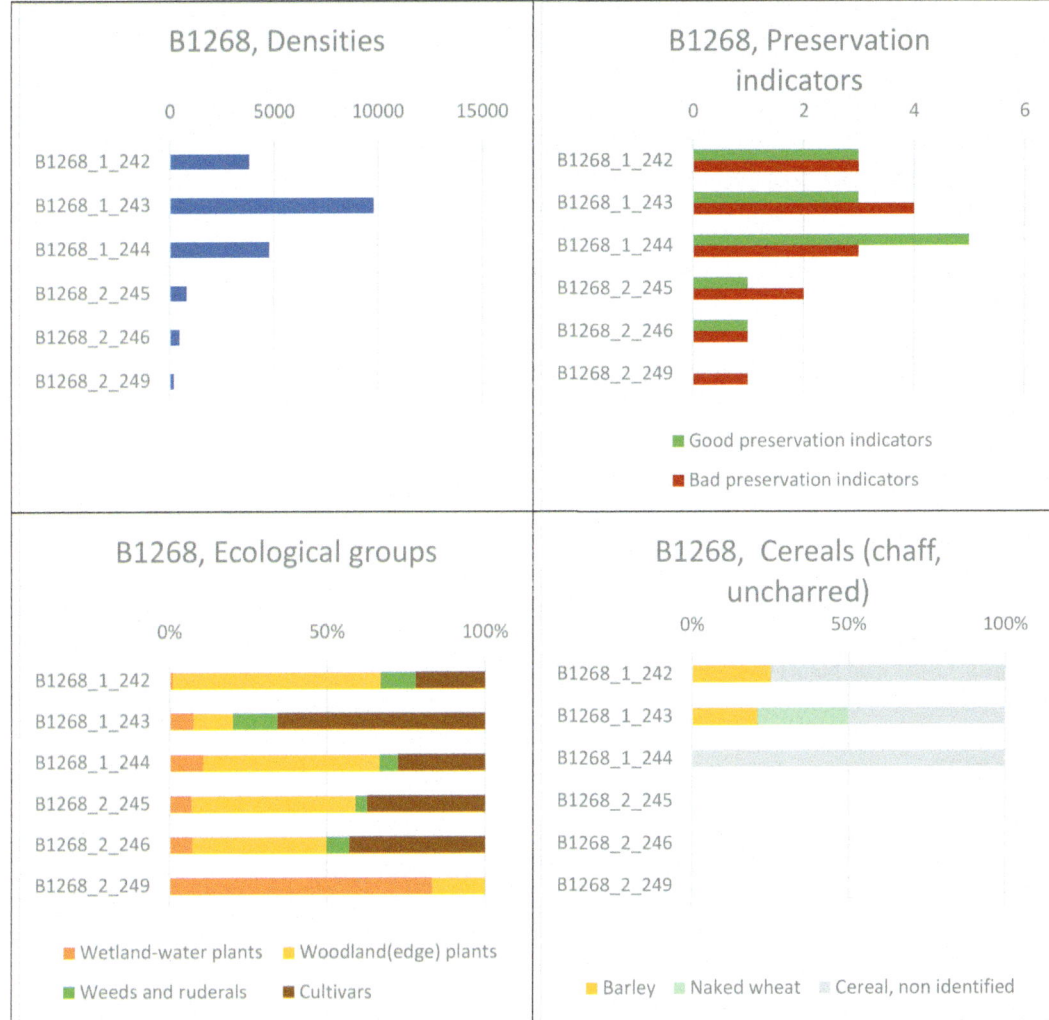

Fig. 16.7 Results of the archaeobotanical analysis of Core 1268 (Authors)

cores. Uncharred cereal chaff remains were only present in the layers with higher density values. Opium poppy was also present in considerable numbers, but no flax remains were recovered. Possible pea pod fragments (Fabaceae) were only found in core B1270.

Core B1273

Core B1273 was taken outside of the island, in the stream of water to the northwest. Two reduced horizons were observed and checked in case they contained remains associated with the occupation of the island. The chronology of the samples is unknown. As expected, the core

mostly yielded molluscs and aquatic plants/algae such as *Chara* sp. and *Najas intermedia/marina*. Indicators of human activity were completely absent. Preservation indicators also point towards bad preservation conditions for these investigated layers.

Core B1274

Core B1274 recorded at least two distinct archaeological phases, one at the base of the core, dated to a late Early Neolithic phase (c. 4700–4600 cal BC), and one towards the centre, dated to the end of the 5th millennium cal BC. In both cases, samples achieved 15,000 r/L or more. The majority of indicators for good preservation could be found in these areas of the core, while indicators for bad preservation dominated at the top of the stratigraphy. Aquatic plants were more abundant towards the base, particularly *Najas* in the samples between the two important archaeological deposits. Cultivars became more important towards the top of the stratigraphy, where also a concentration of charred material was found. In the oldest deposits, only barley and naked wheat were identified. In the uppermost layer, emmer and einkorn were also documented. Timopheevii's wheat was found in two samples of the second (youngest) phase in the core. Flax (seeds and capsules) were abundant and so was opium poppy. *Silene cretica*, a typical weed of flax fields, was also present. Woodland plants were always represented, with relevant amounts of *Corylus avellana*, *Quercus* sp., *Fragaria vesca*, *Rubus* sp. and *Vitis vinifera* subsp. *sylvestris* and needles of *Abies* sp. Other wild plants were dominated by the high presence of Amaranthaceae.

Core B1276

This core is dated to two different phases: 4300–4000 cal BC at the base of the core and 3900–3700 cal BC at the top. Core B1276 (Fig. 16.8) showed extremely good preservation conditions in several samples, mostly towards the bottom and middle part of the core. The densities of macroremains were often well above 10,000 r/L. Intermediate sandy layers have been identified. Aquatic plants were present along the core, except in the uppermost part. *Najas marina/intermedia*, *Trapa natans*, *Polygonum lapathifolium* and *Potamogeton* were dominant. Cultivars and woodland plants usually dominated the sample composition. Among the cultivars, barley and naked wheat were the most relevant, with some presence of glume wheats, particularly towards the middle part of the core. Possible pea pod fragments (Fabaceae) were found in several samples towards the base of the stratigraphy. Charred chaff indicates that both glume wheats and Timopheevii's wheat were well represented from the base of the core. Their abundance suggests crop processing residue accumulation. The assemblages were also rich in gathered wild plants

from woodland areas, including *Quercus* sp., *Physalis alkekengi*, *Corylus avellana*, *Rubus* species, *Fragaria vesca*. Other wild plants were once more dominated by the Amaranthaceae family.

Core B1277

Core B1277 could be similarly dated as Core B1276, but we only have a date for the base of the stratigraphy, which dates human occupation on this part of the island from 4300–4000 cal BC onwards. In core B1277 all samples were rich in finds, but those at the centre of the stratigraphy showed the best preservation conditions and overall density of remains. Cultivars were particularly well represented, with abundant charred cereal chaff remains, but also flax and poppy were found in high numbers. Barley and naked wheat dominated the assemblage, with a progressive better representation of barley towards the top of the stratigraphy. *Silene cretica* was also present in this core, along with abundant remains of the Amaranthaceae family. Aquatic and wetland plants were not particularly abundant, except for *Polygonum lapathifolium*. Among woodland plants, *Abies alba*, *Quercus* and *Fragaria vesca* yielded the highest densities.

16.4 Discussion

16.4.1 Stratigraphic Sequence

The site shows not only a vertical but also a horizontal stratigraphy. While Early Neolithic occupations concentrate at the north of the island, during the late 5th millennium and early 4th millennium cal BC there seems to be a settlement shift, having most activities taken place in the central-southern part of the island. The nature of the layers is different in the different sampled areas. Core B1262 presented more loamy layers, which may have to do also with an advanced state of decay of the organic material in this area, but also probably with the construction technique used consisting in superimposing wooden floors every few decades. The cores at the north-eastern part of the island were more obviously affected by shoreline processes (i.e. more molluscs were

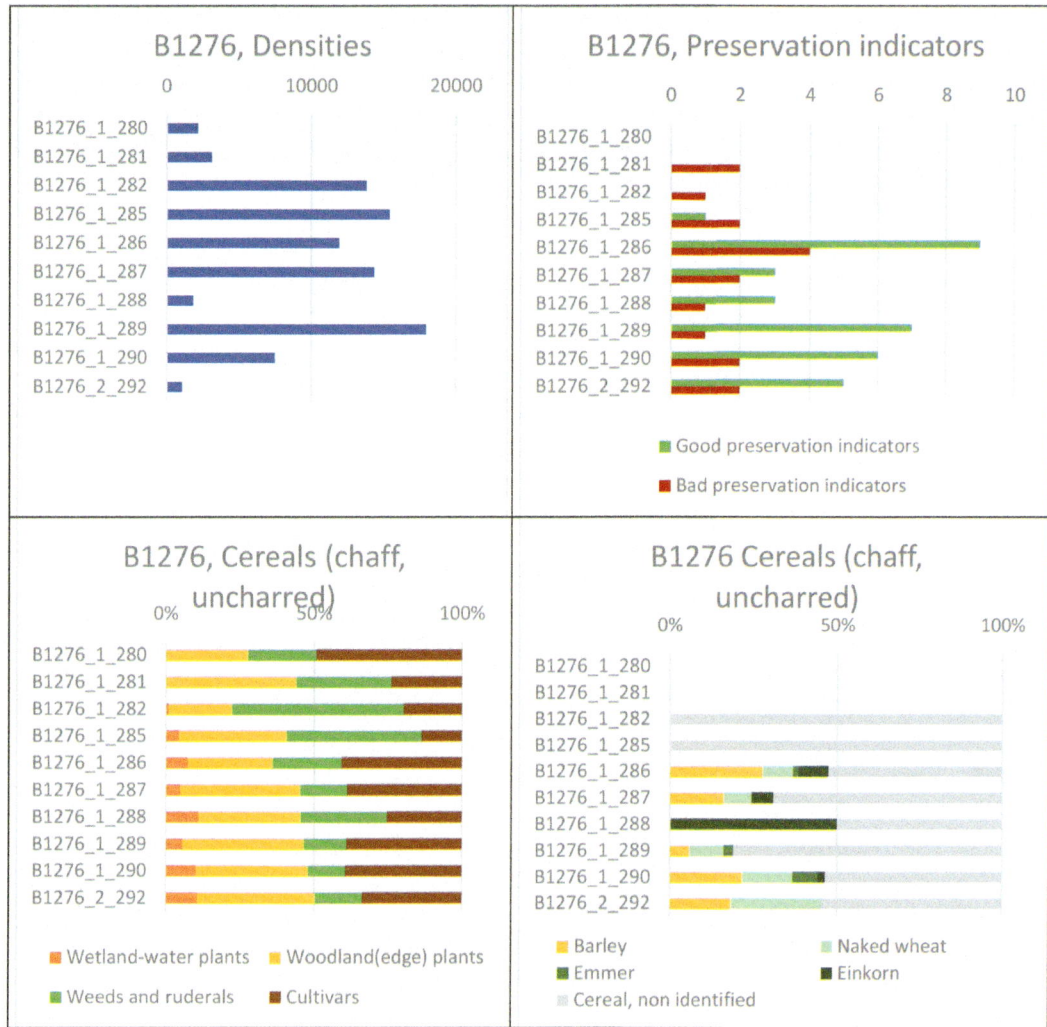

Fig. 16.8 Results of the archaeobotanical analysis of Core B1276 (Authors)

present). The remaining cores yielded layers dated to the later phase of the Early Neolithic (B1274) and mostly to the end of the 5th millennium cal BC and the first quarter of the 4th millennium cal BC. Sandy layers separated more organic-rich deposits. Charred cereal chaff was more abundant in these cores, particularly in deposits dated around 4300 and 4000 BC. Their presence might be interpreted as crop processing residues generated nearby and subsequently charred. No layers could be directly related to an accidental burning of the village or some constructions within it, so the charring of chaff might be best understood as its value as fuel (Smith 1998).

16.4.2 Preservation Quality

Density values and preservation parameters were used to characterise layer preservation quality. As a result, we observed that not all samples were equally well-preserved and not all layers were fully comparable. Higher densities of seed and fruit remains usually coincided with a higher number of indicators for good preservation conditions. Sandy layers usually yielded lower density values and higher numbers of aquatic plants (often *Najas* sp). The fact that these layers appear in all cores suggests that, at times (even if not frequently), the water table must have risen considerably so that sand accumulation occurred also in the inner parts of the island, where aquatic plants could have grown during these periods (or their diaspores transported by water into the settlement). The only core with relatively poor preservation was core B1262, which probably shows the decay generated by the open trench at the centre of the island, where the water table oscillates depending on the weather. Often, samples with small density values and indicators for bad preservation did not yield any waterlogged cereal remains, while these were always present in high abundance when the preservation was good.

Our preliminary analyses show nevertheless that there were well-preserved samples (with density values similar to well-preserved pile-dwelling sites located north of the Alps, (Jacomet 2013; Steiner 2018) in all chronological periods that allow a detailed characterisation of plant resource management at the site for more than 1000 years of occupation.

16.4.3   First Observations on Plant Use at the Site During the Neolithic Period

The volume of sediment dealt within this work is too small to draw firm conclusions on the main crops and gathered plants at the site (Antolín et al. 2017b) but they allow a first overview based on a high number of well-preserved finds. For this discussion, we will follow the periodisation used in Antolín et al. (2022), which does not fully correspond to the local chrono-cultural nomenclature (see Pedrotti et al. 2022). The changes observed are much more obvious regarding cultivated plants than wild plants. The latter seem to be similarly important across the different occupations. A synthesis of the results can be found in Fig. 16.9, but the results will be discussed per phase.

16.4.3.1   Phase 1 (4950–4700 cal BC)
Phase 1 samples involve cores B1262, B1268, B1269 and B1270. Core B1274 also yielded some material dated to the last phase of the Early Neolithic.

The results obtained are quite consistent. The cereal crops are dominated by barley and naked wheat, while oil plants are represented by opium poppy, and pulses may be represented by pea, in the form of pod fragments (identified as Fabaceae). Flax is only found in the latest phase of the Early Neolithic documented in core B1274.

Among gathered plants, we can highlight the presence of *Abies* (as leaf fodder?), *Malus/Pyrus*, *Quercus* sp, *Rubus* species, *Vitis vinifera* subsp. *sylvestris* and *Fragaria vesca* (as gathered fruits). The high amounts of Amaranthaceae in all samples probably indicate that it also was intentionally gathered (occasionally seeds also appeared in a charred state). *Abies* and *Quercus* sp. could have also been used for carpentry (Banchieri et al. 2009). Both wild and domestic resources must have contributed significantly to human diet.

16.4.3.2   Phase 2 (4300–4000 cal BC)
Phase 2 samples concern the upper part of core B1274 and the bottom part of cores B1276 and B1277.

Regarding cultivated plants, glume wheats become more present, including emmer, einkorn and timopheevii's wheat. Barley and naked wheat are still the dominant crops (both considering charred and uncharred remains). Flax and poppy are the main oil plants and peas might be an important crop as well, represented in the form of pod fragments. This speaks for a more diverse crop set than in the Early Neolithic. Crop processing must have taken place on site and chaff may have been used as fuel since concentrations of charred chaff remains were detected.

Gathered plants were also abundantly documented in all cores. *Abies alba*, *Malus/Pyrus*, *Quercus* sp. and Amaranthaceae were always found in significant amounts.

16.4.3.3   Phase 3 (3950–3650 cal BC)
This phase is only represented in the central-upper parts of cores B1276 and B1277. We observe changes in the crop spectrum (only in charred form) that need to be confirmed with additional analyses. It seems that glume wheats are as well represented as (when not more

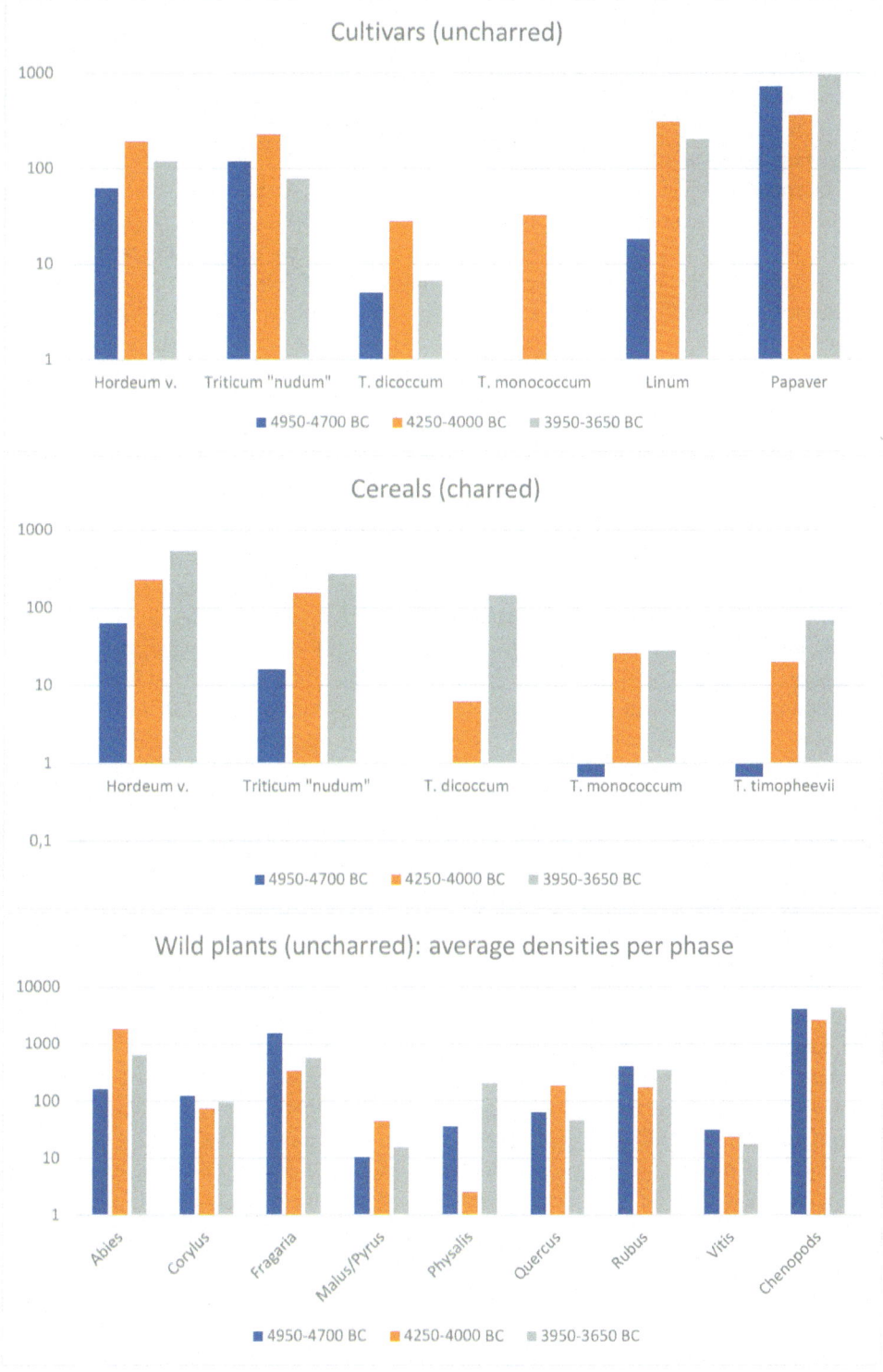

Fig. 16.9 Average densities per phase of the main taxa recovered at Isolino Virginia (Authors)

important than) free-threshing cereals (naked wheat and naked barley) in some samples. This could indicate a shift to a different agricultural system that would involve, for instance, a higher dedication in the routine activity of cereal dehusking. Palynological analyses detect a progressive disappearance of *Abies alba* at the nearby Lake Biandronno c. 4000 cal BC (Castelletti and Motella De Carlo 2017, 47; Drescher Schneider 1990, Fig. 3). In the investigated samples, it was still present in relatively high amounts, suggesting a clear intentionality in its gathering. Other gathered plants remained very well represented in the record.

16.4.4 Isolino Virginia: Looking at East or West?

Early Neolithic agricultural practices in Europe can be clearly separated between the Balkanic-central European farmers and the Western Mediterranean farmers. The latter focus their agriculture on naked wheat and naked barley (de Vareilles et al. 2020), while the former base their nutrition on einkorn and emmer (Kreuz and Marinova 2017). Opium poppy is also better known in the western Mediterranean and central-northern Europe (Salavert et al. 2020). In the light of the results obtained in this study, we can conclude that the crops cultivated at Isolino Virginia are more typical of the Western Mediterranean area. Even though the site might have also been in contact with the north-eastern Neolithic sites in Italy, archaeobotanical research in the area indicates that glume wheats were dominant in most cases (Rottoli and Castgilioni 2009; Reed and Rottoli 2014). This confirms that Isolino Virginia definitely had many aspects in common with sites to the west. This pattern seems to stay true until the end of the 5th millennium cal BC. Similar observations are also supported from other archaeological materials (Pedrotti et al. 2022). The change observed in the increase of glume wheats at the onset of the 4th millennium cal BC might be an indication of eastern influence that could have spread further West across the Alps into Southern France, where einkorn becomes an important crop during the 4th millennium cal BC (Jesus et al. 2021; Bouby et al. 2020).

16.5 Conclusions

The coring programme conducted at Isolino Virginia allowed relevant insights into the history of the settlement and the cultivated plants during the Neolithic period.

The stratigraphy of the site is complex, involving not only different periods, but also a horizontal distribution of the occupations that may require a more systematic coring programme in order for them to be fully understood. We could confirm previous assessments indicating that the Early Neolithic concentrates at the northern part of the island and that during the late 5th millennium cal BC and early 4th millennium cal BC, a shift to the central and southern part of the island becomes clear. Several indicators suggest that the organic deposits found at the site sedimented in a very wet environment, probably submerged in water. This could explain their excellent preservation conditions. A more detailed evaluation of the aquatic and wetland plants present at the site will allow further insights into this topic (e.g. Steiner et al. 2020).

The preservation of archaeobotanical material is very good, comparable to well-preserved sites north of the Alps. Therefore, there is a high potential of a very accurate reconstruction of the diet and economy at the site by increasing the number of samples. Open trenches and the archaeological deposits close to the current surface of the site have probably been going through a constant process of degradation due to water table oscillation.

We could observe that agricultural practices in the Early Neolithic resemble in their main crops, other sites in the Western Mediterranean area. Indeed, the cultivation of naked wheat, barley and opium poppy is rare in northern Italy, where glume wheats are the most important crops at most sites. During the 5th millennium cal BC, this pattern stays the same, although towards the end of the millennium glume wheats become more important and eventually, in the 4th millennium cal BC could even become the main crops at the site, suggesting the implementation of a new farming model, possibly reflecting new networks involving sites in the north-eastern part of Italy.

Acknowledgements We would like to express our gratitude to M. Mainberger, S. Jacomet, S. van Willigen and M. Borrello for their help and advice, and G. Serafini for the topographic work, and the Archaeological and Civic Museum–Villa Mirabello. Thanks also to the *Soprintendenza* of Lombardia. This research took place under the project funded by the Swiss National Science Foundation entitled 'Small seeds for large purposes: an integrated approach to agricultural change and climate during the Neolithic in Western Europe' (AgriChange Project, SNSF Professorship grant number: PP00P1_170515, PI: F. Antolín).

References

Antolín F, Schäfer M (2020 online first) Insect pests of pulse crops and their management in neolithic Europe. Environm Archaeol 1–14. https://doi.org/10.1080/14614103.2020.1713602

Antolín F, Steiner BL, Akeret Ö, Brombacher C, Kühn M, Vandorpe P, Bleicher N, Gross E, Schaeren G, Jacomet S (2017a) Studying the preservation of plant macroremains from waterlogged archaeological deposits for an assessment of layer taphonomy. Rev Palaeobot Palynol 246:120–145. https://doi.org/10.1016/j.revpalbo.2017.06.010

Antolín F, Steiner BL, Jacomet S (2017b) The bigger the better? On sample volume and the representativeness of archaeobotanical data in waterlogged deposits. J Archaeol Sci Rep 12:323–333. https://doi.org/10.1016/j.jasrep.2017.02.008

Antolín F, Martínez-Grau H, Steiner BL, Follmann F, Soteras R, Häberle S, Prats G, Schäfer M, Mainberger M, Hajdas I, Banchieri DG (2022) Neolithic occupations (c. 5200–3400 cal BC) at Isolino Virginia (Lake Varese, Italy) and the onset of the pile-dwelling phenomenon around the Alps. J Archaeol Sci Rep 42, 103375. https://doi.org/10.1016/j.jasrep.2022.103375

Banchieri DG, Rottoli M (2009) Isolino Virginia: una nuova data per la storia del papavero da oppio (Papaver somniferum subsp. somniferum). Sibrium XXV:31–49

Banchieri DG, Martinelli N, Pignatelli O (2009) Nuove indagini sui resti lignei dell'Isolino Virginia. Sibrium XXV:178–184

Banchieri DG, Bini A, Mainberger M (2015) Isolino Virginia, a waterlogged tell in a south pre-alpine lake: preservation and erosion problems. In: Brem H, Ramseyer D, Roulière-Lambert M-J, Schifferdecker F, Schlichtherle H (eds) Archéologie and Érosion—3. Monitoring et mesures de protection pour la sauvegarde des palafittees préhistoriques autour des Alpes. Actes de la troisième Rencontre Internationale Arenenberg et Hemmehofen 8–10 Octobre 2014. Centre jurassien du patrimoine: Mêta-Jura, Lons-le-Saunier, pp 183–190

Banchieri D, Bini A, Rottoli M, Mainberger M (2020) Le Prealpi varesine e l'alimentazione durante la Preistoria. In: Preistoria del cibo. Studi di Preistoria e Protostoria, vol 6. Istituto Italiano di Preistoria e Protostoria, Firenze, pp 187–196

Banchieri DG (2017) Il Neolitico nel territorio di Varese. In: Harari M (ed) Il territorio di Varese in età preistorica e protostorica. Nomos Edizioni, Varese, pp 86–119

Bini A, Zuccoli Bini LZ (2016) Isolino Virginia: dati stratigrafici, datazioni, ^{14}C e sezioni geologiche. Sibrium XXX:97–133

Bouby L, Philippe M, Núria R (2020) Late Neolithic plant subsistence and farming activities on the southern margins of the Massif Central (France). The Holocene 30:599–617. https://doi.org/10.1177/0959683619895

Brombacher C, Jacomet S (1997) Ackerbau, Sammelwirtschaft und Umwelt: Ergebnisse archäobotanischer Untersuchungen. In: Schibler J, Hüster-Plogmann H, Jacomet S, Brombacher C, Gross-Klee E, Rast-Eicher A (eds) Ökonomie und Ökologie neolithischer und bronzezeitlicher Ufersiedlungen am Zürichsee. Ergebnisse der Ausgrabungen Mozartstrasse, Kanalisationssanierungen Seefeld, AKAD/Pressehaus und Mythenschloss in Zürich. Direktion der öffentlichen Bauten des Kantons Zürich, Zürich, Egg

Bronk Ramsey C (2017) Methods for summarizing radiocarbon datasets. Radiocarbon 59:1809–1833. https://doi.org/10.1017/RDC.2017.108

Cappers R, Bekker RM, Jans JEA (2006) Digitale Zadenatlas van Nederland (Digital seed atlas of the Netherlands). Barkhuis Publishing & Groningen University Library, Groningen

Castelletti L, Motella De Carlo S (2017) Il contesto paleoambientale. In: Harari M (ed) Il territorio di Varese in età preistorica e protostorica. Nomos Edizioni, Varese, pp 29–77

de Vareilles A, Bouby L, Jesus A, Martin L, Rottoli M, Vander Linden M, Antolín F (2020) One sea but many routes to Sail: the early maritime dispersal of Neolithic crops from the Aegean to the western Mediterranean. J Archaeol Sci Rep 29:102140

Drescher Schneider R (1990) L'influsso umano sulla vegetazione neolitica del territorio di Varese dedotto dai diagrammi pollinici. In: Biagi P (ed) The Neolithisation of the alpine region. Monografie di "Natura Bresciana", vol 13. Museo civico di scienze naturali, Brescia, pp 91–97

Jacomet S, Brombacher C, Dick M (1989) Archäobotanik am Zürichsee. Ackerbau, Sammelwirtschaft und Umwelt von neolithischen und bronzezeitlichen Seeufersiedlungen im Raum Zürich. Ergebnisse von Untersuchungen pflanzlicher Makroreste der Jahre 1979–1988. Orell Füssli Verlag, Zürich

Jacomet S (2013) Archaeobotany: analyses of plant remains from waterlogged archaeological sites. In: Menotti F, O'Sullivan A (eds) The oxford handbook of Wetland archaeology. Oxford University Press, Oxford

Jesus A, Prats G, Follmann F, Jacomet S, Antolín F (2021) Middle Neolithic farming of open-air sites in SE France: new insights from archaeobotanical investigations of three wells found at Les Bagnoles (L'Isle-sur-la-Sorgue, Dépt. Vaucluse, France). Veg Hist Archaeobotany 30:445–461

Kreuz A, Marinova E (2017) Archaeobotanical evidence of crop growing and diet within the areas of the Karanovo and the Linear pottery cultures: a quantitative and qualitative approach. Veg Hist Archaeobotany 26:639–657

Kreuz A, Schäfer E (2014) Archäobotanisches datenbankprogramm ArboDat 2013. Hessen Archäologie, Wiesbaden

Märkle T, Rösch M (2008) Experiments on the effects of carbonization on some cultivated plant seeds. Veg His Archaeoabotany 17:257–264

Oberdorfer E (2001) Pflanzensoziologische Exkursionsflora für Deutschland und angrenzende Gebiete. Eugen Ulmer Verlag, Stuttgart

Pedrotti A, Poggiani Keller R, Banchieri DG, Longhi C (2022) Il Neolitico in Lombardia. In: Preistoria e Protostoria in Lombardia e Canton Ticino, Milano—Como, 17–21 ottobre 2017, LII Riunione Scientifica. Istituto Italiano di Preistoria e Protostoria, Firenze

Pignatti S (1982) Flora d'Italia. Edagricole, Bologna

Reed K, Rottoli M (2014) L'agricoltura in Friuli e Dalmazia nel Neolitico—Neolithic agriculture in Friuli and Dalmatia. In: Visentini P, Podrug E (eds) Adriatico senza confini: Via die comunicazione e crocevia di popoli nel 6000 a.C. Civici Musei di Udine, Museo Friulano di Storia Naturale, Udine

Reimer PJ, Austin WEN, Bard E, Bayliss A, Blackwell PG, Bronk Ramsey C, Butzin M, Cheng H, Edwards RL, Friedrich M, Grootes PM, Guilderson TP, Hajdas I, Heaton TJ, Hogg AG, Hughen KA, Kromer B, Manning SW, Muscheler R, Palmer JG, Pearson C, Van Der Plicht J, Reimer RW, Richards DA, Scott EM, Southon JR, Turney CSM, Wacker L, Adolphi F, Büntgen U, Capano M, Fahrni SM, Fogtmann-Schulz A, Friedrich R, Köhler P, Kudsk S, Miyake F, Olsen J, Reinig F, Sakamoto M, Sookdeo A, Talamo S (2020) The IntCal20 Northern hemisphere radiocarbon age calibration curve (0–55 cal kBP). Radiocarbon 62:725–757. https://doi.org/10.1017/RDC.2020.41

Rottoli M, Castgilioni E (2009) Prehistory of plant growing and collecting in northern Italy, based on seed remains from the early Neolithic to the Chalcolithic (c. 5600–2100 cal B.C.). Veg Hist Archaeoabotany 18:91–103

Salavert A, Zazzo A, Martin L, Antolín F, Gauthier C, Thil F, Tombret O, Bouby L, Manen C, Mineo M, Mueller-Bieniek A, Piqué R, Rottoli M, Rovira N, Toulemonde F, Vostrovská I (2020) Direct dating reveals the early history of opium poppy in western Europe. Sci Rep 10:20263. https://doi.org/10.1038/s41598-020-76924-3

Smith W (1998) Fuel for thought: archaeobotanical evidence for the use of two alternatives to wood fuel in late antique North Africa. J Mediterr Archaeol 11:191–205

Steiner BL, Antolín F, Jacomet S (2015) Testing of the consistency of the sieving (wash-over) process of waterlogged sediments by multiple operators. J Archaeol Sci Rep 2:310–320

Steiner BL, Alonso N, Grillas P, Jorda C, Piquès G, Tillier M, Rovira N (2020) Languedoc lagoon environments and man: building a modern analogue botanical macroremain database for understanding the role of water and edaphology in sedimentation dynamics of archaeobotanical remains at the Roman port of Lattara (Lattes, France). PLoS ONE 15(6):e0234853. https://doi.org/10.1371/journal.pone.0234853

Steiner BL (2018) Aspects of archaeobotanical methodology applied to the sediments of archaeological wetland deposits. Philosophisch-Naturwissenschaftliche Fakultät, Universität Basel, Basel

Archaeobotanical Investigations at the Mid-5th Millennium BCE Pile-Dwelling Site of Ploča Mičov Grad, Lake Ohrid, North Macedonia

Amy Holguin, Ferran Antolín, Mike Charles, Ana Jesus, Héctor Martínez Grau, Raül Soteras, Bigna L. Steiner, Elizabeth Stroud, and Amy Bogaard

Abstract

Abundant, well-preserved, waterlogged macrobotanical remains were found in a thick cultural organic layer associated with the Late Neolithic pile-dwelling site of Ploča Mičov Grad, Lake Ohrid, North Macedonia. Located in a biogeographically transitional zone between Mediterranean, alpine and continental regions, and on a topographically accessible link between imposing mountain ranges, the site presents a valuable opportunity to explore how, with the movement of people, plant foods and/or ideas, new subsistence strategies were established in the area. Here, we present the first archaeobotanical results from this lakeshore settlement to investigate changing subsistence strategies. We find that during the Late Neolithic occupation phase lasting an estimated 100 years, populations at the site of Ploča Mičov Grad cultivated a range of cereals (particularly einkorn, emmer, and barley), pulses (including lentil, pea, and bitter vetch) and oil-seed crops (flax and opium poppy), alongside a variety of collected fruits and nuts (such as almond, pistachio, blackberries and strawberry). Crop processing techniques are inferred from the partially charred glume bases consistent with singeing of cereal ears to remove awns before dehusking and from the weed seeds which resemble those found in fine-sieving by-products. Using functional weed ecology, we infer that the agrosystem at Ploča Mičov Grad resembles high-input practices suggesting that cultivation was small-scale and labour-intensive. Such inferences are often not possible due to the preservation conditions of terrestrial assemblages, resulting in a lack of extensive weed datasets in the southwestern Balkans and southern Europe more broadly.

A. Holguin (✉) · M. Charles · E. Stroud · A. Bogaard
School of Archaeology, University of Oxford, Oxford, UK
e-mail: amy.holguin@arch.ox.ac.uk

M. Charles
e-mail: michael.charles@arch.ox.ac.uk

E. Stroud
e-mail: elizabeth.stroud@arch.ox.ac.uk

A. Bogaard
e-mail: amy.bogaard@arch.ox.ac.uk

F. Antolín · R. Soteras
Department of Natural Sciences, Germany Archaeological Institute, Berlin, Germany
e-mail: ferran.antolin@dainst.de

R. Soteras
e-mail: raul.soteras@dainst.de

F. Antolín · A. Jesus · H. Martínez Grau · B. L. Steiner
Integrative Prehistory and Archaeological Science (IPAS), Department of Environmental Sciences, University of Basel, Basel, Switzerland
e-mail: hector.martinezgrau@unibas.ch

B. L. Steiner
e-mail: bigna.steiner@unibas.ch

© The Author(s) 2025
A. Ballmer et al. (eds.), *Prehistoric Wetland Sites of Southern Europe*,
Natural Science in Archaeology, https://doi.org/10.1007/978-3-031-52780-7_17

Archaeobotanical research on wetland prehistoric sites from this region is currently limited. Our new work highlights the potential of such sites in this area for better understanding of the spread of agriculture and patterns of plant food use during the Neolithic in the southwestern Balkans.

Keywords

Late Neolithic · Agriculture · Crop processing · Agrosystems · Functional plant ecology · Southwestern Balkans · Prehistoric pile-dwelling · EXPLO

17.1 Introduction

The Neolithic (5th millennium BCE) and Bronze Age (2nd millennium BCE) pile-dwelling site of Ploča Mičov Grad is located on the eastern side of Lake Ohrid, North Macedonia. A c. 1.7-m-thick organic cultural layer with abundant waterlogged organic remains was investigated through the lake sediment cores extracted from the area of the pile-dwelling settlement. Upon completion of the EXPLO project, this will be the first time that such a well-preserved organically rich waterlogged deposit is systematically investigated for plant macroremains in southeastern Europe. In this chapter, we present preliminary results from the archaeobotanical analysis of waterlogged macrobotanical remains from the Late Neolithic (calibrated 2-sigma ages 4600–4300 cal BCE) organic cultural layer (Hafner et al. 2021; see Chap. 6 in this book). The aim of this work is to investigate the spectrum of cultivated and collected plant foods from the site in order to better understand the adoption and adaptation of agriculture and other subsistence strategies in the Neolithic. The southwestern Balkans is a particularly important region for understanding the spread of agriculture due to its diverse landscape with Mediterranean, alpine and continental biogeographical regions, each of which presents different challenges to plant cultivation (Fig. 17.1). Despite this, the southwestern Balkans is an under-researched region. The site of Ploča Mičov Grad presents an exciting

opportunity to contribute to efforts to fill this research gap, due to the great inferential potential of the site's abundant and well-preserved waterlogged plant remains.

17.1.1 The Potential of Waterlogged Assemblages from Prehistoric Wetland Sites

Waterlogged archaeobotanical remains from prehistoric wetland sites offer a rich record of human subsistence. Plant parts, such as pods and leaves, that do not normally survive charring, the most common form of archaeobotanical preservation, can be well preserved in such anaerobic conditions. Waterlogging can also preserve fragile or oily seeds, such as poppy seeds (*Papaver* sp.), which are less likely to survive the charring process than more robust plant remains such as cereal grains (Jacomet et al. 1989; Märkle and Rösch 2008). Colledge and Conolly (2014, 199) found that charred assemblages only preserve approximately 35% of the range of wild taxa found in their waterlogged counterparts. Thus, waterlogged assemblages frequently have both a greater range of plant parts and a greater variety of plant taxa. This highlights the value of wetland sites with waterlogged assemblages for gaining a deeper understanding of the diversity and composition of plant assemblages at prehistoric sites (Jacomet et al. 1989; Brombacher and Jacomet 1997; Hosch and Jacomet 2004). In turn, such material promises great inferential power when investigating patterns of past cultivation and consumption of plant foods (Antolín et al. 2020).

17.1.2 The Importance of Prehistoric Wetland Sites in the Lake Ohrid Basin

Several prehistoric sites with the potential for high levels of organic preservation thanks to waterlogging are known within the Lake Ohrid basin (Westphal et al. 2011; Naumov 2020; Hafner et al. 2021; Chap. 6 in this book).

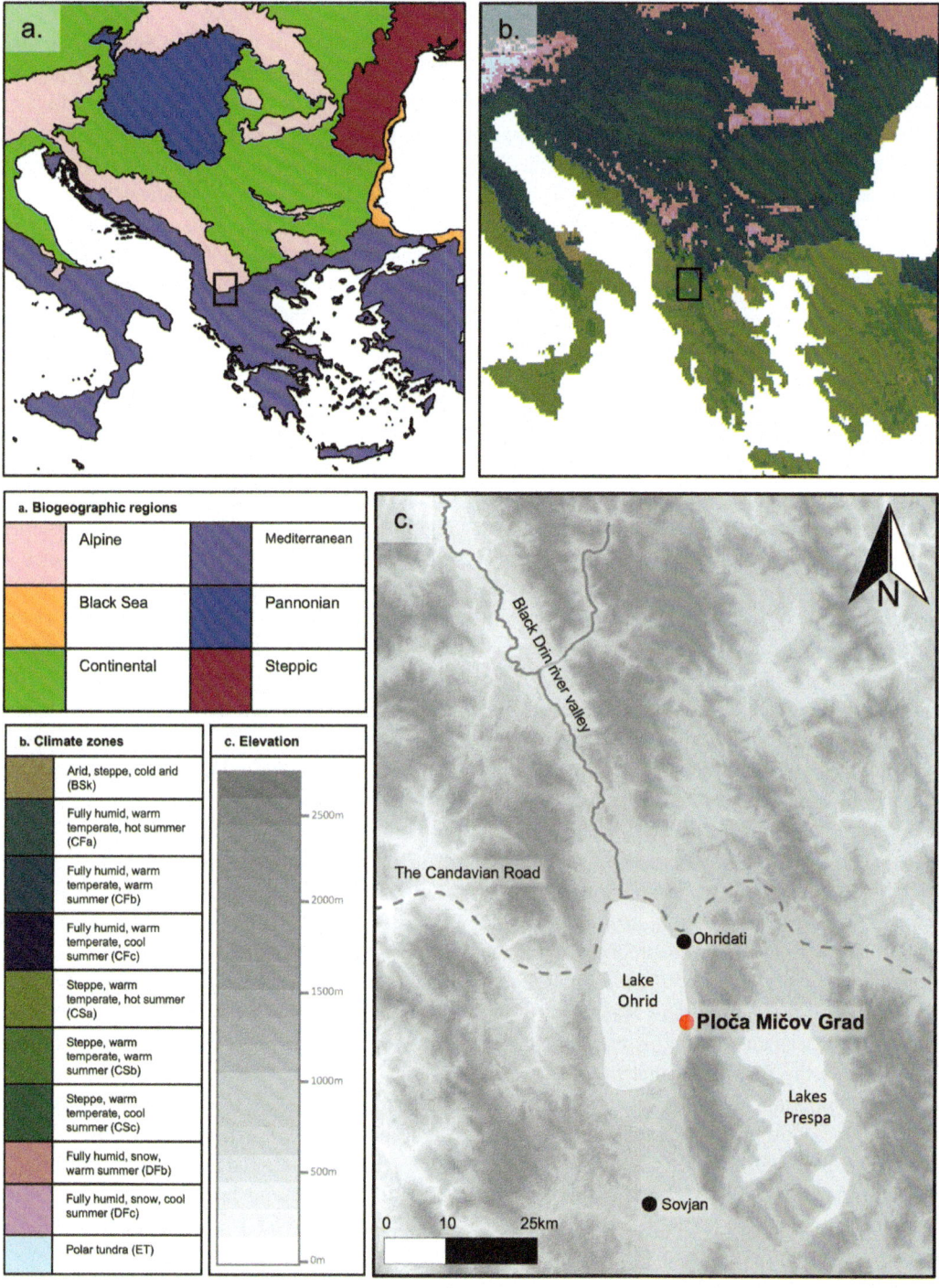

Fig. 17.1 **a** Biogeographic regions in the southwestern Balkans. **b** Köppen-Geiger climate zones of the southwestern Balkans. **c** Topographic map showing the location of the pile dwelling site of Ploča Mičov Grad, the Black Drin River valley, the possible route of the Candavian Road/Via Egnatia, and selected other sites mentioned in the text. Data for biogeographic regions from the European Environment Agency (© Council of Europe (CoE), Directorate-General for Environment (DG ENV), European Environment Agency (2017b)). Köppen-Geiger climate classification after Kottek et al. (2006) downscaling after Rubel et al. (2017) (© Climate Change and Infectious Diseases Group, Institute for Veterinary Public Health, Vetmeduni Vienna, Austria.) Elevation data from USGS Global Multi-resolution Terrain Elevation Data. Possible route of Via Egnatia based on Filipović (2018) and https://www.viaegnatiafoundation. eu/ (last accessed 15/06/2021)

Accompanying Ploča Mičov Grad, on the eastern side of the lake, is the pile-dwelling site of Ohridati, one of the better researched prehistoric sites on the lakeshore. Recovered weights thought to be related to fishing nets and harpoons imply that lacustrine resources were important at the Middle-Late Neolithic site of Ohridati. The distribution of the wooden piles suggests the site was comprised of individual dwellings (Naumov 2020). Both Ploča Mičov Grad and Ohridati show extraordinary levels of organic preservation in the form of vertical wooden piles and so have great archaeobotanical potential, but little research has centred upon the plant foods consumed in the area. Prehistoric sites in this region are of particular interest in investigating changing subsistence practices and understanding the adoption and spread of agriculture across the southwestern Balkans, and into central Europe, as the Lake Ohrid basin lies at a topographically accessible point between biogeographically and ecologically different regions.

The Galičica and Mokra mountain ranges, to the east and west of Lake Ohrid respectively, both have peaks higher than 2000 m, making lowland pathways between the mountains particularly favourable. A prehistoric route crossing the mountain ranges, believed to date back at least to the Iron Age, known as the 'Candavian Road' (*Kandaviski Pat*), runs longitudinally through the Lake Ohrid basin at the northern side of the lake and acted as the predecessor of the Roman road, 'Via Egnatia' (Lolos 2007; Jovanova 2013; Mitrevski 2015; Weissová et al. 2018). The existence of these historically attested routes opens the possibility that the area was a similarly important link between western and eastern zones during earlier prehistoric occupations. The Black Drin River valley towards the northern part of Lake Ohrid further enabled movement between northern and southern regions through the basin (Kuzman 2017; Weissová et al. 2018). Similarities between Neolithic material culture recovered from sites in the Ohrid basin and from pile-dwelling and tell sites further north along the Black Drin River valley, south towards the Korče valley (Kuzman 2017) and to the east in the Pelagonia basin

(Kuzman 2013; Naumov 2020) suggest the interconnectivity of the region, linking the earliest known settlements in the area. Thus, the prehistoric sites on the shore of Lake Ohrid lay at a likely point of passage, enabling the movement of people, materials, and ideas between adjacent areas, due to its topographic accessibility, on the flanks of challenging mountain ranges.

The southwestern Balkans is an area with notable ecological and biogeographic variability which likely influenced when and how different food practices and ideas could have been adopted and adapted (Fig. 17.1). Lake Ohrid is situated within the Pindus Mountains mixed forest ecozone (European Environment Agency 2017a) and, alongside the rest of western North Macedonia, belongs to the 'alpine' biogeographical region of Europe (European Environment Agency 2017b). Regions to the south, such as southern Albania and Greece, are classed as 'Mediterranean' biogeographical regions (European Environment Agency 2017b), featuring the Aegean and Western Turkey sclerophyllous and mixed forests, and the Illyrian deciduous forests ecozones (European Environment Agency 2017a). Areas to the north, such as Kosovo and Serbia, are considered 'continental' biogeographically (European Environment Agency 2017b), with Balkan mixed forests ecozones (European Environment Agency 2017a).

Environments within an area vary locally, and current ecological conditions may not accurately characterise the prehistoric landscape. Nevertheless, the influence of physical geographical factors such as geology, latitude, and altitude, coupled with data indicating relatively unchanged climatic conditions in the region following the Rapid Climate Change event (6550–6050 cal BCE) (Krauß et al. 2018), suggests that biogeographic classifications remain relevant and useful for exploring broad patterns. Different biogeographic regions and ecozones likely fostered different subsistence strategies. Exploring how subsistence strategies at prehistoric sites in the Lake Ohrid basin compare with those at other contemporary alpine sites and sites located in the Mediterranean and continental zones may afford insights into how new ideas or foods were adopted and adapted in different areas, and suggest what balance of biogeographical, ecological, and cultural factors influenced such changes. This is especially so given that the lake lies at this plausible point of connectivity, through which new ideas, people, or foods could have travelled, between the north and south, and the east and west.

17.1.3 The State of Research of Prehistoric Wetland Sites in the Lake Ohrid Region in the Context of the Southwestern Balkans

While prehistoric wetland sites in the Ohrid basin are numerous and have the potential for extraordinary levels of organic preservation, none of these sites have been the subject of a published comprehensive archaeobotanical study or extensive chronological investigation, and so such questions regarding prehistoric subsistence strategies in the area remain under-researched. This echoes the pattern more broadly seen in the southwestern Balkans. Many prehistoric sites are known on riverbeds and lakeshores across the region, such as on the ancient Lake Maliq (Albania), the Panonic Lake (Bosnia), Lake Orestia (Greece), and Great Lake Prespa (North Macedonia, Albania, and Greece), but few are well researched, and doubtlessly, many more remain undiscovered (Naumov 2020). Archaeological interest in wetland sites in this area has only recently grown during the last few decades. For comparison, the waterlogged prehistoric sites in the Alpine forelands (Austria, France, Germany, Italy, Slovenia, and Switzerland) have been researched for the past c. 150 years (Menotti 2004) to the extent that, at present, nearly 1000 such sites have been identified, 111 of which are included on the UNESCO World Heritage list (UNESCO World Heritage Centre 2021) Thus, extensive excavations and new research projects are needed to fill this distinct gap in knowledge surrounding prehistoric wetland sites in the southwestern Balkans.

17.1.4 Aims and Outcomes of Preliminary Archaeobotanical Research at the Site of Ploča Mičov Grad

The research presented in this chapter derives from new excavations at the prehistoric site of Ploča Mičov Grad, on the eastern shore of Lake Ohrid. Current excavations, beginning with a pilot study in 2018 and an extensive campaign in 2019, are part of the ERC Synergy project 'Exploring the dynamics and causes of prehistoric land use change in the cradle of European farming', known as EXPLO, a collaboration of the universities of Bern, Oxford, and the Aristotle University of Thessaloniki, designed to investigate the adoption of agriculture at lakeshore sites in the southwestern Balkans. Recent extensive dendrochronological and radiocarbon analysis by the EXPLO team at the University of Bern suggests that the site reflects multiple settlement phases from the Late Neolithic (5th millennium BCE) to the Bronze Age (2nd millennium BCE) (Hafner et al. 2021; see Chaps. 6 and 14 in this book). The Late Neolithic settlement phase related to the organic cultural layer, sampled through the lake sediment coring, is estimated to have lasted approximately 100 years based on tree felling dates. Hence, the site offers the opportunity to explore changes to subsistence strategies through a well-defined occupation period. Here, we report the archaeobotanical findings from one of the lake cores from Ploča Mičov Grad with plant remains dating to the Late Neolithic. The high levels of preservation and the abundance of the archaeobotanical material from this single core show that the site has great promise for archaeobotany. We present preliminary inferences regarding the site's subsistence strategies and agrosystems and consider how this compares to researched food systems from other lakeshore sites in the southwestern Balkans and southern Europe more broadly. Future work on further lake core samples and bulk samples from an underwater sondage at Ploča Mičov Grad will enable us to consider intra-site variability, perhaps reflecting different activities across the site or different depositional/taphonomic processes, and to identify settlement-wide trends.

17.2 Materials and Methods

17.2.1 The Site of Ploča Mičov Grad, Lake Ohrid, North Macedonia

Ploča Mičov Grad (40.994° N, 20.798° E) lies on the eastern shore of Lake Ohrid, at an altitude of 693 m a.s.l. The Late Neolithic to Bronze Age site is now fully submerged, above which a modern museum is situated with pile-dwelling reconstructions (Fig. 17.1). The lake measures approximately 30 km in length (N–S) and 14 km width (E–W), with an estimated total volume of 50.7 km^3 and a maximum depth of 293 m. At its deepest point, sediment accumulation rates are estimated to be c. 0.5 mm per year (Lacey et al. 2015). The Lake Ohrid basin is located on the border between present-day North Macedonia to the northern and eastern shores, and Albania to the southern and western sides of the lake. Within the Pindus Mountains mixed forest ecozone, this area is characterised by mixed broadleaf forest in lowland zones and coniferous forests at higher altitudes (1200–2500 m) (Dinaric Arc and Balkans Environment Outlook 2010) and a Mediterranean-type climate with warm summers (Kottek et al. 2006). Within the basin, the average annual rainfall is 907 mm and temperature is 11.1 °C, reaching heights of 31.5 °C and lows of − 5.7 °C (Popovska and Bonacci 2007).

17.2.2 Coring, Sampling, and Processing of the Samples

During the 2019 season, 14 cores were extracted using a Niederreiter coring system by the EXPLO palaeoecological coring team, led by Prof. André Lotter. These were arranged along two approximately perpendicular transects (north–south and east–west) centred on the main

settlement area of the visible vertical piles (Fig. 17.2). Plant macrofossil samples were taken for radiocarbon dating from the top and bottom of the visible organic layer from four of these cores, returning calibrated 2-sigma ages ranging from c. 4600 to 4300 cal BCE. Bayesian modelling of these dates alongside felling dates from dendrochronological analysis of the piles suggests that the organic layer present in the cores accumulated across approximately 100 years, between the start of the 45th and the middle of the 44th century BCE (Hafner et al. 2021; see Chaps. 6 and 14 in this book). Eight bulk samples were also taken along a grid from an excavated underwater sondage within the settlement area and will be the subject of further archaeobotanical analysis. One of the four cores, QLM19 0, is the focus of this chapter.

Core QLM19 0 was extracted in 3 parts. The top sub-core measured 60 cm in length, the middle 90 cm, and the bottom 100 cm (Fig. 17.2). A thick organic layer was visible throughout the top and middle sub-cores. The bottom sub-core appeared to be largely lake marl and was left for future macrofossil and pollen analysis. In total, the organic layer measured 1.7 m in thickness.

28 samples were taken from the organic layer. A sample was taken from each distinct stratigraphic horizon. When layers were not apparent, samples were taken at regular intervals (average c. 6 cm) based on depth. 17 samples were taken from the lower part of the organic layer in middle sub-core. These were originally labelled 0.1.1–0.1.17. For the purposes of clarity in this chapter, the samples have been renamed 1–17. A further 11 samples were taken from the upper part of the organic layer in the top sub-core, originally labelled 0.0.1–0.0.11 and renamed 18–28. Sample 1 is the lowermost sample and sample 28 is the uppermost.

The volume of each sample was measured before processing, by calculating the volume of water displaced when a sample was submerged in a beaker (Antolín et al. 2015). The sample volumes ranged from 30 to 200 ml with an average of 93 ml. The total volume of sieved sediment measured 2602 ml. The organic and inorganic

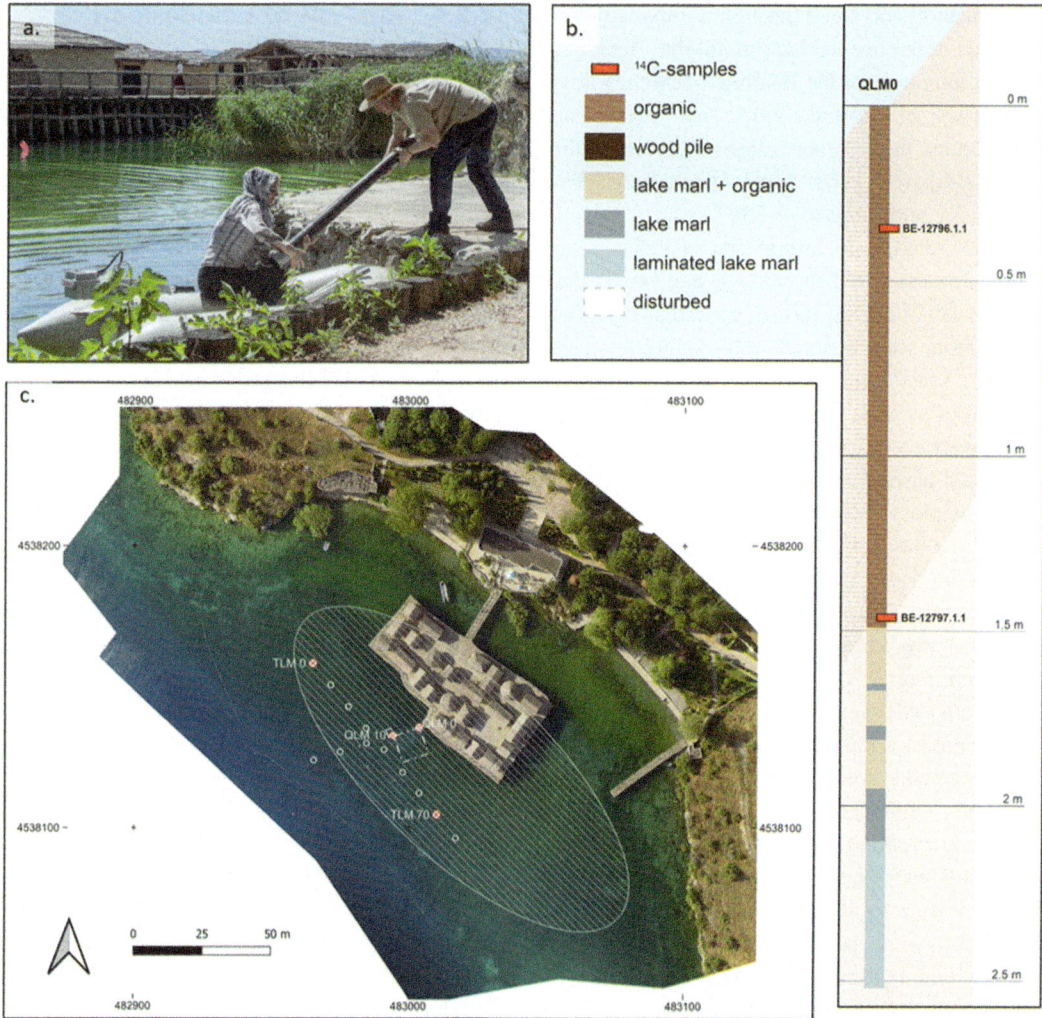

Fig. 17.2 **a** Team members moving a core (R. Soteras). **b** Schematic diagram of drill core QLM19 0, showing key stratigraphic features and the radiocarbon sampling locations (A. Bieri, U Bern, J. Reich and A. Ballmer EXPLO/ U Bern). **c** Aerial orthophoto of the pile-dwelling site of Ploča Mičov Grad with the area containing remains of wooden piles marked by a hatched oval. A modern museum is built above the remains of the site. White circles indicate the locations of core drillings, those with a red cross indicate the cores from which samples were taken for radiocarbon dating (J. Reich, EXPLO/U Bern)

fractions of each sample were sieved through 2 and 0.35 mm meshed sieves using the wash-over method (e.g. Steiner et al. 2015). The < 0.35 mm fraction was retained unanalysed. The 2 mm fraction was analysed in full. If the 0.35 mm fraction sample volume exceeded 14 ml, it was randomly sub-sampled to a volume of approximately 10 ml by grid sampling (e.g. Steiner et al. 2017). No random sub-samples amounted to less than approximately a quarter of the total sample.

17.2.3 Identification and Counting of Botanical Macroremains

Samples were sorted under a low-powered microscope (x6.3-50). The 17 samples from the lower part of the organic layer (1–17) were sorted on site by AB, FA, HMG, ES and AJ. The upper 11 samples (18–28) were sorted at the Institute of Archaeology, University of Oxford, by AH.

Identification was done through comparison with the seed reference collection in the Archaeobotany Laboratory at the Institute of Archaeology, University of Oxford, aided by images and descriptions in reference atlases (Körber-Grohne 1964; Berggren 1969, 1981; Anderberg 1994; Bojňanský and Fargašová 2007).

Items relating to fauna (for instance bones, dung, insects, and fish scales), and some plant items, such as leaves, thorns, charcoal, stems and cereal bran, were counted semi-quantitatively on a 3-part scale from rare to abundant (following Antolín et al. 2017). All other crop material and seeds were recorded quantitatively.

Glume bases were counted if the whole glume base was present. Spikelet forks were counted as two glume bases. For rachis fragments, the upper part of the node (with glume attachment points) was counted. Fragments of apical and embryo ends of charred cereal grains were both counted, as they were rare; if fragments of comparable grains with both ends were present, they were counted together as one. Pulse pod and pistachio (*Pistacia sp.*) fragments were counted if they exceeded 12 mm^2, while acorn (*Quercus* sp.), hazelnut (*Corylus avellana*) and almond (*Amygdalus*) fragments were counted with a threshold of 25 mm^2 due to their larger size. Two pulse cotyledons were counted as one seed. Flax (*Linum usitatissimum*) capsules naturally break into ten segments or 'septa' (Diederichsen and Richards 2003); tips of these septa, or fragments containing two mid-ribs to which a tip could not be connected, were counted. The principle of minimum number of individuals (MNI) was applied visually when counting all other seeds. Some crop items for which it was challenging to estimate an MNI, such as cereal awns and pulse stalks, were recorded as fragment counts.

In the text and figures, counts from subsampled fractions were multiplied to represent the absolute number in the full sample and rounded to a whole number. Counts from the 2 and 0.35 mm fraction were combined.

17.2.4 Analysis of Composition of Samples

Correspondence analysis was performed to explore compositional variation among the samples, using CANOCO 5. This ordination technique arranges samples along two axes based on the extent of their compositional similarity. The horizontal axis is the primary axis and captures the dominant trend in the sample compositions. The position of a sample from the origin reflects the degree of similarity between samples, based on taxon abundance and the combinations of taxa present in the sample. Only taxa occurring in at least 10 samples were included. In cases where the same taxa were represented in uncharred and charred form, they were entered separately into the analysis. In total, 27 taxon categories were included. A logarithmic transformation was used due to the very high numbers of certain taxa in some samples.

Discriminant analysis was performed to investigate how the weed/wild seed characteristics compare to those of known crop processing products and by-products. This approach uses three attributes (seed size, whether seeds are free or enclosed in a head, and whether seeds are easily blown by the wind) to assess the similarity of weed/wild seeds in the archaeobotanical samples to weed seeds in winnowing by-products, coarse-sieving by-products, fine-sieving by-products, and fine-sieving products of manually processed cereals documented on the island of Amorgos, Greece (Jones 1984, 1987). Following Jones (1984, 1987), the samples were classified as unknown cases in a discriminant analysis, using functions extracted to maximise the separation of ethnobotanical crop processing groups using the 'leave one out' option (82% of modern ethnobotanical samples correctly reclassified in IBM SPSS Statistics 26). Results were plotted using R version 3.6.1 (R Core Team 2019).

Discriminant analysis was also employed to explore how the functional ecological traits of the weed/wild seeds compared to those from different modern agricultural regimes, following Bogaard et al. (2016, 2018). This approach considers the specific leaf area, canopy diameter, leaf area per node: leaf thickness, and canopy height of each weed taxon to create a discriminant function distinguishing between weeds commonly found in low labour input agricultural 'fields' and those found in high labour input 'gardens'. This is based on modern survey datasets collected from present-day traditional systems managed under different intensities in Morocco, Spain, France, and Greece (Bogaard et al. 2016, 2018). The presence or absence of weed species in different regimes was used to create a discriminant function that maximally separates out these high and low labour input systems. All archaeobotanical weed/wild taxa identified to species, or at least to a small number of ecologically comparable species (e.g. *Mentha aquatica/arvensis*), were included in the analysis. This totalled 29 taxa. Two of the lower samples (2 and 6) were removed from the analysis since they contained less than 10 weed seeds included in the analysis. The samples were entered into the classification phase of the discriminant analysis, as unknown cases using a discriminant function, extracted using the 'leave one out' option (94% of modern agricultural plots correctly reclassified as 'low input' or 'high input' in IBM SPSS Statistics 26). Results were plotted using R version 3.6.1 (R Core Team 2019).

17.3 Results

17.3.1 Overall Core

The organic layer was visually relatively homogeneous, lacking visible sterile layers interspersed between organic layers as commonly observed in Alpine foreland lakeshore site stratigraphies (e.g. Sipplingen, Lake of Constance (Kolb 1997)). Large fragments of pottery, bone, teeth, molluscs, burnt clay, and stone were recorded within the organic layer, as well as some variation in macroscopic composition that we are continuing to assess in the light of the archaeobotanical results.

The average density of fully quantified plant items per litre sediment throughout the organic layer was 8276. Sample 3, towards the bottom of the organic layer, had the lowest density of 1265 items/litre. Seven samples had more than 10,000 items/litre, the greatest density being 16,780 items/litre in sample 8 (Fig. 17.3).

Crop material was the main botanical category found in the organic layer, comprising 57.8% of all items, followed by fruits and nuts (26.3%), weed/wild taxa (9.7%), oil-seed crops (4.1%), wetland types (2.0%), and finally other wild items (0.4%) (Fig. 17.4). The other wild taxa category largely comprised of arboreal seeds and needles such as fir (*Abies* sp.) and pine (*Pinus* sp.). Counts of wetland taxa (primarily *Chara* sp. and *Najas* sp.) were low throughout the organic layer, but increased notably to the top, making up 34.0% of the total items recovered in the uppermost sample (28). Rodent dung occurred in most samples but small ruminant dung only in a few. Other non-botanical organic remains included fish scales, fish bone, and insect remains.

The waterlogged botanical remains were primarily preserved uncharred (88.8% of all items). However, charred material (11.2% of all items) was still present throughout the samples. All of the main botanical categories discussed included some charred material, except for wetland taxa which were only found uncharred.

17.3.2 Botanical Categories

The largest botanical category, the crop material, included cereal and pulse remains (Fig. 17.5). Uncharred, and in lower numbers charred, cereal chaff items (particularly spikelet forks, glume bases, and rachis) were recovered from all samples. The uncharred glume bases were by far the most dominant group, making up 83.1% of all crop material (Fig. 17.6). These proved challenging to identify to species, as they were frequently distorted. Some uncharred glume

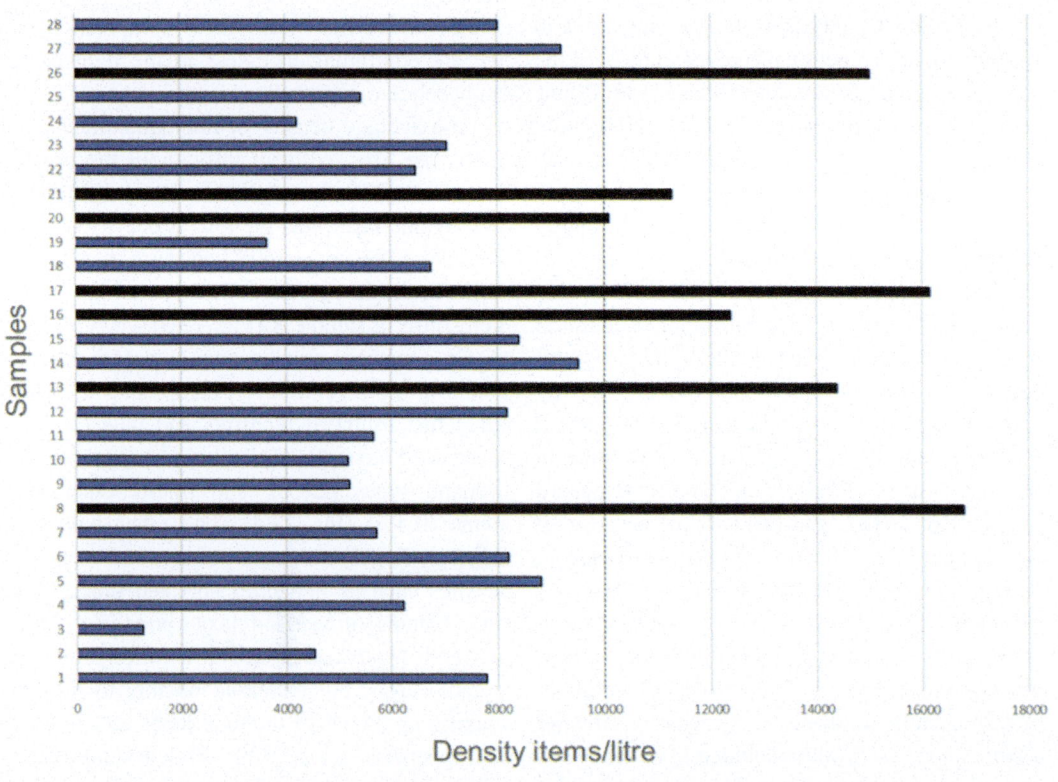

Fig. 17.3 The density of each sample in items per litre. Samples are ordered on the *y*-axis following stratigraphic order, with the uppermost sample (28) at the top of the bar-chart. There are seven samples with over 10,000 items per litre indicated in black (Authors)

bases appeared to be blackened and singed in places, and others appeared lightly charred ('toasted'), being brown in colour and soft (Fig. 17.7). These singed (8.2% of glume bases) and brown (2.5% of glume bases) glume bases retained their original shape to a greater extent and so some of these, alongside some of the charred glume bases (7.0% of glume bases), could be identified to species as einkorn (*Triticum monococcum*, *n* = 395), emmer (*Triticum dicoccum*, *n* = 396) and New Glume Wheat (*Triticum timopheevii* group, *n* = 13). Uncharred and charred rachis nodes were also found making up 3.2% and 1.1% of all crop material, respectively. These were identified as barley (*Hordeum distichon/vulgare*, *n* = 381) and free-threshing wheat rachis (*n* = 43). Work is ongoing to identify barley rachis to species/variety. Across the chaff, einkorn, emmer, and barley were identified in similar numbers. However,

8826 glume bases were unidentifiable to species. The charred cereal chaff generally exhibited a good state of preservation. For example, many of the spikelet forks had a complete internode, and palea and lemma tissues were sometimes preserved still nestled within the spikelet forks. Charred cereal grains were comparatively few (*n* = 108), often less well preserved and fragmented. Some cereal grain fragments (*n* = 22) appeared to have been broken before charring (Valamoti et al. 2021). Fragments of cereal bran were found in all samples.

Uncharred pulse pod was found throughout the organic layer and comprised 4.7% of all crop material, with particularly high concentrations (*n* = 237) in sample 7 in the lower quarter of the organic layer (Fig. 17.6). A small amount of charred pulse pod (0.1% of all crop material) and charred pulse seed, hilum, and testa (0.2% of all crop material) was recovered. The uncharred

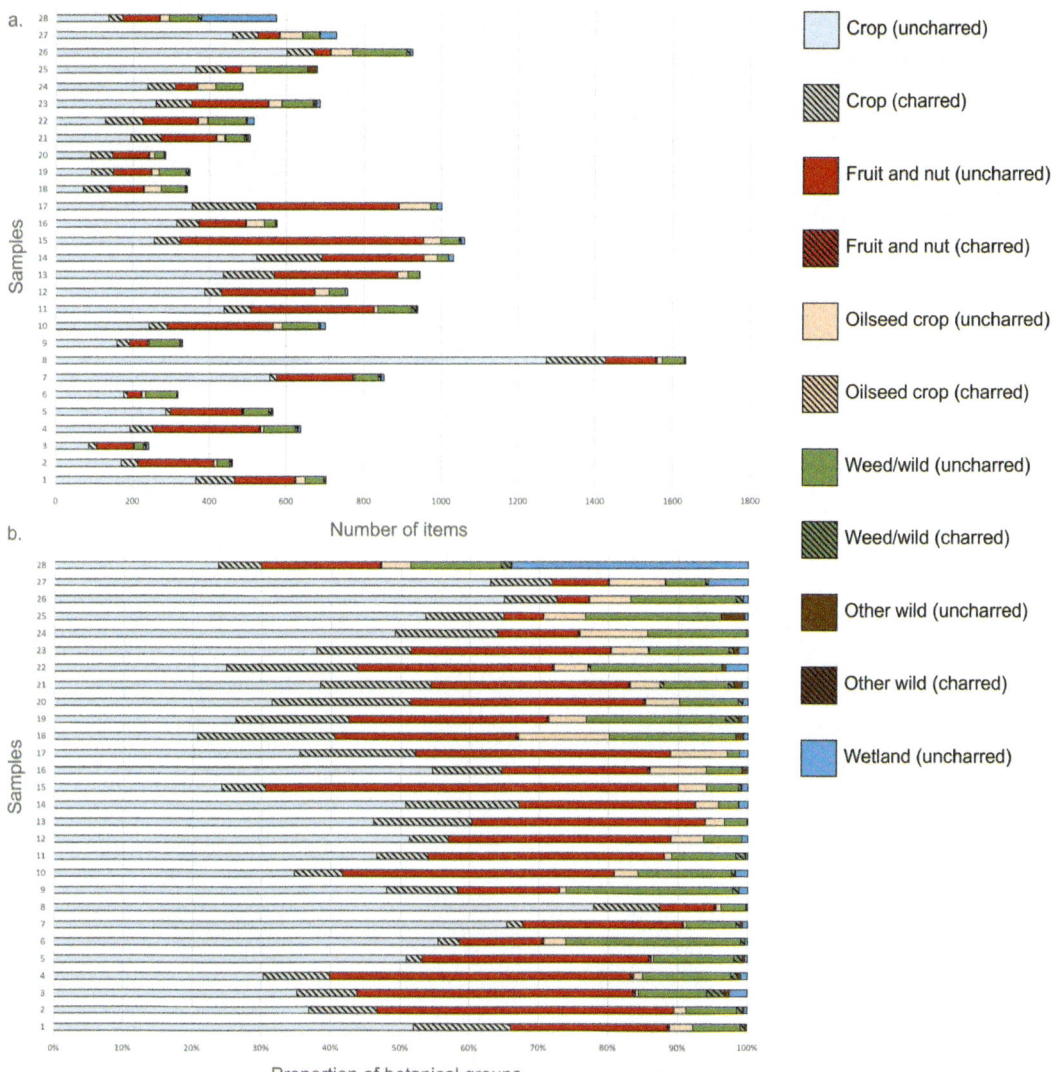

Fig. 17.4 **a** Bar-chart showing the absolute number of items in each major botanical category in each sample. **b** Relative proportion of each botanical category in each sample. Charred material is indicated with a diagonal fill. The most abundant category is the crop remains which here refers to cereal and pulse crops. Counts of wetland taxa are very low, although they are notably most abundant in the uppermost sample (28), likely reflecting taphonomic process (Authors)

pulse pod was mostly found highly fragmented with little discernible shape; further taxonomic identification is ongoing. The charred pulse seeds were identified as bitter vetch (*Vicia ervilia*, $n = 3$), lentil (*Lens culinaris*, $n = 2$) and, pea (*Pisum sativum*, $n = 2$).

Oil-seed crops were also found, 99.1% of which were uncharred (Fig. 17.8). Uncharred opium poppy (*Papaver somniferum*) seeds were found in almost all samples, markedly increasing in the upper half of the organic layer. Sample 18 towards the middle of the organic layer had the largest concentration of seeds ($n = 41$). Uncharred flax seeds (*Linum usitatissimum*) were present in most samples, although they were notably absent in the top 4 samples. The highest numbers could be found in samples towards the middle and at the very bottom of the organic layer.

Fig. 17.5 Crop remains from Ploča Mičov Grad. **a** Uncharred glume base. **b** Singed glume base. **c** Brown *Triticum dicoccum* spikelet fork. **d** Uncharred *Hordeum distichon/vulgare* rachis. **e** Charred *Triticum monococcum* spikelet fork. **f** Charred *Triticum dicoccum* spikelet fork. **g** Charred *Triticum timopheevii* group (new glume wheat) spikelet fork. **h** Charred *Hordeum distichon/vulgare* rachis. **i** Charred *Triticum monococcum* grain. **j** Charred *Triticum dicoccum* apical grain fragment with a pre-charring break. **k** Charred *Hordeum distichon/vulgare* grain. **l** Uncharred pulse pod fragment. **m** Uncharred *Linum usitatissimum* capsule fragment. **n** Uncharred *Linum usitatissimum* seed. **o** Charred *Vicia ervilia* seed. **p** Uncharred *Papaver somniferum* seed. **q** Charred pulse pod tip fragment (R. Soteras and A. Holguin)

Uncharred flax capsule fragments were found in every sample, increasing in number towards the middle and the top of the organic layer. The number of flax seeds and flax capsules did not appear to strongly correlate with each other. A few charred poppy seeds and charred flax capsule fragments were recovered from samples near the top of the organic layer.

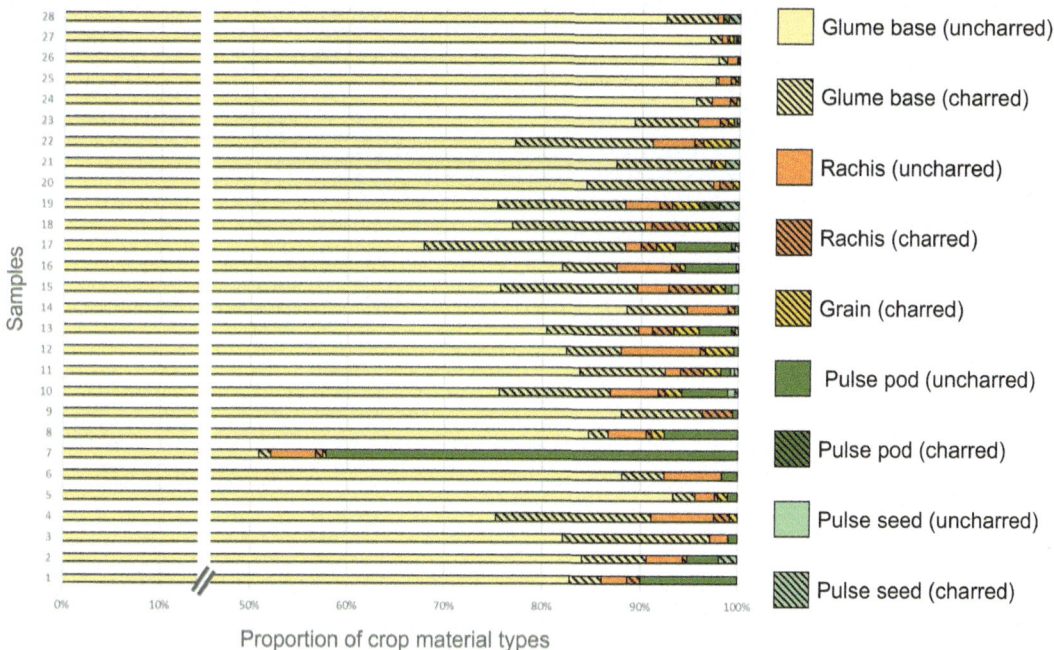

Glume base (uncharred)

Glume base (charred)

Rachis (uncharred)

Rachis (charred)

Grain (charred)

Pulse pod (uncharred)

Pulse pod (charred)

Pulse seed (uncharred)

Pulse seed (charred)

Proportion of crop material types

Fig. 17.6 Relative proportions of crop material in each sample. Material relating to cereals is shown in yellows and oranges and those relating to pulses in green. Charred material is indicated by a diagonal fill. Uncharred glume bases are by far the most abundant crop remain, making up at least 50% of all samples; as such, there is a break in the x-axis (Authors)

Uncharred blackberry (*Rubus fruticosus* agg.) seeds constituted by far the most abundant taxon in the fruits and nuts category. The greatest number of blackberry seeds (*n* = 552) were found in sample 15 towards the middle of the organic layer. Almond (*Amygdalus* sp.), hazelnut (*Corylus avellana*), acorn (*Quercus* sp.), and pistachio (*Pistacia atlantica/terebinthus*) nutshell fragments were found, along with other fruit seeds including strawberry (*Fragaria vesca*), juniper (*Juniperus* sp.), cornelian cherry (*Cornus mas*), fig (*Ficus carica*), and elderberry (*Sambucus* sp.) (Fig. 17.9). Items in this category were predominantly uncharred (99.5%), with a single charred elderberry and a few charred nutshell and blackberry seed fragments recovered.

The majority of weed/wild seeds were uncharred (94.7%). The most numerous uncharred weed/wild seeds were St John's-wort (*Hypericum perforatum*), thyme-leaf sandwort (*Arenaria serpyllifolia*), dodder (*Cuscuta* sp.) fathen (*Chenopodium album*), and black bindweed (*Polygonum convolvulus*). Charred weed/wild taxa included wild oat (*Avena* sp.), campion (*Silene* sp.), and St John's-wort (*Hypericum perforatum*). Some seeds such as lemon balm (*Melissa officinalis*), oregano (*Origanum vulgare*), and horse mint (*Mentha longifolia*) could have been collected as herbs. However, they are counted here as weed/wild seeds due to their low numbers, as they are also commonly found growing wild in the region. The weed/wild taxa in the assemblage notably include some autumn-germinating species (e.g. *Valerianella dentata*) as well as spring-germinating ones (e.g. *Polygonum convolvulus*), based on their flowering time (Bogaard et al. 2001).

17.3.3 Correspondence Analysis

Together axis 1 and 2 of the correspondence analysis accounted for 35.5% of the variance in the samples (Fig. 17.10). The distribution of

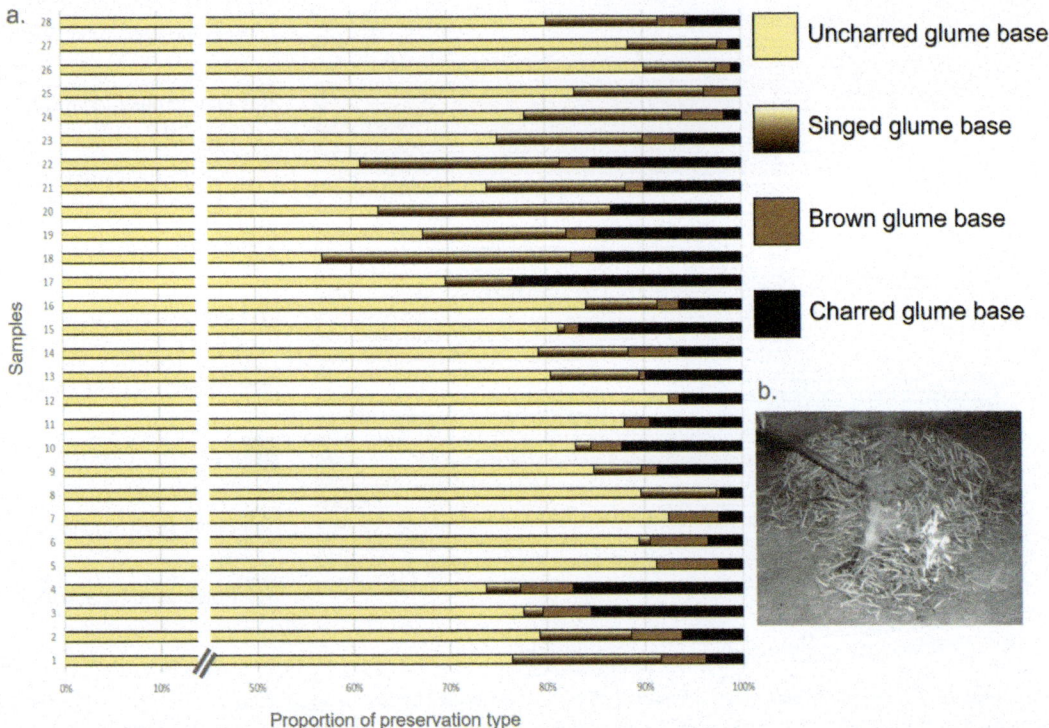

Fig. 17.7 **a** Relative proportions of glume bases in each sample according to preservation type. Glume bases were recovered in uncharred, singed, brown, and fully charred form. Uncharred glume bases are by far the most abundant crop remain, making up at least 50% of all samples; therefore, there is a break in the *x*-axis. **b** Singeing of hulled wheats to scorch awns before or during dehusking in Asturias, Spain (Peña-Chocarro and Zapata 2003, Fig. 6)

samples along axis 1 (the horizontal *x*-axis) roughly corresponds to their depth; samples towards the top of the organic layer are found further towards the left of the scatterplot and those towards the bottom are found further towards the right. Samples further towards the top of the organic layer (with the exception of the uppermost sample, 28) were positioned relatively close together along axis 2 (the vertical *y*-axis), whereas samples further towards the bottom are more dispersed vertically, reflecting greater variation between the compositions of the lower samples.

Proportions of uncharred and charred material per sample do not reflect trends along either axis. Uncharred material dominated all samples. The proportion of charred material fluctuated but did not reflect the position of samples on the axes.

The proportions of different botanical categories, however, did approximately correspond to the position of the samples. Samples with greater amounts of fruits and nuts generally were positioned to the top right of the scatterplot and those with greater amounts of oil-seed crops to the bottom left. The uppermost sample (28) is the only sample positioned high in the top left-hand corner of the scatterplot. It is also the only sample with a large amount of wetland taxa present (Fig. 17.10).

17.3.4 Crop Processing Discriminant Analysis

Almost all of the samples coincide in the discriminant function plot with the distribution of

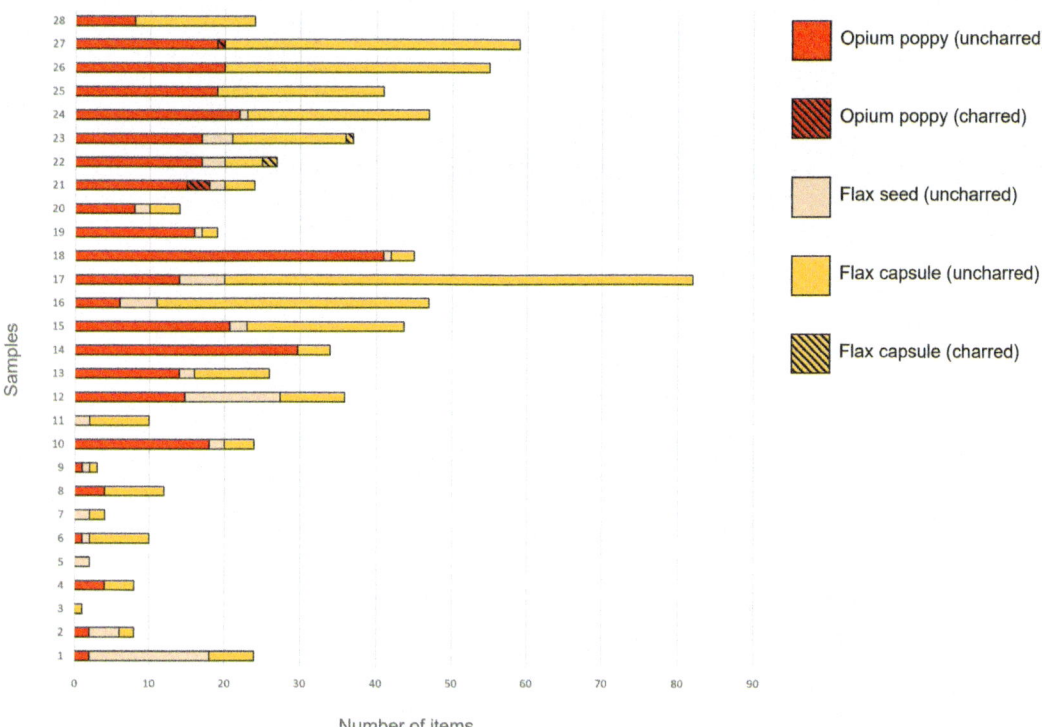

Fig. 17.8 Absolute number of remains from different oil-seed crops (namely opium poppy and flax seeds or capsule fragments in the sample) in each sample. Charred material is shown with a diagonal fill. It is however very rare in this category (Authors)

ethnobotanical samples from known crop processing stages (Fig. 17.11). The archaeobotanical samples largely overlap with the fine-sieving by-product group. A few samples are positioned below this group. Some samples are placed in the region of the plot where fine-sieving products overlap with fine-sieving by-products and winnowing by-products. Samples show some grouping by depth.

17.3.5 Weed Functional Ecology

The discriminant scores of the archaeobotanical samples, in comparison to modern high- and low-input agricultural plots are shown in Fig. 17.12. Most of the samples are intermediate between the low-input and high-input regimes but mostly align with the lower end of the high-input group. There is some grouping by depth but no consistent trend.

17.4 Discussion

Many samples showed a very high density of items per litre with 7 samples exceeding 10,000 items/litre. This is comparable to well-preserved organic cultural layers in lakeshore sites of the Alpine foreland (Steiner et al. 2017). Such high densities mean that data at a high resolution can be collected from the samples and thus the core offers great inferential potential. The lack of sterile layers within the organic layer suggests that it resulted from likely relatively continuous accumulation of material. Loss of some material due to degradation is, however, probable. The thickness of the organic layer, the low number of wetland species, the high number of crop remains, and the finds of cultural artefacts such as pottery all strongly suggest that the organic layer is cultural in origin and reflects repeated activities in the pile-dwelling settlement.

Fig. 17.9 Collected and wild uncharred macrobotanical remains from Ploča Mičov Grad. **a** *Pistacia atlantica/terebinthus* nutshell. **b** *Cornus mas.* **c** *Amygdalus* sp. nutshell. **d** *Sambucus* sp. **e** *Juniperus* sp. **f** *Rubus fruticosus* agg. **g** *Fragaria vesca.* **h** *Ficus carica.* **i** *Abies* sp. needle. **j** *Chenopodium album.* **k** *Hypericum perforatum.* **l** *Origanum vulgare.* **m** *Cuscuta* sp. **n** *Valerianella dentata.* **o** *Polygonum convolvulus* (R. Soteras and A. Holguin)

Variation in the density and the exact compositions of samples shows that the organic layer is not perfectly homogeneous. The evident variation surrounds a continuous theme, with crop remains and gathered fruits and nuts overwhelmingly the most abundant items throughout. Thus, the slightly variable, but cohesive, densities and compositions suggest that the layer accumulated as a result of repeated plant-related activities, across an estimated 100 years, corroborated by evidence through the radiocarbon dates and tree felling dates through dendrochronological analysis (see Chap. 14 in this book).

Glume wheats (einkorn and emmer) are the most abundant cereal taxa, followed by barley. The importance of glume wheats is particularly evident when unidentified glume bases are taken into consideration. Charred grains of einkorn, emmer, and barley were all found. However, these are too few in number to make conclusions about the relative abundance of the different cereals. Emmer, einkorn, and barley are consistently thought to be the most important crops throughout the Neolithic across Europe more broadly. In the Balkans, this can be seen in charred assemblages from the Late Neolithic lakeshore site of Maliq, Albania (Xhuveli and Schultze-Motel 1995) and the Early-Middle Neolithic sites of Bâlgarčevo II, Bulgaria (Marinova and Popova 2008). The Early Neolithic site of Vrbjanska Čuka, situated in the Pelagonia Valley, North Macedonia, shows a similar spectrum of crops, but with a dominance of einkorn (Antolín et al. 2020). New Glume Wheat has been identified at some sites in the Balkans including the Middle-Late Neolithic site of Karanovo, Bulgaria (Kreuz et al. 2005).

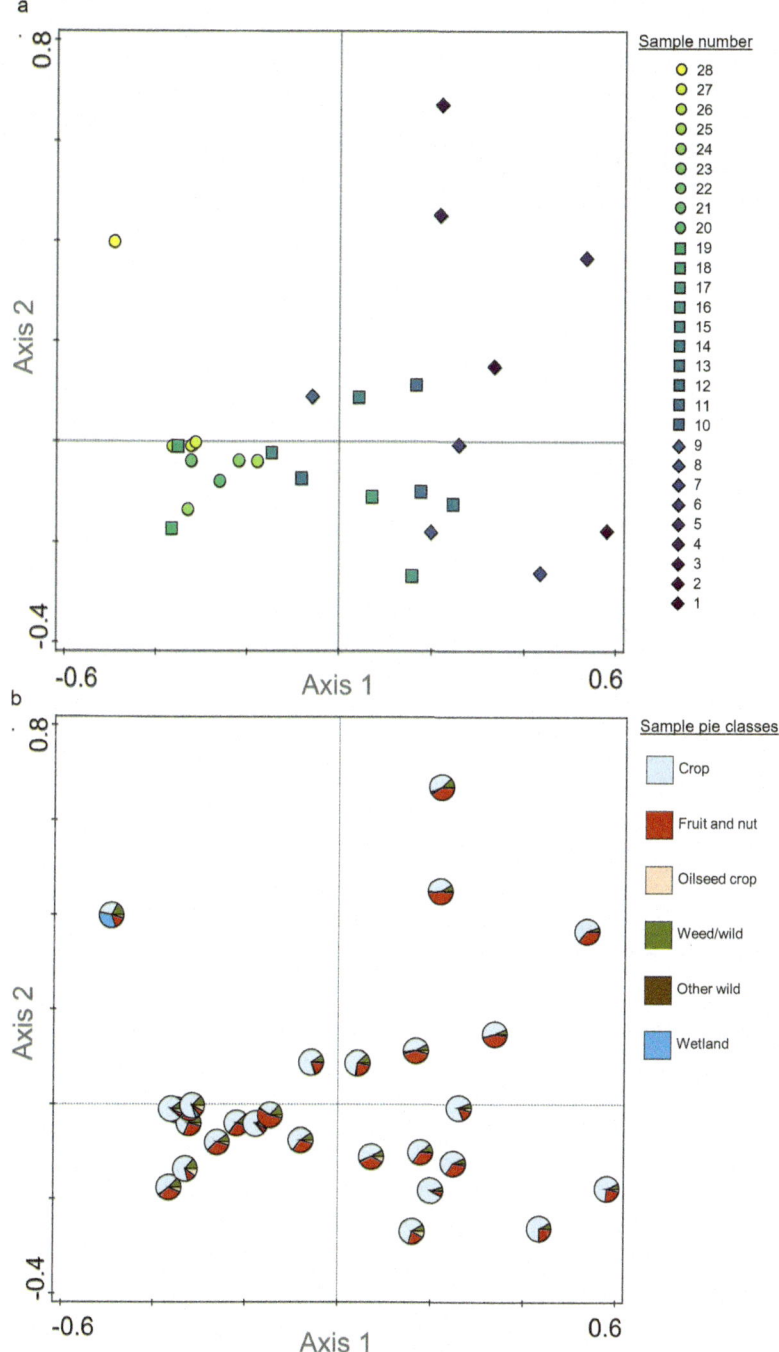

Fig. 17.10 Scatterplots showing the results of the correspondence analysis (axis 1 horizontal and axis 2 vertical). **a** Samples are coloured according to depth with the uppermost sample (28) in yellow and the lowermost sample (1) in dark purple. **b** Samples are shown as pie-charts showing the relative proportions of the major botanical categories (Authors)

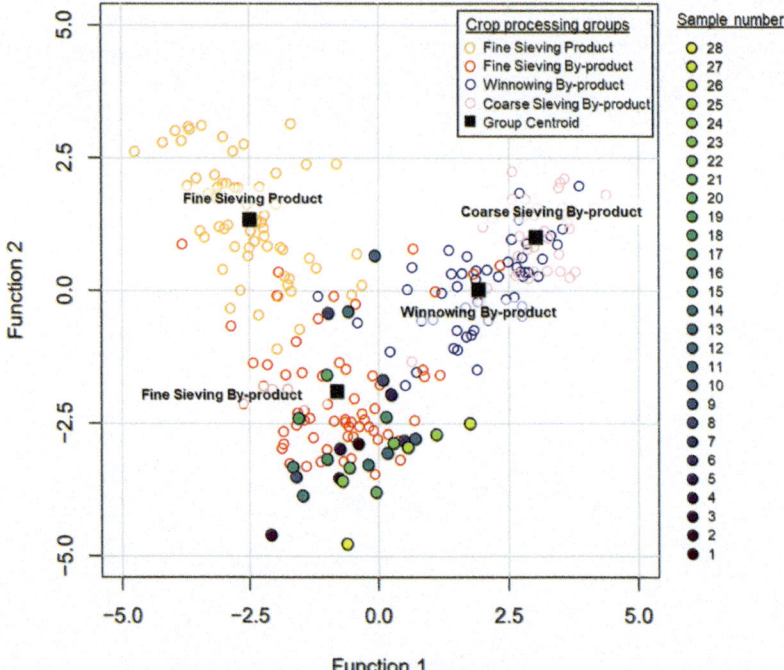

Fig. 17.11 Scatterplot showing the distribution of the archaeobotanical samples from Ploča Mičov Grad along two discriminant functions extracted to distinguish four modern ethnobotanical crop processing sample groups based on physical characteristics of weed seeds. Samples are coloured according to depth with the uppermost sample (28) in yellow and the lowermost sample (1) in dark purple (Authors)

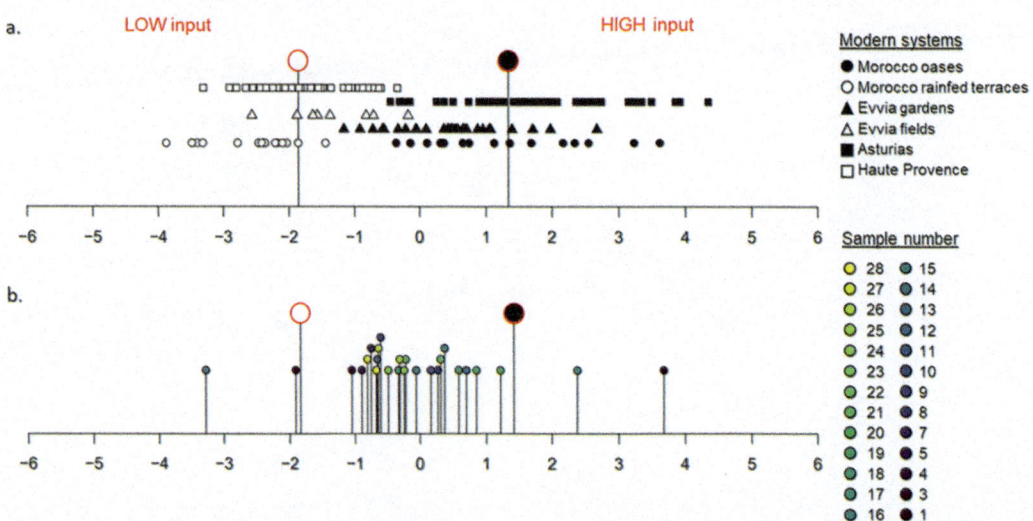

Fig. 17.12 **a** Results of a discriminant analysis separating modern low- and high-input fields based on four functional attributes of the constituent weed taxa. **b** Distribution of the archaeobotanical samples from Ploča Mičov Grad when entered as unknown cases. The group centroids are shown with larger symbols with a red border (Authors)

The range of preservation conditions of glume bases, from uncharred to singed to lightly charred ('brown') to fully charred, is interesting in terms of crop processing methods. Some modern communities practising traditional crop processing methods have been known to singe glume wheats to scorch off awns prior to, or during, dehusking (Peña-Chocarro 1999; Peña-Chocarro and Zapata 2003). Such crop processing practices could explain why we find notable amounts of partially charred glume bases and some charred cereal awns. In terrestrial assemblages, only fully charred material would survive. This gives us a more complete understanding of the contexts or conditions under which cereal chaff may become charred, than could be inferred from the charred assemblage alone.

Lentil, bitter vetch, and pea are commonly found at prehistoric sites in southeastern Europe including Vrbjanska Čuka, North Macedonia (Antolín et al. 2020). Bitter vetch is valued for its low soil fertility requirements and drought tolerance, making it particularly attractive in areas with winter rain and hot summers, as is found in parts of the southern Balkans (Kreuz and Marinova 2017). Both pea and lentil appear important crops at *Linearbandkeramik* sites since the Early Neolithic. However, bitter vetch is not present at such sites until the Bronze Age. This perhaps relates to a climatic barrier preventing the adoption of certain crops in central Europe, and a lesser need for drought-tolerant crops in temperate regions (Colledge and Conolly 2007; Kreuz and Marinova 2017). Kreuz and Marinova (2017) also notes that *Vicia* species may have been a less desirable crop as many require additional processing than for instance peas and lentils, due to the poisonous testa which needs to be removed prior to consumption. Fragments of testa and detached pulse hila, closely resembling those of bitter vetch, were recovered from the samples, and these remains plausibly relate to food processing.

The numbers of charred oil-seed crops, fruits, and nuts and weed seeds are all too few to compare between the composition of the charred and uncharred components in each category. Thus, we are unable as yet to explore if there is a taphonomic or cultural bias relating to preservation type creating different proportions of different taxa or to investigate the possible scenarios that resulted in these remains being charred. These lines of enquiry will become more feasible as additional cores are analysed.

While flax is known from many other sites in southeast Europe, such as at the Early Neolithic sites of Slatina and Kapitan Dimitrievo, western Bulgaria (Marinova 2007), no finds of opium poppy seeds are known in the area during the Neolithic period (Salavert et al. 2020). Other sites in the region such as Vrbjanska Čuka have not yielded evidence for oil-seed crops to date (Antolín et al. 2020). However, this absence may be due to a taphonomic bias, as oil-seed crops, particularly poppy seeds, are unlikely to be preserved by charring (Märkle and Rösch 2008). Opium poppy seeds have been found in abundance in uncharred contexts in Neolithic sites in the pre-alpine (e.g. Isolino Virginia, Italy (Banchieri and Rottoli 2004; see Chap. 16 in this book)) and northern alpine foreland lake dwellings (e.g. Zurich KanSan, Layer 2 (Brombacher and Jacomet 1997; Jacomet 2006)), and from southwestern Europe, such as the Early Neolithic lakeshore sites of La Marmotta, Italy (Rottoli and Pessina 2007; see Chap. 4 in this book) and La Draga, Spain (Antolín 2016; Salavert et al. 2020; see Chap. 11 in this book). Recent developments in geometric morphometric analysis may enable classification of the poppy seeds from Ploča Mičov Grad as domesticated or wild subspecies (Jesus et al. 2021) and therefore contribute to a better understanding of the economic role of the plant at the site.

Many of the gathered fruits and nuts found at Ploča Mičov Grad are well-known from other Neolithic sites in the Balkans and other parts of Europe. In the southwestern Balkans, such gathered fruits and nuts were similarly recovered from charred assemblages from the Early Neolithic Macedonian site of Vrbjanska Čuka (hazelnut, blackberry, elderberry, strawberry, and apple) (Antolín et al. 2020) and the Early Bronze Age phase of Sovjan, Albania (blackberry, fig, elderberry, cornelian cherry, and physalis) (Fouache et al. 2001). However, there was a far greater range and number of collected taxa at Ploča Mičov Grad. This may be due to preservation biases, as found by Colledge and Conolly (2014), that result in less diverse and numerous wild taxa surviving in charred assemblages.

A preliminary inference of the presence of autumn- and spring-germinating weed/wild taxa is that there was at least some autumn sowing, though additional spring sowing cannot be excluded. Similar observations have been made for the Early Neolithic Macedonian site of Vrbjanska Čuka (Mazzucco et al. in press).

The correspondence analysis suggests that there is a trend based on the depth of the samples in the core. However, the proportion of uncharred-versus-charred remains does not explain this pattern. Instead, the composition of the samples, particularly relating to proportions of fruits and nuts and oil-seed crops, appears to explain the variation seen in the samples. Those to the top of the organic layer contain a larger proportion of oil-seed crops, while those towards the bottom have higher proportions of fruits and nuts. The increase in oil-seed crops towards the top of the organic layer particularly relates to an increase in opium poppy seeds and flax capsules, but not flax seeds, which are present in greater numbers in samples towards the bottom. These results will be compared to the three other lake cores and the bulk samples to assess whether these trends are representative of the whole settlement or specific to this core.

The results of the crop processing analysis suggest that the weed/wild taxa have similar attributes to those associated with fine-sieving by-products. This fits with the high amount of glume bases recovered from the samples, as such material would also be removed at this stage with the weed seeds (Hillman 1984; Jones 1984, 1987). This compatibility between the predominant crop and weed/wild composition of the samples suggests that the latter represent the arable weed flora associated with the glume wheats rather than originating from another source such as animal dung (cf. Charles 1998). Though rodent dung was ubiquitous in the samples, larger fragments of possible ruminant dung were a rare occurrence. Some of our

samples are positioned below that of the fine-sieving by-product. This finding is similar to that of other sites in central Europe such as the Early Neolithic site of Vaihingen, Germany (Bogaard 2011), and in southeast Europe such as the Late Bronze Age site of Assiros Toumba, northern Greece (Jones 1987). A possible explanation for this is given by Bogaard (2011) who suggests that different crop fields with different proportions of weed seed types could result in samples positioned outside the main group.

We carried out a functional ecological analysis of the weed/wild assemblage given the likelihood that the weed/wild seeds are arable weed seeds. The samples showed variable scores along the discriminant function, again highlighting variability through the core. The samples broadly align with the lower end of high-input regimes suggesting that the agrosystem at Ploča Mičov Grad resembles modern agrosystems, that are relatively small-scale and labour-intensive. High-input regimes, resembling gardens or cultivated patches, rather than typical modern fields, are associated with higher soil fertility and mechanical disturbance, due to labour-intensive practices such as weeding, tillage, and manuring. Sites from the Neolithic in southwestern Balkans, and southern Europe more broadly, generally do not have a wealth of weed data. As such, a major aim of this project is the collection of weed data for the reconstruction of the agroecology at Ploča Mičov Grad. However, this means that, at present, it is hard to compare these results to contemporary and spatially proximal sites. Variable discriminant scores, reflecting medium to high intensity conditions, can be seen at other Neolithic sites in central Europe such as the forementioned Vaihingen (Bogaard et al. 2016), the Early Neolithic Körös site of Ecsegfalva, Hungary (Bogaard et al. 2007, 2016) and Neolithic Çatalhöyük, Turkey (Green et al. 2019). A range of discriminant scores is not surprising, as this likely reflects the variability typically found in 'real' agrosystems with limited labour and thus shows the expected flexibility and pragmatism involved, influencing when and where labour is focused in such agrosystems (Bogaard et al. 2016).

The degree of similarity between samples shown by these initial analyses will be considered when assessing whether or not to amalgamate adjacent samples taken according to depth rather than visible variation in compositions. This decision will only be taken after analysis of archaeobotanical samples from a further three cores which appear to possess the same organic layer, in order to discern the stratigraphic relationship between samples and between cores. Ongoing work on these samples, including additional cores, aims to further increase our understanding of plant-related activity during the Late Neolithic at the site of Ploča Mičov Grad, providing further data to explore trends in the diversity and abundance of different plant taxa, and strengthen inferences regarding crop processing techniques and agrosystems.

17.5 Conclusions

Waterlogged plant remains from a thick cultural organic layer relating to the lakeshore pile-dwelling site of Ploča Mičov Grad, Lake Ohrid in North Macedonia, offer evidence for repeated plant-related activities during the Late Neolithic across an estimated 100-year occupation from the mid-5th millennium BCE. A range of cereal, pulse, and oil-seed crops were cultivated, together with a collection of varied fruits and nuts. The spectrum of cultivated and collected plants is similar to others in the southwestern Balkans but notably includes a greater number and range of collected plants and oil-seed crops than many terrestrial sites, likely reflecting the broader preservation of waterlogged assemblages. These results will be compared to ongoing work on three further lake core samples and bulk samples from Ploča Mičov Grad to explore intra-site and whole settlement trends. Evidence relating to crop processing techniques comes from partially charred glume bases and the characteristics of the weed/wild seeds, which resemble the arable weeds found in fine-sieving by-products. The results of the functional weed ecological analysis suggest that the agrosystem at Ploča Mičov Grad was small-scale and labour-intensive. This aligns

well with results from other Neolithic sites from southern and central Europe. Comparison to contemporary sites from the southwestern Balkans is limited due to little archaeobotanical research in this region and a lack of large weed datasets. The work presented here alongside ongoing work on further samples from Ploča Mičov Grad and other EXPLO sites aims to contribute to efforts to fill this research gap in the southwestern Balkans, emphasising the potential of prehistoric wetland sites in this area to better understand the adoption and adaptation of agriculture and patterns of plant food use during the Neolithic.

Acknowledgements We would like to thank the anonymous peer reviewer for their constructive comments on this manuscript. This work was funded by a Clarendon doctoral scholarship held by Amy Holguin and by the ERC Synergy project EXPLO (Oxford PI Bogaard, grant no. 810586).

References

Anderberg AL (1994) Atlas of seeds and small fruits of northwest-European plant species. Part 4: Resedaceae—Umbelliferae. Swedish Museum of Natural History, Stockholm

Antolín F (2016) Local, intensive and diverse? Early farmers and plant economy in the North-East of the Iberian Peninsula (5500–2300 cal BC). Barkhuis Publishing, Groningen

Antolín F, Steiner B, Vach W, Jacomet S (2015) What is a litre of sediment? Testing volume measurement techniques for wet sediment and their implications in archaeobotanical analyses at the Late Neolithic lake-dwelling site of Parkhaus Opéra (Zürich, Switzerland). J Archaeol Sci 61:36–44

Antolín F, Steiner B, Akeret Ö, Brombacher C, Kühn M, Vandorpe P, Bleicher N, Gross E, Schaeren G, Jacomet S (2017) Studying the preservation of plant macroremains from waterlogged archaeological deposits for an assessment of layer taphonomy. Rev Palaeobot Palynol 246:120–145

Antolín F, Sabanov A. Naumov G, Soteras R (2020) Crop choice, gathered plants and household activities at the beginnings of farming in the Pelagonia Valley of North Macedonia. Antiquity 94:e21. https://doi.org/10.15184/aqy.2020.119

Banchieri D, Rottoli M (2004) Una nuova data per la storia del papavero da oppio (*Papaver somniferum* subsp. somniferum). Sibrium 25:31–49

Berggren G (1969) Atlas of seeds and small fruits of northwest-European plant species. Part 2: Cyperaceae. Swedish Museum of Natural History, Stockholm

Berggren G (1981) Atlas of seeds and small fruits of northwest-European plant species, with morphological descriptions. Part 3. Saliaceae—Cruciferae. Swedish Museum of Natural History, Stockholm

Bogaard A (2011) Plant use and crop husbandry in an early Neolithic village: Vaihingen an der Enz, Baden-Württemberg. Frankfurter Archäologische Schriften, vol 16. Habel, Bonn

Bogaard A, Jones G, Charles M, Hodgson J (2001) On the archaeobotanical inference of crop sowing time using the FIBS method. J Archaeol Sci 28:1171–1183

Bogaard A, Bending J, Jones G (2007) Archaeo botanical evidence for plant husbandry and use at Ecsegfalva 23. In: Whittle A (ed) The Early Neolithic on the Great Hungarian Plain: investigations of the Körös culture site of Ecsegfalva 23. Hungarian Academy of Sciences, Budapest, Co. Békés, pp 421–445

Bogaard A, Hodgson J, Nitsch E, Jones G, Styring A, Diffey C, Pouncett J, Herbig C, Charles M, Ertuğ F, Tugay O, Filipovic D, Fraser R (2016) Combining functional weed ecology and crop stable isotope ratios to identify cultivation intensity: a comparison of cereal production regimes in Haute Provence, France and Asturias, Spain. Veg Hist Archaeobot 25:57–73. https://doi.org/10.1007/s00334-015-0524-0

Bogaard A, Styring A, Ater M, Hmimsa Y, Green L, Stroud E, Whitlam J, Diffey C, Nitsch E, Charles M, Jones G, Hodgson J (2018) From traditional farming in Morocco to early urban agroecology in northern Mesopotamia: combining present-day arable weed surveys and crop isotope analysis to reconstruct past agrosystems in (semi-)arid regions. Environ Archaeol 23:303–322. https://doi.org/10.1080/14614103.2016.1261217

Bojňanský V, Fargašová A (2007) Atlas of seeds and fruits of Central and East-European flora. The Carpathian Mountains region. Springer, Dordrecht

Brombacher C, Jacomet S (1997) Ackerbau, Sammelwirtschaft und Umwelt: Ergebnisse archäobotanischer Untersuchungen. In: Schibler J, Hüster-Plogmann S, Jacomet S (eds) Ökonomie und Ökologie neolithischer und bronzezeitlicher Ufersiedlungen am Zürichsee. Monographien der Kantonsarchäologie Zürich 20. Fotorotar, Zürich, Egg, pp 220–291

Charles M (1998) Fodder from dung: the recognition and interpretation of dung-derived plant material from archaeological sites. Environ Archaeol 1:111–122

Colledge S, Conolly J (2007) The Neolithisation of the Balkans: a review of the archaeobotanical evidence. In: Spataro M, Biagi P (eds) A short walk through the Balkans: the first farmers of the Carpathian Basin and adjacent regions. Proceedings of the conference held at the institute of archaeology UCL, 20–22 June 2005, Società per La Preistoria e Protostoria Della Regione Friuli-Venezia Giulia. Quaderno, Trieste, pp 25–38

Colledge S, Conolly J (2014) Wild plant use in European Neolithic subsistence economies: a formal assessment of preservation bias in archaeobotanical assemblages and the implications for understanding changes in plant diet breadth. Q Sci Rev 101:193–206. https://doi.org/10.1016/j.quascirev.2014.07.013

Diederichsen A, Richards K (2003) Cultivated flax and the genus Linum L.: taxonomy and germplasm conservation. In: Muir AD, Westcott ND (eds) Flax: the genus Linum, medicinal and aromatic plants—industrial profiles. Taylor & Francis Group, Boca Raton

Dinaric Arc and Balkans Environment Outlook (DABEO) (2010) Proposal on the geographic extent of the Balkans/Dinaric Arc Region for the DABEO reporting process. UNEP/DEWA/GRID, Geneva

European Environment Agency (2017a) Digital map of European ecological regions (WWW document). https://www.eea.europa.eu/data-and-maps/data/digital-map-of-european-ecological-regions. Accessed 25 May 2021

European Environment Agency (2017b) Biogeographical regions in Europe (WWW document). European Environment Agency. https://www.eea.europa.eu/data-and-maps/figures/biogeographical-regions-in-europe-2. Accessed 25 May 2021

Filipović V (2018) Some observations on communications and contacts in the central Balkan and neighbouring regions during the 7th to 5th century BC based on the distribution of weapons. God Cent Za Balk Ispit 47:105–115

Fouache E, Dufaure J-J, Denefle M, Lezine A-M, Lera P, Prendi F, Touchais G (2001) Man and environment around Lake Maliq (southern Albania) during the Late Holocene. Veg Hist Archaeobot 10:79–86

Green L, Charles M, Bogaard A (2019) Exploring the agroecology of Neolithic Çatalhöyük, Central Anatolia: an archaeobotanical approach to agricultural intensity based on functional ecological analysis of arable weed flora. Paleorient 44:29–43

Hafner A, Reich J, Ballmer AT, Bolliger M, Emmenegger L, Fandre J, Francuz J, Hostettler M, Lotter A, Maczkowski A, Naumov G, Nedeljkovic D, Stäheli C, Szidat S, Taneski B, Todorova V, Wieser A, Bogaard A, Kotsakis K, Tinner W (2021) First absolute chronologies of Neolithic and Bronze Age settlements at Lake Ohrid based on dendrochronology and radiocarbon dating. J Archaeol Sci Rep 38:103107. https://doi.org/10.1016/j.jasrep.2021.103107

Hillman G (1984) Interpretation of archaeological plant remains. The application of ethnographic models from Turkey. In: van Zeist W, Casparie WA (eds) Plants and ancient man, studies in palaeoethnobotany. Proceedings of the sixth symposium of the international work group or palaeoethnobotany, Groningen, 30 May–3 June 1983. Symposium of the international work-group for palaeoethnobotany, vol 6. A.A. Balkema, Rotterdam, Boston, pp 1–42

Hosch S, Jacomet S (2004) Ackerbau und Sammelwirtschaft—Ergebnisse der Untersuchung von Samen und Früchten. In: Jacomet S, Leuzinger U, Schibler J (eds) Die Jungsteinzeitliche Seeufersiedlung Arbon Bleiche 3: Umwelt Und Wirtschaft. Huber & Co, Frauenfeld, pp 112–157

Jacomet S (2006) Plant economy of the northern Alpine lake dwellings 3500–2400 cal. BC. Environ Archaeol 11(1): 65–85. https://doi.org/10.1179/174963106x97061

Jacomet S, Brombacher C, Dick M (1989) Archäobotanik am Zürichsee. Ackerbau, Sammelwirtschaft und Umwelt von neolithischen und bronzezeitlichen Seeufersiedlungen im Raum Zürich. Orell Füssli Verlag, Zurich

Jesus A, Bonhomme V, Evin A, Ivorra S, Soteras R, Salavert A, Antolín F, Bouby L (2021) A morphometric approach to track opium poppy domestication. Sci Rep 11:9778

Jones G (1984) Interpretation of archaeological plant remains: ethnographic models from Greece. In: van Zeist W, Casparie WA (eds) Plants and ancient man: studies in palaeoethnobotany: proceedings of the sixth symposium of the international work group for palaeoethnobotany, 30 May–3 June 1983. A.A. Balkema, Rotterdam, Boston, pp 43–61

Jones G (1987) A statistical approach to the archaeological identification of crop processing. J Archaeol Sci 14:311–323. https://doi.org/10.1016/0305-4403(87)90019-7

Jovanova L (2013) Makedonija vo rimskiot period, arheologia. In: Kuzman P, Dimitrova E, Donev J (eds) Makedonija. Mileniumski kulturno-istoriski fakti. Skopje, pp 789–930

Kolb M (1997) Die Seeufersiedlung Sipplingen und die Entwicklung der Horgener Kultur am Bodensee. In: Schlichtherle H (ed) Pfahlbauten rund um die Alpen. Archäologie in Deutschland, Sonderheft 1997. Theiss, Stuttgart, pp 22–28

Körber-Grohne U (1964) Bestimmungsschlüssel für subfossile Juncus-Samen und Gramineen-Früchte. Probleme der Küstenforschung im südlichen Nordseegebiet, vol 7. Lax, Hildesheim

Kottek M, Grieser J, Beck C, Rudolf B, Rubel F (2006) World map of the Köppen-Geiger climate classification updated. Meteorol Z 15:259–263. https://doi.org/10.1127/0941-2948/2006/0130

Krauß R, Marinova E, De Brue H, Weninger B (2018) The rapid spread of early farming from the Aegean into the Balkans via the Sub-Mediterranean-Aegean Vegetation Zone. Q Int 496:24–41. https://doi.org/10.1016/j.quaint.2017.01.019

Kreuz A, Marinova E (2017) Archaeobotanical evidence of crop growing and diet within the areas of the Karanovo and the Linear Pottery Cultures: a quantitative and qualitative approach. Veg Hist Archaeobot 26:639–657. https://doi.org/10.1007/s00334-017-0643-x

Kreuz A, Marinova E, Schäfer E, Wiethold J (2005) A comparison of early Neolithic crop and weed assemblages from the Linearbandkeramik and the Bulgarian Neolithic cultures: differences and similarities. Veg Hist Archaeobot 14:237–258. https://doi.org/10.1007/s00334-005-0080-0

Kuzman P (2013) Praistoriski palafitni naselbi vo Makedonija. In: Kuzman P, Dimitrova E, Donev J (eds) Makedonija. Mileniumski kulturno-istoriski fakti. Skopje, pp 297–430

Kuzman P (2017) Chronological and geographic routes of Ohrid's oldest population. Arheol Inform 1:156–164

Lacey JH, Francke A, Leng MJ, Vane CH, Wagner B (2015) A high-resolution Late Glacial to Holocene record of environmental change in the Mediterranean from Lake Ohrid (Macedonia/Albania). Int J Earth Sci 104:1623–1638. https://doi.org/10.1007/s00531-014-1033-6

Lolos Y (2007) Via Egnatia after Egnatius: imperial policy and inter-regional contacts. Mediterr Hist Rev 22:273–293. https://doi.org/10.1080/09518960802005844

Marinova E (2007) Archaeobotanical data from the early Neolithic of Bulgaria. In: Colledge S, Conolly J, Shennan S (eds) The origins and spread of domestic plants in southwest Asia and Europe. Taylor & Francis Group, Walnut Creek, pp 93–109

Marinova E, Popova T (2008) *Cicer arietinum* (chick pea) in the Neolithic and Chalcolithic of Bulgaria: implications for cultural contacts with the neighbouring regions? Veg Hist Archaeobot 17:73–80. https://doi.org/10.1007/s00334-008-0159-5

Märkle T, Rösch M (2008) Experiments on the effects of carbonization on some cultivated plant seeds. Veg Hist Archaeobot 17:257–263. https://doi.org/10.1007/s00334-008-0165-7

Mazzucco N, Sabanov A, Antolín F, Naumov G, Fidanoski L, Gibaja JF (in press) The spread of agriculture in south-eastern Europe: new data from North Macedonia. Antiquity 96:15–33. https://doi.org/10.15184/aqy.2021.32

Menotti F (2004) Living on the lake in prehistoric Europe: 150 years of lake-dwelling research. Routledge, London, New York

Mitrevski D (2015) Праисторија на Република Македонија (The prehistory in Republic of Macedonia)

Naumov G (2020) Neolithic wetland and lakeside settlements in the Balkans. In: Settling waterscapes in Europe. The archaeology of Neolithic and Bronze Age pile-dwellings. Open series in prehistoric archaeology, vol 1. Propylaeum, Bern, Heidelberg, pp 111–135

Peña-Chocarro L (1999) Prehistoric agriculture in southern Spain during the Neolithic and the Bronze Age: the application of ethnographic models. BAR international series, vol 818. Archaeopress, Oxford

Peña-Chocarro L, Zapata L (2003) Post-harvest processing of hulled wheats. An ethnoarchaeological approach. In: Anderson PC, Cummings LS, Schippers TS, Simonel B (eds) Le traitement des récoltes: un regard sur la diversité, du Néolithique au présent. Actes des XXIIIe rencontres internationales d'archéologie et d'histoire d'Antibes, 17–19 Oct 2002. Éditions APDCA, Antibes

Popovska C, Bonacci O (2007) Basic data on the hydrology of Lakes Ohrid and Prespa. Hydrol Process 21:658–664. https://doi.org/10.1002/hyp.6252

R Core Team (2019) R: alanguage and environment for statistical computing, Vienna, Austria. https://www.R-project.org/

Rottoli M, Pessina A (2007) Neolithic agriculture in Italy: an update of archaeobotanical data with particular emphasis on northern settlements. In: Colledge S, Conolly J, Shennan S (eds) The origins and spread of domestic plants in southwest Asia and Europe. Taylor & Francis Group, Walnut Creek, pp 141–154

Rubel F, Brugger K, Haslinger K, Auer I (2017) The climate of the European Alps: shift of very high resolution Köppen-Geiger climate zones 1800–2100. Meteorol Z 26:115–125. https://doi.org/10.1127/metz/2016/0816

Salavert A, Zazzo A, Martin L, Antolín F, Gauthier C, Thil F, Tombret O, Bouby L, Manen C, Mineo M, Mueller-Bieniek A, Piqué R, Rottoli M, Rovira N, Toulemonde F, Vostrovská I (2020) Direct dating reveals the early history of opium poppy in western Europe. Sci Rep 10:20263. https://doi.org/10.1038/s41598-020-76924-3

Steiner BL, Antolín F, Jacomet S (2015) Testing of the consistency of the sieving (wash-over) process of waterlogged sediments by multiple operators. J Archaeol Sci Rep 2:310–320. https://doi.org/10.1016/j.jasrep.2015.02.012

Steiner BL, Antolín F, Vach W, Jacomet S (2017) Systematic subsampling of large-volume samples in waterlogged sediments. A time-saving strategy or a source of error? Rev Palaeobot Palynol 245:10–27. https://doi.org/10.1016/j.revpalbo.2017.05.013

UNESCO World Heritage Centre (2021) Prehistoric pile dwellings around the Alps (WWW document). UNESCO World Heritage Centre. https://whc.unesco.org/en/list/1363/. Accessed 26 May 2021

Valamoti SM, Petridou C, Berihuete-Azorín M, Stika H-S, Papadopoulou L, Mimi I (2021) Deciphering ancient 'recipes' from charred cereal fragments: an integrated methodological approach using experimental, ethnographic and archaeological evidence. J Archaeol Sci 128:105347

Weissová B, Tušlová P, Ardjanliev P, Verčík M (2018) The frontier studies. Survey of the northern part of the Lake Ohrid Basin, preliminary report on the season 2017. Stud Hercynia 22:99–133

Westphal T, Tegel W, Heußner KU, Lera P, Rittershofer F (2011) Erste dendrochronologische Datierungen historischer Hölzer in Albanien. Archäol Anzeiger 2:75–95

Xhuveli L, Schultze-Motel J (1995) Neolithic cultivated plants from Albania. Veg Hist Archaeobot 4:245–248. https://doi.org/10.1007/BF00235755

Geographical Index

Subject Index

GPSR Compliance

The European Union's (EU) General Product Safety Regulation (GPSR) is a set of rules that requires consumer products to be safe and our obligations to ensure this.

If you have any concerns about our products, you can contact us on ProductSafety@springernature.com

In case Publisher is established outside the EU, the EU authorized representative is:

Springer Nature Customer Service Center GmbH
Europaplatz 3
69115 Heidelberg, Germany

The manufacturer's authorised representative in the EU is Springer
Nature Customer Service Centre GmbH, Europaplatz 3, 69115 Heidelberg,
Germany. If you have any concerns regarding our products, please
contact ProductSafety@springernature.com

Printed and bound by CPI Group (UK) Ltd, Croydon, CR0 4YY
29/04/2026
02099454-0008